Artificial Intelligence
An Introduction

Compiled by
Sybil Foreman

Scribbles

Year of Publication 2018

ISBN : 9789386874016

Book Published by

Scribbles

(An Imprint of Alpha Editions)

email - alphaedis@gmail.com

Produced by: PediaPress GmbH
Limburg an der Lahn
Germany
http://pediapress.com/

Contents

Tools 183

Applications 425

History

History of artificial intelligence

Artificial intelligence
Major goals
Knowledge reasoning
Planning
Machine learning
Natural language processing
Computer vision
Robotics
Artificial general intelligence
Approaches
Symbolic
Deep learning
Bayesian networks
Evolutionary algorithms
Philosophy
Ethics
Existential risk
Turing test
Chinese room
Friendly AI
History
Timeline
Progress
AI winter

Technology
Applications
Projects
Programming languages
Glossary
Glossary

History of computing
Hardware
• Hardware before 1960 • Hardware 1960s to present
Software
• Software • Unix • Free software and open-source software
Computer science
• Artificial intelligence • Compiler construction • Computer science • Operating systems • Programming languages • Prominent pioneers • Software engineering
Modern concepts
• General-purpose CPUs • Graphical user interface • Internet • Laptops • Personal computers • Video games • World Wide Web
By country
• Bulgaria • Poland • Romania • Soviet Bloc • Soviet Union • Yugoslavia
Timeline of computing

• before 1950 • 1950–1979 • 1980–1989 • 1990–1999 • 2000–2009 • 2010–2019 • *more timelines ...*
Glossary of computer science
• **Category**
• v • t • e[1]

The **history of Artificial Intelligence** (**AI**) began in antiquity, with myths, stories and rumors of artificial beings endowed with intelligence or consciousness by master craftsmen; as Pamela McCorduck writes, AI began with "an ancient wish to forge the gods."[2]

The seeds of modern AI were planted by classical philosophers who attempted to describe the process of human thinking as the mechanical manipulation of symbols. This work culminated in the invention of the programmable digital computer in the 1940s, a machine based on the abstract essence of mathematical reasoning. This device and the ideas behind it inspired a handful of scientists to begin seriously discussing the possibility of building an electronic brain.

The field of AI research was founded at a workshop held on the campus of Dartmouth College during the summer of 1956. Those who attended would become the leaders of AI research for decades. Many of them predicted that a machine as intelligent as a human being would exist in no more than a generation and they were given millions of dollars to make this vision come true.

Eventually it became obvious that they had grossly underestimated the difficulty of the project due to computer hardware limitations. In 1973, in response to the criticism of James Lighthill and ongoing pressure from congress, the U.S. and British Governments stopped funding undirected research into artificial intelligence, and the difficult years that followed would later be known as an "AI winter". Seven years later, a visionary initiative by the Japanese Government inspired governments and industry to provide AI with billions of dollars, but by the late 80s the investors became disillusioned by the absence of the needed computer power (hardware) and withdrew funding again.

Investment and interest in AI boomed in the first decades of the 21st century, when machine learning was successfully applied to many problems in academia and industry due to the presence of powerful computer hardware. As in previous "AI summers", some observers (such as Ray Kurzweil[3]) predicted the imminent arrival of artificial general intelligence: a machine with intellectual capabilities that exceed the abilities of human beings.

Precursors

The dream of artificial intelligence was first thought of in Indian philosophies like those of Charvaka, a famous philosophy tradition dating back to 1500 BCE and surviving documents around 600 BCE. McCorduck (2004) writes "artificial intelligence in one form or another is an idea that has pervaded intellectual history, a dream in urgent need of being realized," expressed in humanity's myths, legends, stories, speculation and clockwork automatons.[4]

AI in myth, fiction and speculation

Mechanical men and artificial beings appear in Greek myths, such as the golden robots of Hephaestus and Pygmalion's Galatea. In the Middle Ages, there were rumors of secret mystical or alchemical means of placing mind into matter, such as Jābir ibn Hayyān's *Takwin*, Paracelsus' homunculus and Rabbi Judah Loew's Golem.[5] By the 19th century, ideas about artificial men and thinking machines were developed in fiction, as in Mary Shelley's *Frankenstein* or Karel Čapek's *R.U.R. (Rossum's Universal Robots)*,[6] and speculation, such as Samuel Butler's "Darwin among the Machines."[7] AI has continued to be an important element of science fiction into the present.

Automatons

Realistic humanoid automatons were built by craftsman from every civilization, including Yan Shi, Hero of Alexandria, Al-Jazari[8] , Pierre Jaquet-Droz, and Wolfgang von Kempelen.[9] The oldest known automatons were the sacred statues of ancient Egypt and Greece. The faithful believed that craftsman had imbued these figures with very real minds, capable of wisdom and emotion—Hermes Trismegistus wrote that "by discovering the true nature of the gods, man has been able to reproduce it."[10,11]

Figure 1: *Al-Jazari's programmable automata (1206 CE)*

Formal reasoning

Artificial intelligence is based on the assumption that the process of human thought can be mechanized. The study of mechanical—or "formal"—reasoning has a long history. Chinese, Indian and Greek philosophers all developed structured methods of formal deduction in the first millennium BCE. Their ideas were developed over the centuries by philosophers such as Aristotle (who gave a formal analysis of the syllogism), Euclid (whose *Elements* was a model of formal reasoning), al-Khwārizmī (who developed algebra and gave his name to "algorithm") and European scholastic philosophers such as William of Ockham and Duns Scotus.

Majorcan philosopher Ramon Llull (1232–1315) developed several *logical machines* devoted to the production of knowledge by logical means;[12] Llull described his machines as mechanical entities that could combine basic and undeniable truths by simple logical operations, produced by the machine by mechanical meanings, in such ways as to produce all the possible knowledge.[13] Llull's work had a great influence on Gottfried Leibniz, who redeveloped his ideas.[14]

In the 17th century, Leibniz, Thomas Hobbes and René Descartes explored the possibility that all rational thought could be made as systematic as algebra or geometry.[15] Hobbes famously wrote in *Leviathan*: "reason is nothing but reckoning".[16] Leibniz envisioned a universal language of reasoning (his *characteristica universalis*) which would reduce argumentation to calculation, so that "there would be no more need of disputation between two philosophers than between two accountants. For it would suffice to take their pencils in

Figure 2: *Gottfried Leibniz, who speculated that human reason could be reduced to mechanical calculation*

hand, down to their slates, and to say each other (with a friend as witness, if they liked): *Let us calculate*."[17] These philosophers had begun to articulate the physical symbol system hypothesis that would become the guiding faith of AI research.

In the 20th century, the study of mathematical logic provided the essential breakthrough that made artificial intelligence seem plausible. The foundations had been set by such works as Boole's *The Laws of Thought* and Frege's *Begriffsschrift*. Building on Frege's system, Russell and Whitehead presented a formal treatment of the foundations of mathematics in their masterpiece, the *Principia Mathematica* in 1913. Inspired by Russell's success, David Hilbert challenged mathematicians of the 1920s and 30s to answer this fundamental question: "can all of mathematical reasoning be formalized?" His question was answered by Gödel's incompleteness proof, Turing's machine and Church's Lambda calculus.[18]

Their answer was surprising in two ways. First, they proved that there were, in fact, limits to what mathematical logic could accomplish. But second (and more important for AI) their work suggested that, within these limits, *any* form of mathematical reasoning could be mechanized. The Church-Turing thesis implied that a mechanical device, shuffling symbols as simple as 0 and 1, could imitate any conceivable process of mathematical deduction. The key

Figure 3: *The ENIAC, at the Moore School of Electrical Engineering. This photo has been artificially darkened, obscuring details such as the women who were present and the IBM equipment in use.*[19]

insight was the Turing machine—a simple theoretical construct that captured the essence of abstract symbol manipulation. This invention would inspire a handful of scientists to begin discussing the possibility of thinking machines.[20]

Computer science

Calculating machines were built in antiquity and improved throughout history by many mathematicians, including (once again) philosopher Gottfried Leibniz. In the early 19th century, Charles Babbage designed a programmable computer (the Analytical Engine), although it was never built. Ada Lovelace speculated that the machine "might compose elaborate and scientific pieces of music of any degree of complexity or extent".[21] (She is often credited as the first programmer because of a set of notes she wrote that completely detail a method for calculating Bernoulli numbers with the Engine.)

The first modern computers were the massive code breaking machines of the Second World War (such as Z3, ENIAC and Colossus). The latter two of these machines were based on the theoretical foundation laid by Alan Turing and developed by John von Neumann.

Figure 4: *The IBM 702: a computer used by the first generation of AI researchers.*

The birth of artificial intelligence 1952–1956

In the 1940s and 50s, a handful of scientists from a variety of fields (mathematics, psychology, engineering, economics and political science) began to discuss the possibility of creating an artificial brain. The field of artificial intelligence research was founded as an academic discipline in 1956.

Cybernetics and early neural networks

The earliest research into thinking machines was inspired by a confluence of ideas that became prevalent in the late 1930s, 1940s, and early 1950s. Recent research in neurology had shown that the brain was an electrical network of neurons that fired in all-or-nothing pulses. Norbert Wiener's cybernetics described control and stability in electrical networks. Claude Shannon's information theory described digital signals (i.e., all-or-nothing signals). Alan Turing's theory of computation showed that any form of computation could be described digitally. The close relationship between these ideas suggested that it might be possible to construct an electronic brain.[22]

Examples of work in this vein includes robots such as W. Grey Walter's turtles and the Johns Hopkins Beast. These machines did not use computers, digital electronics or symbolic reasoning; they were controlled entirely by analog circuitry.[23]

Walter Pitts and Warren McCulloch analyzed networks of idealized artificial neurons and showed how they might perform simple logical functions. They were the first to describe what later researchers would call a neural network.[24] One of the students inspired by Pitts and McCulloch was a young Marvin Minsky, then a 24-year-old graduate student. In 1951 (with Dean Edmonds) he built the first neural net machine, the SNARC.[25] Minsky was to become one of the most important leaders and innovators in AI for the next 50 years.

Turing's test

In 1950 Alan Turing published a landmark paper in which he speculated about the possibility of creating machines that think.[26] He noted that "thinking" is difficult to define and devised his famous Turing Test. If a machine could carry on a conversation (over a teleprinter) that was indistinguishable from a conversation with a human being, then it was reasonable to say that the machine was "thinking". This simplified version of the problem allowed Turing to argue convincingly that a "thinking machine" was at least *plausible* and the paper answered all the most common objections to the proposition.[27] The Turing Test was the first serious proposal in the philosophy of artificial intelligence.

Game AI

In 1951, using the Ferranti Mark 1 machine of the University of Manchester, Christopher Strachey wrote a checkers program and Dietrich Prinz wrote one for chess.[28] Arthur Samuel's checkers program, developed in the middle 50s and early 60s, eventually achieved sufficient skill to challenge a respectable amateur.[29] Game AI would continue to be used as a measure of progress in AI throughout its history.

Symbolic reasoning and the Logic Theorist

When access to digital computers became possible in the middle fifties, a few scientists instinctively recognized that a machine that could manipulate numbers could also manipulate symbols and that the manipulation of symbols could well be the essence of human thought. This was a new approach to creating thinking machines.

In 1955, Allen Newell and (future Nobel Laureate) Herbert A. Simon created the "Logic Theorist" (with help from J. C. Shaw). The program would eventually prove 38 of the first 52 theorems in Russell and Whitehead's *Principia Mathematica*, and find new and more elegant proofs for some.[30] Simon said that they had "solved the venerable mind/body problem, explaining how a system composed of matter can have the properties of mind."[31] (This was an early statement of the philosophical position John Searle would later call "Strong AI": that machines can contain minds just as human bodies do.)

Dartmouth Conference 1956: the birth of AI

The Dartmouth Conference of 1956[32] was organized by Marvin Minsky, John McCarthy and two senior scientists: Claude Shannon and Nathan Rochester of IBM. The proposal for the conference included this assertion: "every aspect of learning or any other feature of intelligence can be so precisely described that a machine can be made to simulate it".[33] The participants included Ray Solomonoff, Oliver Selfridge, Trenchard More, Arthur Samuel, Allen Newell and Herbert A. Simon, all of whom would create important programs during the first decades of AI research.[34] At the conference Newell and Simon debuted the "Logic Theorist" and McCarthy persuaded the attendees to accept "Artificial Intelligence" as the name of the field.[35] The 1956 Dartmouth conference was the moment that AI gained its name, its mission, its first success and its major players, and is widely considered the birth of AI.[36]

The golden years 1956–1974

The years after the Dartmouth conference were an era of discovery, of sprinting across new ground. The programs that were developed during this time were, to most people, simply "astonishing":[37] computers were solving algebra word problems, proving theorems in geometry and learning to speak English. Few at the time would have believed that such "intelligent" behavior by machines was possible at all.[38] Researchers expressed an intense optimism in private and in print, predicting that a fully intelligent machine would be built in less than 20 years.[39] Government agencies like DARPA poured money into the new field.

The work

There were many successful programs and new directions in the late 50s and 1960s. Among the most influential were these:

Reasoning as search

Many early AI programs used the same basic algorithm. To achieve some goal (like winning a game or proving a theorem), they proceeded step by step towards it (by making a move or a deduction) as if searching through a maze, backtracking whenever they reached a dead end. This paradigm was called "reasoning as search".[40]

The principal difficulty was that, for many problems, the number of possible paths through the "maze" was simply astronomical (a situation known as a "combinatorial explosion"). Researchers would reduce the search space by using heuristics or "rules of thumb" that would eliminate those paths that were unlikely to lead to a solution.[41]

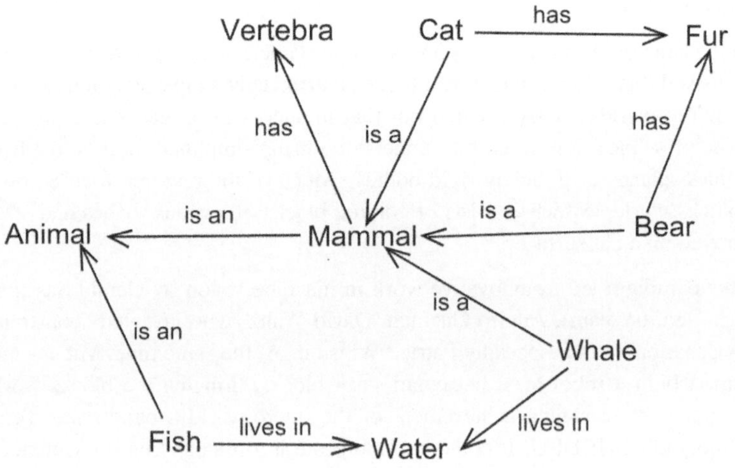

Figure 5: *An example of a semantic network*

Newell and Simon tried to capture a general version of this algorithm in a program called the "General Problem Solver".[42] Other "searching" programs were able to accomplish impressive tasks like solving problems in geometry and algebra, such as Herbert Gelernter's Geometry Theorem Prover (1958) and SAINT, written by Minsky's student James Slagle (1961).[43] Other programs searched through goals and subgoals to plan actions, like the STRIPS system developed at Stanford to control the behavior of their robot Shakey.[44]

Natural language

An important goal of AI research is to allow computers to communicate in natural languages like English. An early success was Daniel Bobrow's program STUDENT, which could solve high school algebra word problems.[45]

A semantic net represents concepts (e.g. "house","door") as nodes and relations among concepts (e.g. "has-a") as links between the nodes. The first AI program to use a semantic net was written by Ross Quillian and the most successful (and controversial) version was Roger Schank's Conceptual dependency theory.

Joseph Weizenbaum's ELIZA could carry out conversations that were so realistic that users occasionally were fooled into thinking they were communicating with a human being and not a program. But in fact, ELIZA had no idea what she was talking about. She simply gave a canned response or repeated back what was said to her, rephrasing her response with a few grammar rules. ELIZA was the first chatterbot.

Micro-worlds

In the late 60s, Marvin Minsky and Seymour Papert of the MIT AI Laboratory proposed that AI research should focus on artificially simple situations known as micro-worlds. They pointed out that in successful sciences like physics, basic principles were often best understood using simplified models like frictionless planes or perfectly rigid bodies. Much of the research focused on a "blocks world," which consists of colored blocks of various shapes and sizes arrayed on a flat surface.[46]

This paradigm led to innovative work in machine vision by Gerald Sussman (who led the team), Adolfo Guzman, David Waltz (who invented "constraint propagation"), and especially Patrick Winston. At the same time, Minsky and Papert built a robot arm that could stack blocks, bringing the blocks world to life. The crowning achievement of the micro-world program was Terry Winograd's SHRDLU. It could communicate in ordinary English sentences, plan operations and execute them.[47]

The optimism

The first generation of AI researchers made these predictions about their work:

- 1958, H. A. Simon and Allen Newell: "within ten years a digital computer will be the world's chess champion" and "within ten years a digital computer will discover and prove an important new mathematical theorem."[48]
- 1965, H. A. Simon: "machines will be capable, within twenty years, of doing any work a man can do."[49]
- 1967, Marvin Minsky: "Within a generation ... the problem of creating 'artificial intelligence' will substantially be solved."[50]
- 1970, Marvin Minsky (in *Life* Magazine): "In from three to eight years we will have a machine with the general intelligence of an average human being."[51]

The money

In June 1963, MIT received a $2.2 million grant from the newly created Advanced Research Projects Agency (later known as DARPA). The money was used to fund project MAC which subsumed the "AI Group" founded by Minsky and McCarthy five years earlier. DARPA continued to provide three million dollars a year until the 70s. DARPA made similar grants to Newell and Simon's program at CMU and to the Stanford AI Project (founded by John McCarthy in 1963). Another important AI laboratory was established at Edinburgh University by Donald Michie in 1965. These four institutions would continue to be the main centers of AI research (and funding) in academia for many years.[52]

The money was proffered with few strings attached: J. C. R. Licklider, then the director of ARPA, believed that his organization should "fund people, not projects!" and allowed researchers to pursue whatever directions might interest them. This created a freewheeling atmosphere at MIT that gave birth to the hacker culture,[53] but this "hands off" approach would not last.

Robotics

In Japan, Waseda University initiated the WABOT project in 1967, and in 1972 completed the WABOT-1, the world's first full-scale intelligent humanoid robot,[54,55] or android. Its limb control system allowed it to walk with the lower limbs, and to grip and transport objects with hands, using tactile sensors. Its vision system allowed it to measure distances and directions to objects using external receptors, artificial eyes and ears. And its conversation system allowed it to communicate with a person in Japanese, with an artificial mouth.[56,57]

The first AI winter 1974–1980

In the 1970s, AI was subject to critiques and financial setbacks. AI researchers had failed to appreciate the difficulty of the problems they faced. Their tremendous optimism had raised expectations impossibly high, and when the promised results failed to materialize, funding for AI disappeared.[58] At the same time, the field of connectionism (or neural nets) was shut down almost completely for 10 years by Marvin Minsky's devastating criticism of perceptrons.[59] Despite the difficulties with public perception of AI in the late 70s, new ideas were explored in logic programming, commonsense reasoning and many other areas.

The problems

In the early seventies, the capabilities of AI programs were limited. Even the most impressive could only handle trivial versions of the problems they were supposed to solve; all the programs were, in some sense, "toys". AI researchers had begun to run into several fundamental limits that could not be overcome in the 1970s. Although some of these limits would be conquered in later decades, others still stymie the field to this day.

- **Limited computer power**: There was not enough memory or processing speed to accomplish anything truly useful. For example, Ross Quillian's successful work on natural language was demonstrated with a vocabulary of only *twenty* words, because that was all that would fit in memory.[60] Hans Moravec argued in 1976 that computers were still millions of times too weak to exhibit intelligence. He suggested an analogy: artificial intelligence requires computer power in the same way that aircraft require horsepower. Below a certain threshold, it's impossible, but, as power increases, eventually it could become easy.[61] With regard to computer vision, Moravec estimated that simply matching the edge and motion detection capabilities of human retina in real time would require a general-purpose computer capable of 10^9 operations/second (1000 MIPS). As of 2011, practical computer vision applications require 10,000 to 1,000,000 MIPS. By comparison, the fastest supercomputer in 1976, Cray-1 (retailing at $5 million to $8 million), was only capable of around 80 to 130 MIPS, and a typical desktop computer at the time achieved less than 1 MIPS.

- **Intractability and the combinatorial explosion**. In 1972 Richard Karp (building on Stephen Cook's 1971 theorem) showed there are many problems that can probably only be solved in exponential time (in the size of the inputs). Finding optimal solutions to these problems requires unimaginable amounts of computer time except when the problems are trivial. This almost certainly meant that many of the "toy" solutions used by AI would probably never scale up into useful systems.[62]
- **Commonsense knowledge and reasoning**. Many important artificial intelligence applications like vision or natural language require simply enormous amounts of information about the world: the program needs to have some idea of what it might be looking at or what it is talking about. This requires that the program know most of the same things about the world that a child does. Researchers soon discovered that this was a truly *vast* amount of information. No one in 1970 could build a database so large and no one knew how a program might learn so much information.[63]
- **Moravec's paradox**: Proving theorems and solving geometry problems is comparatively easy for computers, but a supposedly simple task like recognizing a face or crossing a room without bumping into anything is extremely difficult. This helps explain why research into vision and robotics had made so little progress by the middle 1970s.
- **The frame and qualification problems**. AI researchers (like John McCarthy) who used logic discovered that they could not represent ordinary deductions that involved planning or default reasoning without making changes to the structure of logic itself. They developed new logics (like non-monotonic logics and modal logics) to try to solve the problems.

The end of funding

The agencies which funded AI research (such as the British government, DARPA and NRC) became frustrated with the lack of progress and eventually cut off almost all funding for undirected research into AI. The pattern began as early as 1966 when the ALPAC report appeared criticizing machine translation efforts. After spending 20 million dollars, the NRC ended all support.[64] In 1973, the Lighthill report on the state of AI research in England criticized the utter failure of AI to achieve its "grandiose objectives" and led to the dismantling of AI research in that country.[65] (The report specifically mentioned the combinatorial explosion problem as a reason for AI's failings.)[66] DARPA was deeply disappointed with researchers working on the Speech Understanding Research program at CMU and canceled an annual grant of three million dollars.[67] By 1974, funding for AI projects was hard to find.

Hans Moravec blamed the crisis on the unrealistic predictions of his colleagues. "Many researchers were caught up in a web of increasing exaggeration."[68] However, there was another issue: since the passage of the Mansfield Amendment in 1969, DARPA had been under increasing pressure to fund "mission-oriented direct research, rather than basic undirected research". Funding for the creative, freewheeling exploration that had gone on in the 60s would not come from DARPA. Instead, the money was directed at specific projects with clear objectives, such as autonomous tanks and battle management systems.[69]

Critiques from across campus

Several philosophers had strong objections to the claims being made by AI researchers. One of the earliest was John Lucas, who argued that Gödel's incompleteness theorem showed that a formal system (such as a computer program) could never see the truth of certain statements, while a human being could.[70] Hubert Dreyfus ridiculed the broken promises of the 1960s and critiqued the assumptions of AI, arguing that human reasoning actually involved very little "symbol processing" and a great deal of embodied, instinctive, unconscious "know how".[71,72] John Searle's Chinese Room argument, presented in 1980, attempted to show that a program could not be said to "understand" the symbols that it uses (a quality called "intentionality"). If the symbols have no meaning for the machine, Searle argued, then the machine can not be described as "thinking".[73]

These critiques were not taken seriously by AI researchers, often because they seemed so far off the point. Problems like intractability and commonsense knowledge seemed much more immediate and serious. It was unclear what difference "know how" or "intentionality" made to an actual computer program. Minsky said of Dreyfus and Searle "they misunderstand, and should be ignored."[74] Dreyfus, who taught at MIT, was given a cold shoulder: he later said that AI researchers "dared not be seen having lunch with me."[75] Joseph Weizenbaum, the author of ELIZA, felt his colleagues' treatment of Dreyfus was unprofessional and childish. Although he was an outspoken critic of Dreyfus' positions, he "deliberately made it plain that theirs was not the way to treat a human being."[76]

Weizenbaum began to have serious ethical doubts about AI when Kenneth Colby wrote a "computer program which can conduct psychotherapeutic dialogue" based on ELIZA.[77] Weizenbaum was disturbed that Colby saw a mindless program as a serious therapeutic tool. A feud began, and the situation was not helped when Colby did not credit Weizenbaum for his contribution to the program. In 1976, Weizenbaum published *Computer Power and Human Reason* which argued that the misuse of artificial intelligence has the potential to devalue human life.[78]

Perceptrons and the dark age of connectionism

A perceptron was a form of neural network introduced in 1958 by Frank Rosenblatt, who had been a schoolmate of Marvin Minsky at the Bronx High School of Science. Like most AI researchers, he was optimistic about their power, predicting that "perceptron may eventually be able to learn, make decisions, and translate languages." An active research program into the paradigm was carried out throughout the 1960s but came to a sudden halt with the publication of Minsky and Papert's 1969 book *Perceptrons*. It suggested that there were severe limitations to what perceptrons could do and that Frank Rosenblatt's predictions had been grossly exaggerated. The effect of the book was devastating: virtually no research at all was done in connectionism for 10 years. Eventually, a new generation of researchers would revive the field and thereafter it would become a vital and useful part of artificial intelligence. Rosenblatt would not live to see this, as he died in a boating accident shortly after the book was published.

The neats: logic and symbolic reasoning

Logic was introduced into AI research as early as 1958, by John McCarthy in his Advice Taker proposal. In 1963, J. Alan Robinson had discovered a simple method to implement deduction on computers, the resolution and unification algorithm. However, straightforward implementations, like those attempted by McCarthy and his students in the late 1960s, were especially intractable: the programs required astronomical numbers of steps to prove simple theorems. A more fruitful approach to logic was developed in the 1970s by Robert Kowalski at the University of Edinburgh, and soon this led to the collaboration with French researchers Alain Colmerauer and Philippe Roussel who created the successful logic programming language Prolog. Prolog uses a subset of logic (Horn clauses, closely related to "rules" and "production rules") that permit tractable computation. Rules would continue to be influential, providing a foundation for Edward Feigenbaum's expert systems and the continuing work by Allen Newell and Herbert A. Simon that would lead to Soar and their unified theories of cognition.

Critics of the logical approach noted, as Dreyfus had, that human beings rarely used logic when they solved problems. Experiments by psychologists like Peter Wason, Eleanor Rosch, Amos Tversky, Daniel Kahneman and others provided proof.[79] McCarthy responded that what people do is irrelevant. He argued that what is really needed are machines that can solve problems—not machines that think as people do.[80]

The scruffies: frames and scripts

Among the critics of McCarthy's approach were his colleagues across the country at MIT. Marvin Minsky, Seymour Papert and Roger Schank were trying to solve problems like "story understanding" and "object recognition" that *required* a machine to think like a person. In order to use ordinary concepts like "chair" or "restaurant" they had to make all the same illogical assumptions that people normally made. Unfortunately, imprecise concepts like these are hard to represent in logic. Gerald Sussman observed that "using precise language to describe essentially imprecise concepts doesn't make them any more precise." Schank described their "anti-logic" approaches as "scruffy", as opposed to the "neat" paradigms used by McCarthy, Kowalski, Feigenbaum, Newell and Simon.[81]

In 1975, in a seminal paper, Minsky noted that many of his fellow "scruffy" researchers were using the same kind of tool: a framework that captures all our common sense assumptions about something. For example, if we use the concept of a bird, there is a constellation of facts that immediately come to mind: we might assume that it flies, eats worms and so on. We know these facts are not always true and that deductions using these facts will not be "logical", but these structured sets of assumptions are part of the *context* of everything we say and think. He called these structures "frames". Schank used a version of frames he called "scripts" to successfully answer questions about short stories in English.[82] Many years later object-oriented programming would adopt the essential idea of "inheritance" from AI research on frames.

Boom 1980–1987

In the 1980s a form of AI program called "expert systems" was adopted by corporations around the world and knowledge became the focus of mainstream AI research. In those same years, the Japanese government aggressively funded AI with its fifth generation computer project. Another encouraging event in the early 1980s was the revival of connectionism in the work of John Hopfield and David Rumelhart. Once again, AI had achieved success.

The rise of expert systems

An expert system is a program that answers questions or solves problems about a specific domain of knowledge, using logical rules that are derived from the knowledge of experts. The earliest examples were developed by Edward Feigenbaum and his students. Dendral, begun in 1965, identified compounds from spectrometer readings. MYCIN, developed in 1972, diagnosed infectious blood diseases. They demonstrated the feasibility of the approach.[83]

Expert systems restricted themselves to a small domain of specific knowledge (thus avoiding the commonsense knowledge problem) and their simple design made it relatively easy for programs to be built and then modified once they were in place. All in all, the programs proved to be *useful*: something that AI had not been able to achieve up to this point.[84]

In 1980, an expert system called XCON was completed at CMU for the Digital Equipment Corporation. It was an enormous success: it was saving the company 40 million dollars annually by 1986. Corporations around the world began to develop and deploy expert systems and by 1985 they were spending over a billion dollars on AI, most of it to in-house AI departments. An industry grew up to support them, including hardware companies like Symbolics and Lisp Machines and software companies such as IntelliCorp and Aion.[85]

The knowledge revolution

The power of expert systems came from the expert knowledge they contained. They were part of a new direction in AI research that had been gaining ground throughout the 70s. "AI researchers were beginning to suspect—reluctantly, for it violated the scientific canon of parsimony—that intelligence might very well be based on the ability to use large amounts of diverse knowledge in different ways," writes Pamela McCorduck. "[T]he great lesson from the 1970s was that intelligent behavior depended very much on dealing with knowledge, sometimes quite detailed knowledge, of a domain where a given task lay". Knowledge based systems and knowledge engineering became a major focus of AI research in the 1980s.[86]

The 1980s also saw the birth of Cyc, the first attempt to attack the commonsense knowledge problem directly, by creating a massive database that would contain all the mundane facts that the average person knows. Douglas Lenat, who started and led the project, argued that there is no shortcut — the only way for machines to know the meaning of human concepts is to teach them, one concept at a time, by hand. The project was not expected to be completed for many decades.[87]

Chess playing programs HiTech and Deep Thought defeated chess masters in 1989. Both were developed by Carnegie Mellon University; Deep Thought development paved the way for Deep Blue.

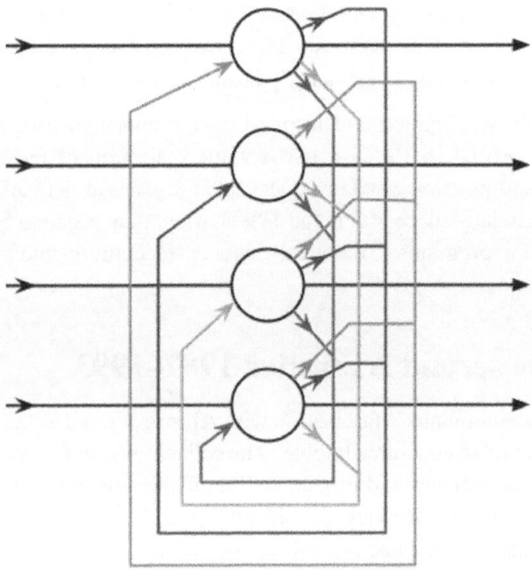

Figure 6: *A Hopfield net with four nodes.*

The money returns: the fifth generation project

In 1981, the Japanese Ministry of International Trade and Industry set aside $850 million for the Fifth generation computer project. Their objectives were to write programs and build machines that could carry on conversations, translate languages, interpret pictures, and reason like human beings.[88] Much to the chagrin of scruffies, they chose Prolog as the primary computer language for the project.

Other countries responded with new programs of their own. The UK began the £350 million Alvey project. A consortium of American companies formed the Microelectronics and Computer Technology Corporation (or "MCC") to fund large scale projects in AI and information technology. DARPA responded as well, founding the Strategic Computing Initiative and tripling its investment in AI between 1984 and 1988.[89]

The revival of connectionism

In 1982, physicist John Hopfield was able to prove that a form of neural network (now called a "Hopfield net") could learn and process information in a completely new way. Around the same time, David Rumelhart popularized a new method for training neural networks called "backpropagation" (discovered years earlier by Paul Werbos). These two discoveries revived the field of connectionism which had been largely abandoned since 1970.

The new field was unified and inspired by the appearance of *Parallel Distributed Processing* in 1986—a two volume collection of papers edited by Rumelhart and psychologist James McClelland. Neural networks would become commercially successful in the 1990s, when they began to be used as the engines driving programs like optical character recognition and speech recognition.

Bust: the second AI winter 1987–1993

The business community's fascination with AI rose and fell in the 1980s in the classic pattern of an economic bubble. The collapse was in the *perception* of AI by government agencies and investors – the field continued to make advances despite the criticism. Rodney Brooks and Hans Moravec, researchers from the related field of robotics, argued for an entirely new approach to artificial intelligence.

AI winter

The term "AI winter" was coined by researchers who had survived the funding cuts of 1974 when they became concerned that enthusiasm for expert systems had spiraled out of control and that disappointment would certainly follow.[90] Their fears were well founded: in the late 1980s and early 1990s, AI suffered a series of financial setbacks.

The first indication of a change in weather was the sudden collapse of the market for specialized AI hardware in 1987. Desktop computers from Apple and IBM had been steadily gaining speed and power and in 1987 they became more powerful than the more expensive Lisp machines made by Symbolics and others. There was no longer a good reason to buy them. An entire industry worth half a billion dollars was demolished overnight.

Eventually the earliest successful expert systems, such as XCON, proved too expensive to maintain. They were difficult to update, they could not learn, they were "brittle" (i.e., they could make grotesque mistakes when given unusual inputs), and they fell prey to problems (such as the qualification problem) that had been identified years earlier. Expert systems proved useful, but only in a few special contexts.[91]

In the late 1980s, the Strategic Computing Initiative cut funding to AI "deeply and brutally." New leadership at DARPA had decided that AI was not "the next wave" and directed funds towards projects that seemed more likely to produce immediate results.

By 1991, the impressive list of goals penned in 1981 for Japan's Fifth Generation Project had not been met. Indeed, some of them, like "carry on a casual conversation" had not been met by 2010.[92] As with other AI projects, expectations had run much higher than what was actually possible.

The importance of having a body: nouvelle AI and embodied reason

In the late 1980s, several researchers advocated a completely new approach to artificial intelligence, based on robotics. They believed that, to show real intelligence, a machine needs to have a *body* — it needs to perceive, move, survive and deal with the world. They argued that these sensorimotor skills are essential to higher level skills like commonsense reasoning and that abstract reasoning was actually the *least* interesting or important human skill (see Moravec's paradox). They advocated building intelligence "from the bottom up."[93]

The approach revived ideas from cybernetics and control theory that had been unpopular since the sixties. Another precursor was David Marr, who had come to MIT in the late 1970s from a successful background in theoretical neuroscience to lead the group studying vision. He rejected all symbolic approaches (*both* McCarthy's logic and Minsky's frames), arguing that AI needed to understand the physical machinery of vision from the bottom up before any symbolic processing took place. (Marr's work would be cut short by leukemia in 1980.)

In a 1990 paper, "Elephants Don't Play Chess,"[94] robotics researcher Rodney Brooks took direct aim at the physical symbol system hypothesis, arguing that symbols are not always necessary since "the world is its own best model. It is always exactly up to date. It always has every detail there is to be known. The trick is to sense it appropriately and often enough." In the 1980s and 1990s, many cognitive scientists also rejected the symbol processing model of the mind and argued that the body was essential for reasoning, a theory called the embodied mind thesis.[95]

AI 1993–2011

The field of AI, now more than a half a century old, finally achieved some of its oldest goals. It began to be used successfully throughout the technology industry, although somewhat behind the scenes. Some of the success was due to increasing computer power and some was achieved by focusing on specific isolated problems and pursuing them with the highest standards of scientific accountability. Still, the reputation of AI, in the business world at least, was less than pristine. Inside the field there was little agreement on the reasons for AI's failure to fulfill the dream of human level intelligence that had captured the imagination of the world in the 1960s. Together, all these factors helped to fragment AI into competing subfields focused on particular problems or approaches, sometimes even under new names that disguised the tarnished pedigree of "artificial intelligence".[96] AI was both more cautious and more successful than it had ever been.

Milestones and Moore's law

On 11 May 1997, Deep Blue became the first computer chess-playing system to beat a reigning world chess champion, Garry Kasparov. The super computer was a specialized version of a framework produced by IBM, and was capable of processing twice as many moves per second as it had during the first match (which Deep Blue had lost), reportedly 200,000,000 moves per second. The event was broadcast live over the internet and received over 74 million hits.

In 2005, a Stanford robot won the DARPA Grand Challenge by driving autonomously for 131 miles along an unrehearsed desert trail.[97] Two years later, a team from CMU won the DARPA Urban Challenge by autonomously navigating 55 miles in an Urban environment while adhering to traffic hazards and all traffic laws.[98] In February 2011, in a Jeopardy! quiz show exhibition match, IBM's question answering system, Watson, defeated the two greatest Jeopardy! champions, Brad Rutter and Ken Jennings, by a significant margin.

These successes were not due to some revolutionary new paradigm, but mostly on the tedious application of engineering skill and on the tremendous power of computers today.[99] In fact, Deep Blue's computer was 10 million times faster than the Ferranti Mark 1 that Christopher Strachey taught to play chess in 1951.[100] This dramatic increase is measured by Moore's law, which predicts that the speed and memory capacity of computers doubles every two years. The fundamental problem of "raw computer power" was slowly being overcome.

Intelligent agents

A new paradigm called "intelligent agents" became widely accepted during the 1990s.[101] Although earlier researchers had proposed modular "divide and conquer" approaches to AI,[102] the intelligent agent did not reach its modern form until Judea Pearl, Allen Newell, Leslie P. Kaelbling, and others brought concepts from decision theory and economics into the study of AI. When the economist's definition of a rational agent was married to computer science's definition of an object or module, the intelligent agent paradigm was complete.

An intelligent agent is a system that perceives its environment and takes actions which maximize its chances of success. By this definition, simple programs that solve specific problems are "intelligent agents", as are human beings and organizations of human beings, such as firms. The intelligent agent paradigm defines AI research as "the study of intelligent agents". This is a generalization of some earlier definitions of AI: it goes beyond studying human intelligence; it studies all kinds of intelligence.[103]

The paradigm gave researchers license to study isolated problems and find solutions that were both verifiable and useful. It provided a common language to describe problems and share their solutions with each other, and with other fields that also used concepts of abstract agents, like economics and control theory. It was hoped that a complete agent architecture (like Newell's SOAR) would one day allow researchers to build more versatile and intelligent systems out of interacting intelligent agents.

"Victory of the neats"

AI researchers began to develop and use sophisticated mathematical tools more than they ever had in the past. There was a widespread realization that many of the problems that AI needed to solve were already being worked on by researchers in fields like mathematics, economics or operations research. The shared mathematical language allowed both a higher level of collaboration with more established and successful fields and the achievement of results which were measurable and provable; AI had become a more rigorous "scientific" discipline. Russell & Norvig (2003) describe this as nothing less than a "revolution" and "the victory of the neats".[104]

Judea Pearl's highly influential 1988 book brought probability and decision theory into AI. Among the many new tools in use were Bayesian networks, hidden Markov models, information theory, stochastic modeling and classical optimization. Precise mathematical descriptions were also developed for "computational intelligence" paradigms like neural networks and evolutionary algorithms.

AI behind the scenes

Algorithms originally developed by AI researchers began to appear as parts of larger systems. AI had solved a lot of very difficult problems[105] and their solutions proved to be useful throughout the technology industry,[106] such as data mining, industrial robotics, logistics, speech recognition,[107] banking software,[108] medical diagnosis and Google's search engine.

The field of AI received little or no credit for these successes in the 1990s and early 2000s. Many of AI's greatest innovations have been reduced to the status of just another item in the tool chest of computer science.[109] Nick Bostrom explains "A lot of cutting edge AI has filtered into general applications, often without being called AI because once something becomes useful enough and common enough it's not labeled AI anymore."

Many researchers in AI in 1990s deliberately called their work by other names, such as informatics, knowledge-based systems, cognitive systems or computational intelligence. In part, this may be because they considered their field to be fundamentally different from AI, but also the new names help to procure funding. In the commercial world at least, the failed promises of the AI Winter continued to haunt AI research into the 2000s, as the New York Times reported in 2005: "Computer scientists and software engineers avoided the term artificial intelligence for fear of being viewed as wild-eyed dreamers."

Where is HAL 9000?

In 1968, Arthur C. Clarke and Stanley Kubrick had imagined that by the year 2001, a machine would exist with an intelligence that matched or exceeded the capability of human beings. The character they created, HAL 9000, was based on a belief shared by many leading AI researchers that such a machine would exist by the year 2001.

In 2001, AI founder Marvin Minsky asked "So the question is why didn't we get HAL in 2001?"[110] Minsky believed that the answer is that the central problems, like commonsense reasoning, were being neglected, while most researchers pursued things like commercial applications of neural nets or genetic algorithms. John McCarthy, on the other hand, still blamed the qualification problem. For Ray Kurzweil, the issue is computer power and, using Moore's Law, he predicted that machines with human-level intelligence will appear by 2029. Jeff Hawkins argued that neural net research ignores the essential properties of the human cortex, preferring simple models that have been successful at solving simple problems. There were many other explanations and for each there was a corresponding research program underway.

Deep learning, big data and artificial general intelligence: 2011–present

In the first decades of the 21st century, access to large amounts of data (known as "big data"), faster computers and advanced machine learning techniques were successfully applied to many problems throughout the economy. In fact, McKinsey Global Institute estimated in their famous paper "Big data: The next frontier for innovation, competition, and productivity" that "by 2009, nearly all sectors in the US economy had at least an average of 200 terabytes of stored data".

By 2016, the market for AI-related products, hardware, and software reached more than 8 billion dollars, and the New York Times reported that interest in AI had reached a "frenzy". The applications of big data began to reach into other fields as well, such as training models in ecology and for various applications in economics. Advances in deep learning (particularly deep convolutional neural networks and recurrent neural networks) drove progress and research in image and video processing, text analysis, and even speech recognition.

Deep learning

Deep learning is a branch of machine learning that models high level abstractions in data by using a deep graph with many processing layers. According to the Universal approximation theorem, deep-ness isn't necessary for a neural network to be able to approximate arbitrary continuous functions. Even so, there are many problems that are common to shallow networks (such as overfitting) that deep networks help avoid. As such, deep neural networks are able to realistically generate much more complex models as compared to their shallow counterparts.

However, deep learning has problems of its own. A common problem for recurrent neural networks is the vanishing gradient problem, which is where gradients passed between layers gradually shrink and literally disappear as they are rounded off to zero. There have been many methods developed to approach this problem, such as Long short-term memory units.

State-of-the-art deep neural network architectures can sometimes even rival human accuracy in fields like computer vision, specifically on things like the MNIST database, and traffic sign recognition.

Language processing engines powered by smart search engines can easily beat humans at answering general trivia questions (such as IBM Watson), and recent developments in deep learning have produced astounding results in competing with humans, in things like Go and Doom (which, being a FPS, has sparked some controversy).

Big Data

Big data refers to a collection of data that cannot be captured, managed, and processed by conventional software tools within a certain time frame. It is a massive amount of decision-making, insight, and process optimization capabilities that require new processing models. In the Big Data Era written by Victor Meyer Schonberg and Kenneth Cooke, big data means that instead of random analysis (sample survey), all data is used for analysis. The 5V characteristics of big data (proposed by IBM): *Volume*, *Velocity*, *Variety*, *Value*, *Veracity*. The strategic significance of big data technology is not to master huge data information, but to specialize in these meaningful data. In other words, if big data is likened to an industry, the key to realizing profitability in this industry is to increase the "Process capability" of the data and realize the "Value added" of the data through "Processing".

Artificial general intelligence

Artificial intelligence is a branch of computer science that attempts to understand the essence of intelligence and produce a new intelligent machine that responds in a manner similar to human intelligence. Research in this area includes robotics, speech recognition, image recognition, Natural language processing and expert systems. Since the birth of artificial intelligence, the theory and technology have become more and more mature, and the application fields have been expanding. It is conceivable that the technological products brought by artificial intelligence in the future will be the "container" of human wisdom. Artificial intelligence can simulate the information process of human consciousness and thinking. Artificial intelligence is not human intelligence, but it can be like human thinking, and it may exceed human intelligence. Artificial general intelligence is also referred to as "**strong AI**",[111] "**full AI**"[112] or as the ability of a machine to perform "general intelligent action".[3] Academic sources reserve "strong AI" to refer to machines capable of experiencing consciousness.

References

<templatestyles src="Template:Refbegin/styles.css" />

- Berlinski, David (2000), *The Advent of the Algorithm*, Harcourt Books, ISBN 0-15-601391-6, OCLC 46890682[113].
- Buchanan, Bruce G. (Winter 2005), "A (Very) Brief History of Artificial Intelligence"[114] (PDF), *AI Magazine*, pp. 53–60, archived from the original[115] (PDF) on 26 December 2007, retrieved 30 August 2007.

- Brooks, Rodney (1990), "Elephants Don't Play Chess"[116] (PDF), *Robotics and Autonomous Systems*, **6**: 3–15, doi: 10.1016/S0921-8890(05)80025-9[117], retrieved 2007-08-30.
- Butler, Samuel (13 June 1863), "Darwin Among the Machines"[118], *the Press, Christchurch, New Zealand*, retrieved 10 October 2008.
- Colby, Kenneth M.; Watt, James B.; Gilbert, John P. (1966), "A Computer Method of Psychotherapy: Preliminary Communication"[119], *The Journal of Nervous and Mental Disease*, vol. 142 no. 2, pp. 148–152, retrieved 17 June 2018.
- Colby, Kenneth M. (September 1974), *Ten Criticisms of Parry*[120] (PDF), Stanford Artificial Intelligence Laboratory, REPORT NO. STAN-CS-74-457, retrieved 17 June 2018.
- CNN (26 July 2006), *AI set to exceed human brain power*[121], CNN.com, retrieved 16 October 2007.
- Copeland, Jack (2000), *Micro-World AI*[122], retrieved 8 October 2008.
- Cordeschi, Roberto (2002), *The Discovery of the Artificial*, Dordrecht: Kluwer..
- Crevier, Daniel (1993), *AI: The Tumultuous Search for Artificial Intelligence*, New York, NY: BasicBooks, ISBN 0-465-02997-3
- Darrach, Brad (20 November 1970), "Meet Shaky, the First Electronic Person", *Life Magazine*, pp. 58–68.
- Doyle, J. (1983), "What is rational psychology? Toward a modern mental philosophy", *AI Magazine*, vol. 4 no. 3, pp. 50–53.
- Dreyfus, Hubert (1965), *Alchemy and AI*, RAND Corporation Memo.
- Dreyfus, Hubert (1972), *What Computers Can't Do*, New York: MIT Press, ISBN 0-06-090613-8, OCLC 5056816[123].
- The Economist (7 June 2007), "Are You Talking to Me?"[124], *The Economist*, retrieved 16 October 2008.
- Feigenbaum, Edward A.; McCorduck, Pamela (1983), *The Fifth Generation: Artificial Intelligence and Japan's Computer Challenge to the World*, Michael Joseph, ISBN 0-7181-2401-4.
- Hawkins, Jeff; Blakeslee, Sandra (2004), *On Intelligence*, New York, NY: Owl Books, ISBN 0-8050-7853-3, OCLC 61273290[125].
- Hebb, D.O. (1949), *The Organization of Behavior*, New York: Wiley, ISBN 0-8058-4300-0, OCLC 48871099[126].
- Hewitt, Carl; Bishop, Peter; Steiger, Richard (1973), *A Universal Modular Actor Formalism for Artificial Intelligence*[127] (PDF), IJCAI, archived from the original[128] (PDF) on 29 December 2009
- Hobbes, Thomas (1651), *Leviathan*.
- Hofstadter, Douglas (1999) [1979], *Gödel, Escher, Bach: an Eternal Golden Braid*, Basic Books, ISBN 0-465-02656-7, OCLC 225590743[129].

- Howe, J. (November 1994), *Artificial Intelligence at Edinburgh University: a Perspective*[130], retrieved 30 August 2007.
- Kolata, G. (1982), "How can computers get common sense?", *Science*, **217** (4566): 1237–1238, Bibcode: 1982Sci...217.1237K[131], doi: 10.1126/science.217.4566.1237[132], PMID 17837639[133].
- Kurzweil, Ray (2005), *The Singularity is Near*, Viking Press, ISBN 0-14-303788-9, OCLC 71826177[134].
- Lakoff, George (1987), *Women, Fire, and Dangerous Things: What Categories Reveal About the Mind*, University of Chicago Press., ISBN 0-226-46804-6.
- Lenat, Douglas; Guha, R. V. (1989), *Building Large Knowledge-Based Systems*, Addison-Wesley, ISBN 0-201-51752-3, OCLC 19981533[135].
- Levitt, Gerald M. (2000), *The Turk, Chess Automaton*, Jefferson, N.C.: McFarland, ISBN 0-7864-0778-6.
- Lighthill, Professor Sir James (1973), "Artificial Intelligence: A General Survey", *Artificial Intelligence: a paper symposium*, Science Research Council
- Lucas, John (1961), "Minds, Machines and Gödel"[136], *Philosophy*, **36** (XXXVI): 112–127, doi: 10.1017/S0031819100057983[137], retrieved 15 October 2008
- Maker, Meg Houston (2006), *AI@50: AI Past, Present, Future*[138], Dartmouth College, archived from the original[139] on 8 October 2008, retrieved 16 October 2008
- Markoff, John (14 October 2005), "Behind Artificial Intelligence, a Squadron of Bright Real People"[140], *The New York Times*, retrieved 16 October 2008
- McCarthy, John; Minsky, Marvin; Rochester, Nathan; Shannon, Claude (31 August 1955), *A Proposal for the Dartmouth Summer Research Project on Artificial Intelligence*[141], retrieved 16 October 2008
- McCarthy, John; Hayes, P. J. (1969), "Some philosophical problems from the standpoint of artificial intelligence", in Meltzer, B. J.; Mitchie, Donald, *Machine Intelligence 4*[142], Edinburgh University Press, pp. 463–502, retrieved 16 October 2008
- McCorduck, Pamela (2004), *Machines Who Think* (2nd ed.), Natick, MA: A. K. Peters, Ltd., ISBN 1-56881-205-1, OCLC 52197627[143].
- McCullough, W. S.; Pitts, W. (1943), "A logical calculus of the ideas immanent in nervous activity", *Bulletin of Mathematical Biophysics*, **5** (4): 115–127, doi: 10.1007/BF02478259[144]
- Menabrea, Luigi Federico; Lovelace, Ada (1843), "Sketch of the Analytical Engine Invented by Charles Babbage"[145], *Scientific Memoirs*, **3**, retrieved 2008-08-29 With notes upon the Memoir by the Translator

- Minsky, Marvin (1967), *Computation: Finite and Infinite Machines*, Englewood Cliffs, N.J.: Prentice-Hall
- Minsky, Marvin; Papert, Seymour (1969), *Perceptrons: An Introduction to Computational Geometry*, The MIT Press, ISBN 0-262-63111-3, OCLC 16924756[146]
- Minsky, Marvin (1974), *A Framework for Representing Knowledge*[147], retrieved 16 October 2008
- Minsky, Marvin (1986), *The Society of Mind*, Simon and Schuster, ISBN 0-671-65713-5, OCLC 223353010[148]
- Minsky, Marvin (2001), *It's 2001. Where Is HAL?*[149], Dr. Dobb's Technetcast, retrieved 8 August 2009
- Moravec, Hans (1976), *The Role of Raw Power in Intelligence*[150], retrieved 16 October 2008
- Moravec, Hans (1988), *Mind Children*, Harvard University Press, ISBN 0-674-57618-7, OCLC 245755104[151]
- NRC (1999), "Developments in Artificial Intelligence", *Funding a Revolution: Government Support for Computing Research*, National Academy Press, ISBN 0-309-06278-0, OCLC 246584055[152]
- Newell, Allen; Simon, H. A. (1963), "GPS: A Program that Simulates Human Thought", in Feigenbaum, E.A.; Feldman, J., *Computers and Thought*, New York: McGraw-Hill, ISBN 0-262-56092-5, OCLC 246968117[153]
- Nick, Martin (2005), *Al Jazari: The Ingenious 13th Century Muslin Mechanic*[154], Al Shindagah, retrieved 16 October 2008.
- O'Connor, Kathleen Malone (1994), *The alchemical creation of life (takwin) and other concepts of Genesis in medieval Islam*[155], University of Pennsylvania, retrieved 2007-01-10
- Olsen, Stefanie (10 May 2004), *Newsmaker: Google's man behind the curtain*[156], CNET, retrieved 17 October 2008.
- Olsen, Stefanie (18 August 2006), *Spying an intelligent search engine*[157], CNET, retrieved 17 October 2008.
- Pearl, J. (1988), *Probabilistic Reasoning in Intelligent Systems: Networks of Plausible Inference*, San Mateo, California: Morgan Kaufmann, ISBN 1-55860-479-0, OCLC 249625842[158].
- Russell, Stuart J.; Norvig, Peter (2003), *Artificial Intelligence: A Modern Approach*[159] (2nd ed.), Upper Saddle River, New Jersey: Prentice Hall, ISBN 0-13-790395-2.
- Poole, David; Mackworth, Alan; Goebel, Randy (1998), *Computational Intelligence: A Logical Approach*[160], Oxford University Press., ISBN 0-19-510270-3.
- Samuel, Arthur L. (July 1959), "Some studies in machine learning using the game of checkers"[161], *IBM Journal of Research and Development*, **3**

(3): 210–219, doi: 10.1147/rd.33.0210[162], retrieved 2007-08-20.

- Searle, John (1980), "Minds, Brains and Programs"[163], *Behavioral and Brain Sciences*, **3** (3): 417–457, doi: 10.1017/S0140525X00005756[164], retrieved May 13, 2009.
- Simon, H. A.; Newell, Allen (1958), "Heuristic Problem Solving: The Next Advance in Operations Research", *Operations Research*, **6**: 1, doi: 10.1287/opre.6.1.1[165].
- Simon, H. A. (1965), *The Shape of Automation for Men and Management*, New York: Harper & Row.
- Skillings, Jonathan (2006), *Newsmaker: Getting machines to think like us*[166], CNET, retrieved 8 October 2008.
- Tascarella, Patty (14 August 2006), "Robotics firms find fundraising struggle, with venture capital shy"[167], *Pittsburgh Business Times*, retrieved 15 Mar 2016.
- Turing, Alan (1936–37), "On Computable Numbers, with an Application to the Entscheidungsproblem"[168], *Proceedings of the London Mathematical Society*, 2 (42): 230–265, doi: 10.1112/plms/s2-42.1.230[169], retrieved 8 October 2008.
- Turing, Alan (October 1950), "Computing Machinery and Intelligence", *Mind*, **LIX** (236): 433–460, doi: 10.1093/mind/LIX.236.433[170], ISSN 0026-4423[171].
- Weizenbaum, Joseph (1976), *Computer Power and Human Reason*, W.H. Freeman & Company, ISBN 0-14-022535-8, OCLC 10952283[172].

<indicator name="good-star"> ⊕ </indicator>

Timeline of artificial intelligence

Artificial intelligence
Major goals
Knowledge reasoning
Planning
Machine learning
Natural language processing
Computer vision
Robotics
Artificial general intelligence
Approaches
Symbolic
Deep learning
Bayesian networks
Evolutionary algorithms
Philosophy
Ethics
Existential risk
Turing test
Chinese room
Friendly AI
History
Timeline
Progress
AI winter
Technology
Applications
Projects
Programming languages
Glossary
Glossary

This is a **timeline of artificial intelligence**.

To 1900

Date	Development
Antiquity	Greek myths of Hephaestus and Pygmalion incorporated the idea of intelligent robots (such as Talos) and artificial beings (such as Galatea and Pandora).
Antiquity	Yan Shi presented King Mu of Zhou with mechanical men.
Antiquity	Sacred mechanical statues built in Egypt and Greece were believed to be capable of wisdom and emotion. Hermes Trismegistus would write "they have *sensus* and *spiritus* ... by discovering the true nature of the gods, man has been able to reproduce it." Mosaic law prohibits the use of automatons in religion.
384 BC–322 BC	Aristotle described the syllogism, a method of formal, mechanical thought.
1st century	Heron of Alexandria created mechanical men and other automatons.
260	Porphyry of Tyros wrote *Isagogê* which categorized knowledge and logic.
∼800	Geber develops the Arabic alchemical theory of *Takwin*, the artificial creation of life in the laboratory, up to and including human life.
1206	Al-Jazari created a programmable orchestra of mechanical human beings.[173]
1275	Ramon Llull, Spanish theologian invents the Ars Magna, a tool for combining concepts mechanically, based on an Arabic astrological tool, the Zairja. The method would be developed further by Gottfried Leibniz in the 17th century.
∼1500	Paracelsus claimed to have created an artificial man out of magnetism, sperm and alchemy.
∼1580	Rabbi Judah Loew ben Bezalel of Prague is said to have invented the Golem, a clay man brought to life.
Early 17th century	René Descartes proposed that bodies of animals are nothing more than complex machines (but that mental phenomena are of a different "substance").
1623	Wilhelm Schickard drew a calculating clock on a letter to Kepler. This will be the first of five unsuccessful attempts at designing a *direct entry* calculating clock in the 17th century (including the designs of Tito Burattini, Samuel Morland and René Grillet)).[174]
1641	Thomas Hobbes published *Leviathan* and presented a mechanical, combinatorial theory of cognition. He wrote "...for reason is nothing but reckoning".[175]
1642	Blaise Pascal invented the mechanical calculator,[176] the first digital calculating machine
1672	Gottfried Leibniz improved the earlier machines, making the Stepped Reckoner to do multiplication and division. He also invented the binary numeral system and envisioned a universal calculus of reasoning (alphabet of human thought) by which arguments could be decided mechanically. Leibniz worked on assigning a specific number to each and every object in the world, as a prelude to an algebraic solution to all possible problems.
1726	Jonathan Swift published *Gulliver's Travels*, which includes this description of the Engine, a machine on the island of Laputa: "a Project for improving speculative Knowledge by practical and mechanical Operations " by using this "Contrivance", "the most ignorant Person at a reasonable Charge, and with a little bodily Labour, may write Books in Philosophy, Poetry, Politicks, Law, Mathematicks, and Theology, with the least Assistance from Genius or study."[177] The machine is a parody of Ars Magna, one of the inspirations of Gottfried Leibniz' mechanism.

1750	Julien Offray de La Mettrie published *L'Homme Machine*, which argued that human thought is strictly mechanical.
1769	Wolfgang von Kempelen built and toured with his chess-playing automaton, The Turk. The Turk was later shown to be a hoax, involving a human chess player.
1818	Mary Shelley published the story of *Frankenstein; or the Modern Prometheus*, a fictional consideration of the ethics of creating sentient beings.
1822–1859	Charles Babbage & Ada Lovelace worked on programmable mechanical calculating machines.
1837	The mathematician Bernard Bolzano made the first modern attempt to formalize semantics.
1854	George Boole set out to "investigate the fundamental laws of those operations of the mind by which reasoning is performed, to give expression to them in the symbolic language of a calculus", inventing Boolean algebra.
1863	Samuel Butler suggested that Darwinian evolution also applies to machines, and speculates that they will one day become conscious and eventually supplant humanity.[178]

1901–1950

Date	Development
1913	Bertrand Russell and Alfred North Whitehead published *Principia Mathematica,* which revolutionized formal logic.
1915	Leonardo Torres y Quevedo built a chess automaton, El Ajedrecista and published speculation about thinking and automata.
1923	Karel Čapek's play *R.U.R. (Rossum's Universal Robots)* opened in London. This is the first use of the word "robot" in English.
1920s and 1930s	Ludwig Wittgenstein and Rudolf Carnap lead philosophy into logical analysis of knowledge. Alonzo Church develops Lambda Calculus to investigate computability using recursive functional notation.
1931	Kurt Gödel showed that sufficiently powerful formal systems, if consistent, permit the formulation of true theorems that are unprovable by any theorem-proving machine deriving all possible theorems from the axioms. To do this he had to build a universal, integer-based programming language, which is the reason why he is sometimes called the "father of theoretical computer science".
1941	Konrad Zuse built the first working program-controlled computers.[179]
1943	Warren Sturgis McCulloch and Walter Pitts publish "A Logical Calculus of the Ideas Immanent in Nervous Activity" (1943), laying foundations for artificial neural networks.
1943	Arturo Rosenblueth, Norbert Wiener and Julian Bigelow coin the term "cybernetics". Wiener's popular book by that name published in 1948.
1945	Game theory which would prove invaluable in the progress of AI was introduced with the 1944 paper, Theory of Games and Economic Behavior by mathematician John von Neumann and economist Oskar Morgenstern.
1945	Vannevar Bush published *As We May Think* (The Atlantic Monthly, July 1945) a prescient vision of the future in which computers assist humans in many activities.

1948	John von Neumann (quoted by E.T. Jaynes) in response to a comment at a lecture that it was impossible for a machine to think: "You insist that there is something a machine cannot do. If you will tell me *precisely* what it is that a machine cannot do, then I can always make a machine which will do just that!". Von Neumann was presumably alluding to the Church-Turing thesis which states that any effective procedure can be simulated by a (generalized) computer.

1950s

Date	Development
1950	Alan Turing proposes the Turing Test as a measure of machine intelligence.[180]
1950	Claude Shannon published a detailed analysis of chess playing as search.
1950	Isaac Asimov published his Three Laws of Robotics.
1951	The first working AI programs were written in 1951 to run on the Ferranti Mark 1 machine of the University of Manchester: a checkers-playing program written by Christopher Strachey and a chess-playing program written by Dietrich Prinz.
1952–1962	Arthur Samuel (IBM) wrote the first game-playing program,[181] for checkers (draughts), to achieve sufficient skill to challenge a respectable amateur. His first checkers-playing program was written in 1952, and in 1955 he created a version that learned to play.[182]
1956	The Dartmouth College summer AI conference is organized by John McCarthy, Marvin Minsky, Nathan Rochester of IBM and Claude Shannon. McCarthy coins the term *artificial intelligence* for the conference.
1956	The first demonstration of the Logic Theorist (LT) written by Allen Newell, J.C. Shaw and Herbert A. Simon (Carnegie Institute of Technology, now Carnegie Mellon University or CMU). This is often called the first AI program, though Samuel's checkers program also has a strong claim.
1957	The General Problem Solver (GPS) demonstrated by Newell, Shaw and Simon while at CMU.
1958	John McCarthy (Massachusetts Institute of Technology or MIT) invented the Lisp programming language.
1958	Herbert Gelernter and Nathan Rochester (IBM) described a theorem prover in geometry that exploits a semantic model of the domain in the form of diagrams of "typical" cases.
1958	Teddington Conference on the Mechanization of Thought Processes was held in the UK and among the papers presented were John McCarthy's *Programs with Common Sense,* Oliver Selfridge's *Pandemonium,* and Marvin Minsky's *Some Methods of Heuristic Programming and Artificial Intelligence.*
1959	John McCarthy and Marvin Minsky founded the MIT AI Lab.
Late 1950s, early 1960s	Margaret Masterman and colleagues at University of Cambridge design semantic nets for machine translation.

1960s

Date	Development
1960s	Ray Solomonoff lays the foundations of a mathematical theory of AI, introducing universal Bayesian methods for inductive inference and prediction.
1960	*Man-Computer Symbiosis* by J.C.R. Licklider.
1961	James Slagle (PhD dissertation, MIT) wrote (in Lisp) the first symbolic integration program, SAINT, which solved calculus problems at the college freshman level.
1961	In *Minds, Machines and Gödel*, John Lucas denied the possibility of machine intelligence on logical or philosophical grounds. He referred to Kurt Gödel's result of 1931: sufficiently powerful formal systems are either inconsistent or allow for formulating true theorems unprovable by any theorem-proving AI deriving all provable theorems from the axioms. Since humans are able to "see" the truth of such theorems, machines were deemed inferior.
1961	Unimation's industrial robot Unimate worked on a General Motors automobile assembly line.
1963	Thomas Evans' program, ANALOGY, written as part of his PhD work at MIT, demonstrated that computers can solve the same analogy problems as are given on IQ tests.
1963	Edward Feigenbaum and Julian Feldman published *Computers and Thought*, the first collection of articles about artificial intelligence.
1963	Leonard Uhr and Charles Vossler published "A Pattern Recognition Program That Generates, Evaluates, and Adjusts Its Own Operators", which described one of the first machine learning programs that could adaptively acquire and modify features and thereby overcome the limitations of simple perceptrons of Rosenblatt
1964	Danny Bobrow's dissertation at MIT (technical report #1 from MIT's AI group, Project MAC), shows that computers can understand natural language well enough to solve algebra word problems correctly.
1964	Bertram Raphael's MIT dissertation on the SIR program demonstrates the power of a logical representation of knowledge for question-answering systems.
1965	J. Alan Robinson invented a mechanical proof procedure, the Resolution Method, which allowed programs to work efficiently with formal logic as a representation language.
1965	Joseph Weizenbaum (MIT) built ELIZA, an interactive program that carries on a dialogue in English language on any topic. It was a popular toy at AI centers on the ARPANET when a version that "simulated" the dialogue of a psychotherapist was programmed.
1965	Edward Feigenbaum initiated Dendral, a ten-year effort to develop software to deduce the molecular structure of organic compounds using scientific instrument data. It was the first expert system.
1966	Ross Quillian (PhD dissertation, Carnegie Inst. of Technology, now CMU) demonstrated semantic nets.
1966	Machine Intelligence[183] workshop at Edinburgh – the first of an influential annual series organized by Donald Michie and others.
1966	Negative report on machine translation kills much work in Natural language processing (NLP) for many years.

1967	Dendral program (Edward Feigenbaum, Joshua Lederberg, Bruce Buchanan, Georgia Sutherland at Stanford University) demonstrated to interpret mass spectra on organic chemical compounds. First successful knowledge-based program for scientific reasoning.
1968	Joel Moses (PhD work at MIT) demonstrated the power of symbolic reasoning for integration problems in the Macsyma program. First successful knowledge-based program in mathematics.
1968	Richard Greenblatt (programmer) at MIT built a knowledge-based chess-playing program, MacHack, that was good enough to achieve a class-C rating in tournament play.
1968	Wallace and Boulton's program, Snob (Comp.J. 11(2) 1968), for unsupervised classification (clustering) uses the Bayesian Minimum Message Length criterion, a mathematical realisation of Occam's razor.
1969	Stanford Research Institute (SRI): Shakey the Robot, demonstrated combining animal locomotion, perception and problem solving.
1969	Roger Schank (Stanford) defined conceptual dependency model for natural language understanding. Later developed (in PhD dissertations at Yale University) for use in story understanding by Robert Wilensky and Wendy Lehnert, and for use in understanding memory by Janet Kolodner.
1969	Yorick Wilks (Stanford) developed the semantic coherence view of language called Preference Semantics, embodied in the first semantics-driven machine translation program, and the basis of many PhD dissertations since such as Bran Boguraev and David Carter at Cambridge.
1969	First International Joint Conference on Artificial Intelligence (IJCAI) held at Stanford.
1969	Marvin Minsky and Seymour Papert publish *Perceptrons*, demonstrating previously unrecognized limits of this feed-forward two-layered structure. This book is considered by some to mark the beginning of the AI winter of the 1970s, a failure of confidence and funding for AI. Nevertheless, significant progress in the field continued (see below).
1969	McCarthy and Hayes started the discussion about the frame problem with their essay, "Some Philosophical Problems from the Standpoint of Artificial Intelligence".

1970s

Date	Development
Early 1970s	Jane Robinson and Don Walker established an influential Natural Language Processing group at SRI.
1970	Jaime Carbonell (Sr.) developed SCHOLAR, an interactive program for computer assisted instruction based on semantic nets as the representation of knowledge.
1970	Bill Woods described Augmented Transition Networks (ATN's) as a representation for natural language understanding.
1970	Patrick Winston's PhD program, ARCH, at MIT learned concepts from examples in the world of children's blocks.
1971	Terry Winograd's PhD thesis (MIT) demonstrated the ability of computers to understand English sentences in a restricted world of children's blocks, in a coupling of his language understanding program, SHRDLU, with a robot arm that carried out instructions typed in English.

1971	Work on the Boyer-Moore theorem prover started in Edinburgh.
1972	Prolog programming language developed by Alain Colmerauer.
1972	Earl Sacerdoti developed one of the first hierarchical planning programs, ABSTRIPS.
1973	The Assembly Robotics Group at University of Edinburgh builds Freddy Robot, capable of using visual perception to locate and assemble models. (See Edinburgh *Freddy* Assembly Robot: a versatile computer-controlled assembly system.)
1973	The Lighthill report gives a largely negative verdict on AI research in Great Britain and forms the basis for the decision by the British government to discontinue support for AI research in all but two universities.
1974	Ted Shortliffe's PhD dissertation on the MYCIN program (Stanford) demonstrated a very practical rule-based approach to medical diagnoses, even in the presence of uncertainty. While it borrowed from DENDRAL, its own contributions strongly influenced the future of expert system development, especially commercial systems.
1975	Earl Sacerdoti developed techniques of partial-order planning in his NOAH system, replacing the previous paradigm of search among state space descriptions. NOAH was applied at SRI International to interactively diagnose and repair electromechanical systems.
1975	Austin Tate developed the Nonlin hierarchical planning system able to search a space of partial plans characterised as alternative approaches to the underlying goal structure of the plan.
1975	Marvin Minsky published his widely read and influential article on Frames as a representation of knowledge, in which many ideas about schemas and semantic links are brought together.
1975	The Meta-Dendral learning program produced new results in chemistry (some rules of mass spectrometry) the first scientific discoveries by a computer to be published in a refereed journal.
Mid-1970s	Barbara Grosz (SRI) established limits to traditional AI approaches to discourse modeling. Subsequent work by Grosz, Bonnie Webber and Candace Sidner developed the notion of "centering", used in establishing focus of discourse and anaphoric references in Natural language processing.
Mid-1970s	David Marr and MIT colleagues describe the "primal sketch" and its role in visual perception.
1976	Douglas Lenat's AM program (Stanford PhD dissertation) demonstrated the discovery model (loosely guided search for interesting conjectures).
1976	Randall Davis demonstrated the power of meta-level reasoning in his PhD dissertation at Stanford.
1978	Tom Mitchell, at Stanford, invented the concept of Version spaces for describing the search space of a concept formation program.
1978	Herbert A. Simon wins the Nobel Prize in Economics for his theory of bounded rationality, one of the cornerstones of AI known as "satisficing".
1978	The MOLGEN program, written at Stanford by Mark Stefik and Peter Friedland, demonstrated that an object-oriented programming representation of knowledge can be used to plan gene-cloning experiments.
1979	Bill VanMelle's PhD dissertation at Stanford demonstrated the generality of MYCIN's representation of knowledge and style of reasoning in his EMYCIN program, the model for many commercial expert system "shells".

1979	Jack Myers and Harry Pople at University of Pittsburgh developed INTERNIST, a knowledge-based medical diagnosis program based on Dr. Myers' clinical knowledge.
1979	Cordell Green, David Barstow, Elaine Kant and others at Stanford demonstrated the CHI system for automatic programming.
1979	The Stanford Cart, built by Hans Moravec, becomes the first computer-controlled, autonomous vehicle when it successfully traverses a chair-filled room and circumnavigates the Stanford AI Lab.
1979	BKG, a backgammon program written by Hans Berliner at CMU, defeats the reigning world champion (in part via luck).
1979	Drew McDermott and Jon Doyle at MIT, and John McCarthy at Stanford begin publishing work on non-monotonic logics and formal aspects of truth maintenance.
Late 1970s	Stanford's SUMEX-AIM resource, headed by Ed Feigenbaum and Joshua Lederberg, demonstrates the power of the ARPAnet for scientific collaboration.

1980s

Date	Development
1980s	Lisp machines developed and marketed. First expert system shells and commercial applications.
1980	First National Conference of the American Association for Artificial Intelligence (AAAI) held at Stanford.
1981	Danny Hillis designs the connection machine, which utilizes Parallel computing to bring new power to AI, and to computation in general. (Later founds Thinking Machines Corporation)
1982	The Fifth Generation Computer Systems project (FGCS), an initiative by Japan's Ministry of International Trade and Industry, begun in 1982, to create a "fifth generation computer" (see history of computing hardware) which was supposed to perform much calculation utilizing massive parallelism.
1983	John Laird and Paul Rosenbloom, working with Allen Newell, complete CMU dissertations on Soar (program).
1983	James F. Allen invents the Interval Calculus, the first widely used formalization of temporal events.
Mid-1980s	Neural Networks become widely used with the Backpropagation algorithm (first described by Paul Werbos in 1974).
1985	The autonomous drawing program, AARON, created by Harold Cohen, is demonstrated at the AAAI National Conference (based on more than a decade of work, and with subsequent work showing major developments).
1986	The team of Ernst Dickmanns at Bundeswehr University of Munich builds the first robot cars, driving up to 55 mph on empty streets.
1986	Barbara Grosz and Candace Sidner create the first computation model of discourse, establishing the field of research.
1987	Marvin Minsky published *The Society of Mind*, a theoretical description of the mind as a collection of cooperating agents. He had been lecturing on the idea for years before the book came out (c.f. Doyle 1983).

1987	Around the same time, Rodney Brooks introduced the subsumption architecture and behavior-based robotics as a more minimalist modular model of natural intelligence; Nouvelle AI.
1987	Commercial launch of generation 2.0 of Alacrity by Alacritous Inc./Allstar Advice Inc. Toronto, the first commercial strategic and managerial advisory system. The system was based upon a forward-chaining, self-developed expert system with 3,000 rules about the evolution of markets and competitive strategies and co-authored by Alistair Davidson and Mary Chung, founders of the firm with the underlying engine developed by Paul Tarvydas. The Alacrity system also included a small financial expert system that interpreted financial statements and models.
1989	Dean Pomerleau at CMU creates ALVINN (An Autonomous Land Vehicle in a Neural Network).

1990s

Date	Development
Early 1990s	TD-Gammon, a backgammon program written by Gerry Tesauro, demonstrates that reinforcement (learning) is powerful enough to create a championship-level game-playing program by competing favorably with world-class players.
1990s	Major advances in all areas of AI, with significant demonstrations in machine learning, intelligent tutoring, case-based reasoning, multi-agent planning, scheduling, uncertain reasoning, data mining, natural language understanding and translation, vision, virtual reality, games, and other topics.
1991	DART scheduling application deployed in the first Gulf War paid back DARPA's investment of 30 years in AI research.[184]
1993	Ian Horswill extended behavior-based robotics by creating Polly, the first robot to navigate using vision and operate at animal-like speeds (1 meter/second).
1993	Rodney Brooks, Lynn Andrea Stein and Cynthia Breazeal started the widely publicized MIT Cog project with numerous collaborators, in an attempt to build a humanoid robot child in just five years.
1993	ISX corporation wins "DARPA contractor of the year" for the Dynamic Analysis and Replanning Tool (DART) which reportedly repaid the US government's entire investment in AI research since the 1950s.
1994	With passengers on board, the twin robot cars VaMP and VITA-2 of Ernst Dickmanns and Daimler-Benz drive more than one thousand kilometers on a Paris three-lane highway in standard heavy traffic at speeds up to 130 km/h. They demonstrate autonomous driving in free lanes, convoy driving, and lane changes left and right with autonomous passing of other cars.
1994	English draughts (checkers) world champion Tinsley resigned a match against computer program Chinook. Chinook defeated 2nd highest rated player, Lafferty. Chinook won the USA National Tournament by the widest margin ever.
1995	"No Hands Across America": A semi-autonomous car drove coast-to-coast across the United States with computer-controlled steering for 2,797 miles (4,501 km) of the 2,849 miles (4,585 km). Throttle and brakes were controlled by a human driver.
1995	One of Ernst Dickmanns' robot cars (with robot-controlled throttle and brakes) drove more than 1000 miles from Munich to Copenhagen and back, in traffic, at up to 120 mph, occasionally executing maneuvers to pass other cars (only in a few critical situations a safety driver took over). Active vision was used to deal with rapidly changing street scenes.

1997	The Deep Blue chess machine (IBM) defeats the (then) world chess champion, Garry Kasparov.
1997	First official RoboCup football (soccer) match featuring table-top matches with 40 teams of interacting robots and over 5000 spectators.
1997	Computer Othello program Logistello defeated the world champion Takeshi Murakami with a score of 6–0.
1998	Tiger Electronics' Furby is released, and becomes the first successful attempt at producing a type of A.I to reach a domestic environment.
1998	Tim Berners-Lee published his Semantic Web Road map paper.
1998	Leslie P. Kaelbling, Michael Littman, and Anthony Cassandra introduce the first method for solving POMDPs offline, jumpstarting widespread use in robotics and automated planning and scheduling
1999	Sony introduces an improved domestic robot similar to a Furby, the AIBO becomes one of the first artificially intelligent "pets" that is also autonomous.
Late 1990s	Web crawlers and other AI-based information extraction programs become essential in widespread use of the World Wide Web.
Late 1990s	Demonstration of an Intelligent room and Emotional Agents at MIT's AI Lab.
Late 1990s	Initiation of work on the Oxygen architecture, which connects mobile and stationary computers in an adaptive network.

2000s

Date	Development
2000	Interactive robopets ("smart toys") become commercially available, realizing the vision of the 18th century novelty toy makers.
2000	Cynthia Breazeal at MIT publishes her dissertation on Sociable machines, describing Kismet (robot), with a face that expresses emotions.
2000	The Nomad robot explores remote regions of Antarctica looking for meteorite samples.
2002	iRobot's Roomba autonomously vacuums the floor while navigating and avoiding obstacles.
2004	OWL Web Ontology Language W3C Recommendation (10 February 2004).
2004	DARPA introduces the DARPA Grand Challenge requiring competitors to produce autonomous vehicles for prize money.
2004	NASA's robotic exploration rovers Spirit and Opportunity autonomously navigate the surface of Mars.
2005	Honda's ASIMO robot, an artificially intelligent humanoid robot, is able to walk as fast as a human, delivering trays to customers in restaurant settings.
2005	Recommendation technology based on tracking web activity or media usage brings AI to marketing. See TiVo Suggestions.
2005	Blue Brain is born, a project to simulate the brain at molecular detail.

2006	The Dartmouth Artificial Intelligence Conference: The Next 50 Years (AI@50) AI@50 (14–16 July 2006)
2007	Philosophical Transactions of the Royal Society, B – Biology, one of the world's oldest scientific journals, puts out a special issue on using AI to understand biological intelligence, titled *Models of Natural Action Selection*
2007	Checkers is solved by a team of researchers at the University of Alberta.
2007	DARPA launches the Urban Challenge for autonomous cars to obey traffic rules and operate in an urban environment.
2009	Google builds self driving car.

2010s

Date	Development
2010	Microsoft launched Kinect for Xbox 360, the first gaming device to track human body movement[185], using just a 3D camera and infra-red detection, enabling users to play their Xbox 360 wirelessly. The award-winning machine learning for human motion capture technology for this device was developed by the Computer Vision group[186] at Microsoft Research, Cambridge.
2011	IBM's Watson computer defeated television game show Jeopardy! champions Rutter and Jennings.
2011–2014	Apple's Siri (2011), Google's Google Now (2012) and Microsoft's Cortana (2014) are smartphone apps that use natural language to answer questions, make recommendations and perform actions.
2013	Robot HRP-2 built by SCHAFT Inc of Japan, a subsidiary of Google, defeats 15 teams to win DARPA's Robotics Challenge Trials. HRP-2 scored 27 out of 32 points in 8 tasks needed in disaster response. Tasks are drive a vehicle, walk over debris, climb a ladder, remove debris, walk through doors, cut through a wall, close valves and connect a hose.
2013	NEIL, the Never Ending Image Learner, is released at Carnegie Mellon University to constantly compare and analyze relationships between different images.
2015	An open letter to ban development and use of autonomous weapons signed by Hawking, Musk, Wozniak and 3,000 researchers in AI and robotics.
2015	Google DeepMind's AlphaGo (version: Fan) defeated 3 time European Go champion 2 dan professional Fan Hui by 5 games to 0.
2016	Google DeepMind's AlphaGo (version: Lee) defeated Lee Sedol 4–1. Lee Sedol is a 9 dan professional Korean Go champion who won 27 major tournaments from 2002 to 2016. Before the match with AlphaGo, Lee Sedol was confident in predicting an easy 5–0 or 4–1 victory.
2017	Asilomar Conference on Beneficial AI was held, to discuss AI ethics and how to bring about beneficial AI while avoiding the existential risk from artificial general intelligence.
2017	Poker AI Libratus individually defeated each of its 4 human opponents—among the best players in the world—at an exceptionally high aggregated winrate, over a statistically significant sample. In contrast to Chess and Go, Poker is an imperfect information game.

2017	Google DeepMind's AlphaGo (version: Master) won 60–0 rounds on two public Go websites including 3 wins against world Go champion Ke Jie.
2017	An OpenAI-machined learned bot played at The International 2017 *Dota 2* tournament in August 2017. It won during a 1v1 demonstration game against professional *Dota 2* player Dendi.
2017	Google DeepMind revealed that AlphaGo Zero—an improved version of AlphaGo—displayed significant performance gains while using far fewer tensor processing units (as compared to AlphaGo Lee; it used same amount of TPU's as AlphaGo Master). Unlike previous versions, which learned the game by observing millions of human moves, AlphaGo Zero learned by playing only against itself. The system then defeated AlphaGo Lee 100 games to zero, and defeated AlphaGo Master 89 to 11. Although unsupervised learning is a step forward, much has yet to be learned about general intelligence. AlphaZero masters chess in 4 hours, defeating the best chess engine, StockFish 8. AlphaZero won 28 out of 100 games, and the remaining 72 games ended in a draw.
2018	Alibaba language processing AI outscores top humans at a Stanford University reading and comprehension test, scoring 82.44 against 82.304 on a set of 100,000 questions.[187]
2018	The European Lab for Learning and Intelligent Systems (*aka* Ellis) proposed as a pan-European competitor to American AI efforts, with the aim of staving off a brain drain of talent, along the lines of CERN after World War II.
2018	Announcement of Google Duplex, a service to allow an AI assistant to book appointments over the phone. The LA Times judges the AI's voice to be a "nearly flawless" imitation of human-sounding speech.

References

- Berlinski, David (2000), *The Advent of the Algorithm*, Harcourt Books
- Buchanan, Bruce G. (2005), "A (Very) Brief History of Artificial Intelligence"[188] (PDF), *AI Magazine*, pp. 53–60, archived from the original[189] (PDF) on 26 September 2007, retrieved 30 August 2007
- Brooks, Rodney (1990), "Elephants Don't Play Chess"[190] (PDF), *Robotics and Autonomous Systems*, **6** (1–2): 3–15, doi: 10.1016/S0921-8890(05)80025-9[191], retrieved 30 August 2007
- Crevier, Daniel (1993), *AI: The Tumultuous Search for Artificial Intelligence*, New York, NY: BasicBooks, ISBN 0-465-02997-3
- Darrach, Brad (20 November 1970), "Meet Shakey, the First Electronic Person", *Life Magazine*, pp. 58–68
- Doyle, J. (1983), "What is rational psychology? Toward a modern mental philosophy", *AI Magazine*, vol. 4 no. 3, pp. 50–53
- Dreyfus, Hubert (1972), *What Computers Can't Do*, MIT Press
- Feigenbaum, Edward A.; McCorduck, Pamela (1983), *The Fifth Generation: Artificial Intelligence and Japan's Computer Challenge to the World*, Michael Joseph, ISBN 0-7181-2401-4
- Hobbes (1651), *Leviathan*
- Hofstadter, Douglas (1980), *Gödel, Escher, Bach: an Eternal Golden Braid*

- Howe, J. (November 1994), *Artificial Intelligence at Edinburgh University: a Perspective*[192], retrieved 30 August 2007
- Kurzweil, Ray (2005), *The Singularity is Near*, Viking Press
- Lakoff, George (1987), *Women, Fire, and Dangerous Things: What Categories Reveal About the Mind*, University of Chicago Press., ISBN 0-226-46804-6
- Lenat, Douglas; Guha, R. V. (1989), *Building Large Knowledge-Based Systems*, Addison-Wesley
- Levitt, Gerald M. (2000), *The Turk, Chess Automaton*, Jefferson, N.C.: McFarland, ISBN 0-7864-0778-6
- Lighthill, Professor Sir James (1973), "Artificial Intelligence: A General Survey", *Artificial Intelligence: a paper symposium*, Science Research Council
- Lucas, John (1961), *Minds, Machines and Gödel*[193]
- McCarthy, John; Minsky, Marvin; Rochester, Nathan; Shannon, Claude (1955), *A Proposal for the Dartmouth Summer Research Project on Artificial Intelligence*[194], archived from the original[195] on 26 August 2007
- McCarthy, John; Hayes, P. J. (1969), "Some philosophical problems from the standpoint of artificial intelligence"[196], *Machine Intelligence*, **4**: 463–502
- McCorduck, Pamela (2004), *Machines Who Think* (2nd ed.), Natick, MA: A. K. Peters, Ltd., ISBN 1-56881-205-1.
- McCullough, W. S.; Pitts, W. (1943), "A logical calculus of the ideas immanent in nervous activity", *Bulletin of Mathematical Biophysics*, **5** (4): 115–127, doi: 10.1007/BF02478259[197]
- Minsky, Marvin (1967), *Computation: Finite and Infinite Machines*, Englewood Cliffs, N.J.: Prentice-Hall
- Minsky, Marvin; Seymour Papert (1969), *Perceptrons: An Introduction to Computational Geometry*, The MIT Press
- Minsky, Marvin (1974), *A Framework for Representing Knowledge*[198]
- Minsky, Marvin (1986), *The Society of Mind*, Simon and Schuster
- Moravec, Hans (1976), *The Role of Raw Power in Intelligence*[199]
- Moravec, Hans (1988), *Mind Children*, Harvard University Press
- United States National Research Council (1999), "Developments in Artificial Intelligence"[200], *Funding a Revolution: Government Support for Computing Research*, National Academy Press, retrieved 30 August 2007
- Newell, Allen; Simon, H. A. (1963), "GPS: A Program that Simulates Human Thought", in Feigenbaum, E.A.; Feldman, J., *Computers and Thought*, New York: McGraw-Hill
- Pearl, J. (1988), *Probabilistic Reasoning in Intelligent Systems: Networks of Plausible Inference*, San Mateo, California: Morgan Kaufmann
- Russell, Stuart J.; Norvig, Peter (2003), *Artificial Intelligence: A Modern*

Approach[201] (2nd ed.), Upper Saddle River, New Jersey: Prentice Hall, ISBN 0-13-790395-2

- Poole, David; Mackworth, Alan; Goebel, Randy (1998), *Computational Intelligence: A Logical Approach*[202], Oxford University Press., ISBN 0-19-510270-3
- Samuel, Arthur L. (July 1959), "Some studies in machine learning using the game of checkers"[203], *IBM Journal of Research and Development*, **3** (3): 210–219, doi: 10.1147/rd.33.0210[204], retrieved 20 August 2007
- Searle, John (1980), "Minds, Brains and Programs"[205], *Behavioral and Brain Sciences*, **3** (3): 417–457, doi: 10.1017/S0140525X00005756[206], archived from the original[207] on 18 January 2010
- Simon, H. A.; Newell, Allen (1958), "Heuristic Problem Solving: The Next Advance in Operations Research", *Operations Research*, **6** (1): 1, doi: 10.1287/opre.6.1.1[208]
- Simon, H. A. (1965), *The Shape of Automation for Men and Management*, New York: Harper & Row
- Turing, Alan (1936–37), "On Computable Numbers, with an Application to the Entscheidungsproblem"[209], *Proceedings of the London Mathematical Society*, 2 (42): 230–265, doi: 10.1112/plms/s2-42.1.230[210]
- Turing, Alan (October 1950), "Computing machinery and intelligence"[211], *Mind*, **LIX** (236): 433–60, doi: 10.1093/mind/LIX.236.433[212], archived from the original[213] on 2 July 2008
- Weizenbaum, Joseph (1976), *Computer Power and Human Reason*, W.H. Freeman & Company

External links

- "Brief History (timeline)"[214], *AI Topics*, Association for the Advancement of Artificial Intelligence
- "Timeline: Building Smarter Machines"[215], *New York Times*, June 24, 2010

Problem

Knowledge representation and reasoning

Knowledge representation and reasoning (**KR**, **KR²**, **KR&R**) is the field of artificial intelligence (AI) dedicated to representing information about the world in a form that a computer system can utilize to solve complex tasks such as diagnosing a medical condition or having a dialog in a natural language. Knowledge representation incorporates findings from psychology about how humans solve problems and represent knowledge in order to design formalisms that will make complex systems easier to design and build. Knowledge representation and reasoning also incorporates findings from logic to automate various kinds of *reasoning*, such as the application of rules or the relations of sets and subsets.

Examples of knowledge representation formalisms include semantic nets, systems architecture, frames, rules, and ontologies. Examples of automated reasoning engines include inference engines, theorem provers, and classifiers.

The KR conference series was established to share ideas and progress on this challenging field.

History

The earliest work in computerized knowledge representation was focused on general problem solvers such as the General Problem Solver (GPS) system developed by Allen Newell and Herbert A. Simon in 1959. These systems featured data structures for planning and decomposition. The system would begin with a goal. It would then decompose that goal into sub-goals and then set out to construct strategies that could accomplish each subgoal.

In these early days of AI, general search algorithms such as A* were also developed. However, the amorphous problem definitions for systems such as GPS meant that they worked only for very constrained toy domains (e.g. the "blocks world"). In order to tackle non-toy problems, AI researchers such as Ed Feigenbaum and Frederick Hayes-Roth realized that it was necessary to focus systems on more constrained problems.

It was the failure of these efforts that led to the cognitive revolution in psychology and to the phase of AI focused on knowledge representation that resulted in expert systems in the 1970s and 80s, production systems, frame languages, etc. Rather than general problem solvers, AI changed its focus to expert systems that could match human competence on a specific task, such as medical diagnosis.

Expert systems gave us the terminology still in use today where AI systems are divided into a Knowledge Base with facts about the world and rules and an inference engine that applies the rules to the knowledge base in order to answer questions and solve problems. In these early systems the knowledge base tended to be a fairly flat structure, essentially assertions about the values of variables used by the rules.

In addition to expert systems, other researchers developed the concept of frame based languages in the mid 1980s. A frame is similar to an object class: It is an abstract description of a category describing things in the world, problems, and potential solutions. Frames were originally used on systems geared toward human interaction, e.g. understanding natural language and the social settings in which various default expectations such as ordering food in a restaurant narrow the search space and allow the system to choose appropriate responses to dynamic situations.

It wasn't long before the frame communities and the rule-based researchers realized that there was synergy between their approaches. Frames were good for representing the real world, described as classes, subclasses, slots (data values) with various constraints on possible values. Rules were good for representing and utilizing complex logic such as the process to make a medical diagnosis. Integrated systems were developed that combined Frames and Rules. One of the most powerful and well known was the 1983 Knowledge Engineering Environment (KEE) from Intellicorp. KEE had a complete rule engine with forward and backward chaining. It also had a complete frame based knowledge base with triggers, slots (data values), inheritance, and message passing. Although message passing originated in the object-oriented community rather than AI it was quickly embraced by AI researchers as well in environments such as KEE and in the operating systems for Lisp machines from Symbolics, Xerox, and Texas Instruments.

The integration of Frames, rules, and object-oriented programming was significantly driven by commercial ventures such as KEE and Symbolics spun off from various research projects. At the same time as this was occurring, there was another strain of research which was less commercially focused and was driven by mathematical logic and automated theorem proving. One of the most influential languages in this research was the KL-ONE language of the mid 80's. KL-ONE was a frame language that had a rigorous semantics, formal definitions for concepts such as an Is-A relation. KL-ONE and languages that were influenced by it such as Loom had an automated reasoning engine that was based on formal logic rather than on IF-THEN rules. This reasoner is called the classifier. A classifier can analyze a set of declarations and infer new assertions, for example, redefine a class to be a subclass or superclass of some other class that wasn't formally specified. In this way the classifier can function as an inference engine, deducing new facts from an existing knowledge base. The classifier can also provide consistency checking on a knowledge base (which in the case of KL-ONE languages is also referred to as an Ontology).

Another area of knowledge representation research was the problem of common sense reasoning. One of the first realizations learned from trying to make software that can function with human natural language was that humans regularly draw on an extensive foundation of knowledge about the real world that we simply take for granted but that is not at all obvious to an artificial agent. Basic principles of common sense physics, causality, intentions, etc. An example is the frame problem, that in an event driven logic there need to be axioms that state things maintain position from one moment to the next unless they are moved by some external force. In order to make a true artificial intelligence agent that can converse with humans using natural language and can process basic statements and questions about the world, it is essential to represent this kind of knowledge. One of the most ambitious programs to tackle this problem was Doug Lenat's Cyc project. Cyc established its own Frame language and had large numbers of analysts document various areas of common sense reasoning in that language. The knowledge recorded in Cyc included common sense models of time, causality, physics, intentions, and many others.

The starting point for knowledge representation is the *knowledge representation hypothesis* first formalized by Brian C. Smith in 1985:

> *Any mechanically embodied intelligent process will be comprised of structural ingredients that a) we as external observers naturally take to represent a propositional account of the knowledge that the overall process exhibits, and b) independent of such external semantic attribution, play a formal but causal and essential role in engendering the behavior that manifests that knowledge.*

Currently one of the most active areas of knowledge representation research are projects associated with the semantic web. The semantic web seeks to add a layer of semantics (meaning) on top of the current Internet. Rather than indexing web sites and pages via keywords, the semantic web creates large ontologies of concepts. Searching for a concept will be more effective than traditional text only searches. Frame languages and automatic classification play a big part in the vision for the future semantic web. The automatic classification gives developers technology to provide order on a constantly evolving network of knowledge. Defining ontologies that are static and incapable of evolving on the fly would be very limiting for Internet-based systems. The classifier technology provides the ability to deal with the dynamic environment of the Internet.

Recent projects funded primarily by the Defense Advanced Research Projects Agency (DARPA) have integrated frame languages and classifiers with markup languages based on XML. The Resource Description Framework (RDF) provides the basic capability to define classes, subclasses, and properties of objects. The Web Ontology Language (OWL) provides additional levels of semantics and enables integration with classification engines.

Overview

Knowledge-representation is the field of artificial intelligence that focuses on designing computer representations that capture information about the world that can be used to solve complex problems.

The justification for knowledge representation is that conventional procedural code is not the best formalism to use to solve complex problems. Knowledge representation makes complex software easier to define and maintain than procedural code and can be used in expert systems.

For example, talking to experts in terms of business rules rather than code lessens the semantic gap between users and developers and makes development of complex systems more practical.

Knowledge representation goes hand in hand with automated reasoning because one of the main purposes of explicitly representing knowledge is to be able to reason about that knowledge, to make inferences, assert new knowledge, etc. Virtually all knowledge representation languages have a reasoning or inference engine as part of the system.

A key trade-off in the design of a knowledge representation formalism is that between expressivity and practicality. The ultimate knowledge representation formalism in terms of expressive power and compactness is First Order Logic (FOL). There is no more powerful formalism than that used by mathematicians to define general propositions about the world. However, FOL has two drawbacks as a knowledge representation formalism: ease of use and practicality of implementation. First order logic can be intimidating even for many software developers. Languages which do not have the complete formal power of FOL can still provide close to the same expressive power with a user interface that is more practical for the average developer to understand. The issue of practicality of implementation is that FOL in some ways is too expressive. With FOL it is possible to create statements (e.g. quantification over infinite sets) that would cause a system to never terminate if it attempted to verify them.

Thus, a subset of FOL can be both easier to use and more practical to implement. This was a driving motivation behind rule-based expert systems. IF-THEN rules provide a subset of FOL but a very useful one that is also very intuitive. The history of most of the early AI knowledge representation formalisms; from databases to semantic nets to theorem provers and production systems can be viewed as various design decisions on whether to emphasize expressive power or computability and efficiency.

In a key 1993 paper on the topic, Randall Davis of MIT outlined five distinct roles to analyze a knowledge representation framework:

- A knowledge representation (KR) is most fundamentally a surrogate, a substitute for the thing itself, used to enable an entity to determine consequences by thinking rather than acting, i.e., by reasoning about the world rather than taking action in it.
- It is a set of ontological commitments, i.e., an answer to the question: In what terms should I think about the world?
- It is a fragmentary theory of intelligent reasoning, expressed in terms of three components: (i) the representation's fundamental conception of intelligent reasoning; (ii) the set of inferences the representation sanctions; and (iii) the set of inferences it recommends.
- It is a medium for pragmatically efficient computation, i.e., the computational environment in which thinking is accomplished. One contribution to this pragmatic efficiency is supplied by the guidance a representation provides for organizing information so as to facilitate making the recommended inferences.
- It is a medium of human expression, i.e., a language in which we say things about the world.

Knowledge representation and reasoning are a key enabling technology for the Semantic web. Languages based on the Frame model with automatic classification provide a layer of semantics on top of the existing Internet. Rather than searching via text strings as is typical today, it will be possible to define logical queries and find pages that map to those queries. The automated reasoning component in these systems is an engine known as the classifier. Classifiers focus on the subsumption relations in a knowledge base rather than rules. A classifier can infer new classes and dynamically change the ontology as new information becomes available. This capability is ideal for the ever-changing and evolving information space of the Internet.

The Semantic web integrates concepts from knowledge representation and reasoning with markup languages based on XML. The Resource Description Framework (RDF) provides the basic capabilities to define knowledge-based objects on the Internet with basic features such as Is-A relations and object properties. The Web Ontology Language (OWL) adds additional semantics and integrates with automatic classification reasoners.

Characteristics

In 1985, Ron Brachman categorized the core issues for knowledge representation as follows:

- Primitives. What is the underlying framework used to represent knowledge? Semantic networks were one of the first knowledge representation primitives. Also, data structures and algorithms for general fast search. In this area, there is a strong overlap with research in data structures and algorithms in computer science. In early systems, the Lisp programming language, which was modeled after the lambda calculus, was often used as a form of functional knowledge representation. Frames and Rules were the next kind of primitive. Frame languages had various mechanisms for expressing and enforcing constraints on frame data. All data in frames are stored in slots. Slots are analogous to relations in entity-relation modeling and to object properties in object-oriented modeling. Another technique for primitives is to define languages that are modeled after First Order Logic (FOL). The most well known example is Prolog, but there are also many special purpose theorem proving environments. These environments can validate logical models and can deduce new theories from existing models. Essentially they automate the process a logician would go through in analyzing a model. Theorem proving technology had some specific practical applications in the areas of software engineering. For example, it is possible to prove that a software program rigidly adheres to a formal logical specification.
- Meta-representation. This is also known as the issue of reflection in computer science. It refers to the capability of a formalism to have access to information about its own state. An example would be the meta-object protocol in Smalltalk and CLOS that gives developers run time access to the class objects and enables them to dynamically redefine the structure of the knowledge base even at run time. Meta-representation means the knowledge representation language is itself expressed in that language. For example, in most Frame based environments all frames would be

instances of a frame class. That class object can be inspected at run time, so that the object can understand and even change its internal structure or the structure of other parts of the model. In rule-based environments, the rules were also usually instances of rule classes. Part of the meta protocol for rules were the meta rules that prioritized rule firing.

- Incompleteness. Traditional logic requires additional axioms and constraints to deal with the real world as opposed to the world of mathematics. Also, it is often useful to associate degrees of confidence with a statement. I.e., not simply say "Socrates is Human" but rather "Socrates is Human with confidence 50%". This was one of the early innovations from expert systems research which migrated to some commercial tools, the ability to associate certainty factors with rules and conclusions. Later research in this area is known as fuzzy logic.
- Definitions and universals vs. facts and defaults. Universals are general statements about the world such as "All humans are mortal". Facts are specific examples of universals such as "Socrates is a human and therefore mortal". In logical terms definitions and universals are about universal quantification while facts and defaults are about existential quantifications. All forms of knowledge representation must deal with this aspect and most do so with some variant of set theory, modeling universals as sets and subsets and definitions as elements in those sets.
- Non-monotonic reasoning. Non-monotonic reasoning allows various kinds of hypothetical reasoning. The system associates facts asserted with the rules and facts used to justify them and as those facts change updates the dependent knowledge as well. In rule based systems this capability is known as a truth maintenance system.
- Expressive adequacy. The standard that Brachman and most AI researchers use to measure expressive adequacy is usually First Order Logic (FOL). Theoretical limitations mean that a full implementation of FOL is not practical. Researchers should be clear about how expressive (how much of full FOL expressive power) they intend their representation to be.
- Reasoning efficiency. This refers to the run time efficiency of the system. The ability of the knowledge base to be updated and the reasoner to develop new inferences in a reasonable period of time. In some ways, this is the flip side of expressive adequacy. In general, the more powerful a representation, the more it has expressive adequacy, the less efficient its automated reasoning engine will be. Efficiency was often an issue, especially for early applications of knowledge representation technology. They were usually implemented in interpreted environments such as Lisp, which were slow compared to more traditional platforms of the time.

Ontology engineering

In the early years of knowledge-based systems the knowledge-bases were fairly small. The knowledge-bases that were meant to actually solve real problems rather than do proof of concept demonstrations needed to focus on well defined problems. So for example, not just medical diagnosis as a whole topic, but medical diagnosis of certain kinds of diseases.

As knowledge-based technology scaled up, the need for larger knowledge bases and for modular knowledge bases that could communicate and integrate with each other became apparent. This gave rise to the discipline of ontology engineering, designing and building large knowledge bases that could be used by multiple projects. One of the leading research projects in this area was the Cyc project. Cyc was an attempt to build a huge encyclopedic knowledge base that would contain not just expert knowledge but common sense knowledge. In designing an artificial intelligence agent, it was soon realized that representing common sense knowledge, knowledge that humans simply take for granted, was essential to make an AI that could interact with humans using natural language. Cyc was meant to address this problem. The language they defined was known as CycL.

After CycL, a number of ontology languages have been developed. Most are declarative languages, and are either frame languages, or are based on first-order logic. Modularity—the ability to define boundaries around specific domains and problem spaces—is essential for these languages because as stated by Tom Gruber, "Every ontology is a treaty- a social agreement among people with common motive in sharing." There are always many competing and differing views that make any general purpose ontology impossible. A general purpose ontology would have to be applicable in any domain and different areas of knowledge need to be unified.[216]

There is a long history of work attempting to build ontologies for a variety of task domains, e.g., an ontology for liquids,[217] the lumped element model widely used in representing electronic circuits (e.g.,[218]), as well as ontologies for time, belief, and even programming itself. Each of these offers a way to see some part of the world.

The lumped element model, for instance, suggests that we think of circuits in terms of components with connections between them, with signals flowing instantaneously along the connections. This is a useful view, but not the only possible one. A different ontology arises if we need to attend to the electrodynamics in the device: Here signals propagate at finite speed and an object (like a resistor) that was previously viewed as a single component with an I/O behavior may now have to be thought of as an extended medium through which an electromagnetic wave flows.

Ontologies can of course be written down in a wide variety of languages and notations (e.g., logic, LISP, etc.); the essential information is not the form of that language but the content, i.e., the set of concepts offered as a way of thinking about the world. Simply put, the important part is notions like connections and components, not the choice between writing them as predicates or LISP constructs.

The commitment made selecting one or another ontology can produce a sharply different view of the task at hand. Consider the difference that arises in selecting the lumped element view of a circuit rather than the electrodynamic view of the same device. As a second example, medical diagnosis viewed in terms of rules (e.g., MYCIN) looks substantially different from the same task viewed in terms of frames (e.g., INTERNIST). Where MYCIN sees the medical world as made up of empirical associations connecting symptom to disease, INTERNIST sees a set of prototypes, in particular prototypical diseases, to be matched against the case at hand.

Further reading

- Ronald J. Brachman; What IS-A is and isn't. An Analysis of Taxonomic Links in Semantic Networks[219]; IEEE Computer, 16 (10); October 1983
- Ronald J. Brachman, Hector J. Levesque *Knowledge Representation and Reasoning*, Morgan Kaufmann, 2004 ISBN 978-1-55860-932-7
- Ronald J. Brachman, Hector J. Levesque (eds) *Readings in Knowledge Representation*, Morgan Kaufmann, 1985, ISBN 0-934613-01-X
- Chein, M., Mugnier, M.-L. (2009), *Graph-based Knowledge Representation: Computational Foundations of Conceptual Graphs*[220], Springer, 2009,ISBN 978-1-84800-285-2.
- Randall Davis, Howard Shrobe, and Peter Szolovits; What Is a Knowledge Representation?[221] AI Magazine, 14(1):17-33,1993
- Ronald Fagin, Joseph Y. Halpern, Yoram Moses, Moshe Y. Vardi *Reasoning About Knowledge*, MIT Press, 1995, ISBN 0-262-06162-7
- Jean-Luc Hainaut, Jean-Marc Hick, Vincent Englebert, Jean Henrard, Didier Roland: Understanding Implementations of IS-A Relations[222]. ER 1996: 42-57
- Hermann Helbig: *Knowledge Representation and the Semantics of Natural Language*[223], Springer, Berlin, Heidelberg, New York 2006
- Frank van Harmelen, Vladimir Lifschitz and Bruce Porter: Handbook of Knowledge Representation[224] 2007.
- Arthur B. Markman: *Knowledge Representation* Lawrence Erlbaum Associates, 1998
- John F. Sowa: *Knowledge Representation*: Logical, Philosophical, and Computational Foundations. Brooks/Cole: New York, 2000

- Adrian Walker, Michael McCord, John F. Sowa, and Walter G. Wilson: *Knowledge Systems and Prolog*, Second Edition, Addison-Wesley, 1990
- Mary-Anne Williams and Hans Rott: "Frontiers in Belief Revision, Kluwer", 2001.

External links

- What is a Knowledge Representation?[225] by Randall Davis and others
- Introduction to Knowledge Modeling[226] by Pejman Makhfi
- Introduction to Description Logics course[227] by Enrico Franconi, Faculty of Computer Science, Free University of Bolzano, Italy
- DATR Lexical knowledge representation language[228]
- Loom Project Home Page[229]
- Description Logic in Practice: A CLASSIC Application[230]
- The Rule Markup Initiative[231]
- Nelements KOS[232] - a non-free 3d knowledge representation system

Wikimedia Commons has media related to ***Knowledge representation***.

Commonsense knowledge (artificial intelligence)

In artificial intelligence research, **commonsense knowledge** consists of facts about the everyday world, such as "Lemons are sour", that all humans are expected to know. The first AI program to address common sense knowledge was Advice Taker in 1959 by John McCarthy .

It is currently an unsolved problem in Artificial General Intelligence and is a focus of the Paul Allen Institute for Artificial Intelligence. Commonsense knowledge can underpin a commonsense reasoning process, to attempt inferences such as "You might bake a cake because you want people to eat the cake." A natural language processing process can be attached to the commonsense knowledge base to allow the knowledge base to attempt to answer commonsense questions about the world.[233] Common sense knowledge also helps to solve problems in the face of incomplete information. Using widely held beliefs about everyday objects, or common sense knowledge, AI systems make common sense assumptions or default assumptions about the unknown similar to the way people do. In an AI system or in english, this is expressed as 'Normally P holds', 'Usually P' or 'Typically P so Assume P'. For example if we know the fact 'tweety is a bird' , because we know the commonly held belief about birds, Typically Birds Fly, without knowing anything else about tweety, we may reasonably assume the fact that 'tweety can fly.' As more knowledge of the world is discovered or learned over time, the AI system can revise its assumptions about tweety using a truth maintenance process. If we later learn that 'tweety is a penguin' then truth maintenance revises this assumption because we also know 'penguins do not fly'.

Commonsense reasoning

Commonsense reasoning simulates the human ability to make presumptions about the type and essence of ordinary situations they encounter every day, including time, missing or incomplete information and cause and effect. The ability to explain cause and effect is an important aspect of explainable AI. Compared with humans, all existing computer programs that attempt human-level AI perform extremely poorly on modern "commonsense reasoning" benchmark tests such as the Winograd Schema Challenge. The problem of attaining human-level competency at "commonsense knowledge" tasks is considered to probably be "AI complete" (that is, solving it would require the ability to synthesize a fully human-level intelligence),[234,235] although some oppose this notion and believe compassionate intelligence is also required for human-level AI.[236] Common sense reasoning has been applied successfully in more limited domains such as automated diagnosis or analysis.

Applications

Around 2013, MIT researchers developed BullySpace, an extension of the commonsense knowledgebase ConceptNet, to catch taunting social media comments. BullySpace included over 200 semantic assertions based around stereotypes, to help the system infer that comments like "Put on a wig and lipstick and be who you really are" are more likely to be an insult if directed at a boy than a girl.

ConceptNet has also been used by chatbots and by computers that compose original fiction. At Lawrence Livermore National Laboratory, common sense knowledge was used in an intelligent software agent to detect violations of a comprehensive nuclear test ban treaty.

Data

As an example, as of 2012 ConceptNet includes these 21 language-independent relations:[237]

- IsA
- UsedFor
- HasA
- CapableOf
- Desires
- CreatedBy ("cake" can be creating by "baking")
- PartOf
- Causes
- LocatedNear
- AtLocation (Somewhere a "cook" can be is a "restaurant")
- DefinedAs
- SymbolOf (*X* represents *Y*)
- ReceivesAction ("cake" can be "eaten")
- HasPrerequisite (*X* can't do *Y* unless *A* does *B*)
- MotivatedByGoal (You would "bake" because you want to "eat")
- CausesDesire ("baking" makes you want to "follow recipe")
- MadeOf
- HasFirstSubevent (The first thing required when you're doing *X* is for entity *Y* to do *Z*)
- HasSubevent ("eat" has subevent "swallow")
- HasLastSubevent

Commonsense knowledge bases

- Cyc
- Open Mind Common Sense (data source) and ConceptNet (datastore and NLP engine)
- ThoughtTreasure
- Mindpixel
- True Knowledge
- DBpediaWikipedia:Citation needed
- Graphiq

Automated planning and scheduling

Automated planning and scheduling, sometimes denoted as simply **AI Planning**, is a branch of artificial intelligence that concerns the realization of strategies or action sequences, typically for execution by intelligent agents, autonomous robots and unmanned vehicles. Unlike classical control and classification problems, the solutions are complex and must be discovered and optimized in multidimensional space. Planning is also related to decision theory.

In known environments with available models, planning can be done offline. Solutions can be found and evaluated prior to execution. In dynamically unknown environments, the strategy often needs to be revised online. Models and policies must be adapted. Solutions usually resort to iterative trial and error processes commonly seen in artificial intelligence. These include dynamic programming, reinforcement learning and combinatorial optimization. Languages used to describe planning and scheduling are often called action languages.

Overview

Given a description of the possible initial states of the world, a description of the desired goals, and a description of a set of possible actions, the planning problem is to synthesise a plan that is guaranteed (when applied to any of the initial states) to generate a state which contains the desired goals (such a state is called a goal state).

The difficulty of planning is dependent on the simplifying assumptions employed. Several classes of planning problems can be identified depending on the properties the problems have in several dimensions.

- Are the actions deterministic or nondeterministic? For nondeterministic actions, are the associated probabilities available?
- Are the state variables discrete or continuous? If they are discrete, do they have only a finite number of possible values?
- Can the current state be observed unambiguously? There can be full observability and partial observability.
- How many initial states are there, finite or arbitrarily many?
- Do actions have a duration?
- Can several actions be taken concurrently, or is only one action possible at a time?
- Is the objective of a plan to reach a designated goal state, or to maximize a reward function?

- Is there only one agent or are there several agents? Are the agents cooperative or selfish? Do all of the agents construct their own plans separately, or are the plans constructed centrally for all agents?

The simplest possible planning problem, known as the Classical Planning Problem, is determined by:

- a unique known initial state,
- durationless actions,
- deterministic actions,
- which can be taken only one at a time,
- and a single agent.

Since the initial state is known unambiguously, and all actions are deterministic, the state of the world after any sequence of actions can be accurately predicted, and the question of observability is irrelevant for classical planning.

Further, plans can be defined as sequences of actions, because it is always known in advance which actions will be needed.

With nondeterministic actions or other events outside the control of the agent, the possible executions form a tree, and plans have to determine the appropriate actions for every node of the tree.

Discrete-time Markov decision processes (MDP) are planning problems with:

- durationless actions,
- nondeterministic actions with probabilities,
- full observability,
- maximization of a reward function,
- and a single agent.

When full observability is replaced by partial observability, planning corresponds to partially observable Markov decision process (POMDP).

If there are more than one agent, we have multi-agent planning, which is closely related to game theory.

Domain Independent Planning

In AI Planning, planners typically input a domain model (a description of a set of possible actions which model the domain) as well as the specific problem to be solved specified by the initial state and goal, in contrast to those in which there is no input domain specified. Such planners are called "Domain Independent" to emphasis the fact that they can solve planning problems from a wide range of domains. Typical examples of domains are block stacking, logistics, workflow management, and robot task planning. Hence a single domain independent planner can be used to solve planning problems in all these various domains. On the other hand, a route planner is typical of a domain specific planner.

Planning Domain Modelling Languages

The most commonly used languages for representing planning domains and specific planning problems, such as STRIPS and PDDL for Classical Planning, are based on state variables. Each possible state of the world is an assignment of values to the state variables, and actions determine how the values of the state variables change when that action is taken. Since a set of state variables induce a state space that has a size that is exponential in the set, planning, similarly to many other computational problems, suffers from the curse of dimensionality and the combinatorial explosion.

An alternative language for describing planning problems is that of hierarchical task networks, in which a set of tasks is given, and each task can be either realized by a primitive action or decomposed into a set of other tasks. This does not necessarily involve state variables, although in more realistic applications state variables simplify the description of task networks.

Algorithms for planning

Classical planning

- forward chaining state space search, possibly enhanced with heuristics
- backward chaining search, possibly enhanced by the use of state constraints (see STRIPS, graphplan)
- partial-order planning

Reduction to other problems

- reduction to the propositional satisfiability problem (satplan).
- reduction to Model checking - both are essentially problems of traversing state spaces, and the classical planning problem corresponds to a subclass of model checking problems.

Temporal planning

Temporal planning can be solved with methods similar to classical planning. The main difference is, because of the possibility of several, temporally overlapping actions with a duration being taken concurrently, that the definition of a state has to include information about the current absolute time and how far the execution of each active action has proceeded. Further, in planning with rational or real time, the state space may be infinite, unlike in classical planning or planning with integer time. Temporal planning is closely related to scheduling problems. Temporal planning can also be understood in terms of timed automata.

Probabilistic planning

Probabilistic planning can be solved with iterative methods such as value iteration and policy iteration, when the state space is sufficiently small. With partial observability, probabilistic planning is similarly solved with iterative methods, but using a representation of the value functions defined for the space of beliefs instead of states.

Preference-based planning

In preference-based planning, the objective is not only to produce a plan but also to satisfy user-specified preferences. A difference to the more common reward-based planning, for example corresponding to MDPs, preferences don't necessarily have a precise numerical value.

Deployment of planning systems

- The Hubble Space Telescope uses a short-term system called SPSS[238] and a long-term planning system called Spike[239].

Further reading

- Vlahavas, I. "Planning and Scheduling"[240]. *EETN*. Archived from the original[241] on 2013-12-22.

External links

- International Conference on Automated Planning and Scheduling[242]

Machine learning

Machine learning and data mining

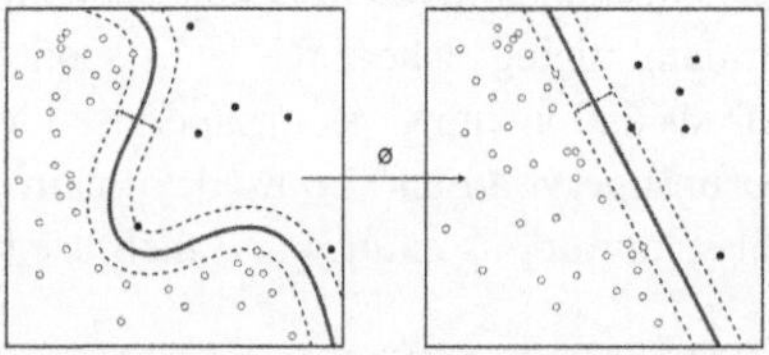

- Machine learning portal

- v
- t
- e[243]

Machine learning is a subset of artificial intelligence in the field of computer science that often uses statistical techniques to give computers the ability to "learn" (i.e., progressively improve performance on a specific task) with data, without being explicitly programmed.[244]

The name *machine learning* was coined in 1959 by Arthur Samuel. Evolved from the study of pattern recognition and computational learning theory in artificial intelligence,[245] machine learning explores the study and construction of algorithms that can learn from and make predictions on data – such algorithms overcome following strictly static program instructions by making data-driven predictions or decisions,:2 through building a model from sample inputs. Machine learning is employed in a range of computing tasks where designing and programming explicit algorithms with good performance is difficult or infeasible; example applications include email filtering, detection of network intruders or malicious insiders working towards a data breach, optical character recognition (OCR), learning to rank, and computer vision.

Machine learning is closely related to (and often overlaps with) computational statistics, which also focuses on prediction-making through the use of computers. It has strong ties to mathematical optimization, which delivers methods, theory and application domains to the field. Machine learning is sometimes conflated with data mining, where the latter subfield focuses more on exploratory data analysis and is known as unsupervised learning.:vii Machine learning can also be unsupervised and be used to learn and establish baseline behavioral profiles for various entities and then used to find meaningful anomalies.

Within the field of data analytics, machine learning is a method used to devise complex models and algorithms that lend themselves to prediction; in commercial use, this is known as predictive analytics. These analytical models allow researchers, data scientists, engineers, and analysts to "produce reliable, repeatable decisions and results" and uncover "hidden insights" through learning from historical relationships and trends in the data.

Overview

Tom M. Mitchell provided a widely quoted, more formal definition of the algorithms studied in the machine learning field: "A computer program is said to learn from experience E with respect to some class of tasks T and performance measure P if its performance at tasks in T, as measured by P, improves with experience E." This definition of the tasks in which machine learning is concerned offers a fundamentally operational definition rather than defining the field in cognitive terms. This follows Alan Turing's proposal in his paper "Computing Machinery and Intelligence", in which the question "Can machines think?" is replaced with the question "Can machines do what we (as thinking entities) can do?". In Turing's proposal the various characteristics that could be possessed by a *thinking machine* and the various implications in constructing one are exposed.

Machine learning tasks

Machine learning tasks are typically classified into two broad categories, depending on whether there is a learning "signal" or "feedback" available to a learning system:

- Supervised learning: The computer is presented with example inputs and their desired outputs, given by a "teacher", and the goal is to learn a general rule that maps inputs to outputs. As special cases, the input signal can be only partially available, or restricted to special feedback:

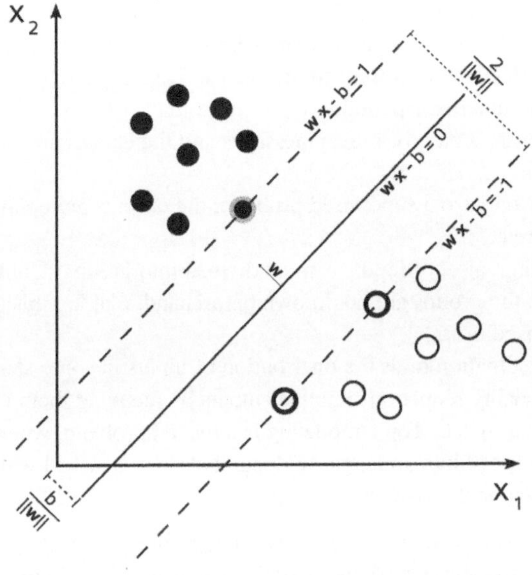

Figure 7: *A support vector machine is a classifier that divides its input space into two regions, separated by a linear boundary. Here, it has learned to distinguish black and white circles.*

- Semi-supervised learning: the computer is given only an incomplete training signal: a training set with some (often many) of the target outputs missing.
- Active learning: the computer can only obtain training labels for a limited set of instances (based on a budget), and also has to optimize its choice of objects to acquire labels for. When used interactively, these can be presented to the user for labeling.
- Reinforcement learning: training data (in form of rewards and punishments) is given only as feedback to the program's actions in a dynamic environment, such as driving a vehicle or playing a game against an opponent.:3

- Unsupervised learning: No labels are given to the learning algorithm, leaving it on its own to find structure in its input. Unsupervised learning can be a goal in itself (discovering hidden patterns in data) or a means towards an end (feature learning).

Machine learning applications

Another categorization of machine learning tasks arises when one considers the desired *output* of a machine-learned system::3

- In classification, inputs are divided into two or more classes, and the learner must produce a model that assigns unseen inputs to one or more (multi-label classification) of these classes. This is typically tackled in a supervised manner. Spam filtering is an example of classification, where the inputs are email (or other) messages and the classes are "spam" and "not spam".
- In regression, also a supervised problem, the outputs are continuous rather than discrete.
- In clustering, a set of inputs is to be divided into groups. Unlike in classification, the groups are not known beforehand, making this typically an unsupervised task.
- Density estimation finds the distribution of inputs in some space.
- Dimensionality reduction simplifies inputs by mapping them into a lower-dimensional space. Topic modeling is a related problem, where a program is given a list of human language documents and is tasked with finding out which documents cover similar topics.

Among other categories of machine learning problems, learning to learn learns its own inductive bias based on previous experience. Developmental learning, elaborated for robot learning, generates its own sequences (also called curriculum) of learning situations to cumulatively acquire repertoires of novel skills through autonomous self-exploration and social interaction with human teachers and using guidance mechanisms such as active learning, maturation, motor synergies, and imitation.

History and relationships to other fields

Arthur Samuel, an American pioneer in the field of computer gaming and artificial intelligence, coined the term "Machine Learning" in 1959 while at IBM.[246] As a scientific endeavour, machine learning grew out of the quest for artificial intelligence. Already in the early days of AI as an academic discipline, some researchers were interested in having machines learn from data. They attempted to approach the problem with various symbolic methods, as well as what were then termed "neural networks"; these were mostly perceptrons and other models that were later found to be reinventions of the generalized linear models of statistics. Probabilistic reasoning was also employed, especially in automated medical diagnosis.[:488]

However, an increasing emphasis on the logical, knowledge-based approach caused a rift between AI and machine learning. Probabilistic systems were plagued by theoretical and practical problems of data acquisition and representation.[:488] By 1980, expert systems had come to dominate AI, and statistics was out of favor. Work on symbolic/knowledge-based learning did continue within AI, leading to inductive logic programming, but the more statistical line of research was now outside the field of AI proper, in pattern recognition and information retrieval.:708–710; 755 Neural networks research had been abandoned by AI and computer science around the same time. This line, too, was continued outside the AI/CS field, as "connectionism", by researchers from other disciplines including Hopfield, Rumelhart and Hinton. Their main success came in the mid-1980s with the reinvention of backpropagation.:25

Machine learning, reorganized as a separate field, started to flourish in the 1990s. The field changed its goal from achieving artificial intelligence to tackling solvable problems of a practical nature. It shifted focus away from the symbolic approaches it had inherited from AI, and toward methods and models borrowed from statistics and probability theory. It also benefited from the increasing availability of digitized information, and the ability to distribute it via the Internet.

Machine learning and data mining often employ the same methods and overlap significantly, but while machine learning focuses on prediction, based on *known* properties learned from the training data, data mining focuses on the discovery of (previously) *unknown* properties in the data (this is the analysis step of knowledge discovery in databases). Data mining uses many machine learning methods, but with different goals; on the other hand, machine learning also employs data mining methods as "unsupervised learning" or as a preprocessing step to improve learner accuracy. Much of the confusion between these two research communities (which do often have separate conferences and separate journals, ECML PKDD being a major exception) comes from the basic assumptions they work with: in machine learning, performance is usually evaluated with respect to the ability to *reproduce known* knowledge, while in knowledge discovery and data mining (KDD) the key task is the discovery of previously *unknown* knowledge. Evaluated with respect to known knowledge, an uninformed (unsupervised) method will easily be outperformed by other supervised methods, while in a typical KDD task, supervised methods cannot be used due to the unavailability of training data.

Machine learning also has intimate ties to optimization: many learning problems are formulated as minimization of some loss function on a training set of examples. Loss functions express the discrepancy between the predictions of the model being trained and the actual problem instances (for example, in classification, one wants to assign a label to instances, and models are trained to correctly predict the pre-assigned labels of a set of examples). The difference between the two fields arises from the goal of generalization: while optimization algorithms can minimize the loss on a training set, machine learning is concerned with minimizing the loss on unseen samples.

Relation to statistics

Machine learning and statistics are closely related fields. According to Michael I. Jordan, the ideas of machine learning, from methodological principles to theoretical tools, have had a long pre-history in statistics. He also suggested the term data science as a placeholder to call the overall field.

Leo Breiman distinguished two statistical modelling paradigms: data model and algorithmic model, wherein "algorithmic model" means more or less the machine learning algorithms like Random forest.

Some statisticians have adopted methods from machine learning, leading to a combined field that they call *statistical learning*.

Theory

A core objective of a learner is to generalize from its experience. Generalization in this context is the ability of a learning machine to perform accurately on new, unseen examples/tasks after having experienced a learning data set. The training examples come from some generally unknown probability distribution (considered representative of the space of occurrences) and the learner has to build a general model about this space that enables it to produce sufficiently accurate predictions in new cases.

The computational analysis of machine learning algorithms and their performance is a branch of theoretical computer science known as computational learning theory. Because training sets are finite and the future is uncertain, learning theory usually does not yield guarantees of the performance of algorithms. Instead, probabilistic bounds on the performance are quite common. The bias–variance decomposition is one way to quantify generalization error.

For the best performance in the context of generalization, the complexity of the hypothesis should match the complexity of the function underlying the data. If the hypothesis is less complex than the function, then the model has underfit the data. If the complexity of the model is increased in response, then the training error decreases. But if the hypothesis is too complex, then the model is subject to overfitting and generalization will be poorer.

In addition to performance bounds, computational learning theorists study the time complexity and feasibility of learning. In computational learning theory, a computation is considered feasible if it can be done in polynomial time. There are two kinds of time complexity results. Positive results show that a certain class of functions can be learned in polynomial time. Negative results show that certain classes cannot be learned in polynomial time.

Approaches

Decision tree learning

Decision tree learning uses a decision tree as a predictive model, which maps observations about an item to conclusions about the item's target value.

Association rule learning

Association rule learning is a method for discovering interesting relations between variables in large databases.

Artificial neural networks

An artificial neural network (ANN) learning algorithm, usually called "neural network" (NN), is a learning algorithm that is vaguely inspired by biological neural networks. Computations are structured in terms of an interconnected group of artificial neurons, processing information using a connectionist approach to computation. Modern neural networks are non-linear statistical data modeling tools. They are usually used to model complex relationships between inputs and outputs, to find patterns in data, or to capture the statistical structure in an unknown joint probability distribution between observed variables.

Deep learning

Falling hardware prices and the development of GPUs for personal use in the last few years have contributed to the development of the concept of deep learning which consists of multiple hidden layers in an artificial neural network. This approach tries to model the way the human brain processes light and sound into vision and hearing. Some successful applications of deep learning are computer vision and speech recognition.[247]

Inductive logic programming

Inductive logic programming (ILP) is an approach to rule learning using logic programming as a uniform representation for input examples, background knowledge, and hypotheses. Given an encoding of the known background knowledge and a set of examples represented as a logical database of facts, an ILP system will derive a hypothesized logic program that entails all positive and no negative examples. Inductive programming is a related field that considers any kind of programming languages for representing hypotheses (and not only logic programming), such as functional programs.

Support vector machines

Support vector machines (SVMs) are a set of related supervised learning methods used for classification and regression. Given a set of training examples, each marked as belonging to one of two categories, an SVM training algorithm builds a model that predicts whether a new example falls into one category or the other.

Clustering

Cluster analysis is the assignment of a set of observations into subsets (called *clusters*) so that observations within the same cluster are similar according to some predesignated criterion or criteria, while observations drawn from different clusters are dissimilar. Different clustering techniques make different assumptions on the structure of the data, often defined by some *similarity metric* and evaluated for example by *internal compactness* (similarity between members of the same cluster) and *separation* between different clusters. Other methods are based on *estimated density* and *graph connectivity*. Clustering is a method of unsupervised learning, and a common technique for statistical data analysis.

Bayesian networks

A Bayesian network, belief network or directed acyclic graphical model is a probabilistic graphical model that represents a set of random variables and their conditional independencies via a directed acyclic graph (DAG). For example, a Bayesian network could represent the probabilistic relationships between diseases and symptoms. Given symptoms, the network can be used to compute the probabilities of the presence of various diseases. Efficient algorithms exist that perform inference and learning.

Reinforcement learning

Reinforcement learning is concerned with how an *agent* ought to take *actions* in an *environment* so as to maximize some notion of long-term *reward*. Reinforcement learning algorithms attempt to find a *policy* that maps *states* of the world to the actions the agent ought to take in those states. Reinforcement learning differs from the supervised learning problem in that correct input/output pairs are never presented, nor sub-optimal actions explicitly corrected.

Representation learning

Several learning algorithms, mostly unsupervised learning algorithms, aim at discovering better representations of the inputs provided during training. Classical examples include principal components analysis and cluster analysis. Representation learning algorithms often attempt to preserve the information in their input but transform it in a way that makes it useful, often as a pre-processing step before performing classification or predictions, allowing reconstruction of the inputs coming from the unknown data generating distribution, while not being necessarily faithful for configurations that are implausible under that distribution.

Manifold learning algorithms attempt to do so under the constraint that the learned representation is low-dimensional. Sparse coding algorithms attempt to do so under the constraint that the learned representation is sparse (has many zeros). Multilinear subspace learning algorithms aim to learn low-dimensional representations directly from tensor representations for multidimensional data, without reshaping them into (high-dimensional) vectors. Deep learning algorithms discover multiple levels of representation, or a hierarchy of features, with higher-level, more abstract features defined in terms of (or generating) lower-level features. It has been argued that an intelligent machine is one that learns a representation that disentangles the underlying factors of variation that explain the observed data.

Similarity and metric learning

In this problem, the learning machine is given pairs of examples that are considered similar and pairs of less similar objects. It then needs to learn a similarity function (or a distance metric function) that can predict if new objects are similar. It is sometimes used in Recommendation systems.

Sparse dictionary learning

In this method, a datum is represented as a linear combination of basis functions, and the coefficients are assumed to be sparse. Let x be a d-dimensional datum, D be a d by n matrix, where each column of D represents a basis function. r is the coefficient to represent x using D. Mathematically, sparse dictionary learning means solving $x \approx Dr$ where r is sparse. Generally speaking, n is assumed to be larger than d to allow the freedom for a sparse representation.

Learning a dictionary along with sparse representations is strongly NP-hard and also difficult to solve approximately. A popular heuristic method for sparse dictionary learning is K-SVD.

Sparse dictionary learning has been applied in several contexts. In classification, the problem is to determine which classes a previously unseen datum belongs to. Suppose a dictionary for each class has already been built. Then a new datum is associated with the class such that it's best sparsely represented by the corresponding dictionary. Sparse dictionary learning has also been applied in image de-noising. The key idea is that a clean image patch can be sparsely represented by an image dictionary, but the noise cannot.[248]

Genetic algorithms

A genetic algorithm (GA) is a search heuristic that mimics the process of natural selection, and uses methods such as mutation and crossover to generate new genotype in the hope of finding good solutions to a given problem. In machine learning, genetic algorithms found some uses in the 1980s and 1990s. Conversely, machine learning techniques have been used to improve the performance of genetic and evolutionary algorithms.

Rule-based machine learning

Rule-based machine learning is a general term for any machine learning method that identifies, learns, or evolves "rules" to store, manipulate or apply, knowledge. The defining characteristic of a rule-based machine learner is the identification and utilization of a set of relational rules that collectively represent the knowledge captured by the system. This is in contrast to other machine learners that commonly identify a singular model that can be universally applied to any instance in order to make a prediction. Rule-based machine learning approaches include learning classifier systems, association rule learning, and artificial immune systems.

Learning classifier systems

Learning classifier systems (LCS) are a family of rule-based machine learning algorithms that combine a discovery component (e.g. typically a genetic algorithm) with a learning component (performing either supervised learning, reinforcement learning, or unsupervised learning). They seek to identify a set of context-dependent rules that collectively store and apply knowledge in a piecewise manner in order to make predictions.

Feature selection approach

Feature selection is the process of selecting an optimal subset of relevant features for use in model construction. It is assumed the data contains some features that are either redundant or irrelevant, and can thus be removed to reduce calculation cost without incurring much loss of information. Common optimality criteria include accuracy, similarity and information measures.

Applications

Applications for machine learning include:

- Agriculture
- Automated theorem proving[249,250]
- Adaptive websitesWikipedia:Citation needed
- Affective computing
- Bioinformatics
- Brain–machine interfaces
- Cheminformatics
- Classifying DNA sequences
- Computational anatomy
- Computer Networks
- Telecommunication
- Computer vision, including object recognition
- Detecting credit-card fraud
- General game playing[251]
- Information retrieval
- Internet fraud detection
- Linguistics
- Marketing
- Machine learning control
- Machine perception
- Medical diagnosis
- Economics
- Insurance
- Natural language processing
- Natural language understanding[252]
- Optimization and metaheuristic
- Online advertising
- Recommender systems
- Robot locomotion
- Search engines
- Sentiment analysis (or opinion mining)
- Multimodal sentiment analysis
- Sequence mining
- Software engineering
- Speech and handwriting recognition
- Space mapping
- Financial market analysis
- Structural health monitoring
- Syntactic pattern recognition

- Time series forecasting
- User behavior analytics
- Translation

In 2006, the online movie company Netflix held the first "Netflix Prize" competition to find a program to better predict user preferences and improve the accuracy on its existing Cinematch movie recommendation algorithm by at least 10%. A joint team made up of researchers from AT&T Labs-Research in collaboration with the teams Big Chaos and Pragmatic Theory built an ensemble model to win the Grand Prize in 2009 for $1 million.[253] Shortly after the prize was awarded, Netflix realized that viewers' ratings were not the best indicators of their viewing patterns ("everything is a recommendation") and they changed their recommendation engine accordingly.

In 2010 The Wall Street Journal wrote about the firm Rebellion Research and their use of Machine Learning to predict the financial crisis.

In 2012, co-founder of Sun Microsystems Vinod Khosla predicted that 80% of medical doctors jobs would be lost in the next two decades to automated machine learning medical diagnostic software.

In 2014, it has been reported that a machine learning algorithm has been applied in Art History to study fine art paintings, and that it may have revealed previously unrecognized influences between artists.[254]

Model assessments

Although machine learning has been very transformative in some fields, effective machine learning is difficult because finding patterns is hard and often not enough training data are available; as a result, machine-learning programs often fail to deliver.

Classification machine learning models can be validated by accuracy estimation techniques like the Holdout method, which splits the data into a training and test sets (conventionally 2/3 training set and 1/3 test set designation) and evaluates the performance of the training model on the test set. In comparison, the k-fold-cross-validation method randomly splits the data into k subsets where the k - 1 instances of the data subsets are used to train the model while the kth subset instance is used to test the predictive ability of the training model. In addition to the holdout and cross-validation methods, bootstrap, which samples n instances with replacement from the dataset, can be used to assess model accuracy.

In addition to overall accuracy, investigators frequently report sensitivity and specificity meaning True Positive Rate (TPR) and True Negative Rate (TNR) respectively. Similarly, investigators sometimes report the False Positive Rate

(FPR) as well as the False Negative Rate (FNR). However, these rates are ratios that fail to reveal their numerators and denominators. The Total Operating Characteristic (TOC) is an effective method to express a model's diagnostic ability. TOC shows the numerators and denominators of the previously mentioned rates, thus TOC provides more information than the commonly used Receiver Operating Characteristic (ROC) and ROC's associated Area Under the Curve (AUC).

Ethics

Machine learning poses a host of ethical questions. Systems which are trained on datasets collected with biases may exhibit these biases upon use (algorithmic bias), thus digitizing cultural prejudices. For example, using job hiring data from a firm with racist hiring policies may lead to a machine learning system duplicating the bias by scoring job applicants against similarity to previous successful applicants. Responsible collection of data and documentation of algorithmic rules used by a system thus is a critical part of machine learning.

Because language contains biases, machines trained on language *corpora* will necessarily also learn bias.

Other forms of ethical challenges, not related to personal biases, are more seen in health care. There are concerns among health care professionals that these systems might not be designed in the public's interest, but as income generating machines. This is especially true in the United States where there is a perpetual ethical dilemma of improving health care, but also increasing profits. For example, the algorithms could be designed to provide patients with unnecessary tests or medication in which the algorithm's proprietary owners hold stakes in. There is huge potential for machine learning in health care to provide professionals a great tool to diagnose, medicate, and even plan recovery paths for patients, but this will not happen until the personal biases mentioned previously, and these "greed" biases are addressed.

Software

Software suites containing a variety of machine learning algorithms include the following :

Free and open-source software

- CNTK
- Deeplearning4j
- ELKI
- H2O
- Mahout
- Mallet
- MLPACK
- MXNet
- OpenNN
- Orange
- scikit-learn
- Shogun
- Spark MLlib
- TensorFlow
- Torch / PyTorch
- Weka / MOA
- Yooreeka

Proprietary software with free and open-source editions

- KNIME
- RapidMiner

Proprietary software

- Amazon Machine Learning
- Angoss KnowledgeSTUDIO
- Ayasdi
- IBM Data Science Experience
- Google Prediction API
- IBM SPSS Modeler
- KXEN Modeler
- LIONsolver
- Mathematica
- MATLAB
- Microsoft Azure Machine Learning
- Neural Designer
- NeuroSolutions
- Oracle Data Mining
- Oracle AI Platform Cloud Service
- RCASE

- SAP Leonardo Machine Learning
- SAS Enterprise Miner
- SequenceL
- Splunk
- STATISTICA Data Miner

Journals

- *Journal of Machine Learning Research*
- *Machine Learning*
- *Neural Computation*

Conferences

- Conference on Neural Information Processing Systems
- International Conference on Machine Learning
- International Conference on Learning Representations
- Open Data Science Conference (ODSC)[255]

Further reading

<templatestyles src="Template:Refbegin/styles.css" />

- Nils J. Nilsson, *Introduction to Machine Learning*[256].
- Trevor Hastie, Robert Tibshirani and Jerome H. Friedman (2001). *The Elements of Statistical Learning*[257], Springer. ISBN 0-387-95284-5.
- Pedro Domingos (September 2015), The Master Algorithm, Basic Books, ISBN 978-0-465-06570-7
- Ian H. Witten and Eibe Frank (2011). *Data Mining: Practical machine learning tools and techniques* Morgan Kaufmann, 664pp., ISBN 978-0-12-374856-0.
- Ethem Alpaydin (2004). *Introduction to Machine Learning*, MIT Press, ISBN 978-0-262-01243-0.
- David J. C. MacKay. *Information Theory, Inference, and Learning Algorithms*[258] Cambridge: Cambridge University Press, 2003. ISBN 0-521-64298-1
- Richard O. Duda, Peter E. Hart, David G. Stork (2001) *Pattern classification* (2nd edition), Wiley, New York, ISBN 0-471-05669-3.
- Christopher Bishop (1995). *Neural Networks for Pattern Recognition*, Oxford University Press. ISBN 0-19-853864-2.
- Stuart Russell & Peter Norvig, (2002). *Artificial Intelligence – A Modern Approach*. Prentice Hall, ISBN 0-136-04259-7.

- Ray Solomonoff, *An Inductive Inference Machine*, IRE Convention Record, Section on Information Theory, Part 2, pp. 56–62, 1957.
- Ray Solomonoff, " An Inductive Inference Machine[259]" A privately circulated report from the 1956 Dartmouth Summer Research Conference on AI.

External links

Wikimedia Commons has media related to ***Machine learning***.

- International Machine Learning Society[260]
- Popular online course by Andrew Ng, at Coursera[261]. It uses GNU Octave. The course is a free version of Stanford University's actual course taught by Ng, whose lectures are also available for free[262].
- mloss[263] is an academic database of open-source machine learning software.
- Machine Learning Crash Course[264] by Google. This is a free course on machine learning through the use of TensorFlow
- Popular book on Machine Intelligence in Design Automation[265].
- Draft copy of Andrew Ng's book[266].
- Online copy of deep learning book[267] by Ian Goodfellow, Yoshua Bengio and Aaron Courville.

Natural language processing

Natural language processing (**NLP**) is an area of computer science and artificial intelligence concerned with the interactions between computers and human (natural) languages, in particular how to program computers to process and analyze large amounts of natural language data.

Challenges in natural language processing frequently involve speech recognition, natural language understanding, and natural language generation.

History

The history of natural language processing generally started in the 1950s, although work can be found from earlier periods. In 1950, Alan Turing published an article titled "Computing Machinery and Intelligence" which proposed what is now called the Turing test as a criterion of intelligence.

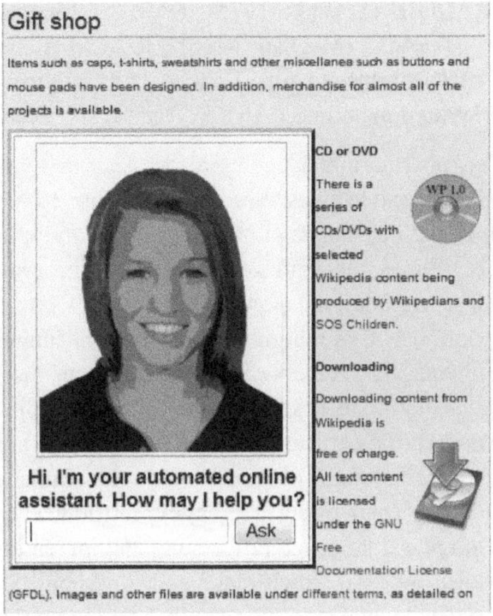

Figure 8: *An automated online assistant providing customer service on a web page, an example of an application where natural language processing is a major component.*[268]

The Georgetown experiment in 1954 involved fully automatic translation of more than sixty Russian sentences into English. The authors claimed that within three or five years, machine translation would be a solved problem. However, real progress was much slower, and after the ALPAC report in 1966, which found that ten-year-long research had failed to fulfill the expectations, funding for machine translation was dramatically reduced. Little further research in machine translation was conducted until the late 1980s, when the first statistical machine translation systems were developed.

Some notably successful natural language processing systems developed in the 1960s were SHRDLU, a natural language system working in restricted "blocks worlds" with restricted vocabularies, and ELIZA, a simulation of a Rogerian psychotherapist, written by Joseph Weizenbaum between 1964 and 1966. Using almost no information about human thought or emotion, ELIZA sometimes provided a startlingly human-like interaction. When the "patient" exceeded the very small knowledge base, ELIZA might provide a generic response, for example, responding to "My head hurts" with "Why do you say your head hurts?".

During the 1970s, many programmers began to write "conceptual ontologies", which structured real-world information into computer-understandable data.

Examples are MARGIE (Schank, 1975), SAM (Cullingford, 1978), PAM (Wilensky, 1978), TaleSpin (Meehan, 1976), QUALM (Lehnert, 1977), Politics (Carbonell, 1979), and Plot Units (Lehnert 1981). During this time, many chatterbots were written including PARRY, Racter, and Jabberwacky.

Up to the 1980s, most natural language processing systems were based on complex sets of hand-written rules. Starting in the late 1980s, however, there was a revolution in natural language processing with the introduction of machine learning algorithms for language processing. This was due to both the steady increase in computational power (see Moore's law) and the gradual lessening of the dominance of Chomskyan theories of linguistics (e.g. transformational grammar), whose theoretical underpinnings discouraged the sort of corpus linguistics that underlies the machine-learning approach to language processing.[269] Some of the earliest-used machine learning algorithms, such as decision trees, produced systems of hard if-then rules similar to existing hand-written rules. However, part-of-speech tagging introduced the use of hidden Markov models to natural language processing, and increasingly, research has focused on statistical models, which make soft, probabilistic decisions based on attaching real-valued weights to the features making up the input data. The cache language models upon which many speech recognition systems now rely are examples of such statistical models. Such models are generally more robust when given unfamiliar input, especially input that contains errors (as is very common for real-world data), and produce more reliable results when integrated into a larger system comprising multiple subtasks.

Many of the notable early successes occurred in the field of machine translation, due especially to work at IBM Research, where successively more complicated statistical models were developed. These systems were able to take advantage of existing multilingual textual corpora that had been produced by the Parliament of Canada and the European Union as a result of laws calling for the translation of all governmental proceedings into all official languages of the corresponding systems of government. However, most other systems depended on corpora specifically developed for the tasks implemented by these systems, which was (and often continues to be) a major limitation in the success of these systems. As a result, a great deal of research has gone into methods of more effectively learning from limited amounts of data.

Recent research has increasingly focused on unsupervised and semi-supervised learning algorithms. Such algorithms are able to learn from data that has not been hand-annotated with the desired answers, or using a combination of annotated and non-annotated data. Generally, this task is much more difficult than supervised learning, and typically produces less accurate results for a given amount of input data. However, there is an enormous amount of non-annotated data available (including, among other things, the entire content of the World Wide Web), which can often make up for the inferior results if the algorithm used has a low enough time complexity to be practical, which some such as Chinese Whispers do.[270]

In recent years, there has been a flurry of results showing deep learning techniques[271,272] achieving state-of-the-art results in many natural language tasks, for example in language modeling,[273] parsing,[274,275] and many others.

Statistical natural language processing (SNLP)

Since the so-called "statistical revolution"[276,277] in the late 1980s and mid 1990s, much natural language processing research has relied heavily on machine learning.

Formerly, many language-processing tasks typically involved the direct hand coding of rules,[278,279] which is not in general robust to natural language variation. The machine-learning paradigm calls instead for using statistical inference to automatically learn such rules through the analysis of large *corpora* of typical real-world examples (a *corpus* (plural, "corpora") is a set of documents, possibly with human or computer annotations).

Many different classes of machine learning algorithms have been applied to natural language processing tasks. These algorithms take as input a large set of "features" that are generated from the input data. Some of the earliest-used algorithms, such as decision trees, produced systems of hard if-then rules similar to the systems of hand-written rules that were then common. Increasingly, however, research has focused on statistical models, which make soft, probabilistic decisions based on attaching real-valued weights to each input feature. Such models have the advantage that they can express the relative certainty of many different possible answers rather than only one, producing more reliable results when such a model is included as a component of a larger system.

Systems based on machine-learning algorithms have many advantages over hand-produced rules:

- The learning procedures used during machine learning automatically focus on the most common cases, whereas when writing rules by hand it is often not at all obvious where the effort should be directed.
- Automatic learning procedures can make use of statistical inference algorithms to produce models that are robust to unfamiliar input (e.g. containing words or structures that have not been seen before) and to erroneous input (e.g. with misspelled words or words accidentally omitted). Generally, handling such input gracefully with hand-written rules—or more generally, creating systems of hand-written rules that make soft decisions—is extremely difficult, error-prone and time-consuming.

- Systems based on automatically learning the rules can be made more accurate simply by supplying more input data. However, systems based on hand-written rules can only be made more accurate by increasing the complexity of the rules, which is a much more difficult task. In particular, there is a limit to the complexity of systems based on hand-crafted rules, beyond which the systems become more and more unmanageable. However, creating more data to input to machine-learning systems simply requires a corresponding increase in the number of man-hours worked, generally without significant increases in the complexity of the annotation process.

Major evaluations and tasks

The following is a list of some of the most commonly researched tasks in natural language processing. Note that some of these tasks have direct real-world applications, while others more commonly serve as subtasks that are used to aid in solving larger tasks.

Though natural language processing tasks are closely intertwined, they are frequently subdivided into categories for convenience. A coarse division is given below.

Syntax

Lemmatization

Morphological segmentation

Separate words into individual morphemes and identify the class of the morphemes. The difficulty of this task depends greatly on the complexity of the morphology (i.e. the structure of words) of the language being considered. English has fairly simple morphology, especially inflectional morphology, and thus it is often possible to ignore this task entirely and simply model all possible forms of a word (e.g. "open, opens, opened, opening") as separate words. In languages such as Turkish or Meitei,[280] a highly agglutinated Indian language, however, such an approach is not possible, as each dictionary entry has thousands of possible word forms.

Part-of-speech tagging

Given a sentence, determine the part of speech for each word. Many words, especially common ones, can serve as multiple parts of speech. For example, "book" can be a noun ("the book on the table") or verb ("to book a flight"); "set" can be a noun, verb or adjective; and "out" can be any of at least five different parts of speech. Some languages have more such ambiguity than others.Wikipedia:Accuracy dispute#Disputed statement Languages with little inflectional morphology, such as English, are particularly prone to such ambiguity. Chinese is prone to such ambiguity because it is a tonal language during verbalization. Such inflection is not readily conveyed via the entities employed within the orthography to convey intended meaning.

Parsing

(see also: Stochastic grammar) Determine the parse tree (grammatical analysis) of a given sentence. The grammar for natural languages is ambiguous and typical sentences have multiple possible analyses. In fact, perhaps surprisingly, for a typical sentence there may be thousands of potential parses (most of which will seem completely nonsensical to a human). There are two primary types of parsing, Dependency Parsing and Constituency Parsing. Dependency Parsing focuses on the relationships between words in a sentence (marking things like Primary Objects and predicates), whereas Constituency Parsing focuses on building out the Parse Tree using a Probabilistic Context-Free Grammar (PCFG).

Sentence breaking (also known as sentence boundary disambiguation)

Given a chunk of text, find the sentence boundaries. Sentence boundaries are often marked by periods or other punctuation marks, but these same characters can serve other purposes (e.g. marking abbreviations).

Stemming

Word segmentation

Separate a chunk of continuous text into separate words. For a language like English, this is fairly trivial, since words are usually separated by spaces. However, some written languages like Chinese, Japanese and Thai do not mark word boundaries in such a fashion, and in those languages text segmentation is a significant task requiring knowledge of the vocabulary and morphology of words in the language.

Terminology extraction

The goal of terminology extraction is to automatically extract relevant terms from a given corpus.

Semantics

Lexical semantics
What is the computational meaning of individual words in context?

Machine translation
Automatically translate text from one human language to another. This is one of the most difficult problems, and is a member of a class of problems colloquially termed "AI-complete", i.e. requiring all of the different types of knowledge that humans possess (grammar, semantics, facts about the real world, etc.) in order to solve properly.

Named entity recognition (NER)
Given a stream of text, determine which items in the text map to proper names, such as people or places, and what the type of each such name is (e.g. person, location, organization). Note that, although capitalization can aid in recognizing named entities in languages such as English, this information cannot aid in determining the type of named entity, and in any case is often inaccurate or insufficient. For example, the first word of a sentence is also capitalized, and named entities often span several words, only some of which are capitalized. Furthermore, many other languages in non-Western scripts (e.g. Chinese or Arabic) do not have any capitalization at all, and even languages with capitalization may not consistently use it to distinguish names. For example, German capitalizes all nouns, regardless of whether they are names, and French and Spanish do not capitalize names that serve as adjectives.

Natural language generation
Convert information from computer databases or semantic intents into readable human language.

Natural language understanding
Convert chunks of text into more formal representations such as first-order logic structures that are easier for computer programs to manipulate. Natural language understanding involves the identification of the intended semantic from the multiple possible semantics which can be derived from a natural language expression which usually takes the form of organized notations of natural language concepts. Introduction and creation of language metamodel and ontology are efficient however empirical solutions. An explicit formalization of natural language semantics without confusions with implicit assumptions such as closed-world assumption (CWA) vs. open-world assumption, or subjective Yes/No vs. objective True/False is expected for the construction of a basis of semantics formalization.[281]

Optical character recognition (OCR)

Given an image representing printed text, determine the corresponding text.

Question answering

Given a human-language question, determine its answer. Typical questions have a specific right answer (such as "What is the capital of Canada?"), but sometimes open-ended questions are also considered (such as "What is the meaning of life?"). Recent works have looked at even more complex questions.[282]

Recognizing Textual entailment

Given two text fragments, determine if one being true entails the other, entails the other's negation, or allows the other to be either true or false.[283]

Relationship extraction

Given a chunk of text, identify the relationships among named entities (e.g. who is married to whom).

Sentiment analysis (see also multimodal sentiment analysis)

Extract subjective information usually from a set of documents, often using online reviews to determine "polarity" about specific objects. It is especially useful for identifying trends of public opinion in the social media, for the purpose of marketing.

Topic segmentation and recognition

Given a chunk of text, separate it into segments each of which is devoted to a topic, and identify the topic of the segment.

Word sense disambiguation

Many words have more than one meaning; we have to select the meaning which makes the most sense in context. For this problem, we are typically given a list of words and associated word senses, e.g. from a dictionary or from an online resource such as WordNet.

Discourse

Automatic summarization

Produce a readable summary of a chunk of text. Often used to provide summaries of text of a known type, such as articles in the financial section of a newspaper.

Coreference resolution

Given a sentence or larger chunk of text, determine which words ("mentions") refer to the same objects ("entities"). Anaphora resolution is a specific example of this task, and is specifically concerned with matching up pronouns with the nouns or names to which they refer. The more general task of coreference resolution also includes identifying so-called "bridging relationships" involving referring expressions. For example, in a sentence such as "He entered John's house through the front door", "the front door" is a referring expression and the bridging relationship to be identified is the fact that the door being referred to is the front door of John's house (rather than of some other structure that might also be referred to).

Discourse analysis

This rubric includes a number of related tasks. One task is identifying the discourse structure of connected text, i.e. the nature of the discourse relationships between sentences (e.g. elaboration, explanation, contrast). Another possible task is recognizing and classifying the speech acts in a chunk of text (e.g. yes-no question, content question, statement, assertion, etc.).

Speech

Speech recognition

Given a sound clip of a person or people speaking, determine the textual representation of the speech. This is the opposite of text to speech and is one of the extremely difficult problems colloquially termed "AI-complete" (see above). In natural speech there are hardly any pauses between successive words, and thus speech segmentation is a necessary subtask of speech recognition (see below). Note also that in most spoken languages, the sounds representing successive letters blend into each other in a process termed coarticulation, so the conversion of the analog signal to discrete characters can be a very difficult process.

Speech segmentation

Given a sound clip of a person or people speaking, separate it into words. A subtask of speech recognition and typically grouped with it.

Text-to-speech

Further reading

<templatestyles src="Template:Refbegin/styles.css" />

- Bates, M (1995). "Models of natural language understanding"[284]. *Proceedings of the National Academy of Sciences of the United States of America*. **92** (22): 9977–9982. doi: 10.1073/pnas.92.22.9977[285]. PMC 40721[284] .
- Steven Bird, Ewan Klein, and Edward Loper (2009). *Natural Language Processing with Python*. O'Reilly Media. ISBN 978-0-596-51649-9.
- Daniel Jurafsky and James H. Martin (2008). *Speech and Language Processing*, 2nd edition. Pearson Prentice Hall. ISBN 978-0-13-187321-6.
- Mohamed Zakaria Kurdi (2016). *Natural Language Processing and Computational Linguistics: speech, morphology, and syntax*, Volume 1. ISTE-Wiley. ISBN 978-1848218482.
- Mohamed Zakaria Kurdi (2017). *Natural Language Processing and Computational Linguistics: semantics, discourse, and applications*, Volume 2. ISTE-Wiley. ISBN 978-1848219212.
- Christopher D. Manning, Prabhakar Raghavan, and Hinrich Schütze (2008). *Introduction to Information Retrieval*. Cambridge University Press. ISBN 978-0-521-86571-5. Official html and pdf versions available without charge.[286]
- Christopher D. Manning and Hinrich Schütze (1999). *Foundations of Statistical Natural Language Processing*. The MIT Press. ISBN 978-0-262-13360-9.
- David M. W. Powers and Christopher C. R. Turk (1989). *Machine Learning of Natural Language*. Springer-Verlag. ISBN 978-0-387-19557-5.

External links

- Media related to Natural language processing at Wikimedia Commons

Machine perception

Machine perception is the capability of a computer system to interpret data in a manner that is similar to the way humans use their senses to relate to the world around them. The basic method that the computers take in and respond to their environment is through the attached hardware. Until recently input was limited to a keyboard, or a mouse, but advances in technology, both in hardware and software, have allowed computers to take in sensory input in a way similar to humans.

Machine perception allows the computer to use this sensory input, as well as conventional computational means of gathering information, to gather information with greater accuracy and to present it in a way that is more comfortable for the user. These include computer vision, machine hearing, and machine touch.

The end goal of machine perception is to give machines the ability to see, feel and perceive the world as humans do and therefore for them to be able to explain in a human way why they are making their decisions, to warn us when it is failing and more importantly, the reason why it is failing.

Machine vision

Computer vision is a field that includes methods for acquiring, processing, analyzing, and understanding images and, in general, high-dimensional data from the real world in order to produce numerical or symbolic information, e.g., in the forms of decisions. Computer vision has many applications already in use today such as facial recognition, geographical modeling, and even aesthetic judgment.

Machine hearing

Machine hearing, also known as machine listening or computer audition, is the ability of a computer or machine to take in and process sound data such as music or speech. This area has a wide range of application including music recording and compression, speech synthesis, and speech recognition. Moreover, this technology allows the machine to replicate the human brain's ability to selectively focus in a specific sound against many other competing sounds and background noise. This particular ability is called "auditory scene analysis". The technology enables the machine to segment several streams occurring at the same time. Many commonly used devices such as a smartphones, voice translators, and cars make use of some form of machine hearing.

Machine touch

Machine touch is an area of machine perception where tactile information is processed by a machine or computer. Applications include tactile perception of surface properties and dexterity whereby tactile information can enable intelligent reflexes and interaction with the environment.

Computer vision

Computer vision is an interdisciplinary field that deals with how computers can be made for gaining high-level understanding from digital images or videos. From the perspective of engineering, it seeks to automate tasks that the human visual system can do.

Computer vision tasks include methods for acquiring, processing, analyzing and understanding digital images, and extraction of high-dimensional data from the real world in order to produce numerical or symbolic information, *e.g.*, in the forms of decisions. Understanding in this context means the transformation of visual images (the input of the retina) into descriptions of the world that can interface with other thought processes and elicit appropriate action. This image understanding can be seen as the disentangling of symbolic information from image data using models constructed with the aid of geometry, physics, statistics, and learning theory.

As a scientific discipline, computer vision is concerned with the theory behind artificial systems that extract information from images. The image data can take many forms, such as video sequences, views from multiple cameras, or multi-dimensional data from a medical scanner. As a technological discipline, computer vision seeks to apply its theories and models for the construction of computer vision systems.

Sub-domains of computer vision include scene reconstruction, event detection, video tracking, object recognition, 3D pose estimation, learning, indexing, motion estimation, and image restoration.

Definition

Computer vision is an interdisciplinary field that deals with how computers can be made for gaining high-level understanding from digital images or videos. From the perspective of engineering, it seeks to automate tasks that the human visual system can do. "Computer vision is concerned with the automatic extraction, analysis and understanding of useful information from a single image or a sequence of images. It involves the development of a theoretical and algorithmic basis to achieve automatic visual understanding."[287] As a scientific discipline, computer vision is concerned with the theory behind artificial systems that extract information from images. The image data can take many forms, such as video sequences, views from multiple cameras, or multi-dimensional data from a medical scanner. As a technological discipline, computer vision seeks to apply its theories and models for the construction of computer vision systems.

History

In the late 1960s, computer vision began at universities that were pioneering artificial intelligence. It was meant to mimic the human visual system, as a stepping stone to endowing robots with intelligent behavior. In 1966, it was believed that this could be achieved through a summer project, by attaching a camera to a computer and having it "describe what it saw".

What distinguished computer vision from the prevalent field of digital image processing at that time was a desire to extract three-dimensional structure from images with the goal of achieving full scene understanding. Studies in the 1970s formed the early foundations for many of the computer vision algorithms that exist today, including extraction of edges from images, labeling of lines, non-polyhedral and polyhedral modeling, representation of objects as interconnections of smaller structures, optical flow, and motion estimation.

The next decade saw studies based on more rigorous mathematical analysis and quantitative aspects of computer vision. These include the concept of scale-space, the inference of shape from various cues such as shading, texture and focus, and contour models known as snakes. Researchers also realized that many of these mathematical concepts could be treated within the same optimization framework as regularization and Markov random fields. By the 1990s, some of the previous research topics became more active than the others. Research in projective 3-D reconstructions led to better understanding of camera calibration. With the advent of optimization methods for camera calibration, it was realized that a lot of the ideas were already explored in bundle adjustment theory from the field of photogrammetry. This led to methods for sparse 3-D reconstructions of scenes from multiple images. Progress was made on the dense stereo correspondence problem and further multi-view stereo techniques. At the same time, variations of graph cut were used to solve image segmentation. This decade also marked the first time statistical learning techniques were used in practice to recognize faces in images (see Eigenface). Toward the end of the 1990s, a significant change came about with the increased interaction between the fields of computer graphics and computer vision. This included image-based rendering, image morphing, view interpolation, panoramic image stitching and early light-field rendering.

Recent work has seen the resurgence of feature-based methods, used in conjunction with machine learning techniques and complex optimization frameworks.

Related fields

Artificial Intelligence

Areas of artificial intelligence deal with autonomous planning or deliberation for robotical systems to navigate through an environment. A detailed understanding of these environments is required to navigate through them. Information about the environment could be provided by a computer vision system, acting as a vision sensor and providing high-level information about the environment and the robot.

Artificial intelligence and computer vision share other topics such as pattern recognition and learning techniques. Consequently, computer vision is sometimes seen as a part of the artificial intelligence field or the computer science field in general.

Solid-state Physics

Solid-state physics is another field that is closely related to computer vision. Most computer vision systems rely on image sensors, which detect electromagnetic radiation, which is typically in the form of either visible or infra-red light. The sensors are designed using quantum physics. The process by which light interacts with surfaces is explained using physics. Physics explains the behavior of optics which are a core part of most imaging systems. Sophisticated image sensors even require quantum mechanics to provide a complete understanding of the image formation process. Also, various measurement problems in physics can be addressed using computer vision, for example motion in fluids.

Neurobiology

A third field which plays an important role is neurobiology, specifically the study of the biological vision system. Over the last century, there has been an extensive study of eyes, neurons, and the brain structures devoted to processing of visual stimuli in both humans and various animals. This has led to a coarse, yet complicated, description of how "real" vision systems operate in order to solve certain vision related tasks. These results have led to a subfield within computer vision where artificial systems are designed to mimic the processing and behavior of biological systems, at different levels of complexity. Also, some of the learning-based methods developed within computer vision (*e.g.* neural net and deep learning based image and feature analysis and classification) have their background in biology.

Some strands of computer vision research are closely related to the study of biological vision – indeed, just as many strands of AI research are closely tied with research into human consciousness, and the use of stored knowledge to interpret, integrate and utilize visual information. The field of biological vision studies and models the physiological processes behind visual perception in humans and other animals. Computer vision, on the other hand, studies and describes the processes implemented in software and hardware behind artificial vision systems. Interdisciplinary exchange between biological and computer vision has proven fruitful for both fields.

Signal Processing

Yet another field related to computer vision is signal processing. Many methods for processing of one-variable signals, typically temporal signals, can be extended in a natural way to processing of two-variable signals or multi-variable signals in computer vision. However, because of the specific nature of images there are many methods developed within computer vision which have no counterpart in processing of one-variable signals. Together with the multi-dimensionality of the signal, this defines a subfield in signal processing as a part of computer vision.

Beside the above-mentioned views on computer vision, many of the related research topics can also be studied from a purely mathematical point of view. For example, many methods in computer vision are based on statistics, optimization or geometry. Finally, a significant part of the field is devoted to the implementation aspect of computer vision; how existing methods can be realized in various combinations of software and hardware, or how these methods can be modified in order to gain processing speed without losing too much performance.

The fields most closely related to computer vision are image processing, image analysis and machine vision. There is a significant overlap in the range of techniques and applications that these cover. This implies that the basic techniques that are used and developed in these fields are similar, something which can be interpreted as there is only one field with different names. On the other hand, it appears to be necessary for research groups, scientific journals, conferences and companies to present or market themselves as belonging specifically to one of these fields and, hence, various characterizations which distinguish each of the fields from the others have been presented.

Computer graphics produces image data from 3D models, computer vision often produces 3D models from image data . There is also a trend towards a combination of the two disciplines, *e.g.*, as explored in augmented reality.

The following characterizations appear relevant but should not be taken as universally accepted:

- Image processing and image analysis tend to focus on 2D images, how to transform one image to another, *e.g.*, by pixel-wise operations such as contrast enhancement, local operations such as edge extraction or noise removal, or geometrical transformations such as rotating the image. This characterization implies that image processing/analysis neither require assumptions nor produce interpretations about the image content.
- Computer vision includes 3D analysis from 2D images. This analyzes the 3D scene projected onto one or several images, *e.g.*, how to reconstruct structure or other information about the 3D scene from one or several images. Computer vision often relies on more or less complex assumptions about the scene depicted in an image.
- Machine vision is the process of applying a range of technologies & methods to provide imaging-based automatic inspection, process control and robot guidance in industrial applications. Machine vision tends to focus on applications, mainly in manufacturing, *e.g.*, vision based robots and systems for vision based inspection, measurement, or picking (such as Bin Picking[288]). This implies that image sensor technologies and control theory often are integrated with the processing of image data to control a robot and that real-time processing is emphasised by means of efficient implementations in hardware and software. It also implies that the external conditions such as lighting can be and are often more controlled in machine vision than they are in general computer vision, which can enable the use of different algorithms.
- There is also a field called imaging which primarily focus on the process of producing images, but sometimes also deals with processing and analysis of images. For example, medical imaging includes substantial work on the analysis of image data in medical applications.
- Finally, pattern recognition is a field which uses various methods to extract information from signals in general, mainly based on statistical approaches and artificial neural networks. A significant part of this field is devoted to applying these methods to image data.

Photogrammetry also overlaps with computer vision, e.g., stereophotogrammetry vs. computer stereo vision.

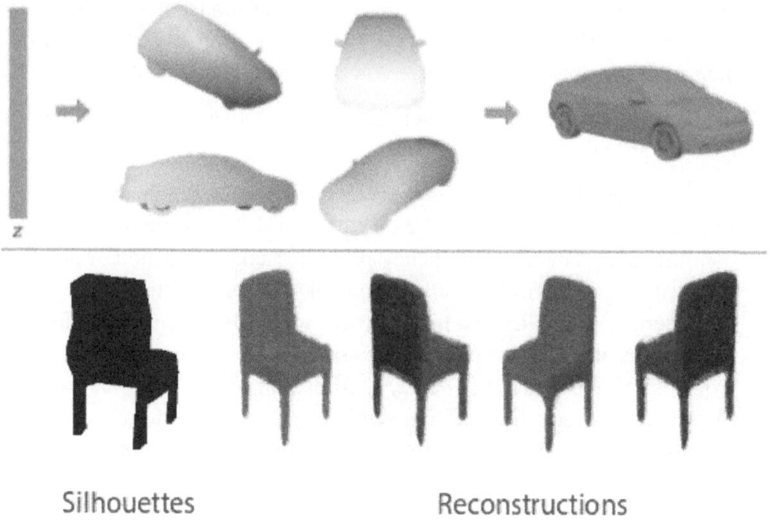

Figure 9: *Learning 3D shapes has been a challenging task in computer vision. Recent advances in deep learning has enabled researchers to build models that are able to generate and reconstruct 3D shapes from single or multi-view depth maps or silhouettes seamlessly and efficiently*

Applications

Applications range from tasks such as industrial machine vision systems which, say, inspect bottles speeding by on a production line, to research into artificial intelligence and computers or robots that can comprehend the world around them. The computer vision and machine vision fields have significant overlap. Computer vision covers the core technology of automated image analysis which is used in many fields. Machine vision usually refers to a process of combining automated image analysis with other methods and technologies to provide automated inspection and robot guidance in industrial applications. In many computer vision applications, the computers are pre-programmed to solve a particular task, but methods based on learning are now becoming increasingly common. Examples of applications of computer vision include systems for:

- Automatic inspection, *e.g.*, in manufacturing applications;
- Assisting humans in identification tasks, e.g., a species identification system;
- Controlling processes, *e.g.*, an industrial robot;
- Detecting events, *e.g.*, for visual surveillance or people counting;

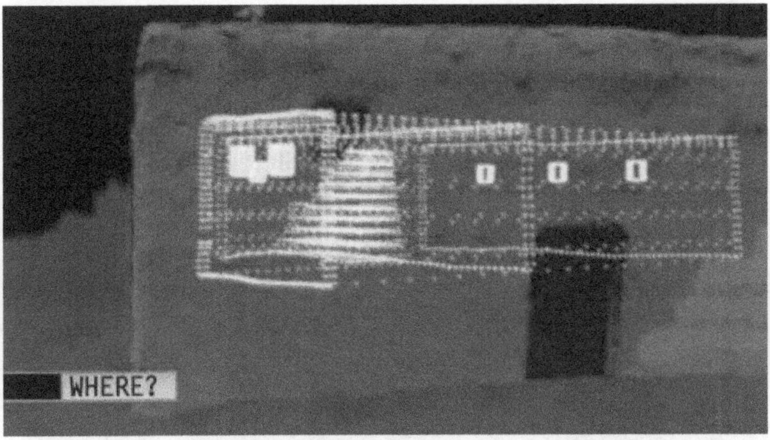

Figure 10: *DARPA's Visual Media Reasoning concept video*

- Interaction, *e.g.*, as the input to a device for computer-human interaction;
- Modeling objects or environments, *e.g.*, medical image analysis or topographical modeling;
- Navigation, *e.g.*, by an autonomous vehicle or mobile robot; and
- Organizing information, *e.g.*, for indexing databases of images and image sequences.

One of the most prominent application fields is medical computer vision, or medical image processing, characterized by the extraction of information from image data to diagnose a patient. An example of this is detection of tumours, arteriosclerosis or other malign changes; measurements of organ dimensions, blood flow, etc. are another example. It also supports medical research by providing new information: *e.g.*, about the structure of the brain, or about the quality of medical treatments. Applications of computer vision in the medical area also includes enhancement of images interpreted by humans—ultrasonic images or X-ray images for example—to reduce the influence of noise.

A second application area in computer vision is in industry, sometimes called machine vision, where information is extracted for the purpose of supporting a manufacturing process. One example is quality control where details or final products are being automatically inspected in order to find defects. Another example is measurement of position and orientation of details to be picked up by a robot arm. Machine vision is also heavily used in agricultural process to remove undesirable food stuff from bulk material, a process called optical sorting.

Figure 11: *Artist's concept of a Mars Exploration Rover, an example of an unmanned land-based vehicle. Notice the stereo cameras mounted on top of the rover.*

Military applications are probably one of the largest areas for computer vision. The obvious examples are detection of enemy soldiers or vehicles and missile guidance. More advanced systems for missile guidance send the missile to an area rather than a specific target, and target selection is made when the missile reaches the area based on locally acquired image data. Modern military concepts, such as "battlefield awareness", imply that various sensors, including image sensors, provide a rich set of information about a combat scene which can be used to support strategic decisions. In this case, automatic processing of the data is used to reduce complexity and to fuse information from multiple sensors to increase reliability.

One of the newer application areas is autonomous vehicles, which include submersibles, land-based vehicles (small robots with wheels, cars or trucks), aerial vehicles, and unmanned aerial vehicles (UAV). The level of autonomy ranges from fully autonomous (unmanned) vehicles to vehicles where computer vision based systems support a driver or a pilot in various situations. Fully autonomous vehicles typically use computer vision for navigation, i.e. for knowing where it is, or for producing a map of its environment (SLAM) and for detecting obstacles. It can also be used for detecting certain task specific events, *e.g.*, a UAV looking for forest fires. Examples of supporting systems are obstacle warning systems in cars, and systems for autonomous landing of aircraft.

Several car manufacturers have demonstrated systems for autonomous driving of cars, but this technology has still not reached a level where it can be put on the market. There are ample examples of military autonomous vehicles ranging from advanced missiles, to UAVs for recon missions or missile guidance. Space exploration is already being made with autonomous vehicles using computer vision, *e.g.*, NASA's Mars Exploration Rover and ESA's ExoMars Rover.

Other application areas include:

- Support of visual effects creation for cinema and broadcast, *e.g.*, camera tracking (matchmoving).
- Surveillance.
- Tracking and counting organisms in the biological sciences

Typical tasks

Each of the application areas described above employ a range of computer vision tasks; more or less well-defined measurement problems or processing problems, which can be solved using a variety of methods. Some examples of typical computer vision tasks are presented below.

Computer vision tasks include methods for acquiring, processing, analyzing and understanding digital images, and extraction of high-dimensional data from the real world in order to produce numerical or symbolic information, *e.g.*, in the forms of decisions. Understanding in this context means the transformation of visual images (the input of the retina) into descriptions of the world that can interface with other thought processes and elicit appropriate action. This image understanding can be seen as the disentangling of symbolic information from image data using models constructed with the aid of geometry, physics, statistics, and learning theory.

Recognition

The classical problem in computer vision, image processing, and machine vision is that of determining whether or not the image data contains some specific object, feature, or activity. Different varieties of the recognition problem are described in the literature:Wikipedia:Citation needed

- **Object recognition** (also called **object classification**) – one or several pre-specified or learned objects or object classes can be recognized, usually together with their 2D positions in the image or 3D poses in the scene. Blippar, Google Goggles and LikeThat provide stand-alone programs that illustrate this functionality.

- **Identification** – an individual instance of an object is recognized. Examples include identification of a specific person's face or fingerprint, identification of handwritten digits, or identification of a specific vehicle.
- **Detection** – the image data are scanned for a specific condition. Examples include detection of possible abnormal cells or tissues in medical images or detection of a vehicle in an automatic road toll system. Detection based on relatively simple and fast computations is sometimes used for finding smaller regions of interesting image data which can be further analyzed by more computationally demanding techniques to produce a correct interpretation.

Currently, the best algorithms for such tasks are based on convolutional neural networks. An illustration of their capabilities is given by the ImageNet Large Scale Visual Recognition Challenge; this is a benchmark in object classification and detection, with millions of images and hundreds of object classes. Performance of convolutional neural networks, on the ImageNet tests, is now close to that of humans.[289] The best algorithms still struggle with objects that are small or thin, such as a small ant on a stem of a flower or a person holding a quill in their hand. They also have trouble with images that have been distorted with filters (an increasingly common phenomenon with modern digital cameras). By contrast, those kinds of images rarely trouble humans. Humans, however, tend to have trouble with other issues. For example, they are not good at classifying objects into fine-grained classes, such as the particular breed of dog or species of bird, whereas convolutional neural networks handle this with ease.

Several specialized tasks based on recognition exist, such as:

- **Content-based image retrieval** – finding all images in a larger set of images which have a specific content. The content can be specified in different ways, for example in terms of similarity relative a target image (give me all images similar to image X), or in terms of high-level search criteria given as text input (give me all images which contains many houses, are taken during winter, and have no cars in them).
- **Pose estimation** – estimating the position or orientation of a specific object relative to the camera. An example application for this technique would be assisting a robot arm in retrieving objects from a conveyor belt in an assembly line situation or picking parts from a bin.
- **Optical character recognition** (OCR) – identifying characters in images of printed or handwritten text, usually with a view to encoding the text in a format more amenable to editing or indexing (*e.g.* ASCII).
- **2D Code reading** Reading of 2D codes such as data matrix and QR codes.
- **Facial recognition**
- **Shape Recognition Technology** (SRT) in people counter systems differentiating human beings (head and shoulder patterns) from objects

Figure 12: *Computer vision for people counter purposes in public places, malls, shopping centres*

Motion analysis

Several tasks relate to motion estimation where an image sequence is processed to produce an estimate of the velocity either at each points in the image or in the 3D scene, or even of the camera that produces the images . Examples of such tasks are:

- **Egomotion** – determining the 3D rigid motion (rotation and translation) of the camera from an image sequence produced by the camera.
- **Tracking** – following the movements of a (usually) smaller set of interest points or objects (*e.g.*, vehicles, humans or other organisms) in the image sequence.
- **Optical flow** – to determine, for each point in the image, how that point is moving relative to the image plane, i.e., its apparent motion. This motion is a result both of how the corresponding 3D point is moving in the scene and how the camera is moving relative to the scene.

Scene reconstruction

Given one or (typically) more images of a scene, or a video, scene reconstruction aims at computing a 3D model of the scene. In the simplest case the model can be a set of 3D points. More sophisticated methods produce a complete 3D surface model. The advent of 3D imaging not requiring motion or scanning, and related processing algorithms is enabling rapid advances in this field. Grid-based 3D sensing can be used to acquire 3D images from multiple angles. Algorithms are now available to stitch multiple 3D images together into point clouds and 3D models .

Image restoration

The aim of image restoration is the removal of noise (sensor noise, motion blur, etc.) from images. The simplest possible approach for noise removal is various types of filters such as low-pass filters or median filters. More sophisticated methods assume a model of how the local image structures look like, a model which distinguishes them from the noise. By first analysing the image data in terms of the local image structures, such as lines or edges, and then controlling the filtering based on local information from the analysis step, a better level of noise removal is usually obtained compared to the simpler approaches.

An example in this field is inpainting.

System methods

The organization of a computer vision system is highly application dependent. Some systems are stand-alone applications which solve a specific measurement or detection problem, while others constitute a sub-system of a larger design which, for example, also contains sub-systems for control of mechanical actuators, planning, information databases, man-machine interfaces, etc. The specific implementation of a computer vision system also depends on if its functionality is pre-specified or if some part of it can be learned or modified during operation. Many functions are unique to the application. There are, however, typical functions which are found in many computer vision systems.

- **Image acquisition** – A digital image is produced by one or several image sensors, which, besides various types of light-sensitive cameras, include range sensors, tomography devices, radar, ultra-sonic cameras, etc. Depending on the type of sensor, the resulting image data is an ordinary 2D image, a 3D volume, or an image sequence. The pixel values typically correspond to light intensity in one or several spectral bands (gray images or colour images), but can also be related to various physical measures, such as depth, absorption or reflectance of sonic or electromagnetic waves, or nuclear magnetic resonance.
- **Pre-processing** – Before a computer vision method can be applied to image data in order to extract some specific piece of information, it is usually necessary to process the data in order to assure that it satisfies certain assumptions implied by the method. Examples are
  - Re-sampling in order to assure that the image coordinate system is correct.
  - Noise reduction in order to assure that sensor noise does not introduce false information.
  - Contrast enhancement to assure that relevant information can be detected.

  - Scale space representation to enhance image structures at locally appropriate scales.
- **Feature extraction** – Image features at various levels of complexity are extracted from the image data. Typical examples of such features are
  - Lines, edges and ridges.
  - Localized interest points such as corners, blobs or points.

  More complex features may be related to texture, shape or motion.
- **Detection/segmentation** – At some point in the processing a decision is made about which image points or regions of the image are relevant for further processing. Examples are
  - Selection of a specific set of interest points
  - Segmentation of one or multiple image regions which contain a specific object of interest.
  - Segmentation of image into nested scene architecture comprised foreground, object groups, single objects or salient object parts (also referred to as spatial-taxon scene hierarchy)[290]
- **High-level processing** – At this step the input is typically a small set of data, for example a set of points or an image region which is assumed to contain a specific object. The remaining processing deals with, for example:
  - Verification that the data satisfy model-based and application specific assumptions.
  - Estimation of application specific parameters, such as object pose or object size.
  - Image recognition – classifying a detected object into different categories.
  - Image registration – comparing and combining two different views of the same object.
- **Decision making** Making the final decision required for the application, for example:
  - Pass/fail on automatic inspection applications
  - Match / no-match in recognition applications
  - Flag for further human review in medical, military, security and recognition applications

Image-understanding systems

Image-understanding systems (IUS) include three levels of abstraction as follows: Low level includes image primitives such as edges, texture elements, or regions; intermediate level includes boundaries, surfaces and volumes; and high level includes objects, scenes, or events. Many of these requirements are really topics for further research.

The representational requirements in the designing of IUS for these levels are: representation of prototypical concepts, concept organization, spatial knowledge, temporal knowledge, scaling, and description by comparison and differentiation.

While inference refers to the process of deriving new, not explicitly represented facts from currently known facts, control refers to the process that selects which of the many inference, search, and matching techniques should be applied at a particular stage of processing. Inference and control requirements for IUS are: search and hypothesis activation, matching and hypothesis testing, generation and use of expectations, change and focus of attention, certainty and strength of belief, inference and goal satisfaction.

Hardware

There are many kinds of computer vision systems, nevertheless all of them contain these basic elements: a power source, at least one image acquisition device (i.e. camera, ccd, etc.), a processor as well as control and communication cables or some kind of wireless interconnection mechanism. In addition, a practical vision system contains software, as well as a display in order to monitor the system. Vision systems for inner spaces, as most industrial ones, contain an illumination system and may be placed in a controlled environment. Furthermore, a completed system includes many accessories like camera supports, cables and connectors.

Most computer vision systems use visible-light cameras passively viewing a scene at frame rates of at most 60 frames per second (usually far slower).

A few computer vision systems use image acquisition hardware with active illumination or something other than visible light or both. For example, a structured-light 3D scanner, a thermographic camera, a hyperspectral imager, radar imaging, a lidar scanner, a magnetic resonance image, a side-scan sonar, a synthetic aperture sonar, or etc. Such hardware captures "images" that are then processed often using the same computer vision algorithms used to process visible-light images.

While traditional broadcast and consumer video systems operate at a rate of 30 frames per second, advances in digital signal processing and consumer graphics hardware has made high-speed image acquisition, processing, and display possible for real-time systems on the order of hundreds to thousands of frames per second. For applications in robotics, fast, real-time video systems are critically important and often can simplify the processing needed for certain algorithms. When combined with a high-speed projector, fast image acquisition allows 3D measurement and feature tracking to be realised.

Egocentric vision systems are composed of a wearable camera that automatically take pictures from a first-person perspective.

As of 2016, vision processing units are emerging as a new class of processor, to complement CPUs and graphics processing units (GPUs) in this role.

References

Further reading

- David Marr (1982). *Vision*. W. H. Freeman and Company. ISBN 0-7167-1284-9.
- Azriel Rosenfeld; Avinash Kak (1982). *Digital Picture Processing*. Academic Press. ISBN 0-12-597301-2.
- Barghout, Lauren; Lawrence W. Lee (2003). *Perceptual information processing system*. U.S. Patent Application 10/618,543. ISBN 0-262-08159-8.
- Berthold Klaus Paul Horn (1986). *Robot Vision*. MIT Press. ISBN 0-262-08159-8.
- Michael C.Fairhurst (1988). *Computer Vision for robotic systems*. Prentice Hall. ISBN 0-13-166919-2.
- Olivier Faugeras (1993). *Three-Dimensional Computer Vision, A Geometric Viewpoint*. MIT Press. ISBN 0-262-06158-9.
- Tony Lindeberg (1994). *Scale-Space Theory in Computer Vision*[291]. Springer. ISBN 0-7923-9418-6.
- James L. Crowley and Henrik I. Christensen (Eds.) (1995). *Vision as Process*. Springer-Verlag. ISBN 3-540-58143-X.
- Gösta H. Granlund; Hans Knutsson (1995). *Signal Processing for Computer Vision*. Kluwer Academic Publisher. ISBN 0-7923-9530-1.
- Reinhard Klette; Karsten Schluens; Andreas Koschan (1998). *Computer Vision – Three-Dimensional Data from Images*[292]. Springer, Singapore. ISBN 981-3083-71-9.
- Emanuele Trucco; Alessandro Verri (1998). *Introductory Techniques for 3-D Computer Vision*. Prentice Hall. ISBN 0-13-261108-2.
- Bernd Jähne (2002). *Digital Image Processing*. Springer. ISBN 3-540-67754-2.
- Richard Hartley and Andrew Zisserman (2003). *Multiple View Geometry in Computer Vision*. Cambridge University Press. ISBN 0-521-54051-8.
- Gérard Medioni; Sing Bing Kang (2004). *Emerging Topics in Computer Vision*. Prentice Hall. ISBN 0-13-101366-1.
- R. Fisher; K Dawson-Howe; A. Fitzgibbon; C. Robertson; E. Trucco (2005). *Dictionary of Computer Vision and Image Processing*. John Wiley. ISBN 0-470-01526-8.

- Nikos Paragios and Yunmei Chen and Olivier Faugeras (2005). *Handbook of Mathematical Models in Computer Vision*[293]. Springer. ISBN 0-387-26371-3.
- Wilhelm Burger; Mark J. Burge (2007). *Digital Image Processing: An Algorithmic Approach Using Java*[294]. Springer. ISBN 1-84628-379-5.
- Pedram Azad; Tilo Gockel; Rüdiger Dillmann (2008). *Computer Vision – Principles and Practice*[295]. Elektor International Media BV. ISBN 0-905705-71-8.
- Richard Szeliski (2010). *Computer Vision: Algorithms and Applications*[296]. Springer-Verlag. ISBN 978-1848829343.
- J. R. Parker (2011). *Algorithms for Image Processing and Computer Vision (2nd ed.)*[297]. Wiley. ISBN 978-0470643853.
- Richard J. Radke (2013). *Computer Vision for Visual Effects*. Cambridge University Press. ISBN 978-0-521-76687-6.

External links

- USC Iris computer vision conference list[298]
- Computer vision papers on the web[299] A complete list of papers of the most relevant computer vision conferences.
- Computer Vision Online[300] News, source code, datasets and job offers related to computer vision.
- Keith Price's Annotated Computer Vision Bibliography[301]
- CVonline[302] Bob Fisher's Compendium of Computer Vision.
- British Machine Vision Association[303] Supporting computer vision research within the UK via the BMVC and MIUA conferences, *Annals of the BMVA* (open-source journal), BMVA Summer School and one-day meetings

Speech recognition

Speech recognition is the inter-disciplinary sub-field of computational linguistics that develops methodologies and technologies that enables the recognition and translation of spoken language into text by computers. It is also known as **automatic speech recognition** (**ASR**), **computer speech recognition** or **speech to text** (**STT**). It incorporates knowledge and research in the linguistics, computer science, and electrical engineering fields.

Some speech recognition systems require "training" (also called "enrollment") where an individual speaker reads text or isolated vocabulary into the system. The system analyzes the person's specific voice and uses it to fine-tune the recognition of that person's speech, resulting in increased accuracy. Systems that do not use training are called "speaker independent" systems. Systems that use training are called "speaker dependent".

Speech recognition applications include voice user interfaces such as voice dialing (e.g. "Call home"), call routing (e.g. "I would like to make a collect call"), domotic appliance control, search (e.g. find a podcast where particular words were spoken), simple data entry (e.g., entering a credit card number), preparation of structured documents (e.g. a radiology report), speech-to-text processing (e.g., word processors or emails), and aircraft (usually termed direct voice input).

The term *voice recognition* or *speaker identification* refers to identifying the speaker, rather than what they are saying. Recognizing the speaker can simplify the task of translating speech in systems that have been trained on a specific person's voice or it can be used to authenticate or verify the identity of a speaker as part of a security process.

From the technology perspective, speech recognition has a long history with several waves of major innovations. Most recently, the field has benefited from advances in deep learning and big data. The advances are evidenced not only by the surge of academic papers published in the field, but more importantly by the worldwide industry adoption of a variety of deep learning methods in designing and deploying speech recognition systems. These speech industry players include Google, Microsoft, IBM, Baidu, Apple, Amazon, Nuance, SoundHound, iFLYTEK many of which have publicized the core technology in their speech recognition systems as being based on deep learning.

History

Early work

In 1952 three Bell Labs researchers built a system for single-speaker digit recognition. Their system worked by locating the formants in the power spectrum of each utterance. The 1950s era technology was limited to single-speaker systems with vocabularies of around ten words.

Gunnar Fant developed the source-filter model of speech production and published it in 1960, which proved to be a useful model of speech production.

Unfortunately, funding at Bell Labs dried up for several years when, in 1969, the influential John Pierce wrote an open letter that was critical of speech recognition research. Pierce defunded speech recognition research at Bell Labs where no research on speech recognition was done until Pierce retired and James L. Flanagan took over.

Raj Reddy was the first person to take on continuous speech recognition as a graduate student at Stanford University in the late 1960s. Previous systems required the users to make a pause after each word. Reddy's system was designed to issue spoken commands for the game of chess.

Also around this time Soviet researchers invented the dynamic time warping (DTW) algorithm and used it to create a recognizer capable of operating on a 200-word vocabulary. The DTW algorithm processed the speech signal by dividing it into short frames, e.g. 10ms segments, and processing each frame as a single unit. Although DTW would be superseded by later algorithms, the technique of dividing the signal into frames would carry on. Achieving speaker independence was a major unsolved goal of researchers during this time period.

In 1971, DARPA funded five years of speech recognition research through its Speech Understanding Research program with ambitious end goals including a minimum vocabulary size of 1,000 words. It was thought that speech *understanding* would be key to making progress in speech *recognition*, although that later proved to not be true. BBN, IBM, Carnegie Mellon and Stanford Research Institute all participated in the program. The government funding revived speech recognition research that had been largely abandoned in the United States after John Pierce's letter.

Despite the fact that CMU's Harpy system met the original goals of the program, many predictions turned out to be nothing more than hype, disappointing DARPA administrators. This disappointment led to DARPA not continuing the funding. Several innovations happened during this time, such as the invention of beam search for use in CMU's Harpy system.[304] The field also benefited from the discovery of several algorithms in other fields such as linear predictive coding and cepstral analysis.

In 1972, the IEEE Acoustics, Speech, and Signal Processing group held a conference in Newton, Massachusetts. Four years later, the first ICASSP was held in Philadelphia, which since then has been a major venue for the publication of research on speech recognition.

During the late 1960s Leonard Baum developed the mathematics of Markov chains at the Institute for Defense Analysis. A decade later, at CMU, Raj Reddy's students James Baker and Janet M. Baker began using the Hidden Markov Model (HMM) for speech recognition. James Baker had learned about HMMs from a summer job at the Institute of Defense Analysis during his undergraduate education. The use of HMMs allowed researchers to combine different sources of knowledge, such as acoustics, language, and syntax, in a unified probabilistic model.

Under Fred Jelinek's lead, IBM created a voice activated typewriter called Tangora, which could handle a 20,000 word vocabulary by the mid 1980s. Jelinek's statistical approach put less emphasis on emulating the way the human brain processes and understands speech in favor of using statistical modeling techniques like HMMs. (Jelinek's group independently discovered the application of HMMs to speech.) This was controversial with linguists since HMMs are too simplistic to account for many common features of human languages. However, the HMM proved to be a highly useful way for modeling speech and replaced dynamic time warping to become the dominant speech recognition algorithm in the 1980s. IBM had a few competitors including Dragon Systems founded by James and Janet M. Baker in 1982. The 1980s also saw the introduction of the n-gram language model. Katz introduced the back-off model in 1987, which allowed language models to use multiple length n-grams. During the same time, also CSELT was using HMM (the diphonies were studied since 1980) to recognize language like Italian.[305,306,307] At the same time, CSELT led a series of European projects (Esprit I, II), and summarized the state-of-the-art in a book, later (2013) reprinted.[308]

Much of the progress in the field is owed to the rapidly increasing capabilities of computers. At the end of the DARPA program in 1976, the best computer available to researchers was the PDP-10 with 4 MB ram. Using these computers it could take up to 100 minutes to decode just 30 seconds of speech. A few decades later, researchers had access to tens of thousands of times as much computing power. As the technology advanced and computers got faster, researchers began tackling harder problems such as larger vocabularies, speaker independence, noisy environments and conversational speech. In particular, this shifting to more difficult tasks has characterized DARPA funding of speech recognition since the 1980s. For example, progress was made on speaker independence first by training on a larger variety of speakers and then later by doing explicit speaker adaptation during decoding. Further reductions in word error rate came as researchers shifted acoustic models to be discriminative instead of using maximum likelihood estimation.

In the mid-Eighties new speech recognition microprocessors were released: for example RIPAC, an independent-speaker recognition (for continuous speech) chip tailored for telephone services, was presented in the Netherlands in 1986. It was designed by CSELT/Elsag and manufactured by SGS.

Practical speech recognition

The 1990s saw the first introduction of commercially successful speech recognition technologies. Two of the earliest products were Dragon Dictate, a consumer product released in 1990 and originally priced at $9,000, and a recognizer from Kurzweil Applied Intelligence released in 1987. AT&T deployed the Voice Recognition Call Processing service in 1992 to route telephone calls without the use of a human operator. The technology was developed by Lawrence Rabiner and others at Bell Labs. By this point, the vocabulary of the typical commercial speech recognition system was larger than the average human vocabulary. Raj Reddy's former student, Xuedong Huang, developed the Sphinx-II system at CMU. The Sphinx-II system was the first to do speaker-independent, large vocabulary, continuous speech recognition and it had the best performance in DARPA's 1992 evaluation. Handling continuous speech with a large vocabulary was a major milestone in the history of speech recognition. Huang went on to found the speech recognition group at Microsoft in 1993. Raj Reddy's student Kai-Fu Lee joined Apple where, in 1992, he helped develop a speech interface prototype for the Apple computer known as Casper.

Lernout & Hauspie, a Belgium-based speech recognition company, acquired several other companies, including Kurzweil Applied Intelligence in 1997 and Dragon Systems in 2000. The L&H speech technology was used in the Windows XP operating system. L&H was an industry leader until an accounting scandal brought an end to the company in 2001. The speech technology from L&H was bought by ScanSoft which became Nuance in 2005. Apple originally licensed software from Nuance to provide speech recognition capability to its digital assistant Siri.

In the 2000s DARPA sponsored two speech recognition programs: Effective Affordable Reusable Speech-to-Text (EARS) in 2002 and Global Autonomous Language Exploitation (GALE). Four teams participated in the EARS program: IBM, a team led by BBN with LIMSI and Univ. of Pittsburgh, Cambridge University, and a team composed of ICSI, SRI and University of Washington. EARS funded the collection of the Switchboard telephone speech corpus containing 260 hours of recorded conversations from over 500 speakers.

The GALE program focused on Arabic and Mandarin broadcast news speech. Google's first effort at speech recognition came in 2007 after hiring some researchers from Nuance. The first product was GOOG-411, a telephone based directory service. The recordings from GOOG-411 produced valuable data that helped Google improve their recognition systems. Google voice search is now supported in over 30 languages.

In the United States, the National Security Agency has made use of a type of speech recognition for keyword spotting since at least 2006. This technology allows analysts to search through large volumes of recorded conversations and isolate mentions of keywords. Recordings can be indexed and analysts can run queries over the database to find conversations of interest. Some government research programs focused on intelligence applications of speech recognition, e.g. DARPA's EARS's program and IARPA's Babel program.

Modern systems

In the early 2000s, speech recognition was still dominated by traditional approaches such as Hidden Markov Models combined with feedforward artificial neural networks.[309] Today, however, many aspects of speech recognition have been taken over by a deep learning method called Long short-term memory (LSTM), a recurrent neural network published by Sepp Hochreiter & Jürgen Schmidhuber in 1997. LSTM RNNs avoid the vanishing gradient problem and can learn "Very Deep Learning" tasks that require memories of events that happened thousands of discrete time steps ago, which is important for speech. Around 2007, LSTM trained by Connectionist Temporal Classification (CTC)[310] started to outperform traditional speech recognition in certain applications.[311] In 2015, Google's speech recognition reportedly experienced a dramatic performance jump of 49% through CTC-trained LSTM, which is now available through Google Voice to all smartphone users.[312]

The use of deep feedforward (non-recurrent) networks for acoustic modeling was introduced during later part of 2009 by Geoffrey Hinton and his students at University of Toronto and by Li Deng and colleagues at Microsoft Research, initially in the collaborative work between Microsoft and University of Toronto which was subsequently expanded to include IBM and Google (hence "The shared views of four research groups" subtitle in their 2012 review paper).[313] A Microsoft research executive called this innovation "the most dramatic change in accuracy since 1979". In contrast to the steady incremental improvements of the past few decades, the application of deep learning decreased word error rate by 30%. This innovation was quickly adopted across the field. Researchers have begun to use deep learning techniques for language modeling as well.

In the long history of speech recognition, both shallow form and deep form (e.g. recurrent nets) of artificial neural networks had been explored for many years during 1980s, 1990s and a few years into the 2000s.[314,315,316] But these methods never won over the non-uniform internal-handcrafting Gaussian mixture model/Hidden Markov model (GMM-HMM) technology based on generative models of speech trained discriminatively. A number of key difficulties had been methodologically analyzed in the 1990s, including gradient diminishing[317] and weak temporal correlation structure in the neural predictive models.[318] All these difficulties were in addition to the lack of big training data and big computing power in these early days. Most speech recognition researchers who understood such barriers hence subsequently moved away from neural nets to pursue generative modeling approaches until the recent resurgence of deep learning starting around 2009–2010 that had overcome all these difficulties. Hinton et al. and Deng et al. reviewed part of this recent history about how their collaboration with each other and then with colleagues across four groups (University of Toronto, Microsoft, Google, and IBM) ignited a renaissance of applications of deep feedforward neural networks to speech recognition.[319,320]

Models, methods, and algorithms

Both acoustic modeling and language modeling are important parts of modern statistically-based speech recognition algorithms. Hidden Markov models (HMMs) are widely used in many systems. Language modeling is also used in many other natural language processing applications such as document classification or statistical machine translation.

Hidden Markov models

Modern general-purpose speech recognition systems are based on Hidden Markov Models. These are statistical models that output a sequence of symbols or quantities. HMMs are used in speech recognition because a speech signal can be viewed as a piecewise stationary signal or a short-time stationary signal. In a short time-scale (e.g., 10 milliseconds), speech can be approximated as a stationary process. Speech can be thought of as a Markov model for many stochastic purposes.

Another reason why HMMs are popular is because they can be trained automatically and are simple and computationally feasible to use. In speech recognition, the hidden Markov model would output a sequence of n-dimensional real-valued vectors (with n being a small integer, such as 10), outputting one of these every 10 milliseconds. The vectors would consist of cepstral coefficients, which are obtained by taking a Fourier transform of a short time window of speech and decorrelating the spectrum using a cosine transform, then taking the first (most significant) coefficients. The hidden Markov model will tend to have in each state a statistical distribution that is a mixture of diagonal covariance Gaussians, which will give a likelihood for each observed vector. Each word, or (for more general speech recognition systems), each phoneme, will have a different output distribution; a hidden Markov model for a sequence of words or phonemes is made by concatenating the individual trained hidden Markov models for the separate words and phonemes.

Described above are the core elements of the most common, HMM-based approach to speech recognition. Modern speech recognition systems use various combinations of a number of standard techniques in order to improve results over the basic approach described above. A typical large-vocabulary system would need context dependency for the phonemes (so phonemes with different left and right context have different realizations as HMM states); it would use cepstral normalization to normalize for different speaker and recording conditions; for further speaker normalization it might use vocal tract length normalization (VTLN) for male-female normalization and maximum likelihood linear regression (MLLR) for more general speaker adaptation. The features would have so-called delta and delta-delta coefficients to capture speech dynamics and in addition might use heteroscedastic linear discriminant analysis (HLDA); or might skip the delta and delta-delta coefficients and use splicing and an LDA-based projection followed perhaps by heteroscedastic linear discriminant analysis or a global semi-tied co variance transform (also known as maximum likelihood linear transform, or MLLT). Many systems use so-called discriminative training techniques that dispense with a purely statistical approach to HMM parameter estimation and instead optimize some classification-related measure of the training data. Examples are maximum mutual information (MMI), minimum classification error (MCE) and minimum phone error (MPE).

Decoding of the speech (the term for what happens when the system is presented with a new utterance and must compute the most likely source sentence) would probably use the Viterbi algorithm to find the best path, and here there is a choice between dynamically creating a combination hidden Markov model, which includes both the acoustic and language model information, and combining it statically beforehand (the finite state transducer, or FST, approach).

A possible improvement to decoding is to keep a set of good candidates instead of just keeping the best candidate, and to use a better scoring function (re scoring) to rate these good candidates so that we may pick the best one according to this refined score. The set of candidates can be kept either as a list (the N-best list approach) or as a subset of the models (a lattice). Re scoring is usually done by trying to minimize the Bayes risk (or an approximation thereof):

Instead of taking the source sentence with maximal probability, we try to take the sentence that minimizes the expectancy of a given loss function with regards to all possible transcriptions (i.e., we take the sentence that minimizes the average distance to other possible sentences weighted by their estimated probability). The loss function is usually the Levenshtein distance, though it can be different distances for specific tasks; the set of possible transcriptions is, of course, pruned to maintain tractability. Efficient algorithms have been devised to re score lattices represented as weighted finite state transducers with edit distances represented themselves as a finite state transducer verifying certain assumptions.

Dynamic time warping (DTW)-based speech recognition

Dynamic time warping is an approach that was historically used for speech recognition but has now largely been displaced by the more successful HMM-based approach.

Dynamic time warping is an algorithm for measuring similarity between two sequences that may vary in time or speed. For instance, similarities in walking patterns would be detected, even if in one video the person was walking slowly and if in another he or she were walking more quickly, or even if there were accelerations and deceleration during the course of one observation. DTW has been applied to video, audio, and graphics – indeed, any data that can be turned into a linear representation can be analyzed with DTW.

A well-known application has been automatic speech recognition, to cope with different speaking speeds. In general, it is a method that allows a computer to find an optimal match between two given sequences (e.g., time series) with certain restrictions. That is, the sequences are "warped" non-linearly to match each other. This sequence alignment method is often used in the context of hidden Markov models.

Neural networks

Neural networks emerged as an attractive acoustic modeling approach in ASR in the late 1980s. Since then, neural networks have been used in many aspects of speech recognition such as phoneme classification, isolated word recognition, audiovisual speech recognition, audiovisual speaker recognition and speaker adaptation.

In contrast to HMMs, neural networks make no assumptions about feature statistical properties and have several qualities making them attractive recognition models for speech recognition. When used to estimate the probabilities of a speech feature segment, neural networks allow discriminative training in a natural and efficient manner. Few assumptions on the statistics of input features are made with neural networks. However, in spite of their effectiveness in classifying short-time units such as individual phonemes and isolated words,[321] neural networks are rarely successful for continuous recognition tasks, largely because of their lack of ability to model temporal dependencies.

However, recently LSTM Recurrent Neural Networks (RNNs)[322] and Time Delay Neural Networks(TDNN's) have been used which have been shown to be able to identify latent temporal dependencies and use this information to perform the task of speech recognition.

Deep Neural Networks and Denoising Autoencoders were also being experimented with to tackle this problem in an effective manner.

Due to the inability of feedforward Neural Networks to model temporal dependencies, an alternative approach is to use neural networks as a pre-processing e.g. feature transformation, dimensionality reduction, for the HMM based recognition.

Deep feedforward and recurrent neural networks

A deep feedforward neural network (DNN) is an artificial neural network with multiple hidden layers of units between the input and output layers. Similar to shallow neural networks, DNNs can model complex non-linear relationships. DNN architectures generate compositional models, where extra layers enable composition of features from lower layers, giving a huge learning capacity and thus the potential of modeling complex patterns of speech data.

A success of DNNs in large vocabulary speech recognition occurred in 2010 by industrial researchers, in collaboration with academic researchers, where large output layers of the DNN based on context dependent HMM states constructed by decision trees were adopted.[323] See comprehensive reviews of this development and of the state of the art as of October 2014 in the recent Springer book from Microsoft Research. See also the related background of automatic speech recognition and the impact of various machine learning paradigms including notably deep learning in recent overview articles.

One fundamental principle of deep learning is to do away with hand-crafted feature engineering and to use raw features. This principle was first explored successfully in the architecture of deep autoencoder on the "raw" spectrogram or linear filter-bank features,[324] showing its superiority over the Mel-Cepstral features which contain a few stages of fixed transformation from spectrograms. The true "raw" features of speech, waveforms, have more recently been shown to produce excellent larger-scale speech recognition results.

End-to-end automatic speech recognition

Since 2014, there has been much research interest in "end-to-end" ASR. Traditional phonetic-based (i.e., all HMM-based model) approaches required separate components and training for the pronunciation, acoustic and language model. End-to-end models jointly learn all the components of the speech recognizer. This is valuable since it simplifies the training process and deployment process. For example, a n-gram language model is required for all HMM-based systems, and a typical n-gram language model often takes several gigabytes in memory making them impractical to deploy on mobile devices. Consequently, modern commercial ASR systems from Google and Apple (as of 2017) are deployed on the cloud and require a network connection as opposed to the device locally.

The first attempt of end-to-end ASR was with Connectionist Temporal Classification (CTC) based systems introduced by Alex Graves of Google DeepMind and Navdeep Jaitly of the University of Toronto in 2014. The model consisted of recurrent neural networks and a CTC layer. Jointly, the RNN-CTC model learns the pronunciation and acoustic model together, however it is incapable of learning the language due to conditional independence assumptions similar to a HMM. Consequently, CTC models can directly learn to map speech acoustics to English characters, but the models make many common spelling mistakes and must rely on a separate language model to clean up the transcripts. Later, Baidu expanded on the work with extremely large datasets and demonstrated some commercial success in Chinese Mandarin and English. In 2016, University of Oxford presented LipNet, the first end-to-end sentence-level lip reading model, using spatiotemporal convolutions coupled with an RNN-CTC architecture, surpassing human-level performance in a restricted grammar dataset.

An alternative approach to CTC-based models are attention-based models. Attention-based ASR models were introduced simultaneously by Chan et al. of Carnegie Mellon University and Google Brain and Bahdanaua et al. of the University of Montreal in 2016. The model named "Listen, Attend and Spell" (LAS), literally "listens" to the acoustic signal, pays "attention" to different parts of the signal and "spells" out the transcript one character at a time. Unlike CTC-based models, attention-based models do not have conditional-independence assumptions and can learn all the components of a speech recognizer including the pronunciation, acoustic and language model directly. This means, during deployment, there is no need to carry around a language model making it very practical for deployment onto applications with limited memory. By the end of 2016, the attention-based models have seen considerable success including outperforming the CTC models (with or without an external language model). Various extensions have been proposed since the original LAS model. Latent Sequence Decompositions (LSD) was proposed by Carnegie Mellon University, MIT and Google Brain to directly emit sub-word units which are more natural than English characters; University of Oxford and Google DeepMind extended LAS to "Watch, Listen, Attend and Spell" (WLAS) to handle lip reading surpassing human-level performance.

Applications

In-car systems

Typically a manual control input, for example by means of a finger control on the steering-wheel, enables the speech recognition system and this is signalled to the driver by an audio prompt. Following the audio prompt, the system has a "listening window" during which it may accept a speech input for recognition.Wikipedia:Citation needed

Simple voice commands may be used to initiate phone calls, select radio stations or play music from a compatible smartphone, MP3 player or music-loaded flash drive. Voice recognition capabilities vary between car make and model. Some of the most recentWikipedia:Manual of Style/Dates and numbers#Chronological items car models offer natural-language speech recognition in place of a fixed set of commands, allowing the driver to use full sentences and common phrases. With such systems there is, therefore, no need for the user to memorize a set of fixed command words.Wikipedia:Citation needed

Health care

Medical documentation

In the health care sector, speech recognition can be implemented in front-end or back-end of the medical documentation process. Front-end speech recognition is where the provider dictates into a speech-recognition engine, the recognized words are displayed as they are spoken, and the dictator is responsible for editing and signing off on the document. Back-end or deferred speech recognition is where the provider dictates into a digital dictation system, the voice is routed through a speech-recognition machine and the recognized draft document is routed along with the original voice file to the editor, where the draft is edited and report finalized. Deferred speech recognition is widely used in the industry currently.

One of the major issues relating to the use of speech recognition in healthcare is that the American Recovery and Reinvestment Act of 2009 (ARRA)

provides for substantial financial benefits to physicians who utilize an EMR according to "Meaningful Use" standards. These standards require that a substantial amount of data be maintained by the EMR (now more commonly referred to as an Electronic Health Record or EHR). The use of speech recognition is more naturally suited to the generation of narrative text, as part of a radiology/pathology interpretation, progress note or discharge summary: the ergonomic gains of using speech recognition to enter structured discrete data (e.g., numeric values or codes from a list or a controlled vocabulary) are relatively minimal for people who are sighted and who can operate a keyboard and mouse.

A more significant issue is that most EHRs have not been expressly tailored to take advantage of voice-recognition capabilities. A large part of the clinician's interaction with the EHR involves navigation through the user interface using menus, and tab/button clicks, and is heavily dependent on keyboard and mouse: voice-based navigation provides only modest ergonomic benefits. By contrast, many highly customized systems for radiology or pathology dictation implement voice "macros", where the use of certain phrases – e.g., "normal report", will automatically fill in a large number of default values and/or generate boilerplate, which will vary with the type of the exam – e.g., a chest X-ray vs. a gastrointestinal contrast series for a radiology system.

As an alternative to this navigation by hand, cascaded use of speech recognition and information extraction has been studied as a way to fill out a handover form for clinical proofing and sign-off. The results are encouraging, and the paper also opens data, together with the related performance benchmarks and some processing software, to the research and development community for studying clinical documentation and language-processing.

Therapeutic use

Prolonged use of speech recognition software in conjunction with word processors has shown benefits to short-term-memory restrengthening in brain AVM patients who have been treated with resection. Further research needs to be conducted to determine cognitive benefits for individuals whose AVMs have been treated using radiologic techniques.Wikipedia:Citation needed

Military

High-performance fighter aircraft

Substantial efforts have been devoted in the last decade to the test and evaluation of speech recognition in fighter aircraft. Of particular note have been the US program in speech recognition for the Advanced Fighter Technology Integration (AFTI)/F-16 aircraft (F-16 VISTA), the program in France for Mirage aircraft, and other programs in the UK dealing with a variety of aircraft platforms. In these programs, speech recognizers have been operated successfully in fighter aircraft, with applications including: setting radio frequencies, commanding an autopilot system, setting steer-point coordinates and weapons release parameters, and controlling flight display.

Working with Swedish pilots flying in the JAS-39 Gripen cockpit, Englund (2004) found recognition deteriorated with increasing g-loads. The report also concluded that adaptation greatly improved the results in all cases and that the introduction of models for breathing was shown to improve recognition scores significantly. Contrary to what might have been expected, no effects of the broken English of the speakers were found. It was evident that spontaneous speech caused problems for the recognizer, as might have been expected. A restricted vocabulary, and above all, a proper syntax, could thus be expected to improve recognition accuracy substantially.

The Eurofighter Typhoon, currently in service with the UK RAF, employs a speaker-dependent system, requiring each pilot to create a template. The system is not used for any safety-critical or weapon-critical tasks, such as weapon release or lowering of the undercarriage, but is used for a wide range of other cockpit functions. Voice commands are confirmed by visual and/or aural feedback. The system is seen as a major design feature in the reduction of pilot workload, and even allows the pilot to assign targets to his aircraft with two simple voice commands or to any of his wingmen with only five commands.

Speaker-independent systems are also being developed and are under test for the F35 Lightning II (JSF) and the Alenia Aermacchi M-346 Master lead-in fighter trainer. These systems have produced word accuracy scores in excess of 98%.

Helicopters

The problems of achieving high recognition accuracy under stress and noise pertain strongly to the helicopter environment as well as to the jet fighter environment. The acoustic noise problem is actually more severe in the helicopter environment, not only because of the high noise levels but also because the helicopter pilot, in general, does not wear a facemask, which would reduce acoustic noise in the microphone. Substantial test and evaluation programs have been carried out in the past decade in speech recognition systems applications in helicopters, notably by the U.S. Army Avionics Research and Development Activity (AVRADA) and by the Royal Aerospace Establishment (RAE) in the UK. Work in France has included speech recognition in the Puma helicopter. There has also been much useful work in Canada. Results have been encouraging, and voice applications have included: control of communication radios, setting of navigation systems, and control of an automated target handover system.

As in fighter applications, the overriding issue for voice in helicopters is the impact on pilot effectiveness. Encouraging results are reported for the AVRADA tests, although these represent only a feasibility demonstration in a test environment. Much remains to be done both in speech recognition and in overall speech technology in order to consistently achieve performance improvements in operational settings.

Training air traffic controllers

Training for air traffic controllers (ATC) represents an excellent application for speech recognition systems. Many ATC training systems currently require a person to act as a "pseudo-pilot", engaging in a voice dialog with the trainee controller, which simulates the dialog that the controller would have to conduct with pilots in a real ATC situation. Speech recognition and synthesis techniques offer the potential to eliminate the need for a person to act as pseudo-pilot, thus reducing training and support personnel. In theory, Air controller tasks are also characterized by highly structured speech as the primary output of the controller, hence reducing the difficulty of the speech recognition task should be possible. In practice, this is rarely the case. The FAA document 7110.65 details the phrases that should be used by air traffic controllers. While this document gives less than 150 examples of such phrases, the number of phrases supported by one of the simulation vendors speech recognition systems is in excess of 500,000.

The USAF, USMC, US Army, US Navy, and FAA as well as a number of international ATC training organizations such as the Royal Australian Air Force and Civil Aviation Authorities in Italy, Brazil, and Canada are currently using ATC simulators with speech recognition from a number of different vendors.Wikipedia:Citation needed

Telephony and other domains

ASR is now commonplace In the field of telephony, and is becoming more widespread in the field of computer gaming and simulation. Despite the high level of integration with word processing in general personal computing. However, ASR in the field of document production has not seen the expectedWikipedia:Manual of Style/Words to watch#Unsupported attributions increases in use.

The improvement of mobile processor speeds has made speech recognition practical in smartphones. Speech is used mostly as a part of a user interface, for creating predefined or custom speech commands. Leading software vendors in this field are: Google, Microsoft Corporation (Microsoft Voice Command), Digital Syphon (Sonic Extractor), LumenVox, Nuance Communications (Nuance Voice Control), Voci Technologies, VoiceBox Technology, Speech Technology Center, Vito Technologies (VITO Voice2Go), Speereo Software (Speereo Voice Translator), Verbyx VRX[325] and SVOX.

Usage in education and daily life

For language learning, speech recognition can be useful for learning a second language. It can teach proper pronunciation, in addition to helping a person develop fluency with their speaking skills.

Students who are blind (see Blindness and education) or have very low vision can benefit from using the technology to convey words and then hear the computer recite them, as well as use a computer by commanding with their voice, instead of having to look at the screen and keyboard.

Students who are physically disabled or suffer from Repetitive strain injury/other injuries to the upper extremities can be relieved from having to worry about handwriting, typing, or working with scribe on school assignments by using speech-to-text programs. They can also utilize speech recognition technology to freely enjoy searching the Internet or using a computer at home without having to physically operate a mouse and keyboard.

Speech recognition can allow students with learning disabilities to become better writers. By saying the words aloud, they can increase the fluidity of their writing, and be alleviated of concerns regarding spelling, punctuation, and other mechanics of writing. Also, see Learning disability.

Use of voice recognition software, in conjunction with a digital audio recorder and a personal computer running word-processing software has proven to be positive for restoring damaged short-term-memory capacity, in stroke and craniotomy individuals.

People with disabilities

People with disabilities can benefit from speech recognition programs. For individuals that are Deaf or Hard of Hearing, speech recognition software is used to automatically generate a closed-captioning of conversations such as discussions in conference rooms, classroom lectures, and/or religious services.

Speech recognition is also very useful for people who have difficulty using their hands, ranging from mild repetitive stress injuries to involve disabilities that preclude using conventional computer input devices. In fact, people who used the keyboard a lot and developed RSI became an urgent early market for speech recognition.[326] Speech recognition is used in deaf telephony, such as voicemail to text, relay services, and captioned telephone. Individuals with learning disabilities who have problems with thought-to-paper communication (essentially they think of an idea but it is processed incorrectly causing it to end up differently on paper) can possibly benefit from the software but the technology is not bug proof. Also the whole idea of speak to text can be hard for intellectually disabled person's due to the fact that it is rare that anyone tries to learn the technology to teach the person with the disability.[327]

This type of technology can help those with dyslexia but other disabilities are still in question. The effectiveness of the product is the problem that is hindering it being effective. Although a kid may be able to say a word depending on how clear they say it the technology may think they are saying another word and input the wrong one. Giving them more work to fix, causing them to have to take more time with fixing the wrong word.

Further applications

- Aerospace (e.g. space exploration, spacecraft, etc.) NASA's Mars Polar Lander used speech recognition technology from Sensory, Inc. in the Mars Microphone on the Lander
- Automatic subtitling with speech recognition
- Automatic translation
- Court reporting (Realtime Speech Writing)
- eDiscovery (Legal discovery)
- Hands-free computing: Speech recognition computer user interface
- Home automation
- Interactive voice response
- Mobile telephony, including mobile email
- Multimodal interaction
- Pronunciation evaluation in computer-aided language learning applications
- Robotics

- Speech-to-text reporter (transcription of speech into text, video captioning, Court reporting)
- Telematics (e.g. vehicle Navigation Systems)
- Transcription (digital speech-to-text)
- Video games, with *Tom Clancy's EndWar* and *Lifeline* as working examples
- Virtual assistant (e.g. Apple's Siri)

Performance

The performance of speech recognition systems is usually evaluated in terms of accuracy and speed.[328] Accuracy is usually rated with word error rate (WER), whereas speed is measured with the real time factor. Other measures of accuracy include Single Word Error Rate (SWER) and Command Success Rate (CSR).

Speech recognition by machine is a very complex problem, however. Vocalizations vary in terms of accent, pronunciation, articulation, roughness, nasality, pitch, volume, and speed. Speech is distorted by a background noise and echoes, electrical characteristics. Accuracy of speech recognition may vary with the following:[329]Wikipedia:Citation needed

- Vocabulary size and confusability
- Speaker dependence versus independence
- Isolated, discontinuous or continuous speech
- Task and language constraints
- Read versus spontaneous speech
- Adverse conditions

Accuracy

As mentioned earlier in this article, accuracy of speech recognition may vary depending on the following factors:

- Error rates increase as the vocabulary size grows:

e.g. the 10 digits "zero" to "nine" can be recognized essentially perfectly, but vocabulary sizes of 200, 5000 or 100000 may have error rates of 3%, 7% or 45% respectively.

- Vocabulary is hard to recognize if it contains confusable words:

e.g. the 26 letters of the English alphabet are difficult to discriminate because they are confusable words (most notoriously, the E-set: "B, C, D, E, G, P, T, V, Z"); an 8% error rate is considered good for this vocabulary.Wikipedia:Citation needed

- Speaker dependence vs. independence:

A speaker-dependent system is intended for use by a single speaker.
A speaker-independent system is intended for use by any speaker (more difficult).

- Isolated, Discontinuous or continuous speech

With isolated speech, single words are used, therefore it becomes easier to recognize the speech.
With discontinuous speech full sentences separated by silence are used, therefore it becomes easier to recognize the speech as well as with isolated speech.
With continuous speech naturally spoken sentences are used, therefore it becomes harder to recognize the speech, different from both isolated and discontinuous speech.

- Task and language constraints

e.g. Querying application may dismiss the hypothesis "The apple is red."
e.g. Constraints may be semantic; rejecting "The apple is angry."
e.g. Syntactic; rejecting "Red is apple the."
Constraints are often represented by a grammar.

- Read vs. Spontaneous Speech

When a person reads it's usually in a context that has been previously prepared, but when a person uses spontaneous speech, it is difficult to recognize the speech because of the disfluencies (like "uh" and "um", false starts, incomplete sentences, stuttering, coughing, and laughter) and limited vocabulary.

- Adverse conditions

Environmental noise (e.g. Noise in a car or a factory)
Acoustical distortions (e.g. echoes, room acoustics)
Speech recognition is a multi-levelled pattern recognition task.

- Acoustical signals are structured into a hierarchy of units;

e.g. Phonemes, Words, Phrases, and Sentences;

- Each level provides additional constraints;

e.g. Known word pronunciations or legal word sequences, which can compensate for errors or uncertainties at lower level;

- This hierarchy of constraints are exploited;

By combining decisions probabilistically at all lower levels, and making more deterministic decisions only at the highest level, speech recognition by a machine is a process broken into several phases. Computationally, it is a problem in which a sound pattern has to be recognized or classified into a category that represents a meaning to a human. Every acoustic signal can be broken in smaller more basic sub-signals. As the more complex sound signal is broken into the smaller sub-sounds, different levels are created, where at the top level we have complex sounds, which are made of simpler sounds on lower level, and going to lower levels even more, we create more basic and shorter and simpler sounds. The lowest level, where the sounds are the most fundamental, a machine would check for simple and more probabilistic rules of what sound should represent. Once these sounds are put together into more complex sound on upper level, a new set of more deterministic rules should predict what new complex sound should represent. The most upper level of a deterministic rule should figure out the meaning of complex expressions. In order to expand our knowledge about speech recognition we need to take into a consideration neural networks. There are four steps of neural network approaches:

- Digitize the speech that we want to recognize

For telephone speech the sampling rate is 8000 samples per second;

- Compute features of spectral-domain of the speech (with Fourier transform);

computed every 10 ms, with one 10 ms section called a frame;

Analysis of four-step neural network approaches can be explained by further information. Sound is produced by air (or some other medium) vibration, which we register by ears, but machines by receivers. Basic sound creates a wave which has 2 descriptions; Amplitude (how strong is it), and frequency (how often it vibrates per second).

The sound waves can be digitized: Sample a strength at short intervals like in picture aboveWikipedia:Naming conventions (geographic names) to get bunch of numbers that approximate at each time step the strength of a wave. Collection of these numbers represent analog wave. This new wave is digital. Sound waves are complicated because they superimpose one on top of each other. Like the waves would. This way they create odd-looking waves. For example, if there are two waves that interact with each other we can add them which creates new odd-looking wave.

- Neural network classifies features into phonetic-based categories;

Given basic sound blocks that a machine digitized, one has a bunch of numbers which describe a wave and waves describe words. Each frame has a unit block of sound, which are broken into basic sound waves and represented by numbers which, after Fourier Transform, can be statistically evaluated to set to which class of sounds it belongs. The nodes in the figure on a slide represent a feature of a sound in which a feature of a wave from the first layer of nodes to the second layer of nodes based on statistical analysis. This analysis depends on programmer's instructions. At this point, a second layer of nodes represents higher level features of a sound input which is again statistically evaluated to see what class they belong to. Last level of nodes should be output nodes that tell us with high probability what original sound really was.

- Search to match the neural-network output scores for the best word, to determine the word that was most likely uttered.

Security concerns

Speech recognition can become a means of attack, theft, or accidental operation. For example, activation words like "Alexa" spoken in an audio or video broadcast can cause devices in homes and offices to start listening for input inappropriately, or possibly take an unwanted action. Voice-controlled devices are also accessible to visitors to the building, or even those outside the building if they can be heard inside. Attackers may be able to gain access to personal information, like calendar, address book contents, private messages, and documents. They may also be able to impersonate the user to send messages or make online purchases.

Two attacks have been demonstrated that use artificial sounds. One transmits ultrasound and attempt to send commands without nearby people noticing. The other adds small, inaudible distortions to other speech or music that are specially crafted to confuse the specific speech recognition system into recognizing music as speech, or to make what sounds like one command to a human sound like a different command to the system.

Further information

Conferences and journals

Popular speech recognition conferences held each year or two include SpeechTEK and SpeechTEK Europe, ICASSP, Interspeech/Eurospeech, and the IEEE ASRU. Conferences in the field of natural language processing, such as ACL, NAACL, EMNLP, and HLT, are beginning to include papers on speech processing. Important journals include the IEEE Transactions on Speech and Audio Processing (later renamed IEEE Transactions on Audio,

Speech and Language Processing and since Sept 2014 renamed IEEE/ACM Transactions on Audio, Speech and Language Processing—after merging with an ACM publication), Computer Speech and Language, and Speech Communication.

Books

Books like "Fundamentals of Speech Recognition" by Lawrence Rabiner can be useful to acquire basic knowledge but may not be fully up to date (1993). Another good source can be "Statistical Methods for Speech Recognition" by Frederick Jelinek and "Spoken Language Processing (2001)" by Xuedong Huang etc. More up to date are "Computer Speech", by Manfred R. Schroeder, second edition published in 2004, and "Speech Processing: A Dynamic and Optimization-Oriented Approach" published in 2003 by Li Deng and Doug O'Shaughnessey. The recently updated textbook of "Speech and Language Processing (2008)" by Jurafsky and Martin presents the basics and the state of the art for ASR. Speaker recognition also uses the same features, most of the same front-end processing, and classification techniques as is done in speech recognition. A most recent comprehensive textbook, "Fundamentals of Speaker Recognition" is an in depth source for up to date details on the theory and practice. A good insight into the techniques used in the best modern systems can be gained by paying attention to government sponsored evaluations such as those organised by DARPA (the largest speech recognition-related project ongoing as of 2007 is the GALE project, which involves both speech recognition and translation components).

A good and accessible introduction to speech recognition technology and its history is provided by the general audience book "The Voice in the Machine. Building Computers That Understand Speech" by Roberto Pieraccini (2012).

The most recent book on speech recognition is "Automatic Speech Recognition: A Deep Learning Approach" (Publisher: Springer) written by D. Yu and L. Deng published near the end of 2014, with highly mathematically-oriented technical detail on how deep learning methods are derived and implemented in modern speech recognition systems based on DNNs and related deep learning methods. A related book, published earlier in 2014, "Deep Learning: Methods and Applications" by L. Deng and D. Yu provides a less technical but more methodology-focused overview of DNN-based speech recognition during 2009–2014, placed within the more general context of deep learning applications including not only speech recognition but also image recognition, natural language processing, information retrieval, multimodal processing, and multitask learning.

Software

In terms of freely available resources, Carnegie Mellon University's Sphinx toolkit is one place to start to both learn about speech recognition and to start experimenting. Another resource (free but copyrighted) is the HTK book (and the accompanying HTK toolkit). For more recent and state-of-the-art techniques, Kaldi toolkit can be used.Wikipedia:Citation needed

A Demo[330] of an on-line speech recognizer is available on Cobalt's webpage.

For more software resources, see List of speech recognition software.

Further reading

- Pieraccini, Roberto. *The Voice in the Machine. Building Computers That Understand Speech.* The MIT Press. ISBN 978-0262016858.
- Woelfel, Matthias; McDonough, John. *Distant Speech Recognition.* Wiley. ISBN 978-0470517048.
- Karat, Clare-Marie; Vergo, John; Nahamoo, David (2007). "Conversational Interface Technologies". In Sears, Andrew; Jacko, Julie A. *The Human-Computer Interaction Handbook: Fundamentals, Evolving Technologies, and Emerging Applications (Human Factors and Ergonomics).* Lawrence Erlbaum Associates Inc. ISBN 978-0-8058-5870-9.
- Cole, Ronald; Mariani, Joseph; Uszkoreit, Hans; Varile, Giovanni Battista; Zaenen, Annie; Zampolli; Zue, Victor, eds. (1997). *Survey of the state of the art in human language technology.* Cambridge Studies In Natural Language Processing. XII–XIII. Cambridge University Press. ISBN 0-521-59277-1.
- Junqua, J.-C.; Haton, J.-P. (1995). *Robustness in Automatic Speech Recognition: Fundamentals and Applications.* Kluwer Academic Publishers. ISBN 978-0-7923-9646-8.
- Pirani, Giancarlo, ed. (2013). *Advanced algorithms and architectures for speech understanding.* Springer Science & Business Media. ISBN 978-3-642-84341-9.

External links

- Signer, Beat and Hoste, Lode: *SpeeG2: A Speech- and Gesture-based Interface for Efficient Controller-free Text Entry*[331], In Proceedings of ICMI 2013, 15th International Conference on Multimodal Interaction, Sydney, Australia, December 2013
- Speech Technology[332] at Curlie (based on DMOZ)

Affective computing

Affective computing (sometimes called **artificial emotional intelligence**, or **emotion AI**) is the study and development of systems and devices that can recognize, interpret, process, and simulate human affects. It is an interdisciplinary field spanning computer science, psychology, and cognitive science. While the origins of the field may be traced as far back as to early philosophical inquiries into emotion,[333] the more modern branch of computer science originated with Rosalind Picard's 1995 paper[334] on affective computing. A motivation for the research is the ability to simulate empathy. The machine should interpret the emotional state of humans and adapt its behavior to them, giving an appropriate response to those emotions.

The difference between sentiment analysis and affective analysis is that the latter detects the different emotions instead of identifying only the polarity of the phrase.

Areas

Detecting and recognizing emotional information

Detecting emotional information begins with passive sensors which capture data about the user's physical state or behavior without interpreting the input. The data gathered is analogous to the cues humans use to perceive emotions in others. For example, a video camera might capture facial expressions, body posture, and gestures, while a microphone might capture speech. Other sensors detect emotional cues by directly measuring physiological data, such as skin temperature and galvanic resistance.

Recognizing emotional information requires the extraction of meaningful patterns from the gathered data. This is done using machine learning techniques that process different modalities, such as speech recognition, natural language processing, or facial expression detection, and produce either a labels (i.e. 'confused') or coordinates in a valence-arousal space.

Emotion in machines

Another area within affective computing is the design of computational devices proposed to exhibit either innate emotional capabilities or that are capable of convincingly simulating emotions. A more practical approach, based on current technological capabilities, is the simulation of emotions in conversational agents in order to enrich and facilitate interactivity between human and machine. While human emotions are often associated with surges in hormones and other neuropeptides, emotions in machines might be associated with abstract states associated with progress (or lack of progress) in autonomous learning systemsWikipedia:Citation needed. In this view, affective emotional states correspond to time-derivatives (perturbations) in the learning curve of an arbitrary learning system.Wikipedia:Citation needed

Marvin Minsky, one of the pioneering computer scientists in artificial intelligence, relates emotions to the broader issues of machine intelligence stating in *The Emotion Machine* that emotion is "not especially different from the processes that we call 'thinking.'"

Technologies

In cognitive science and neuroscience, there have been two leading models describing how humans perceive and classify emotion: the continuous model and the categorical model. The continuous model defines each facial expression of emotion as a feature vector in a face space. This model explains, for example, how expressions of emotion can be seen at different intensities. In contrast, the categorical model consists of C classifiers, each tuned to a specific emotion category. This model explains, among other findings, why the images in a morphing sequence between a happy and a surprise face are perceived as either happy or surprise but not something in between.

These approaches have one major flaw in common: they can only detect one emotion from an image, this is generally done by a winner takes it all method. Yet, every day we can perceive more than one emotional category from a single image. Both the categorical and continuous model are unable to identify multiple emotions, so a new way to model it is to consider new categories as the overlap of a small set of categories. A detailed study related to this topic is given in "A model of the perception of facial expressions of emotion by humans: research overview and perspectives".

The following sections consider the possible features which can be used for the task of emotion recognition.

Emotional speech

Various changes in the autonomic nervous system can indirectly alter a person's speech, and affective technologies can leverage this information to recognize emotion. For example, speech produced in a state of fear, anger, or joy becomes fast, loud, and precisely enunciated, with a higher and wider range in pitch, whereas emotions such as tiredness, boredom, or sadness tend to generate slow, low-pitched, and slurred speech.[335] Some emotions have been found to be more easily computationally identified, such as anger or approval.

Emotional speech processing technologies recognize the user's emotional state using computational analysis of speech features. Vocal parameters and prosodic features such as pitch variables and speech rate can be analyzed through pattern recognition techniques.[336,337]

Speech analysis is an effective method of identifying affective state, having an average reported accuracy of 70 to 80% in recent research. These systems tend to outperform average human accuracy (approximately 60%) but are less accurate than systems which employ other modalities for emotion detection, such as physiological states or facial expressions. However, since many speech characteristics are independent of semantics or culture, this technique is considered to be a promising route for further research.

Algorithms

The process of speech/text affect detection requires the creation of a reliable database, knowledge base, or vector space model, broad enough to fit every need for its application, as well as the selection of a successful classifier which will allow for quick and accurate emotion identification.

Currently, the most frequently used classifiers are linear discriminant classifiers (LDC), k-nearest neighbor (k-NN), Gaussian mixture model (GMM), support vector machines (SVM), artificial neural networks (ANN), decision tree algorithms and hidden Markov models (HMMs). Various studies showed that choosing the appropriate classifier can significantly enhance the overall performance of the system. The list below gives a brief description of each algorithm:

- LDC – Classification happens based on the value obtained from the linear combination of the feature values, which are usually provided in the form of vector features.
- k-NN – Classification happens by locating the object in the feature space, and comparing it with the k nearest neighbors (training examples). The majority vote decides on the classification.
- GMM – is a probabilistic model used for representing the existence of subpopulations within the overall population. Each sub-population is described using the mixture distribution, which allows for classification of observations into the sub-populations.[338]
- SVM – is a type of (usually binary) linear classifier which decides in which of the two (or more) possible classes, each input may fall into.
- ANN – is a mathematical model, inspired by biological neural networks, that can better grasp possible non-linearities of the feature space.
- Decision tree algorithms – work based on following a decision tree in which leaves represent the classification outcome, and branches represent the conjunction of subsequent features that lead to the classification.

- HMMs – a statistical Markov model in which the states and state transitions are not directly available to observation. Instead, the series of outputs dependent on the states are visible. In the case of affect recognition, the outputs represent the sequence of speech feature vectors, which allow the deduction of states' sequences through which the model progressed. The states can consist of various intermediate steps in the expression of an emotion, and each of them has a probability distribution over the possible output vectors. The states' sequences allow us to predict the affective state which we are trying to classify, and this is one of the most commonly used techniques within the area of speech affect detection.

It is proved that having enough acoustic evidence available the emotional state of a person can be classified by a set of majority voting classifiers. The proposed set of classifiers is based on three main classifiers: kNN, C4.5 and SVM-RBF Kernel. This set achieves better performance than each basic classifier taken separately. It is compared with two other sets of classifiers: one-against-all (OAA) multiclass SVM with Hybrid kernels and the set of classifiers which consists of the following two basic classifiers: C5.0 and Neural Network. The proposed variant achieves better performance than the other two sets of classifiers.

Databases

The vast majority of present systems are data-dependent. This creates one of the biggest challenges in detecting emotions based on speech, as it implicates choosing an appropriate database used to train the classifier. Most of the currently possessed data was obtained from actors and is thus a representation of archetypal emotions. Those so-called acted databases are usually based on the Basic Emotions theory (by Paul Ekman), which assumes the existence of six basic emotions (anger, fear, disgust, surprise, joy, sadness), the others simply being a mix of the former ones.[339] Nevertheless, these still offer high audio quality and balanced classes (although often too few), which contribute to high success rates in recognizing emotions.

However, for real life application, naturalistic data is preferred. A naturalistic database can be produced by observation and analysis of subjects in their natural context. Ultimately, such database should allow the system to recognize emotions based on their context as well as work out the goals and outcomes of the interaction. The nature of this type of data allows for authentic real life implementation, due to the fact it describes states naturally occurring during the human-computer interaction (HCI).

Despite the numerous advantages which naturalistic data has over acted data, it is difficult to obtain and usually has low emotional intensity. Moreover, data obtained in a natural context has lower signal quality, due to surroundings noise and distance of the subjects from the microphone. The first attempt to produce such database was the FAU Aibo Emotion Corpus for CEICES (Combining Efforts for Improving Automatic Classification of Emotional User States), which was developed based on a realistic context of children (age 10-13) playing with Sony's Aibo robot pet. Likewise, producing one standard database for all emotional research would provide a method of evaluating and comparing different affect recognition systems.

Speech descriptors

The complexity of the affect recognition process increases with the number of classes (affects) and speech descriptors used within the classifier. It is, therefore, crucial to select only the most relevant features in order to assure the ability of the model to successfully identify emotions, as well as increasing the performance, which is particularly significant to real-time detection. The range of possible choices is vast, with some studies mentioning the use of over 200 distinct features. It is crucial to identify those that are redundant and undesirable in order to optimize the system and increase the success rate of correct emotion detection. The most common speech characteristics are categorized into the following groups.

1. Frequency characteristics
   - Accent shape – affected by the rate of change of the fundamental frequency.
   - Average pitch – description of how high/low the speaker speaks relative to the normal speech.
   - Contour slope – describes the tendency of the frequency change over time, it can be rising, falling or level.
   - Final lowering – the amount by which the frequency falls at the end of an utterance.
   - Pitch range – measures the spread between the maximum and minimum frequency of an utterance.
2. Time-related features:
   - Speech rate – describes the rate of words or syllables uttered over a unit of time
   - Stress frequency – measures the rate of occurrences of pitch accented utterances
3. Voice quality parameters and energy descriptors:
   - Breathiness – measures the aspiration noise in speech
   - Brilliance – describes the dominance of high Or low frequencies In the speech
   - Loudness – measures the amplitude of the speech waveform, translates to the energy of an utterance

- Pause Discontinuity – describes the transitions between sound and silence
- Pitch Discontinuity – describes the transitions of the fundamental frequency.

Facial affect detection

The detection and processing of facial expression are achieved through various methods such as optical flow, hidden Markov models, neural network processing or active appearance models. More than one modalities can be combined or fused (multimodal recognition, e.g. facial expressions and speech prosody, facial expressions and hand gestures, or facial expressions with speech and text for multimodal data and metadata analysis) to provide a more robust estimation of the subject's emotional state.

Facial expression databases

Creation of an emotion database is a difficult and time-consuming task. However, database creation is an essential step in the creation of a system that will recognize human emotions. Most of the publicly available emotion databases include posed facial expressions only. In posed expression databases, the participants are asked to display different basic emotional expressions, while in spontaneous expression database, the expressions are natural. Spontaneous emotion elicitation requires significant effort in the selection of proper stimuli which can lead to a rich display of intended emotions. Secondly, the process involves tagging of emotions by trained individuals manually which makes the databases highly reliable. Since perception of expressions and their intensity is subjective in nature, the annotation by experts is essential for the purpose of validation.

Researchers work with three types of databases, such as a database of peak expression images only, a database of image sequences portraying an emotion from neutral to its peak, and video clips with emotional annotations. Many facial expression databases have been created and made public for expression recognition purpose. Two of the widely used databases are CK+ and JAFFE.

Emotion classification

By doing cross-cultural research in Papua New Guinea, on the Fore Tribesmen, at the end of the 1960s, Paul Ekman proposed the idea that facial expressions of emotion are not culturally determined, but universal. Thus, he suggested that they are biological in origin and can, therefore, be safely and correctly categorized. He therefore officially put forth six basic emotions, in 1972:

- Anger

- Disgust
- Fear
- Happiness
- Sadness
- Surprise

However, in the 1990s Ekman expanded his list of basic emotions, including a range of positive and negative emotions not all of which are encoded in facial muscles. The newly included emotions are:

1. Amusement
2. Contempt
3. Contentment
4. Embarrassment
5. Excitement
6. Guilt
7. Pride in achievement
8. Relief
9. Satisfaction
10. Sensory pleasure
11. Shame

Facial Action Coding System

Defining expressions in terms of muscle actions A system has been conceived in order to formally categorize the physical expression of emotions. The central concept of the Facial Action Coding System, or FACS, as created by Paul Ekman and Wallace V. Friesen in 1978[340] are action units (AU). They are, basically, a contraction or a relaxation of one or more muscles. However, as simple as this concept may seem, it is enough to form the base of a complex and devoid of interpretation emotional identification system.

By identifying different facial cues, scientists are able to map them to their corresponding action unit code. Consequently, they have proposed the following classification of the six basic emotions, according to their action units ("+" here mean "and"):

Emotion	Action units
Happiness	6+12
Sadness	1+4+15
Surprise	1+2+5B+26
Fear	1+2+4+5+20+26
Anger	4+5+7+23

Disgust	9+15+16
Contempt	R12A+R14A

Challenges in facial detection

As with every computational practice, in affect detection by facial processing, some obstacles need to be surpassed, in order to fully unlock the hidden potential of the overall algorithm or method employed. The accuracy of modeling and tracking has been an issue, especially in the incipient stages of affective computing. As hardware evolves, as new discoveries are made and new practices introduced, this lack of accuracy fades, leaving behind noise issues. However, methods for noise removal exist including neighborhood averaging, linear Gaussian smoothing, median filtering, or newer methods such as the Bacterial Foraging Optimization Algorithm.[341,342]

It is generally known that the degree of accuracy in facial recognition (not affective state recognition) has not been brought to a level high enough to permit its widespread efficient use across the world (there have been many attempts, especially by law enforcement, which failed at successfully identifying criminals). Without improving the accuracy of hardware and software used to scan faces, progress is very much slowed down.

Other challenges include

- The fact that posed expressions, as used by most subjects of the various studies, are not natural, and therefore not 100% accurate.
- The lack of rotational movement freedom. Affect detection works very well with frontal use, but upon rotating the head more than 20 degrees, "there've been problems".[343]

Body gesture

Gestures could be efficiently used as a means of detecting a particular emotional state of the user, especially when used in conjunction with speech and face recognition. Depending on the specific action, gestures could be simple reflexive responses, like lifting your shoulders when you don't know the answer to a question, or they could be complex and meaningful as when communicating with sign language. Without making use of any object or surrounding environment, we can wave our hands, clap or beckon. On the other hand, when using objects, we can point at them, move, touch or handle these. A computer should be able to recognize these, analyze the context and respond in a meaningful way, in order to be efficiently used for Human-Computer Interaction.

There are many proposed methods[344] to detect the body gesture. Some literature differentiates 2 different approaches in gesture recognition: a 3D model based and an appearance-based. The foremost method makes use of 3D information of key elements of the body parts in order to obtain several important parameters, like palm position or joint angles. On the other hand, Appearance-based systems use images or videos to for direct interpretation. Hand gestures have been a common focus of body gesture detection, apparentnessWikipedia:Vagueness methods and 3-D modeling methods are traditionally used.

Physiological monitoring

This could be used to detect a user's emotional state by monitoring and analyzing their physiological signs. These signs range from their pulse and heart rate to the minute contractions of the facial muscles. This area of research is still in relative infancy as there seems to be more of a drive towards affect recognition through facial inputs. Nevertheless, this area is gaining momentum and we are now seeing real products which implement the techniques. The three main physiological signs that can be analyzed are blood volume pulse, galvanic skin response, facial electromyography.

Blood volume pulse

Overview

A subject's blood volume pulse (BVP) can be measured by a process called photoplethysmography, which produces a graph indicating blood flow through the extremities.[345] The peaks of the waves indicate a cardiac cycle where the heart has pumped blood to the extremities. If the subject experiences fear or is startled, their heart usually 'jumps' and beats quickly for some time, causing the amplitude of the cardiac cycle to increase. This can clearly be seen on a photoplethysmograph when the distance between the trough and the peak of the wave has decreased. As the subject calms down, and as the body's inner core expands, allowing more blood to flow back to the extremities, the cycle will return to normal.

Methodology

Infra-red light is shone on the skin by special sensor hardware, and the amount of light reflected is measured. The amount of reflected and transmitted light correlates to the BVP as light is absorbed by hemoglobin which is found richly in the blood stream.

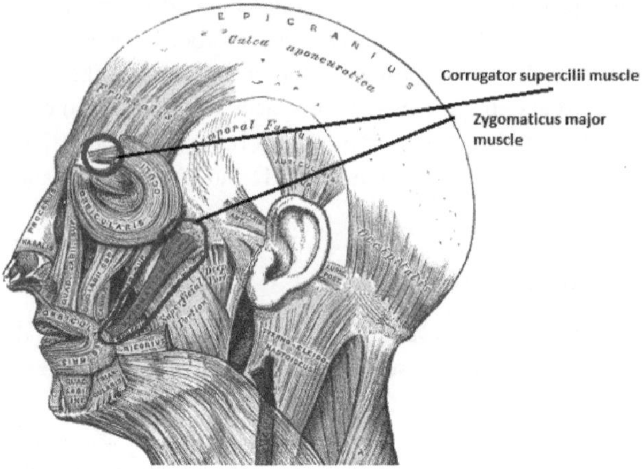

Figure 13: *The corrugator supercilii muscle and zygomaticus major muscle are the 2 main muscles used for measuring the electrical activity, in facial electromyography*

Disadvantages

It can be cumbersome to ensure that the sensor shining an infra-red light and monitoring the reflected light is always pointing at the same extremity, especially seeing as subjects often stretch and readjust their position whilst using a computer. There are other factors which can affect one's blood volume pulse. As it is a measure of blood flow through the extremities, if the subject feels hot, or particularly cold, then their body may allow more, or less, blood to flow to the extremities, all of this regardless of the subject's emotional state.

Facial electromyography

Facial electromyography is a technique used to measure the electrical activity of the facial muscles by amplifying the tiny electrical impulses that are generated by muscle fibers when they contract.[346] The face expresses a great deal of emotion, however, there are two main facial muscle groups that are usually studied to detect emotion: The corrugator supercilii muscle, also known as the 'frowning' muscle, draws the brow down into a frown, and therefore is the best test for negative, unpleasant emotional response.. The zygomaticus major muscle is responsible for pulling the corners of the mouth back when you smile, and therefore is the muscle used to test for a positive emotional response.

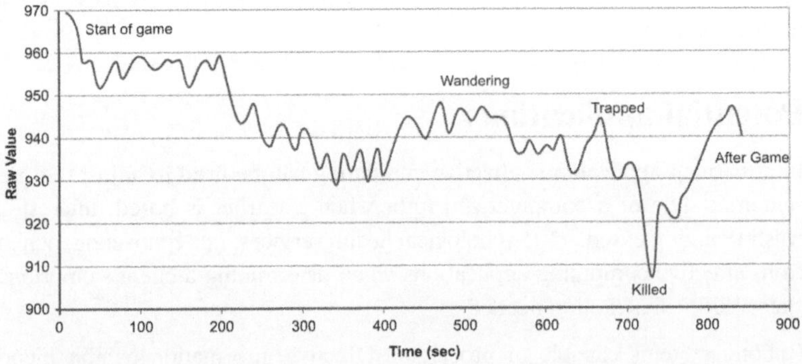

Figure 14: *Here we can see a plot of skin resistance measured using GSR and time whilst the subject played a video game. There are several peaks that are clear in the graph, which suggests that GSR is a good method of differentiating between an aroused and a non-aroused state. For example, at the start of the game where there is usually not much exciting game play, there is a high level of resistance recorded, which suggests a low level of conductivity and therefore less arousal. This is in clear contrast with the sudden trough where the player is killed as one is usually very stressed and tense as their character is killed in the game*

Galvanic skin response

Galvanic skin response (GSR) is a measure of skin conductivity, which is dependent on how moist the skin is. As the sweat glands produce this moisture and the glands are controlled by the body's nervous system, there is a correlation between GSR and the arousal state of the body. The more aroused a subject is, the greater the skin conductivity and GSR reading.

It can be measured using two small silver chloride electrodes placed somewhere on the skin and applying a small voltage between them. The conductance is measured by a sensor. To maximize comfort and reduce irritation the electrodes can be placed on the feet, which leaves the hands fully free to interface with the keyboard and mouse.

Visual aesthetics

Aesthetics, in the world of art and photography, refers to the principles of the nature and appreciation of beauty. Judging beauty and other aesthetic qualities is a highly subjective task. Computer scientists at Penn State treat the challenge of automatically inferring the aesthetic quality of pictures using their visual content as a machine learning problem, with a peer-rated on-line photo sharing website as a data source.[347] They extract certain visual features based on the intuition that they can discriminate between aesthetically pleasing and displeasing images.

Potential applications

In e-learning applications, affective computing can be used to adjust the presentation style of a computerized tutor when a learner is bored, interested, frustrated, or pleased.[348] Psychological health services, i.e. counseling, benefit from affective computing applications when determining a client's emotional state.Wikipedia:Citation needed

Robotic systems capable of processing affective information exhibit higher flexibility while one works in uncertain or complex environments. Companion devices, such as digital pets, use affective computing abilities to enhance realism and provide a higher degree of autonomy.Wikipedia:Citation needed

Social robots, as well as a growing number of robots used in health care benefit from emotional awareness because they can better judge users' and patient's emotional states and alter their actions/programming appropriately. This is especially important in those countries with growing aging populations and/or a lack of younger workers to address their needs.

Other potential applications are centered around social monitoring. For example, a car can monitor the emotion of all occupants and engage in additional safety measures, such as alerting other vehicles if it detects the driver to be angry.Wikipedia:Citation needed Affective computing has potential applications in human-computer interaction, such as affective mirrors allowing the user to see how he or she performs; emotion monitoring agents sending a warning before one sends an angry email; or even music players selecting tracks based on mood.

One idea put forth by the Romanian researcher Dr. Nicu Sebe in an interview is the analysis of a person's face while they are using a certain product (he mentioned ice cream as an example). Companies would then be able to use such analysis to infer whether their product will or will not be well received by the respective market.

One could also use affective state recognition in order to judge the impact of a TV advertisement through a real-time video recording of that person and through the subsequent study of his or her facial expression. Averaging the results obtained on a large group of subjects, one can tell whether that commercial (or movie) has the desired effect and what the elements which interest the watcher most are.

Affective computing is also being applied to the development of communicative technologies for use by people with autism.[349]

Video games

Affective video games can access their players' emotional states through biofeedback devices. A particularly simple form of biofeedback is available through gamepads that measure the pressure with which a button is pressed: this has been shown to correlate strongly with the players' level of arousal; at the other end of the scale are brain–computer interfaces. Affective games have been used in medical research to support the emotional development of autistic children.

Cognitivist vs. interactional approaches

Within the field of human-computer interaction, Rosalind Picard's cognitivist or "information model" concept of emotion has been criticized by and contrasted with the "post-cognitivist" or "interactional" pragmatist approach taken by Kirsten Boehner and others which views emotion as inherently social.

Picard's focus is human-computer interaction, and her goal for affective computing is to "give computers the ability to recognize, express, and in some cases, 'have' emotions". In contrast, the interactional approach seeks to help "people to understand and experience their own emotions" and to improve computer-mediated interpersonal communication. It does not necessarily seek to map emotion into an objective mathematical model for machine interpretation, but rather let humans make sense of each other's emotional expressions in open-ended ways that might be ambiguous, subjective, and sensitive to context.[:284]Wikipedia:AUDIENCE

Picard's critics describe her concept of emotion as "objective, internal, private, and mechanistic". They say it reduces emotion to a discrete psychological signal occurring inside the body that can be measured and which is an input to cognition, undercutting the complexity of emotional experience.[:280:278]

The interactional approach asserts that though emotion has biophysical aspects, it is "culturally grounded, dynamically experienced, and to some degree constructed in action and interaction".[:276] Put another way, it considers "emotion as a social and cultural product experienced through our interactions".

Sources

- Hudlicka, Eva (2003). "To feel or not to feel: The role of affect in human-computer interaction". *International Journal of Human-Computer Studies*. **59** (1–2): 1–32. CiteSeerX 10.1.1.180.6429[350] . doi: 10.1016/s1071-5819(03)00047-8[351].
- Scherer, Klaus R; Banziger, T; Roesch, Etienne B (2010). *A blueprint for affective computing: a sourcebook*. Oxford: Oxford University Press.

External links

- Affective Computing Research Group at the MIT Media Laboratory[352]
- Computational Emotion Group at USC[353]
- Emotive Computing Group at the University of Memphis[354]
- 2011 International Conference on Affective Computing and Intelligent Interaction[355]
- Brain, Body and Bytes: Psychophysiological User Interaction[356] *CHI 2010 Workshop* (10-15, April 2010)
- International Journal of Synthetic Emotions[357]
- IEEE Transactions on Affective Computing[358] *(TAC)*
- Renu Nagpal, Pooja Nagpal and Sumeet Kaur, "Hybrid Technique for Human Face Emotion Detection" International Journal of Advanced Computer Science and Applications(IJACSA), 1(6), 2010[359]
- openSMILE: popular state-of-the-art open-source toolkit for large-scale feature extraction for affect recognition and computational paralinguistics[360]

Artificial general intelligence

Artificial general intelligence (**AGI**) is the intelligence of a machine that could successfully perform any intellectual task that a human being can. It is a primary goal of some artificial intelligence research and a common topic in science fiction and future studies. Artificial general intelligence is also referred to as "**strong AI**",[361] "**full AI**"[362] or as the ability of a machine to perform "general intelligent action".[363] Academic sources reserve "strong AI" to refer to machines capable of experiencing consciousness.

Some references emphasize a distinction between strong AI and "applied AI"[364] (also called "narrow AI" or "weak AI"[365]): the use of software to study or accomplish specific problem solving or reasoning tasks. Weak AI, in contrast to strong AI, does not attempt to perform the full range of human cognitive abilities.

As of 2017, over forty organizations worldwide are doing active research on AGI.

Requirements

Various criteria for intelligence have been proposed (most famously the Turing test) but to date, there is no definition that satisfies everyone.[366] However, there *is* wide agreement among artificial intelligence researchers that intelligence is required to do the following:[367]

- reason, use strategy, solve puzzles, and make judgments under uncertainty;
- represent knowledge, including commonsense knowledge;
- plan;
- learn;
- communicate in natural language;
- and integrate all these skills towards common goals.

Other important capabilities include the ability to sense (e.g. see) and the ability to act (e.g. move and manipulate objects) in the world where intelligent behaviour is to be observed.[368] This would include an ability to detect and respond to hazard. Many interdisciplinary approaches to intelligence (e.g. cognitive science, computational intelligence and decision making) tend to emphasise the need to consider additional traits such as imagination (taken as the ability to form mental images and concepts that were not programmed in) and autonomy.[369] Computer based systems that exhibit many of these capabilities do exist (e.g. see computational creativity, automated reasoning, decision support system, robot, evolutionary computation, intelligent agent), but not yet at human levels.

Tests for confirming human-level AGI

The Turing Test (*Turing*)

A machine and a human both converse sight unseen with a second human, who must evaluate which of the two is the machine, which passes the test if it can fool the evaluator a significant fraction of the time.

The Coffee Test (*Wozniak*)

A machine is required to enter an average American home and figure out how to make coffee: find the coffee machine, find the coffee, add water, find a mug, and brew the coffee by pushing the proper buttons.

The Robot College Student Test (*Goertzel*)

A machine enrolls in a university, taking and passing the same classes that humans would, and obtaining a degree.

The Employment Test (*Nilsson*)

A machine works an economically important job, performing at least as well as humans in the same job.

The flat pack furniture test (*Tony Severyns*)

A machine is required to unpack and assemble an item of flat-packed furniture. It has to read the instructions and assemble the item as described, correctly installing all fixtures.

Problems requiring AGI to solve

The most difficult problems for computers are informally known as "AI-complete" or "AI-hard", implying that solving them is equivalent to the general aptitude of human intelligence, or strong AI, beyond the capabilities of a purpose-specific algorithm.[370]

AI-complete problems are hypothesised to include general computer vision, natural language understanding, and dealing with unexpected circumstances while solving any real world problem.[371]

AI-complete problems cannot be solved with current computer technology alone, and also require human computation. This property can be useful to test for the presence of humans, as with CAPTCHAs, and for computer security to repel brute-force attacks.[372,373]

Classical AI

Modern AI research began in the mid 1950s. The first generation of AI researchers was convinced that artificial general intelligence was possible and that it would exist in just a few decades. As AI pioneer Herbert A. Simon wrote in 1965: "machines will be capable, within twenty years, of doing any work a man can do."[374] Their predictions were the inspiration for Stanley Kubrick and Arthur C. Clarke's character HAL 9000, who accurately embodied what AI researchers believed they could create by the year 2001. Of note is the fact that AI pioneer Marvin Minsky was a consultant[375] on the project of making HAL 9000 as realistic as possible according to the consensus predictions of the time; Crevier quotes him as having said on the subject in 1967, "Within a generation ... the problem of creating 'artificial intelligence' will substantially be solved,"[376] although Minsky states that he was misquoted.Wikipedia:Citation needed

However, in the early 1970s, it became obvious that researchers had grossly underestimated the difficulty of the project. Funding agencies became skeptical of strong AI (ASI) and put researchers under increasing pressure to produce useful "applied AI".[377] As the 1980s began, Japan's Fifth Generation Computer Project revived interest in strong AI (ASI), setting out a ten-year timeline that included strong AI (ASI) goals like "carry on a casual conversation".[378] In response to this and the success of expert systems, both industry and government pumped money back into the field.[379] However, confidence in AI spectacularly collapsed in the late 1980s, and the goals of the Fifth Generation Computer Project were never fulfilled. For the second time in 20 years, AI researchers who had predicted the imminent achievement of strong AI (ASI) had been shown to be fundamentally mistaken. By the 1990s, AI researchers had gained a reputation for making vain promises. They became reluctant to make predictions at all[380] and to avoid any mention of "human level" artificial intelligence for fear of being labeled "wild-eyed dreamer[s]."[381]

Current narrow AI research

In the 1990s and early 21st century, mainstream AI has achieved far greater commercial success and academic respectability by focusing on specific sub-problems where they can produce verifiable results and commercial applications, such as artificial neural networks, computer vision or data mining. These "applied AI" systems are now used extensively throughout the technology industry, and research in this vein is very heavily funded in both academia and industry.

Most mainstream AI researchers hope that strong AI can be developed by combining the programs that solve various sub-problems using an integrated agent architecture, cognitive architecture or subsumption architecture. Hans Moravec wrote in 1988:

> *"I am confident that this bottom-up route to artificial intelligence will one day meet the traditional top-down route more than half way, ready to provide the real world competence and the commonsense knowledge that has been so frustratingly elusive in reasoning programs. Fully intelligent machines will result when the metaphorical golden spike is driven uniting the two efforts."*

However, even this fundamental philosophy has been disputed; for example, Stevan Harnad of Princeton concluded his 1990 paper on the Symbol Grounding Hypothesis by stating:

> *"The expectation has often been voiced that "top-down" (symbolic) approaches to modeling cognition will somehow meet "bottom-up" (sensory) approaches somewhere in between. If the grounding considerations in this paper are valid, then this expectation is hopelessly modular and there is really only one viable route from sense to symbols: from the ground up. A free-floating symbolic level like the software level of a computer will never be reached by this route (or vice versa) – nor is it clear why we should even try to reach such a level, since it looks as if getting there would just amount to uprooting our symbols from their intrinsic meanings (thereby merely reducing ourselves to the functional equivalent of a programmable computer)."*

Artificial general intelligence research

Artificial general intelligence[382] (AGI) describes research that aims to create machines capable of general intelligent action. The term was introduced by Mark Gubrud in 1997 in a discussion of the implications of fully automated military production and operations. The research objective is much older, for example Doug Lenat's Cyc project (that began in 1984), and Allen Newell's Soar project are regarded as within the scope of AGI. AGI research activity in 2006 was described by Pei Wang and Ben Goertzel[383] as "producing publications and preliminary results". As yet, most AI researchers have devoted little attention to AGI, with some claiming that intelligence is too complex to be completely replicated in the near term. However, a small number of computer scientists are active in AGI research, and many of this group are contributing to a series of AGI conferences. The research is extremely diverse and often pioneering in nature. In the introduction to his book,[382] Goertzel says that estimates of the time needed before a truly flexible AGI is built vary from 10 years to over a century, but the consensus in the AGI research community seems to be that the timeline discussed by Ray Kurzweil in *The Singularity is Near* (i.e. between 2015 and 2045) is plausible.[384] Most mainstream AI researchers doubt that progress will be this rapid.Wikipedia:Citation needed Organizations explicitly pursuing AGI include the Swiss AI lab IDSIA,Wikipedia:Citation needed Nnaisense, Vicarious, Maluuba, the OpenCog Foundation, Adaptive AI, LIDA, and Numenta and the associated Redwood Neuroscience Institute. In addition, organizations such as the Machine Intelligence Research Institute and OpenAI have been founded to influence the development path of AGI. Finally, projects such as the Human Brain Project have the goal of building a functioning simulation of the human brain. A 2017 survey of AGI categorized forty-five known "active R&D projects" that explicitly or implicitly (through published research) research AGI, with the largest three being DeepMind, the Human Brain Project, and OpenAI.[385]

Processing power needed to simulate a brain

Whole brain emulation

A popular approach discussed to achieving general intelligent action is whole brain emulation. A low-level brain model is built by scanning and mapping a biological brain in detail and copying its state into a computer system or another computational device. The computer runs a simulation model so faithful to the original that it will behave in essentially the same way as the original brain, or for all practical purposes, indistinguishably.[386] Whole brain emulation is discussed in computational neuroscience and neuroinformatics, in the context of brain simulation for medical research purposes. It is discussed in

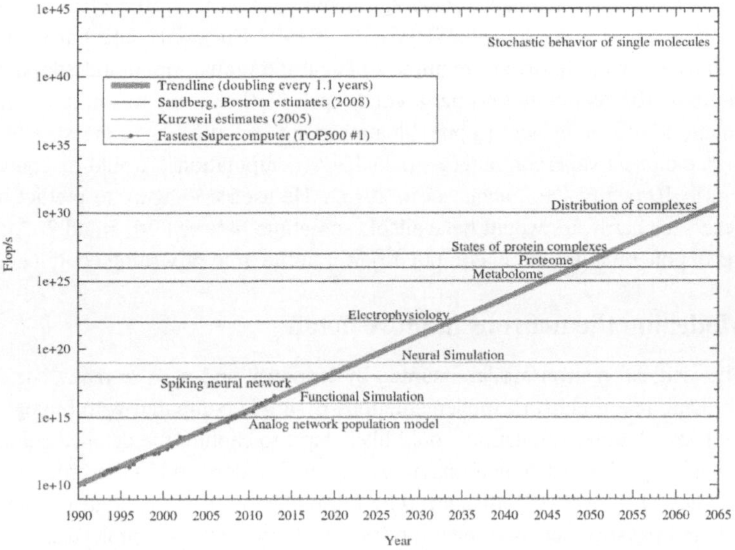

Figure 15: *Estimates of how much processing power is needed to emulate a human brain at various levels (from Ray Kurzweil, and Anders Sandberg and Nick Bostrom), along with the fastest supercomputer from TOP500 mapped by year. Note the logarithmic scale and exponential trendline, which assumes the computational capacity doubles every 1.1 years. Kurzweil believes that mind uploading will be possible at neural simulation, while the Sandberg, Bostrom report is less certain about where consciousness arises.*[387]

artificial intelligence research[384] as an approach to strong AI. Neuroimaging technologies that could deliver the necessary detailed understanding are improving rapidly, and futurist Ray Kurzweil in the book *The Singularity Is Near* predicts that a map of sufficient quality will become available on a similar timescale to the required computing power.

Early estimates

For low-level brain simulation, an extremely powerful computer would be required. The human brain has a huge number of synapses. Each of the 10^{11} (one hundred billion) neurons has on average 7,000 synaptic connections to other neurons. It has been estimated that the brain of a three-year-old child has about 10^{15} synapses (1 quadrillion). This number declines with age, stabilizing by adulthood. Estimates vary for an adult, ranging from 10^{14} to 5×10^{14} synapses (100 to 500 trillion).[388] An estimate of the brain's processing power, based on a simple switch model for neuron activity, is around 10^{14} (100 trillion) synaptic updates per second (SUPS).[389] In 1997 Kurzweil looked at various estimates for the hardware required to equal the human brain and adopted a figure of 10^{16} computations per second (cps).[390] (For comparison, if a "computation" was equivalent to one "floating point operation" – a measure used to rate current supercomputers – then 10^{16} "computations" would be equivalent to 10 petaFLOPS, achieved in 2011). He used this figure to predict the necessary hardware would be available sometime between 2015 and 2025, if the exponential growth in computer power at the time of writing continued.

Modelling the neurons in more detail

The artificial neuron model assumed by Kurzweil and used in many current artificial neural network implementations is simple compared with biological neurons. A brain simulation would likely have to capture the detailed cellular behaviour of biological neurons, presently only understood in the broadest of outlines. The overhead introduced by full modeling of the biological, chemical, and physical details of neural behaviour (especially on a molecular scale) would require computational powers several orders of magnitude larger than Kurzweil's estimate. In addition the estimates do not account for glial cells, which are at least as numerous as neurons, and which may outnumber neurons by as much as 10:1, and are now known to play a role in cognitive processes.

Current research

There are some research projects that are investigating brain simulation using more sophisticated neural models, implemented on conventional computing architectures. The Artificial Intelligence System project implemented non-real time simulations of a "brain" (with 10^{11} neurons) in 2005. It took 50 days on a cluster of 27 processors to simulate 1 second of a model. The Blue Brain project used one of the fastest supercomputer architectures in the world, IBM's Blue Gene platform, to create a real time simulation of a single rat neocortical column consisting of approximately 10,000 neurons and 10^8 synapses in 2006. A longer term goal is to build a detailed, functional simulation of the physiological processes in the human brain: "It is not impossible to build a human brain and we can do it in 10 years," Henry Markram, director of the Blue Brain Project said in 2009 at the TED conference in Oxford.[391] There have also been controversial claims to have simulated a cat brain. Neuro-silicon interfaces have been proposed as an alternative implementation strategy that may scale better.[392]

Hans Moravec addressed the above arguments ("brains are more complicated", "neurons have to be modeled in more detail") in his 1997 paper "When will computer hardware match the human brain?". He measured the ability of existing software to simulate the functionality of neural tissue, specifically the retina. His results do not depend on the number of glial cells, nor on what kinds of processing neurons perform where.

Complications and criticisms of AI approaches based on simulation

A fundamental criticism of the simulated brain approach derives from embodied cognition where human embodiment is taken as an essential aspect of human intelligence. Many researchers believe that embodiment is necessary to ground meaning.[393] If this view is correct, any fully functional brain model will need to encompass more than just the neurons (i.e., a robotic body). Goertzel[384] proposes virtual embodiment (like Second Life), but it is not yet known whether this would be sufficient.

Desktop computers using microprocessors capable of more than 10^9 cps (Kurzweil's non-standard unit "computations per second", see above) have been available since 2005. According to the brain power estimates used by Kurzweil (and Moravec), this computer should be capable of supporting a simulation of a bee brain, but despite some interest[394] no such simulation exists Wikipedia:Citation needed. There are at least three reasons for this:

- Firstly, the neuron model seems to be oversimplified (see next section).
- Secondly, there is insufficient understanding of higher cognitive processes[395] to establish accurately what the brain's neural activity, observed using techniques such as functional magnetic resonance imaging, correlates with.
- Thirdly, even if our understanding of cognition advances sufficiently, early simulation programs are likely to be very inefficient and will, therefore, need considerably more hardware.
- Fourthly, the brain of an organism, while critical, may not be an appropriate boundary for a cognitive model. To simulate a bee brain, it may be necessary to simulate the body, and the environment. The Extended Mind thesis formalizes the philosophical concept, and research into cephalopods has demonstrated clear examples of a decentralized system.

In addition, the scale of the human brain is not currently well-constrained. One estimate puts the human brain at about 100 billion neurons and 100 trillion synapses.[396] Another estimate is 86 billion neurons of which 16.3 billion are in the cerebral cortex and 69 billion in the cerebellum.[397] Glial cell synapses are currently unquantified but are known to be extremely numerous.

Artificial consciousness research

Although the role of consciousness in strong AI/AGI is debatable, many AGI researchers[398] regard research that investigates possibilities for implementing consciousness as vital. In an early effort Igor Aleksander[399] argued that the principles for creating a conscious machine already existed but that it would take forty years to train such a machine to understand language.

Relationship to "strong AI"

In 1980, philosopher John Searle coined the term "strong AI" as part of his Chinese room argument. He wanted to distinguish between two different hypotheses about artificial intelligence:[400]

- An artificial intelligence system can *think* and have a *mind*. (The word "mind" has a specific meaning for philosophers, as used in "the mind body problem" or "the philosophy of mind".)
- An artificial intelligence system can (only) *act like* it thinks and has a mind.

The first one is called "the *strong* AI hypothesis" and the second is "the *weak* AI hypothesis" because the first one makes the *stronger* statement: it assumes something special has happened to the machine that goes beyond all its abilities that we can test. Searle referred to the "strong AI hypothesis" as "strong AI". This usage is also common in academic AI research and textbooks.[401]

The weak AI hypothesis is equivalent to the hypothesis that artificial general intelligence is possible. According to Russell and Norvig, "Most AI researchers take the weak AI hypothesis for granted, and don't care about the strong AI hypothesis."[402]

In contrast to Searle, Kurzweil uses the term "strong AI" to describe any artificial intelligence system that acts like it has a mind, regardless of whether a philosopher would be able to determine if it *actually* has a mind or not.

Possible explanations for the slow progress of AI research

Since the launch of AI research in 1956, the growth of this field has slowed down over time and has stalled the aims of creating machines skilled with intelligent action at the human level.[403] A possible explanation for this delay is that computers lack a sufficient scope of memory or processing power.[403] In addition, the level of complexity that connects to the process of AI research may also limit the progress of AI research.[403]

While most AI researchers believe that strong AI can be achieved in the future, there are some individuals like Hubert Dreyfus and Roger Penrose that deny the possibility of achieving AI.[403] John McCarthy was one of various computer scientists who believe human-level AI will be accomplished, but a date cannot accurately be predicted.[404]

Conceptual limitations are another possible reason for the slowness in AI research.[403] AI researchers may need to modify the conceptual framework of their discipline in order to provide a stronger base and contribution to the quest of achieving strong AI. As William Clocksin wrote in 2003: "the framework starts from Weizenbaum's observation that intelligence manifests itself only relative to specific social and cultural contexts".[403]

Furthermore, AI researchers have been able to create computers that can perform jobs that are complicated for people to do, but conversely they have struggled to develop a computer that is capable of carrying out tasks that are simple for humans to do Wikipedia:AUDIENCE.[403] A problem that is described by David Gelernter is that some people assume that thinking and reasoning are equivalent.[405] However, the idea of whether thoughts and the creator of those thoughts are isolated individually has intrigued AI researchers.[405]

The problems that have been encountered in AI research over the past decades have further impeded the progress of AI. The failed predictions that have been promised by AI researchers and the lack of a complete understanding of human behaviors have helped diminish the primary idea of human-level AI.[384] Although the progress of AI research has brought both improvement and disappointment, most investigators have established optimism about potentially achieving the goal of AI in the 21st century.[384]

Other possible reasons have been proposed for the lengthy research in the progress of strong AI. The intricacy of scientific problems and the need to fully understand the human brain through psychology and neurophysiology have limited many researchers from emulating the function of the human brain into a computer hardware.[406] Many researchers tend to underestimate any doubt that is involved with future predictions of AI, but without taking those issues seriously can people then overlook solutions to problematic questions.[384]

Clocksin says that a conceptual limitation that may impede the progress of AI research is that people may be using the wrong techniques for computer programs and implementation of equipment.[403] When AI researchers first began to aim for the goal of artificial intelligence, a main interest was human reasoning.[407] Researchers hoped to establish computational models of human knowledge through reasoning and to find out how to design a computer with a specific cognitive task.[407]

The practice of abstraction, which people tend to redefine when working with a particular context in research, provides researchers with a concentration on just a few concepts.[407] The most productive use of abstraction in AI research comes from planning and problem solving.[407] Although the aim is to increase the speed of a computation, the role of abstraction has posed questions about the involvement of abstraction operators.[408]

A possible reason for the slowness in AI relates to the acknowledgement by many AI researchers that heuristics is a section that contains a significant breach between computer performance and human performance.[406] The specific functions that are programmed to a computer may be able to account for many of the requirements that allow it to match human intelligence. These explanations are not necessarily guaranteed to be the fundamental causes for the delay in achieving strong AI, but they are widely agreed by numerous researchers.

There have been many AI researchers that debate over the idea whether machines should be created with emotions. There are no emotions in typical models of AI and some researchers say programming emotions into machines allows them to have a mind of their own.[403] Emotion sums up the experiences of humans because it allows them to remember those experiences.[405] David Gelernter writes, "No computer will be creative unless it can simulate all the nuances of human emotion."[405] This concern about emotion has posed problems for AI researchers and it connects to the concept of strong AI as its research progresses into the future.

Consciousness

There are other aspects of the human mind besides intelligence that are relevant to the concept of strong AI which play a major role in science fiction and the ethics of artificial intelligence:

- consciousness: To have subjective experience and thought.[409]
- self-awareness: To be aware of oneself as a separate individual, especially to be aware of one's own thoughts.
- sentience: The ability to "feel" perceptions or emotions subjectively.
- sapience: The capacity for wisdom.

These traits have a moral dimension, because a machine with this form of strong AI may have legal rights, analogous to the rights of non-human animals. Also, Bill Joy, among others, argues a machine with these traits may be a threat to human life or dignity. It remains to be shown whether any of these traits are necessary for strong AI. The role of consciousness is not clear, and currently there is no agreed test for its presence. If a machine is built with a device that simulates the neural correlates of consciousness, would it automatically have self-awareness? It is also possible that some of these properties, such as sentience, naturally emerge from a fully intelligent machine, or that it becomes natural to *ascribe* these properties to machines once they begin to act in a way that is clearly intelligent. For example, intelligent action may be sufficient for sentience, rather than the other way around.

In science fiction, AGI is associated with traits such as consciousness, sentience, sapience, and self-awareness observed in living beings. However, according to philosopher John Searle, it is an open question whether general intelligence is sufficient for consciousness. "Strong AI" (as defined above by Ray Kurzweil) should not be confused with Searle's "'strong AI hypothesis". The strong AI hypothesis is the claim that a computer which behaves as intelligently as a person must also necessarily have a mind and consciousness. AGI refers only to the amount of intelligence that the machine displays, with or without a mind.

Controversies and dangers

Feasibility

Opinions vary both on *whether* and *when* artificial general intelligence will arrive. At one extreme, AI pioneer Herbert A. Simon wrote in 1965: "machines will be capable, within twenty years, of doing any work a man can do"; obviously this prediction failed to come true. Microsoft co-founder Paul Allen believes that such intelligence is unlikely in the 21st century because it would require "unforeseeable and fundamentally unpredictable breakthroughs" and a "scientifically deep understanding of cognition". Writing in The Guardian, roboticist Alan Winfield claimed the gulf between modern computing and human-level artificial intelligence is as wide as the gulf between current space flight and practical faster-than-light spaceflight. Optimism that AGI is feasible waxes and wanes, and may have seen a resurgence in the 2010s. Four polls conducted in 2012 and 2013 suggested that the median guess among experts for when AGI would arrive was 2040 to 2050, depending on the poll.[410]

Potential threat to human existence

The creation of artificial general intelligence may have repercussions so great and so complex that it may not be possible to forecast what will come afterwards. Thus the event in the hypothetical future of achieving strong AI is called the technological singularity, because theoretically one cannot see past it. But this has not stopped philosophers and researchers from guessing what the smart computers or robots of the future may do, including forming a utopia by being our friends or overwhelming us in an AI takeover. The latter potentiality is particularly disturbing as it poses an existential risk for mankind.

Self-replicating machines

Smart computers or robots would be able to design and produce improved versions of themselves[411]. A growing population of intelligent robots could conceivably outcompete inferior humans in job markets, in business, in science, in politics (pursuing robot rights), and technologically, sociologically (by acting as one), and militarily.

Emergent superintelligence

If research into strong AI produced sufficiently intelligent software, it would be able to reprogram and improve itself – a feature called "recursive self-improvement". It would then be even better at improving itself, and would probably continue doing so in a rapidly increasing cycle, leading to an intelligence explosion and the emergence of superintelligence. Such an intelligence would not have the limitations of human intellect, and might be able to invent or discover almost anything.

Hyper-intelligent software might not necessarily decide to support the continued existence of mankind, and might be extremely difficult to stop. This topic has also recently begun to be discussed in academic publications as a real source of risks to civilization, humans, and planet Earth.

One proposal to deal with this is to make sure that the first generally intelligent AI is a friendly AI that would then endeavor to ensure that subsequently developed AIs were also nice to us. But friendly AI is harder to create than plain AGI, and therefore it is likely, in a race between the two, that non-friendly AI would be developed first. Also, there is no guarantee that friendly AI would remain friendly, or that its progeny would also all be good.

References

<templatestyles src="Template:Refbegin/styles.css" />

- Halal, William E. "TechCast Article Series: The Automation of Thought"[412] (PDF). Archived from the original[413] (PDF) on 6 June 2013.
- Aleksander, Igor (1996), *Impossible Minds*, World Scientific Publishing Company, ISBN 978-1-86094-036-1
- Omohundro, Steve (2008), *The Nature of Self-Improving Artificial Intelligence*, presented and distributed at the 2007 Singularity Summit, San Francisco, CA.
- Sandberg, Anders; Boström, Nick (2008), *Whole Brain Emulation: A Roadmap*[414] (PDF), Technical Report #2008-3, Future of Humanity Institute, Oxford University, retrieved 5 April 2009

- Azevedo FA, Carvalho LR, Grinberg LT, et al. (April 2009), "Equal numbers of neuronal and nonneuronal cells make the human brain an isometrically scaled-up primate brain"[415], *The Journal of Comparative Neurology*, **513** (5): 532–41, doi: 10.1002/cne.21974[416], PMID 19226510[417], retrieved 2013-09-04
- Berglas, Anthony (2008), *Artificial Intelligence will Kill our Grandchildren*[418], retrieved 2008-06-13
- Chalmers, David (1996), *The Conscious Mind*, Oxford University Press.
- Clocksin, William (Aug 2003), "Artificial intelligence and the future", *Philosophical Transactions of the Royal Society A*, **361** (1809): 1721–1748, Bibcode: 2003RSPTA.361.1721C[419], doi: 10.1098/rsta.2003.1232[420], PMID 12952683[421].
- Crevier, Daniel (1993), *AI: The Tumultuous Search for Artificial Intelligence*, New York, NY: BasicBooks, ISBN 0-465-02997-3
- Darrach, Brad (20 November 1970), "Meet Shakey, the First Electronic Person", *Life Magazine*, pp. 58–68.
- Drachman, D (2005), "Do we have brain to spare?", *Neurology*, **64** (12): 2004–5, doi: 10.1212/01.WNL.0000166914.38327.BB[422], PMID 15985565[423].
- Feigenbaum, Edward A.; McCorduck, Pamela (1983), *The Fifth Generation: Artificial Intelligence and Japan's Computer Challenge to the World*, Michael Joseph, ISBN 0-7181-2401-4
- Gelernter, David (2010), *Dream-logic, the Internet and Artificial Thought*[424], retrieved 25 July 2010
- Goertzel, Ben; Pennachin, Cassio, eds. (2006), *Artificial General Intelligence*[425] (PDF), Springer, ISBN 3-540-23733-X, archived from the original[426] (PDF) on 20 March 2013
- Goertzel, Ben; Wang, Pei (2006), *Introduction: Aspects of Artificial General Intelligence*[427] (PDF)
- Goertzel, Ben (Dec 2007), "Human-level artificial general intelligence and the possibility of a technological singularity: a reaction to Ray Kurzweil's The Singularity Is Near, and McDermott's critique of Kurzweil"[428], *Artificial Intelligence*, **171** (18, Special Review Issue): 1161–1173, doi: 10.1016/j.artint.2007.10.011[429], retrieved 1 April 2009.
- Gubrud, Mark (November 1997), "Nanotechnology and International Security"[430], *Fifth Foresight Conference on Molecular Nanotechnology*, retrieved 7 May 2011
- Holte, RC; Choueiry, BY (2003), "Abstraction and reformulation in artificial intelligence", *Philosophical Transactions of the Royal Society B*, **358** (1435): 1197–1204, doi: 10.1098/rstb.2003.1317[431], PMC 1693218[432] 🔓, PMID 12903653[433].

- Howe, J. (November 1994), *Artificial Intelligence at Edinburgh University : a Perspective*[434], retrieved 30 August 2007
- Johnson, Mark (1987), *The body in the mind*, Chicago, ISBN 0-226-40317-3
- Kurzweil, Ray (2005), *The Singularity is Near*, Viking Press
- Lighthill, Professor Sir James (1973), "Artificial Intelligence: A General Survey", *Artificial Intelligence: a paper symposium*, Science Research Council
- Luger, George; Stubblefield, William (2004), *Artificial Intelligence: Structures and Strategies for Complex Problem Solving*[435] (5th ed.), The Benjamin/Cummings Publishing Company, Inc., p. 720, ISBN 0-8053-4780-1
- McCarthy, John (Oct 2007), "From here to human-level AI", *Artificial Intelligence*, **171** (18): 1174–1182, doi: 10.1016/j.artint.2007.10.009[436].
- McCorduck, Pamela (2004), *Machines Who Think*[437] (2nd ed.), Natick, MA: A. K. Peters, Ltd., ISBN 1-56881-205-1
- Moravec, Hans (1976), *The Role of Raw Power in Intelligence*[438]
- Moravec, Hans (1988), *Mind Children*, Harvard University Press
- Nagel (1974), "What Is it Like to Be a Bat"[439] (PDF), *Philosophical Review*: 435–50..
- Newell, Allen; Simon, H. A. (1963), "GPS: A Program that Simulates Human Thought", in Feigenbaum, E.A.; Feldman, J., *Computers and Thought*, New York: McGraw-Hill
- Newell, Allen; Simon, H. A. (1976). "Computer Science as Empirical Inquiry: Symbols and Search"[440]. *Communications of the ACM*. **19** (3): 113–126. doi: 10.1145/360018.360022[441].
- Nilsson, Nils (1998), *Artificial Intelligence: A New Synthesis*, Morgan Kaufmann Publishers, ISBN 978-1-55860-467-4
- Russell, Stuart J.; Norvig, Peter (2003), *Artificial Intelligence: A Modern Approach*[442] (2nd ed.), Upper Saddle River, New Jersey: Prentice Hall, ISBN 0-13-790395-2
- NRC (1999), "Developments in Artificial Intelligence"[443], *Funding a Revolution: Government Support for Computing Research*, National Academy Press
- Poole, David; Mackworth, Alan; Goebel, Randy (1998), *Computational Intelligence: A Logical Approach*[444], New York: Oxford University Press
- Searle, John (1980), "Minds, Brains and Programs"[445], *Behavioral and Brain Sciences*, **3** (3): 417–457, doi: 10.1017/S0140525X00005756[446], archived from the original[447] on 18 January 2010
- Simon, H. A. (1965), *The Shape of Automation for Men and Management*, New York: Harper & Row

- Sutherland, J.G. (1990), "Holographic Model of Memory, Learning, and Expression", *International Journal of Neural Systems*, **1–3**: 256–267.
- Williams RW, Herrup K (1988), "The control of neuron number", *Annual Review of Neuroscience*, **11**: 423–53, doi: 10.1146/annurev.ne.11.030188.002231[448], PMID 3284447[449].
- de Vega, Manuel; Glenberg, Arthur; Graesser, Arthur, eds. (2008), *Symbols and Embodiment: Debates on meaning and cognition*, Oxford University Press, ISBN 0-19-921727-0
- Yudkowsky, Eliezer (2006), Goertzel, Ben; Pennachin, Cassio, eds., "Artificial General Intelligence"[450] (PDF), *Annual Review of Psychology*, Springer, **49**: 585–612, doi: 10.1146/annurev.psych.49.1.585[451], ISBN 3-540-23733-X, PMID 9496632[452], archived from the original[453] (PDF) on 11 April 2009
- Zucker, Jean-Daniel (July 2003), "A grounded theory of abstraction in artificial intelligence", *Philosophical Transactions of the Royal Society B*, **358** (1435): 1293–1309, doi: 10.1098/rstb.2003.1308[454], PMC 1693211[455] 🔓, PMID 12903672[456].
- Yudkowsky, Eliezer (2008), "Artificial Intelligence as a Positive and Negative Factor in Global Risk", *Global Catastrophic Risks*, Oxford University Press, Bibcode: 2008gcr..book..303Y[457].

External links

- The AGI portal maintained by Pei Wang[458]
- The Genesis Group at MIT's CSAIL[459] — Modern research on the computations that underlay human intelligence
- OpenCog – open source project to develop a human-level AI[460]
- Simulating logical human thought[461]
- The Game of Intelligent Design[462]— Online game that promotes the development of artificial general intelligence
- What Do We Know about AI Timelines?[463] — Literature review

AI-complete

In the field of artificial intelligence, the most difficult problems are informally known as **AI-complete** or **AI-hard**, implying that the difficulty of these computational problems is equivalent to that of solving the central artificial intelligence problem—making computers as intelligent as people, or strong AI.[464] To call a problem AI-complete reflects an attitude that it would not be solved by a simple specific algorithm.

AI-complete problems are hypothesised to include computer vision, natural language understanding, and dealing with unexpected circumstances while solving any real world problem.[465]

Currently, AI-complete problems cannot be solved with modern computer technology alone, but would also require human computation. This property can be useful, for instance to test for the presence of humans as with CAPTCHAs, and for computer security to circumvent brute-force attacks.[466,467]

History

The term was coined by Fanya Montalvo by analogy with NP-complete and NP-hard in complexity theory, which formally describes the most famous class of difficult problems. Early uses of the term are in Erik Mueller's 1987 Ph.D. dissertation[468] and in Eric Raymond's 1991 Jargon File.[469]

AI-complete problems

AI-complete problems are hypothesised to include:

- Bongard problems
- Computer vision (and subproblems such as object recognition)
- Natural language understanding (and subproblems such as text mining, machine translation, and word sense disambiguation[470])
- Dealing with unexpected circumstances while solving any real world problem, whether it's navigation or planning or even the kind of reasoning done by expert systems.

Machine translation

To translate accurately, a machine must be able to understand the text. It must be able to follow the author's argument, so it must have some ability to reason. It must have extensive world knowledge so that it knows what is being discussed — it must at least be familiar with all the same commonsense facts that the average human translator knows. Some of this knowledge is in the form of facts that can be explicitly represented, but some knowledge is unconscious and closely tied to the human body: for example, the machine may need to understand how an ocean makes one *feel* to accurately translate a specific metaphor in the text. It must also model the authors' goals, intentions, and emotional states to accurately reproduce them in a new language. In short, the machine is required to have wide variety of human intellectual skills, including reason, commonsense knowledge and the intuitions that underlie motion and manipulation, perception, and social intelligence. Machine translation, therefore, is believed to be AI-complete: it may require strong AI to be done as well as humans can do it.

Software brittleness

Current AI systems can solve very simple and/or restricted versions of AI-complete problems, but never in their full generality. When AI researchers attempt to "scale up" their systems to handle more complicated, real world situations, the programs tend to become excessively brittle without commonsense knowledge or a rudimentary understanding of the situation: they fail as unexpected circumstances outside of its original problem context begin to appear. When human beings are dealing with new situations in the world, they are helped immensely by the fact that they know what to expect: they know what all things around them are, why they are there, what they are likely to do and so on. They can recognize unusual situations and adjust accordingly. A machine without strong AI has no other skills to fall back on.

Formalization

Computational complexity theory deals with the relative computational difficulty of computable functions. By definition it does not cover problems whose solution is unknown or has not been characterised formally. Since many AI problems have no formalisation yet, conventional complexity theory does not allow the definition of AI-completeness.

To address this problem, a complexity theory for AI has been proposed.[471] It is based on a model of computation that splits the computational burden between a computer and a human: one part is solved by computer and the other part solved by human. This is formalised by a **human-assisted Turing machine**. The formalisation defines algorithm complexity, problem complexity and reducibility which in turn allows equivalence classes to be defined.

The complexity of executing an algorithm with a human-assisted Turing machine is given by a pair $\langle \Phi_H, \Phi_M \rangle$, where the first element represents the complexity of the human's part and the second element is the complexity of the machine's part.

Results

The complexity of solving the following problems with a human-assisted Turing machine is:

- Optical character recognition for printed text: $\langle O(1), poly(n) \rangle$
- Turing test:
  - for an n -sentence conversation where the oracle remembers the conversation history (persistent oracle): $\langle O(n), O(n) \rangle$
  - for an n -sentence conversation where the conversation history must be retransmitted: $\langle O(n), O(n^2) \rangle$
  - for an n -sentence conversation where the conversation history must be retransmitted and the person takes linear time to read the query: $\langle O(n^2), O(n^2) \rangle$
- ESP game: $\langle O(n), O(n) \rangle$
- Image labelling (based on the Arthur–Merlin protocol): $\langle O(n), O(n) \rangle$
- Image classification: human only: $\langle O(n), O(n) \rangle$, and with less reliance on the human: $\langle O(\log n), O(n \log n) \rangle$.

Approaches

Cybernetics

Cybernetics is a transdisciplinary approach for exploring regulatory systems—their structures, constraints, and possibilities. Norbert Wiener defined cybernetics in 1948 as "the scientific study of control and communication in the animal and the machine." In the 21st century, the term is often used in a rather loose way to imply "control of any system using technology." In other words, it is the scientific study of how humans, animals and machines control and communicate with each other.

Cybernetics is applicable when a system being analyzed incorporates a closed signaling loop—originally referred to as a "circular causal" relationship—that is, where action by the system generates some change in its environment and that change is reflected in the system in some manner (feedback) that triggers a system change. Cybernetics is relevant to, for example, mechanical, physical, biological, cognitive, and social systems. The essential goal of the broad field of cybernetics is to understand and define the functions and processes of systems that have goals and that participate in circular, causal chains that move from action to sensing to comparison with desired goal, and again to action. Its focus is how anything (digital, mechanical or biological) processes information, reacts to information, and changes or can be changed to better accomplish the first two tasks. Cybernetics includes the study of feedback, black boxes and derived concepts such as communication and control in living organisms, machines and organizations including self-organization.

Concepts studied by cyberneticists include, but are not limited to: learning, cognition, adaptation, social control, emergence, convergence, communication, efficiency, efficacy, and connectivity. In cybernetics these concepts (otherwise already objects of study in other disciplines such as biology and engineering) are abstracted from the context of the specific organism or device.

The word *cybernetics* comes from Greek κυβερνητική (*cybernēticḗ*), meaning "governance", i.e., all that are pertinent to κυβερνάω (*cybernáō*), the latter meaning "to steer, navigate or govern", hence κυβέρνησις (*cybérnēsis*), meaning "government", is the government while κυβερνήτης (*cybernḗtēs*) is the governor or "helmperson" of the "ship". Contemporary cybernetics began as an interdisciplinary study connecting the fields of control systems, electrical network theory, mechanical engineering, logic modeling, evolutionary biology, neuroscience, anthropology, and psychology in the 1940s, often attributed to the Macy Conferences. During the second half of the 20th century cybernetics evolved in ways that distinguish first-order cybernetics (about observed systems) from second-order cybernetics (about observing systems).[472] More recently there is talk about a third-order cybernetics (doing in ways that embraces first and second-order).

Studies in cybernetics provide a means for examining the design and function of any system, including social systems such as business management and organizational learning, including for the purpose of making them more efficient and effective. Fields of study which have influenced or been influenced by cybernetics include game theory, system theory (a mathematical counterpart to cybernetics), perceptual control theory, sociology, psychology (especially neuropsychology, behavioral psychology, cognitive psychology), philosophy, architecture, and organizational theory.[473] System dynamics, originated with applications of electrical engineering control theory to other kinds of simulation models (especially business systems) by Jay Forrester at MIT in the 1950s, is a related field.

Definitions

Cybernetics has been defined in a variety of ways, by a variety of people, from a variety of disciplines. Cybernetician Stuart Umpleby reports some notable definitions:

- "Science concerned with the study of systems of any nature which are capable of receiving, storing and processing information so as to use it for control."—A. N. Kolmogorov
- "'The art of steersmanship': deals with all forms of behavior in so far as they are regular, or determinate, or reproducible: stands to the real machine – electronic, mechanical, neural, or economic – much as geometry stands to real object in our terrestrial space; offers a method for the scientific treatment of the system in which complexity is outstanding and too important to be ignored."—W. Ross Ashby
- "A branch of mathematics dealing with problems of control, recursiveness, and information, focuses on forms and the patterns that connect."—Gregory Bateson

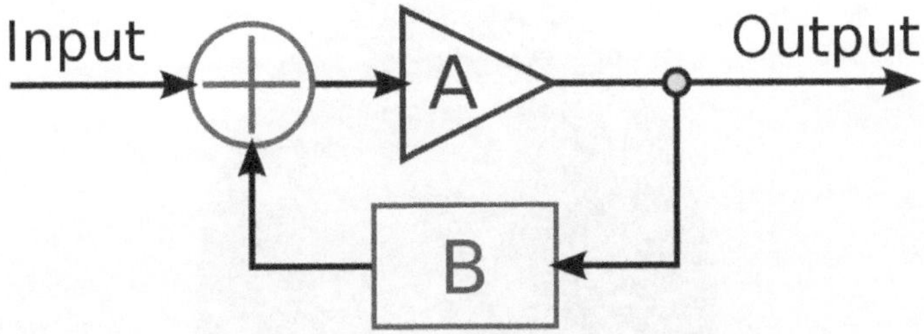

Figure 16: *Simple feedback model. AB < 0 for negative feedback.*

- "The art of securing efficient operation [lit.: the art of effective action]."—Louis Couffignal[474,475]
- "The art of effective organization."—Stafford Beer
- "The art and science of manipulating defensible metaphors" (with relevance to constructivist epistemology. The author later extended the definition to include information flows "in all media", from stars to brains.)—Gordon Pask
- "The art of creating equilibrium in a world of constraints and possibilities."—Ernst von Glasersfeld
- "The science and art of understanding." – Humberto Maturana
- "The ability to cure all temporary truth of eternal triteness."—Herbert Brun

Other notable definitions include:

- "The science and art of the understanding of understanding."—Rodney E. Donaldson, the first president of the American Society for Cybernetics
- "The control of an automaton's feedback loop."—Link Starbureiy
- "A way of thinking about ways of thinking of which it is one."—Larry Richards
- "The art of interaction in dynamic networks."—Roy Ascott
- "The study of systems and processes that interact with themselves and produce themselves from themselves."—Louis Kauffman, President of the American Society for Cybernetics[476]

Etymology

The term *cybernetics* stems from κυβερνήτης (*cybernḗtēs*) "steersman, governor, pilot, or rudder". As with the ancient Greek pilot, independence of thought is important in cybernetics.[477] French physicist and mathematician

Figure 17: *James Watt*

André-Marie Ampère first coined the word "cybernetique" in his 1834 essay *Essai sur la philosophie des sciences* to describe the science of civil government.[478] The term was borrowed by Norbert Wiener, in his book *Cybernetics*, to define the study of control and communication in the animal and the machine.

History

Roots of cybernetic theory

The word *cybernetics* was first used in the context of "the study of self-governance" by Plato in *The Alcibiades* to signify the governance of people. The word 'cybernétique' was also used in 1834 by the physicist André-Marie Ampère (1775–1836) to denote the sciences of government in his classification system of human knowledge.

The first artificial automatic regulatory system was a water clock, invented by the mechanician Ktesibios; based on a tank which poured water into a reservoir before using it to run the mechanism, it used a cone-shaped float to monitor the level of the water in its reservoir and adjust the rate of flow of the water accordingly to maintain a constant level of water in the reservoir. This was the first artificial truly automatic self-regulatory device that required no outside intervention between the feedback and the controls of the mechanism. Although they considered this part of engineering (the use of the term *cybernetics* is much posterior), Ktesibios and others such as Heron and Su Song are considered to be some of the first to study cybernetic principles.

The study of *teleological mechanisms* (from the Greek τέλος or *télos* for *end*, *goal*, or *purpose*) in machines with *corrective feedback* dates from as far back as the late 18th century when James Watt's steam engine was equipped with a governor (1775–1800), a centrifugal feedback valve for controlling the speed of the engine. Alfred Russel Wallace identified this as the principle of evolution in his famous 1858 paper.[479] In 1868 James Clerk Maxwell published a theoretical article on governors, one of the first to discuss and refine the principles of self-regulating devices. Jakob von Uexküll applied the feedback mechanism via his model of functional cycle (*Funktionskreis*) in order to explain animal behaviour and the origins of meaning in general.

Early 20th century

Contemporary cybernetics began as an interdisciplinary study connecting the fields of control systems, electrical network theory, mechanical engineering, logic modeling, evolutionary biology and neuroscience in the 1940s; the ideas are also related to the biological work of Ludwig von Bertalanffy in General Systems Theory. Electronic control systems originated with the 1927 work of Bell Telephone Laboratories engineer Harold S. Black on using negative feedback to control amplifiers.

Early applications of negative feedback in electronic circuits included the control of gun mounts and radar antenna during World War II. The founder of System Dynamics, Jay Forrester, worked with Gordon S. Brown during WWII as a graduate student at the Servomechanisms Laboratory at MIT to develop electronic control systems for the U.S. Navy. Forrester later applied these ideas to social organizations, such as corporations and cities and became an original organizer of the MIT School of Industrial Management at the MIT Sloan School of Management.

W. Edwards Deming, the Total Quality Management guru for whom Japan named its top post-WWII industrial prize, was an intern at Bell Telephone Labs in 1927 and may have been influenced by network theory; Deming made "Understanding Systems" one of the four pillars of what he described as "Profound Knowledge" in his book *The New Economics*.

Numerous papers spearheaded the coalescing of the field. In 1935 Russian physiologist P. K. Anokhin published a book in which the concept of feedback ("back afferentation") was studied. The study and mathematical modelling of regulatory processes became a continuing research effort and two key articles were published in 1943: "Behavior, Purpose and Teleology" by Arturo Rosenblueth, Norbert Wiener, and Julian Bigelow; and the paper "A Logical Calculus of the Ideas Immanent in Nervous Activity" by Warren McCulloch and Walter Pitts.

In 1936, Ştefan Odobleja published "Phonoscopy and the clinical semiotics". In 1937, he participated in the IX International Congress of Military Medicine with "Demonstration de phonoscopie"; in the paper he disseminated a prospectus announcing his future work, "Psychologie consonantiste", the most important of his writings, where he lays the theoretical foundations of generalized cybernetics. The book, published in Paris by *Librairie Maloine* (vol. I in 1938 and vol. II in 1939), contains almost 900 pages and includes 300 figures in the text. The author wrote at the time that "this book is ... a table of contents, an index or a dictionary of psychology, [for] a ... great Treatise of Psychology that should contain 20–30 volumes". Due to the beginning of World War II, the publication went unnoticed (the first Romanian edition of this work did not appear until 1982).

Cybernetics as a discipline was firmly established by Norbert Wiener, McCulloch, Arturo Rosenblueth and others, such as W. Ross Ashby, mathematician Alan Turing, and W. Grey Walter (one of the first to build autonomous robots as an aid to the study of animal behaviour). In the spring of 1947, Wiener was invited to a congress on harmonic analysis, held in Nancy (France was an important geographical locus of early cybernetics together with the US and UK); the event was organized by the Bourbaki, a French scientific society, and mathematician Szolem Mandelbrojt (1899–1983), uncle of the world-famous mathematician Benoît Mandelbrot. During this stay in France, Wiener received the offer to write a manuscript on the unifying character of this part of applied mathematics, which is found in the study of Brownian motion and in telecommunication engineering. The following summer, back in the United States, Wiener decided to introduce the neologism *cybernetics*, coined to denote the study of "teleological mechanisms", into his scientific theory: it was popularized through his book *Cybernetics: Or Control and Communication in the Animal and the Machine* (MIT Press/John Wiley and Sons, NY, 1948). In the UK this became the focus for the Ratio Club.

In the early 1940s John von Neumann, although better known for his work in mathematics and computer science, did contribute a unique and unusual addition to the world of cybernetics: von Neumann cellular automata, and their logical follow up, the von Neumann Universal Constructor. The result of these deceptively simple thought-experiments was the concept of self replication, which cybernetics adopted as a core concept. The concept that the same properties of genetic reproduction applied to social memes, living cells, and

Figure 18: *John von Neumann*

even computer viruses is further proof of the somewhat surprising universality of cybernetic study.

In 1950, Wiener popularized the social implications of cybernetics, drawing analogies between automatic systems (such as a regulated steam engine) and human institutions in his best-selling *The Human Use of Human Beings: Cybernetics and Society* (Houghton-Mifflin).

In the Soviet Union "bourgeois" cybernetics was initially considered a "pseudoscience" and "ideological weapon" of "imperialist reactionaries" (Soviet Philosophical Dictionary, 1954) and later criticised as a narrow form of cybernetics.[480] In the mid to late 1950s Viktor Glushkov and others salvaged the reputation of the field. Soviet cybernetics incorporated much of what became known as computer science in the West.

While not the only instance of a research organization focused on cybernetics, the Biological Computer Lab at the University of Illinois at Urbana-Champaign, under the direction of Heinz von Foerster, was a major center of cybernetic research for almost 20 years, beginning in 1958.

Split from artificial intelligence

Artificial intelligence (AI) was founded as a distinct discipline at the Dartmouth workshop. After some uneasy coexistence, AI gained funding and prominence. Consequently, cybernetic sciences such as the study of artificial neural networks were downplayed; the discipline shifted into the world of social sciences and therapy.

Prominent cyberneticians during this period include Gregory Bateson and Aksel Berg.

New cybernetics

In the 1970s, new cyberneticians emerged in multiple fields, but especially in biology. The ideas of Maturana, Varela and Atlan, according to Jean-Pierre Dupuy (1986) "realized that the cybernetic metaphors of the program upon which molecular biology had been based rendered a conception of the autonomy of the living being impossible. Consequently, these thinkers were led to invent a new cybernetics, one more suited to the organizations which mankind discovers in nature - organizations he has not himself invented".[481] However, during the 1980s the question of whether the features of this new cybernetics could be applied to social forms of organization remained open to debate.

In political science, Project Cybersyn attempted to introduce a cybernetically controlled economy during the early 1970s. In the 1980s, according to Harries-Jones (1988) "unlike its predecessor, the new cybernetics concerns itself with the interaction of autonomous political actors and subgroups, and the practical and reflexive consciousness of the subjects who produce and reproduce the structure of a political community. A dominant consideration is that of recursiveness, or self-reference of political action both with regards to the expression of political consciousness and with the ways in which systems build upon themselves".[482]

One characteristic of the emerging new cybernetics considered in that time by Felix Geyer and Hans van der Zouwen, according to Bailey (1994),[483] was "that it views information as constructed and reconstructed by an individual interacting with the environment. This provides an epistemological foundation of science, by viewing it as observer-dependent. Another characteristic of the new cybernetics is its contribution towards bridging the *micro-macro gap*. That is, it links the individual with the society". Another characteristic noted was the "transition from classical cybernetics to the new cybernetics [that] involves a transition from classical problems to new problems. These shifts in thinking involve, among others, (a) a change from emphasis on the system being steered to the system doing the steering, and the factor which guides the steering decisions; and (b) new emphasis on communication between several systems which are trying to steer each other".

Recent endeavors into the true focus of cybernetics, systems of control and emergent behavior, by such related fields as game theory (the analysis of group interaction), systems of feedback in evolution, and metamaterials (the study of materials with properties beyond the Newtonian properties of their constituent atoms), have led to a revived interest in this increasingly relevant field.

Cybernetics and economic systems

Part of a series on
Economic systems
• **Business and economics portal**
• v • t • e[484]

The design of self-regulating control systems for a real-time planned economy was explored by Viktor Glushkov in the former Soviet Union during the 1960s. By the time information technology was developed enough to enable feasible economic planning based on computers, the Soviet Union and eastern bloc countries began moving away from planning and eventually collapsed.

More recent proposals for socialism involve "New Socialism", outlined by the computer scientists Paul Cockshott and Allin Cottrell, where computers determine and manage the flows and allocation of resources among socially-owned enterprises.[485]

Subdivisions of the field

Cybernetics is sometimes used as a generic term, which serves as an umbrella for many systems-related scientific fields.

Figure 19: *ASIMO uses sensors and sophisticated algorithms to avoid obstacles and navigate stairs.*

Basic cybernetics

Cybernetics studies systems of control as a concept, attempting to discover the basic principles underlying such things as

- Artificial intelligence
- Computer vision
- Control systems
- Conversation theory
- Emergence
- Interactions of actors theory
- Learning organization
- Robotics
- Second-order cybernetics
- Self-organization in cybernetics

In biology

Cybernetics in biology is the study of cybernetic systems present in biological organisms, primarily focusing on how animals adapt to their environment, and how information in the form of genes is passed from generation to generation.[486] There is also a secondary focus on combining artificial systems with biological systems.Wikipedia:Citation needed A notable application to the biology world would be that, in 1955, the physicist George Gamow published a prescient article in *Scientific American* called "Information transfer in the living cell", and cybernetics gave biologists Jacques Monod and François Jacob a language for formulating their early theory of gene regulatory networks in the 1960s.

- Autopoiesis
- Biocybernetics
- Bioengineering
- Bionics
- Ecology
- Heterostasis
- Homeostasis
- Medical cybernetics
- Neuroscience
- Synthetic biology
- Systems biology
- Practopoiesis

In computer science

Computer science directly applies the concepts of cybernetics to the control of devices and the analysis of information.

- Cellular automaton
- Decision support systems
- Design patterns
- Robotics
- Simulation

In engineering

Cybernetics in engineering is used to analyze cascading failures and system accidents, in which the small errors and imperfections in a system can generate disasters. Other topics studied include:

- Adaptive systems
- Biomedical engineering
- Engineering cybernetics
- Ergonomics
- Systems engineering

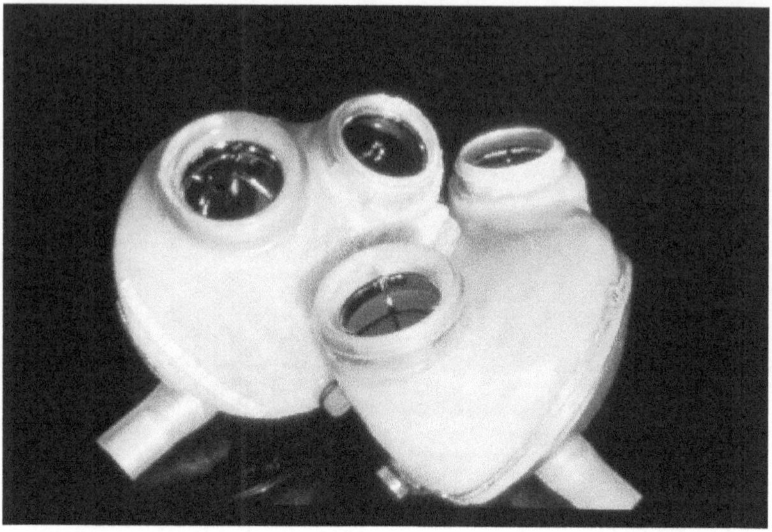

Figure 20: *An artificial heart, a product of biomedical engineering.*

In management

- Autonomous agency theory
- Entrepreneurial cybernetics
- Management cybernetics
- Operations research
- Organizational cybernetics
- Systems engineering
- Viable system theory

In mathematics

Mathematical Cybernetics focuses on the factors of information, interaction of parts in systems, and the structure of systems.

- Control theory
- Dynamical system
- Information theory
- Systems theory

In psychology

- Attachment theory
- Behavioral cybernetics
- Cognitive psychology
- Consciousness
- Embodied cognition
- Human-robot interaction
- Mind-body problem
- Perceptual control theory
- *Psycho-Cybernetics*
- Psychovector analysis
- Systems psychology

In sociology

By examining group behavior through the lens of cybernetics, sociologists can seek the reasons for such spontaneous events as smart mobs and riots, as well as how communities develop rules such as etiquette by consensus without formal discussionWikipedia:Citation needed. Affect Control Theory explains role behavior, emotions, and labeling theory in terms of homeostatic maintenance of sentiments associated with cultural categories. The most comprehensive attempt ever made in the social sciences to increase cybernetics in a generalized theory of society was made by Talcott Parsons. In this way, cybernetics establishes the basic hierarchy in Parsons' AGIL paradigm, which is the ordering system-dimension of his action theory. These and other cybernetic models in sociology are reviewed in a book edited by McClelland and Fararo.[487]

- Affect control theory
- Memetics Wikipedia:Citation needed
- Sociocybernetics

In education

A model of cybernetics in Education was introduced by Gihan Sami Soliman; an educational consultant, as a project idea to be implemented with the help of two team members in Sinai. The Sinai Sustainability Cybernetics Center announced as a semi-finalist project by MIT annual competition 2013.[488] The project idea proposed relating education to sustainable development through an IMS project that applies a multiple educational program related to the original natural self-healing system of life on earth. Education, sustainable development, social justice disciplines interact in a causal circular relationship that education would contribute to the development of the local community in

Sinai village, on both sustainability and social responsibility levels while the community itself provides a unique learning environment that will contribute to the development of the educational program in a closed signaling loop.

In art

Nicolas Schöffer's *CYSP I* (1956) was perhaps the first artwork to explicitly employ cybernetic principles (CYSP is an acronym that joins the first two letters of the words "CYbernetic" and "SPatiodynamic"). The prominent and influential Cybernetic Serendipity exhibition was held at the ICA in 1968 curated by Jasia Reichardt, including Schöffer's *CYSP I* and Gordon Pask's *Colloquy of Mobiles* installation. Pask's reflections on *Colloquy* connected it to his earlier *Musicolour* installation and to he what he termed "aesthetically potent environments", a concept that connected this artistic work to his concerns with teaching and learning.[489] The artist Roy Ascott elaborated an extensive theory of cybernetic art in "Behaviourist Art and the Cybernetic Vision" (*Cybernetica*, Journal of the International Association for Cybernetics (Namur), Volume IX, No.4, 1966; Volume X No.1, 1967) and in "The Cybernetic Stance: My Process and Purpose" (*Leonardo* Vol 1, No 2, 1968). Art historian Edward A. Shanken has written about the history of art and cybernetics in essays including "Cybernetics and Art: Cultural Convergence in the 1960s"[490] and *From Cybernetics to Telematics: The Art, Pedagogy, and Theory of Roy Ascott* (2003), which traces the trajectory of Ascott's work from cybernetic art to telematic art (art using computer networking as its medium, a precursor to net.art.)

- Telematic art
- Interactive art
- Systems art

In architecture and design

Cybernetics was an influence on thinking in architecture and design in the decades after the Second World War. Ashby and Pask were drawn on by design theorists such as Horst Rittel, Christopher Alexander[491] and Bruce Archer.[492] Pask was a consultant to Nicholas Negroponte's Architecture Machine Group, forerunner of the MIT Media Lab, and collaborated with architect Cedric Price and theatre director Joan Littlewood on the influential Fun Palace project during the 1960s.[493] Pask's 1950s Musicolour installation was the inspiration for John and Julia Frazer's work on Price's Generator project.[494] There has been a resurgence of interest in cybernetics and systems thinking amongst designers in recent decades, in relation to developments in technology and increasingly complex design challenges.[495] Figures such as Klaus Krippendorff, Paul Pangaro and Ranulph Glanville have made significant contributions to both cybernetics and design research. The connections between the two fields have come to be understood less in terms of application and more as reflections of each other.[496]

In Earth system science

Geocybernetics aims to study and control the complex co-evolution of ecosphere and anthroposphere,[497] for example, for dealing with planetary problems such as anthropogenic global warming.[498] Geocybernetics applies a dynamical systems perspective to Earth system analysis. It provides a theoretical framework for studying the implications of following different sustainability paradigms on co-evolutionary trajectories of the planetary socio-ecological system to reveal attractors in this system, their stability, resilience and reachability. Concepts such as tipping points in the climate system, planetary boundaries, the safe operating space for humanity and proposals for manipulating Earth system dynamics on a global scale such as geoengineering have been framed in the language of geocybernetic Earth system analysis.

In sport

A model of cybernetics in Sport was introduced by Yuri Verkhoshansky and Mel C. Siff in 1999 in their book *Supertraining*.

In law

As a form of regulation, cybernetics has been always close to law, specially in regulation and legal sciences, through the next topics:

- Organizations and superorganisms
- Ontology, logic and artificial intelligence
- Complex adaptive systems
- Smart contracts
- Control systems
- Self-organization in cybernetics
- Cyberethics
- Regulation
- Consensus systems
- Metagovernment

Related fields

Complexity science

Complexity science attempts to understand the nature of complex systems.

- Complex adaptive system
- Complex systems
- Complexity theory

Biomechatronics

Biomechatronics relates to linking mechatronics to biological organisms, leading to systems that conform to A. N. Kolmogorov's definition of Cybernetics: "Science concerned with the study of systems of any nature which are capable of receiving, storing and processing information so as to use it for control".Wikipedia:Citation needed From this perspective mechatronics are considered technical cybernetics or engineering cybernetics.

Further reading

- Arbib, Michael A. (1987). *Brains, machines, and mathematics* (2nd ed.). New York: Springer-Verlag. ISBN 978-0387965390.
- Arbib, Michael A. (1972). *The Metaphorical Brain*. Wiley. ISBN 0-471-03249-2.
- Ascott, Roy (1967). Behaviourist Art and the Cybernetic Vision. *Cybernetica*, Journal of the International Association for Cybernetics (Namur), 10, pp. 25–56
- Ashby, William Ross (1956). *An introduction to cybernetics*[499] (PDF). Chapman & Hall. Retrieved 3 June 2012.
- Beer, Stafford (1974). *Designing freedom*. Chichester, West Sussex, England: Wiley. ISBN 978-0471951650.
- François, Charles (1999). " Systemics and cybernetics in a historical perspective[500]". In: *Systems Research and Behavioral Science*. Vol 16, pp. 203–219 (1999)
- George, F. H. (1971). *Cybernetics*. Teach Yourself Books. ISBN 0-340-05941-9.
- Gerovitch, Slava (2002). *From newspeak to cyberspeak : a history of Soviet cybernetics*[501]. Cambridge, Massachusetts [u.a.]: MIT Press. ISBN 0262-07232-7.
- Heims, Steve Joshua (1993). *Constructing a social science for post-war America : the cybernetics group, 1946-1953* (1st ed.). Cambridge, Massachusetts u.a.: MIT Press. ISBN 9780262581233.

- Helvey, T.C. (1971). *The age of information; an interdisciplinary survey of cybernetics*. Englewood Cliffs, N.J.: Educational Technology Publications. ISBN 9780877780083.
- Heylighen, Francis, and Cliff Joslyn (2002). " Cybernetics and Second Order Cybernetics[502]", in: R.A. Meyers (ed.), *Encyclopedia of Physical Science & Technology* (3rd ed.), Vol. 4, (Academic Press, San Diego), p. 155-169.
- Hyötyniemi, Heikki (2006). *Neocybernetics in Biological Systems*[503]. Espoo: Helsinki University of Technology, Control Engineering Laboratory.
- Ilgauds, Hans Joachim (1980), *Norbert Wiener*, Leipzig.
- Johnston, John (2008). *The allure of machinic life : cybernetics, artificial life, and the new AI*. Cambridge, Massachusetts: MIT Press. ISBN 978-0-262-10126-4.
- Medina, Eden (2011). *Cybernetic revolutionaries : technology and politics in Allende's Chile*. Cambridge, Massachusetts: MIT Press. ISBN 978-0-262-01649-0.
- Pangaro, Paul. "Cybernetics — A Definition"[504].
- Pask, Gordon (1972). "Cybernetics"[505]. *Encyclopædia Britannica*.
- Patten, Bernard C.; Odum, Eugene P. (December 1981). "The Cybernetic Nature of Ecosystems". *The American Naturalist*. The University of Chicago Press. **118** (6): 886–895. doi: 10.1086/283881[506]. JSTOR 2460822?[507].
- Pekelis, V. (1974). *Cybernetics A to Z*. Moscow: Mir Publishers.
- Pickering, Andrew (2010). *The cybernetic brain : sketches of another future* ([Online-Ausg.] ed.). Chicago: University of Chicago Press. ISBN 978-0226667898.
- Umpleby, Stuart (1989). "The science of cybernetics and the cybernetics of science"[508], in: *Cybernetics and Systems", Vol. 21, No. 1, (1990), pp. 109–121.*
- von Foerster, Heinz, (1995), Ethics and Second-Order Cybernetics[509].
- Wiener, Norbert (1948). Hermann & Cie, ed. *Cybernetics; or, Control and communication in the animal and the machine*[510]. Paris: Technology Press. Retrieved 3 June 2012.
- Wiener, Norbert (1950). *Cybernetics and Society: The Human Use of Human Beings*. Houghton Mifflin.

External links

General

- Norbert Wiener and Stefan Odobleja - A Comparative Analysis[511]
- Reading List for Cybernetics[512]
- *Principia Cybernetica Web*[513]

- Web Dictionary of Cybernetics and Systems[514]
- Glossary Slideshow (136 slides)[515]
- "Basics of Cybernetics"[516]. Archived from the original[517] on 2010-08-11. Retrieved 2016-01-23.
- What is Cybernetics? Livas short introductory videos[518] on YouTube

Music

- Mailman, Joshua B. 2012. "Seven Metaphors for (Music) Listening: DRAMaTIC"[519] in *Journal of Sonic Studies* vol. 2.
- Mailman, Joshua B. 2016. "Cybernetic Phenomenology of Music, Embodied Speculative Realism, and Aesthetics Driven Techné for Spontaneous Audio-visual Expression"[520] in *Perspectives of New Music* vol. 54, no. 1: 5-95.

Societies

- American Society for Cybernetics[521]
- IEEE Systems, Man, & Cybernetics Society[522]
- International Society for Cybernetics and Systems Research[523]
- The Cybernetics Society[524]

Computational neuroscience

Computational neuroscience (also known as **theoretical neuroscience** or **mathematical neuroscience**) is a branch of neuroscience which employs mathematical models, theoretical analysis and abstractions of the brain to understand the principles that govern the development, structure, physiology and cognitive abilities of the nervous system.[525]

Computational neuroscience focuses on the description of biologically plausible neurons (and neural systems) and their physiology and dynamics, and it is therefore not concerned with biologically unrealistic disciplines such as connectionism, machine learning, artificial neural networks, artificial intelligence and computational learning theory.

In theory, computational neuroscience would be a sub-field of theoretical neuroscience which employs computational simulations to validate and solve the mathematical models. However, since the biologically plausible mathematical models formulated in neuroscience are in most cases too complex to be solved analytically, the two terms are essentially synonyms and are used interchangeably. The term mathematical neuroscience is also used sometimes, to stress the quantitative nature of the field.

The mathematical models formulated in computational neuroscience are useful since they capture the essential features of the biological system at multiple spatial-temporal scales, from membrane currents, proteins, and chemical coupling to network oscillations, columnar and topographic architecture, and learning and memory. Furthermore, these computational models frame hypotheses that can be directly tested by biological or psychological experiments.

History

The term "computational neuroscience" was introduced by Eric L. Schwartz, who organized a conference, held in 1985 in Carmel, California, at the request of the Systems Development Foundation to provide a summary of the current status of a field which until that point was referred to by a variety of names, such as neural modeling, brain theory and neural networks. The proceedings of this definitional meeting were published in 1990 as the book *Computational Neuroscience*. The first open international meeting focused on Computational Neuroscience was organized by James M. Bower and John Miller in San Francisco, California in 1989 and has continued each year since as the annual CNS meeting. The first graduate educational program in computational neuroscience was organized as the Computational and Neural Systems Ph.D. program at the California Institute of Technology in 1985.

The early historical roots of the field can be traced to the work of people such as Louis Lapicque, Hodgkin & Huxley, Hubel & Wiesel, and David Marr, to name a few. Lapicque introduced the integrate and fire model of the neuron in a seminal article published in 1907. This model is still popular today for artificial neural networks studies because of its simplicity (see a recent review).

About 40 years later, Hodgkin & Huxley developed the voltage clamp and created the first biophysical model of the action potential. Hubel & Wiesel discovered that neurons in the primary visual cortex, the first cortical area to process information coming from the retina, have oriented receptive fields and are organized in columns. David Marr's work focused on the interactions between neurons, suggesting computational approaches to the study of how functional groups of neurons within the hippocampus and neocortex interact, store, process, and transmit information. Computational modeling of biophysically realistic neurons and dendrites began with the work of Wilfrid Rall, with the first multicompartmental model using cable theory.

Major topics

Research in computational neuroscience can be roughly categorized into several lines of inquiry. Most computational neuroscientists collaborate closely with experimentalists in analyzing novel data and synthesizing new models of biological phenomena.

Single-neuron modeling

Even single neurons have complex biophysical characteristics and can perform computations (e.g.). Hodgkin and Huxley's original model only employed two voltage-sensitive currents (Voltage sensitive ion channels are glycoprotein molecules which extend through the lipid bilayer, allowing ions to traverse under certain conditions through the axolemma), the fast-acting sodium and the inward-rectifying potassium. Though successful in predicting the timing and qualitative features of the action potential, it nevertheless failed to predict a number of important features such as adaptation and shunting. Scientists now believe that there are a wide variety of voltage-sensitive currents, and the implications of the differing dynamics, modulations, and sensitivity of these currents is an important topic of computational neuroscience.

The computational functions of complex dendrites are also under intense investigation. There is a large body of literature regarding how different currents interact with geometric properties of neurons.

Some models are also tracking biochemical pathways at very small scales such as spines or synaptic clefts.

There are many software packages, such as GENESIS and NEURON, that allow rapid and systematic *in silico* modeling of realistic neurons. Blue Brain, a project founded by Henry Markram from the École Polytechnique Fédérale de Lausanne, aims to construct a biophysically detailed simulation of a cortical column on the Blue Gene supercomputer.

Modeling the richness of biophysical properties on the single-neuron scale can supply mechanisms that serve as the building blocks for network dynamics. However, detailed neuron descriptions are computationally expensive and this can handicap the pursuit of realistic network investigations, where many neurons need to be simulated. As a result, researchers that study large neural circuits typically represent each neuron and synapse with an artificially simple model, ignoring much of the biological detail. Hence there is a drive to produce simplified neuron models that can retain significant biological fidelity at a low computational overhead. Algorithms have been developed to produce faithful, faster running, simplified surrogate neuron models from computationally expensive, detailed neuron models.

Development, axonal patterning, and guidance

Computational neuroscience aims to address a wide array of questions. How do axons and dendrites form during development? How do axons know where to target and how to reach these targets? How do neurons migrate to the proper position in the central and peripheral systems? How do synapses form? We know from molecular biology that distinct parts of the nervous system release distinct chemical cues, from growth factors to hormones that modulate and influence the growth and development of functional connections between neurons.

Theoretical investigations into the formation and patterning of synaptic connection and morphology are still nascent. One hypothesis that has recently garnered some attention is the *minimal wiring hypothesis*, which postulates that the formation of axons and dendrites effectively minimizes resource allocation while maintaining maximal information storage.[526]

Sensory processing

Early models of sensory processing understood within a theoretical framework are credited to Horace Barlow. Somewhat similar to the minimal wiring hypothesis described in the preceding section, Barlow understood the processing of the early sensory systems to be a form of efficient coding, where the neurons encoded information which minimized the number of spikes. Experimental and computational work have since supported this hypothesis in one form or another.

Current research in sensory processing is divided among a biophysical modelling of different subsystems and a more theoretical modelling of perception. Current models of perception have suggested that the brain performs some form of Bayesian inference and integration of different sensory information in generating our perception of the physical world.

Memory and synaptic plasticity

Earlier models of memory are primarily based on the postulates of Hebbian learning. Biologically relevant models such as Hopfield net have been developed to address the properties of associative (also known as "content-addressable") style of memory that occur in biological systems. These attempts are primarily focusing on the formation of medium- and long-term memory, localizing in the hippocampus. Models of working memory, relying on theories of network oscillations and persistent activity, have been built to capture some features of the prefrontal cortex in context-related memory.

One of the major problems in neurophysiological memory is how it is maintained and changed through multiple time scales. Unstable synapses are easy to train but also prone to stochastic disruption. Stable synapses forget less easily, but they are also harder to consolidate. One recent computational hypothesis involves cascades of plasticity that allow synapses to function at multiple time scales. Stereochemically detailed models of the acetylcholine receptor-based synapse with the Monte Carlo method, working at the time scale of microseconds, have been built. It is likely that computational tools will contribute greatly to our understanding of how synapses function and change in relation to external stimulus in the coming decades.

Behaviors of networks

Biological neurons are connected to each other in a complex, recurrent fashion. These connections are, unlike most artificial neural networks, sparse and usually specific. It is not known how information is transmitted through such sparsely connected networks, although specific areas of the brain, such as the Visual cortex, are understood in some detail. It is also unknown what the computational functions of these specific connectivity patterns are, if any.

The interactions of neurons in a small network can be often reduced to simple models such as the Ising model. The statistical mechanics of such simple systems are well-characterized theoretically. There has been some recent evidence that suggests that dynamics of arbitrary neuronal networks can be reduced to pairwise interactions. It is not known, however, whether such descriptive dynamics impart any important computational function. With the emergence of two-photon microscopy and calcium imaging, we now have powerful experimental methods with which to test the new theories regarding neuronal networks.

In some cases the complex interactions between *inhibitory* and *excitatory* neurons can be simplified using mean field theory, which gives rise to the population model of neural networks. While many neurotheorists prefer such models with reduced complexity, others argue that uncovering structural functional relations depends on including as much neuronal and network structure as possible. Models of this type are typically built in large simulation platforms like GENESIS or NEURON. There have been some attempts to provide unified methods that bridge and integrate these levels of complexity.

Cognition, discrimination, and learning

Computational modeling of higher cognitive functions has only recently-Wikipedia:Manual of Style/Dates and numbers#Chronological items begun. Experimental data comes primarily from single-unit recording in primates. The frontal lobe and parietal lobe function as integrators of information from multiple sensory modalities. There are some tentative ideas regarding how simple mutually inhibitory functional circuits in these areas may carry out biologically relevant computation.

The brain seems to be able to discriminate and adapt particularly well in certain contexts. For instance, human beings seem to have an enormous capacity for memorizing and recognizing faces. One of the key goals of computational neuroscience is to dissect how biological systems carry out these complex computations efficiently and potentially replicate these processes in building intelligent machines.

The brain's large-scale organizational principles are illuminated by many fields, including biology, psychology, and clinical practice. Integrative neuroscience attempts to consolidate these observations through unified descriptive models and databases of behavioral measures and recordings. These are the bases for some quantitative modeling of large-scale brain activity.

The Computational Representational Understanding of Mind (CRUM) is another attempt at modeling human cognition through simulated processes like acquired rule-based systems in decision making and the manipulation of visual representations in decision making.

Consciousness

One of the ultimate goals of psychology/neuroscience is to be able to explain the everyday experience of conscious life. Francis Crick and Christof Koch made some attempts to formulate a consistent framework for future work in neural correlates of consciousness (NCC), though much of the work in this field remains speculative.

Computational clinical neuroscience

Computational Clinical Neuroscience is a field that brings together experts in neuroscience, neurology, psychiatry, decision sciences and computational modeling to quantitatively define and investigate problems in neurological and psychiatric diseases, and to train scientists and clinicians that wish to apply these models to diagnosis and treatment.

Bibliography

- Chklovskii DB (2004). "Synaptic connectivity and neuronal morphology: two sides of the same coin". *Neuron.* **43** (5): 609–17. doi: 10.1016/j.neuron.2004.08.012[527]. PMID 15339643[528].
- Sejnowski, Terrence J.; Churchland, Patricia Smith (1992). *The computational brain.* Cambridge, Mass: MIT Press. ISBN 0-262-03188-4.

- Gerstner, W.; Kistler, W.; Naud, R.; Paninski, L. (2014). *Neuronal Dynamics*. Cambridge, UK: Cambridge University Press. ISBN 9781107447615.
- Abbott, L. F.; Dayan, Peter (2001). *Theoretical neuroscience: computational and mathematical modeling of neural systems*. Cambridge, Mass: MIT Press. ISBN 0-262-04199-5.
- Eliasmith, Chris; Anderson, Charles H. (2003). *Neural engineering: Representation, computation, and dynamics in neurobiological systems*. Cambridge, Mass: MIT Press. ISBN 0-262-05071-4.
- Hodgkin AL, Huxley AF (28 August 1952). "A quantitative description of membrane current and its application to conduction and excitation in nerve"[529]. *J. Physiol.* **117** (4): 500–44. doi: 10.1113/jphysiol.1952.sp004764[530]. PMC 1392413[531]. PMID 12991237[532].
- William Bialek; Rieke, Fred; David Warland; Rob de Ruyter van Steveninck (1999). *Spikes: exploring the neural code*. Cambridge, Mass: MIT. ISBN 0-262-68108-0.
- Schutter, Erik de (2001). *Computational neuroscience: realistic modeling for experimentalists*. Boca Raton: CRC. ISBN 0-8493-2068-2.
- Sejnowski, Terrence J.; Hemmen, J. L. van (2006). *23 problems in systems neuroscience*. Oxford [Oxfordshire]: Oxford University Press. ISBN 0-19-514822-3.
- Michael A. Arbib; Shun-ichi Amari; Prudence H. Arbib (2002). *The Handbook of Brain Theory and Neural Networks*. Cambridge, Massachusetts: The MIT Press. ISBN 0-262-01197-2.

External links

Wikimedia Commons has media related to ***Computational neuroscience***.

Journals

- Journal of Mathematical Neuroscience[533]
- Journal of Computational Neuroscience[534]
- Neural Computation[535]
- Cognitive Neurodynamics[536]
- Frontiers in Computational Neuroscience[537]
- PLoS Computational Biology[538]
- Frontiers in Neuroinformatics[539]

Software

- BRIAN, a Python based simulator
- Budapest Reference Connectome, web based 3D visualization tool to browse connections in the human brain
- Emergent, neural simulation software.
- GENESIS, a general neural simulation system.

Conferences

- Computational and Systems Neuroscience (COSYNE) – a computational neuroscience meeting with a systems neuroscience focus.
- Annual Computational Neuroscience Meeting (CNS)[540]– a yearly computational neuroscience meeting.
- Neural Information Processing Systems (NIPS)[541]– a leading annual conference covering other machine learning topics as well.
- International Conference on Cognitive Neurodynamics (ICCN)[542]– a yearly conference.
- UK Mathematical Neurosciences Meeting[543]– a new yearly conference, focused on mathematical aspects.
- The NeuroComp Conference[544]– a yearly computational neuroscience conference (France).
- Bernstein Conference on Computational Neuroscience (BCCN)[545]– a yearly conference in Germany, organized by the Bernstein Network for Computational Neuroscience[546].
- AREADNE Conferences[547]– a biennial meeting that includes theoretical and experimental results, held in even years in Santorini, Greece.

Websites

- Encyclopedia of Computational Neuroscience[548], part of Scholarpedia, an online expert curated encyclopedia on computational neuroscience and dynamical systems

Symbolic artificial intelligence

Symbolic artificial intelligence is the term for the collection of all methods in artificial intelligence research that are based on high-level "symbolic" (human-readable) representations of problems, logic and search. Symbolic AI was the dominant paradigm of AI research from the mid-1950s until the late 1980s.Wikipedia:Citing sourcesWikipedia:Citing sources

John Haugeland gave the name **GOFAI** ("Good Old-Fashioned Artificial Intelligence") to symbolic AI in his 1985 book *Artificial Intelligence: The Very Idea*, which explored the philosophical implications of artificial intelligence research. In robotics the analogous term is **GOFR** ("Good Old-Fashioned Robotics").

The approach is based on the assumption that many aspects of intelligence can be achieved by the manipulation of symbols, an assumption defined as the "physical symbol systems hypothesis" by Allen Newell and Herbert A. Simon in the middle 1960s.

The most successful form of symbolic AI is expert systems, which use a network of production rules. Production rules connect symbols in a relationship similar to an If-Then statement. The expert system processes the rules to make deductions and to determine what additional information it needs, i.e. what questions to ask, using human-readable symbols.

Opponents of the symbolic approach include roboticists such as Rodney Brooks, who aims to produce autonomous robots without symbolic representation (or with only minimal representation) and computational intelligence researchers, who apply techniques such as neural networks and optimization to solve problems in machine learning and control engineering.

Symbolic AI was intended to produce general, human-like intelligence in a machine, whereas most modern research is directed at specific sub-problems. Research into general intelligence is now studied in the sub-field of artificial general intelligence.

Machines were initially designed to formulate outputs based on the inputs that were represented by symbols. Symbols are used when the input is definite and falls under certainty. But when there is uncertainty involved, for example in formulating predictions, the representation is done using "fuzzy logic". This can be seen in artificial neural networks.

Tools

Evolutionary computation

Part of a series on
Evolutionary biology
• Evolutionary biology portal • Category • *Book* • Related topics
• v • t • e[549]

In computer science, **evolutionary computation** is a family of algorithms for global optimization inspired by biological evolution, and the subfield of artificial intelligence and soft computing studying these algorithms. In technical terms, they are a family of population-based trial and error problem solvers with a metaheuristic or stochastic optimization character.

In evolutionary computation, an initial set of candidate solutions is generated and iteratively updated. Each new generation is produced by stochastically removing less desired solutions, and introducing small random changes. In biological terminology, a population of solutions is subjected to natural selection (or artificial selection) and mutation. As a result, the population will gradually evolve to increase in fitness, in this case the chosen fitness function of the algorithm.

Evolutionary computation techniques can produce highly optimized solutions in a wide range of problem settings, making them popular in computer science. Many variants and extensions exist, suited to more specific families of problems and data structures. Evolutionary computation is also sometimes used in evolutionary biology as an *in silico* experimental procedure to study common aspects of general evolutionary processes.

History

The use of evolutionary principles for automated problem solving originated in the 1950s. It was not until the 1960s that three distinct interpretations of this idea started to be developed in three different places.

Evolutionary programming was introduced by Lawrence J. Fogel in the US, while John Henry Holland called his method a *genetic algorithm*. In Germany Ingo Rechenberg and Hans-Paul Schwefel introduced *evolution strategies*. These areas developed separately for about 15 years. From the early nineties on they are unified as different representatives ("dialects") of one technology, called *evolutionary computing*. Also in the early nineties, a fourth stream following the general ideas had emerged – *genetic programming*. Since the 1990s, nature-inspired algorithms are becoming an increasingly significant part of the evolutionary computation.

These terminologies denote the field of evolutionary computing and consider evolutionary programming, evolution strategies, genetic algorithms, and genetic programming as sub-areas.

Simulations of evolution using evolutionary algorithms and artificial life started with the work of Nils Aall Barricelli in the 1960s, and was extended by Alex Fraser, who published a series of papers on simulation of artificial selection. Artificial evolution became a widely recognised optimisation method as a result of the work of Ingo Rechenberg in the 1960s and early 1970s, who used evolution strategies to solve complex engineering problems. Genetic algorithms in particular became popular through the writing of John Holland. As academic interest grew, dramatic increases in the power of computers allowed practical applications, including the automatic evolution of computer programs. Evolutionary algorithms are now used to solve multi-dimensional problems more efficiently than software produced by human designers, and also to optimise the design of systems.[550]

Techniques

Evolutionary computing techniques mostly involve metaheuristic optimization algorithms. Broadly speaking, the field includes:

- Ant colony optimization
- Artificial immune systems
- Artificial life (also see digital organism)
- Cultural algorithms
- Differential evolution
- Dual-phase evolution
- Estimation of distribution algorithms
- Evolutionary algorithms
- Evolutionary programming
- Evolution strategy
- Gene expression programming
- Genetic algorithm
- Genetic programming
- Grammatical evolution
- Learnable evolution model
- Learning classifier systems
- Memetic algorithms
- Particle swarm optimization
- Self-organization such as self-organizing maps, competitive learning
- Swarm intelligence

Evolutionary algorithms

Evolutionary algorithms form a subset of evolutionary computation in that they generally only involve techniques implementing mechanisms inspired by biological evolution such as reproduction, mutation, recombination, natural selection and survival of the fittest. Candidate solutions to the optimization problem play the role of individuals in a population, and the cost function determines the environment within which the solutions "live" (see also fitness function). Evolution of the population then takes place after the repeated application of the above operators.

In this process, there are two main forces that form the basis of evolutionary systems: **Recombination** and **mutation** create the necessary diversity and thereby facilitate novelty, while **selection** acts as a force increasing quality.

Many aspects of such an evolutionary process are stochastic. Changed pieces of information due to recombination and mutation are randomly chosen. On the other hand, selection operators can be either deterministic, or stochastic.

In the latter case, individuals with a higher fitness have a higher chance to be selected than individuals with a lower fitness, but typically even the weak individuals have a chance to become a parent or to survive.

Evolutionary algorithms and biology

Genetic algorithms deliver methods to model biological systems and systems biology that are linked to the theory of dynamical systems, since they are used to predict the future states of the system. This is just a vivid (but perhaps misleading) way of drawing attention to the orderly, well-controlled and highly structured character of development in biology.

However, the use of algorithms and informatics, in particular of computational theory, beyond the analogy to dynamical systems, is also relevant to understand evolution itself.

This view has the merit of recognizing that there is no central control of development; organisms develop as a result of local interactions within and between cells. The most promising ideas about program-development parallels seem to us to be ones that point to an apparently close analogy between processes within cells, and the low-level operation of modern computers[551]. Thus, biological systems are like computational machines that process input information to compute next states, such that biological systems are closer to a computation than classical dynamical system .

Furthermore, following concepts from computational theory, micro processes in biological organisms are fundamentally incomplete and undecidable (completeness (logic)), implying that “there is more than a crude metaphor behind the analogy between cells and computers .

The analogy to computation extends also to the relationship between inheritance systems and biological structure, which is often thought to reveal one of the most pressing problems in explaining the origins of life.

Practitioners

The list of active researchers is naturally dynamic and non-exhaustive. A network analysis of the community was published in 2007.

- Thomas Bäck
- Wolfgang Banzhaf
- B V Babu
- Kalyanmoy Deb
- Kenneth A De Jong
- Gusz Eiben

- Anna I Esparcia
- Peter J. Fleming
- David B. Fogel
- Stephanie Forrest
- David E. Goldberg
- Emma Hart
- John Henry Holland
- Theo Jansen
- John Koza
- Zbigniew Michalewicz
- Melanie Mitchell
- Peter Nordin
- Gabriela Ochoa
- Una-May O'Reilly
- Riccardo Poli
- Ingo Rechenberg
- Marc Schoenauer
- Hans-Paul Schwefel
- Jim Smith
- Gloria Townsend
- Xin Yao

Conferences

The main conferences in the evolutionary computation area include

- ACM Genetic and Evolutionary Computation Conference (GECCO),
- IEEE Congress on Evolutionary Computation (CEC),
- EvoStar, which comprises four conferences: EuroGP, EvoApplications, EvoCOP and EvoMUSART,
- Parallel Problem Solving from Nature (PPSN).

Weblinks

- Article in the Stanford Encyclopedia of Philosophy about Biological Information (English)[552]

Bibliography

- Th. Bäck, D.B. Fogel, and Z. Michalewicz (Editors), Handbook of Evolutionary Computation[553], 1997, ISBN 0750303921

- Th. Bäck and H.-P. Schwefel. An overview of evolutionary algorithms for parameter optimization. Evolutionary Computation, 1(1):1–23, 1993.
- W. Banzhaf, P. Nordin, R.E. Keller, and F.D. Francone. Genetic Programming — An Introduction. Morgan Kaufmann, 1998.
- S. Cagnoni, et al., Real-World Applications of Evolutionary Computing[554], Springer-Verlag Lecture Notes in Computer Science, Berlin, 2000.
- R. Chiong, Th. Weise, Z. Michalewicz (Editors), Variants of Evolutionary Algorithms for Real-World Applications[555], Springer, 2012, ISBN 3642234232
- K. A. De Jong, Evolutionary computation: a unified approach. MIT Press, Cambridge MA, 2006
- A. E. Eiben and M. Schoenauer, Evolutionary computing[556], Information Processing Letters, 82(1): 1–6, 2002.
- A. E. Eiben and J.E. Smith, Introduction to Evolutionary Computing[557], Springer, First edition, 2003, ISBN 3-540-40184-9,
- D. B. Fogel. Evolutionary Computation. Toward a New Philosophy of Machine Intelligence. IEEE Press, Piscataway, NJ, 1995.
- L. J. Fogel, A. J. Owens, and M. J. Walsh. Artificial Intelligence through Simulated Evolution. New York: John Wiley, 1966.
- D. E. Goldberg. Genetic algorithms in search, optimization and machine learning. Addison Wesley, 1989.
- J. H. Holland. Adaptation in natural and artificial systems. University of Michigan Press, Ann Arbor, 1975.
- P. Hingston, L. Barone, and Z. Michalewicz (Editors), Design by Evolution, Natural Computing Series[558], 2008, Springer, ISBN 3540741097
- J. R. Koza. Genetic Programming: On the Programming of Computers by means of Natural Evolution. MIT Press, Massachusetts, 1992.
- F.J. Lobo, C.F. Lima, Z. Michalewicz (Editors), Parameter Setting in Evolutionary Algorithms[559], Springer, 2010, ISBN 3642088929
- Z. Michalewicz, Genetic Algorithms + Data Structures – Evolution Programs[560], 1996, Springer, ISBN 3540606769
- Z. Michalewicz and D.B. Fogel, How to Solve It: Modern Heuristics[561], Springer, 2004, ISBN 978-3-540-22494-5
- I. Rechenberg. Evolutionstrategie: Optimierung Technischer Systeme nach Prinzipien des Biologischen Evolution. Fromman-Hozlboog Verlag, Stuttgart, 1973. (in German)
- H.-P. Schwefel. Numerical Optimization of Computer Models. John Wiley & Sons, New-York, 1981. 1995 – 2nd edition.
- D. Simon. Evolutionary Optimization Algorithms[562]. Wiley, 2013.
- M. Sipper, W. Fu, K. Ahuja, and J. H. Moore, Investigating the Parameter Space of Evolutionary Algorithms[563], BioData Mining, 11:2, 2018.

- Y. Zhang and S. Li. " PSA: A novel optimization algorithm based on survival rules of porcellio scaber[564]".

Mathematical optimization

In mathematics, computer science and operations research, **mathematical optimization** or **mathematical programming**, alternatively spelled *optimisation*, is the selection of a best element (with regard to some criterion) from some set of available alternatives.[565]

In the simplest case, an optimization problem consists of maximizing or minimizing a real function by systematically choosing input values from within an allowed set and computing the value of the function. The generalization of optimization theory and techniques to other formulations constitutes a large area of applied mathematics. More generally, optimization includes finding "best available" values of some objective function given a defined domain (or input), including a variety of different types of objective functions and different types of domains.

Optimization problems

An optimization problem can be represented in the following way:

Given: a function $f: A \to \mathbb{R}$ from some set A to the real numbers

Sought: an element $\mathbf{x}_0 \in A$ such that $f(\mathbf{x}_0) \leq f(\mathbf{x})$ for all $\mathbf{x} \in A$ ("minimization") or such that $f(\mathbf{x}_0) \geq f(\mathbf{x})$ for all $\mathbf{x} \in A$ ("maximization").

Such a formulation is called an **optimization problem** or a **mathematical programming problem** (a term not directly related to computer programming, but still in use for example in linear programming – see History below). Many real-world and theoretical problems may be modeled in this general framework. Problems formulated using this technique in the fields of physics and computer vision may refer to the technique as **energy minimization**, speaking of the value of the function f as representing the energy of the system being modeled.

Typically, A is some subset of the Euclidean space $\mathbb{R}^n$, often specified by a set of *constraints*, equalities or inequalities that the members of A have to satisfy. The domain A of f is called the *search space* or the *choice set*, while the elements of A are called *candidate solutions* or *feasible solutions*.

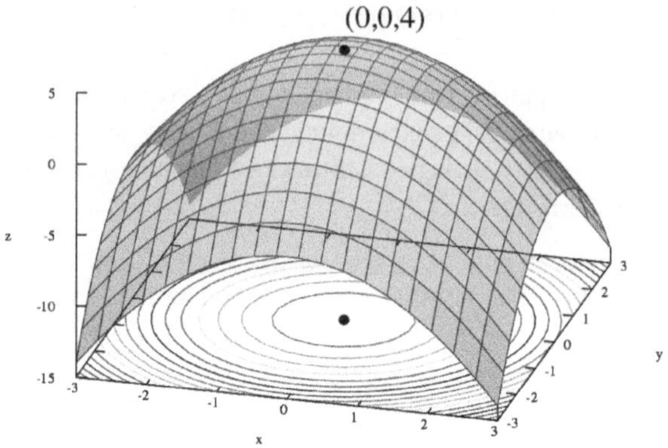

Figure 21: *Graph of a paraboloid given by* $z = f(x, y) = -(x^2 + y^2) + 4$. *The global maximum at* $(x, y, z) = (0, 0, 4)$ *is indicated by a blue dot.*

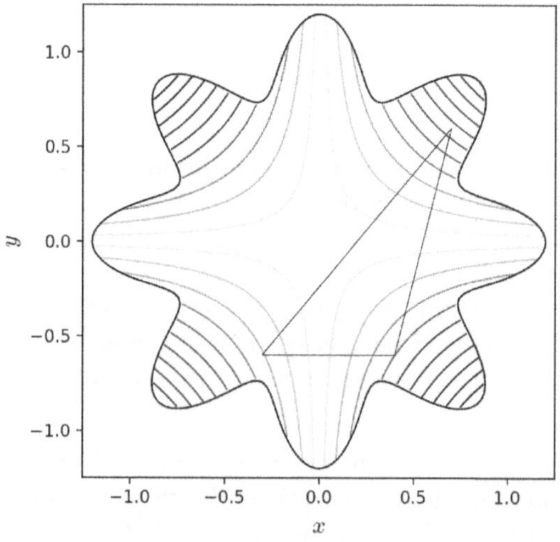

Figure 22: *Nelder-Mead minimum search of Simionescu's function. Simplex vertices are ordered by their value, with 1 having the lowest (best) value.*

The function f is called, variously, an **objective function**, a **loss function** or **cost function** (minimization),[566] a **utility function** or **fitness function** (maximization), or, in certain fields, an **energy function** or **energy functional**. A feasible solution that minimizes (or maximizes, if that is the goal) the objective function is called an *optimal solution*.

In mathematics, conventional optimization problems are usually stated in terms of minimization.

A *local minimum* $\mathbf{x}^*$ is defined as an element for which there exists some $\delta > 0$ such that

> for all $\mathbf{x} \in A$ where $\|\mathbf{x} - \mathbf{x}^*\| \leq \delta$, the expression $f(\mathbf{x}^*) \leq f(\mathbf{x})$ holds;

that is to say, on some region around $\mathbf{x}^*$ all of the function values are greater than or equal to the value at that element. Local maxima are defined similarly.

While a local minimum is at least as good as any nearby elements, a global minimum is at least as good as every feasible element. Generally, unless both the objective function and the feasible region are convex in a minimization problem, there may be several local minima. In a convex problem, if there is a local minimum that is interior (not on the edge of the set of feasible elements), it is also the global minimum, but a nonconvex problem may have more than one local minimum not all of which need be global minima.

A large number of algorithms proposed for solving nonconvex problems – including the majority of commercially available solvers – are not capable of making a distinction between locally optimal solutions and globally optimal solutions, and will treat the former as actual solutions to the original problem. Global optimization is the branch of applied mathematics and numerical analysis that is concerned with the development of deterministic algorithms that are capable of guaranteeing convergence in finite time to the actual optimal solution of a nonconvex problem.

Notation

Optimization problems are often expressed with special notation. Here are some examples:

Minimum and maximum value of a function

Consider the following notation:

$$\min_{x\in\mathbb{R}} (x^2 + 1)$$

This denotes the minimum value of the objective function $x^2 + 1$, when choosing x from the set of real numbers $\mathbb{R}$. The minimum value in this case is 1, occurring at $x = 0$.

Similarly, the notation

$$\max_{x\in\mathbb{R}} 2x$$

asks for the maximum value of the objective function $2x$, where x may be any real number. In this case, there is no such maximum as the objective function is unbounded, so the answer is "infinity" or "undefined".

Optimal input arguments

Consider the following notation:

$$\underset{x\in(-\infty,-1]}{\arg\min} \; x^2 + 1,$$

or equivalently

$$\underset{x}{\arg\min} \; x^2 + 1, \text{ subject to: } x \in (-\infty, -1].$$

This represents the value (or values) of the argument x in the interval $(-\infty, -1]$ that minimizes (or minimize) the objective function $x^2 + 1$ (the actual minimum value of that function is not what the problem asks for). In this case, the answer is $x = -1$, since $x = 0$ is infeasible, i.e. does not belong to the feasible set.

Similarly,

$$\underset{x\in[-5,5],\; y\in\mathbb{R}}{\arg\max} \; x\cos(y),$$

or equivalently

$$\underset{x,\; y}{\arg\max} \; x\cos(y), \text{ subject to: } x \in [-5, 5], \; y \in \mathbb{R},$$

represents the (x, y) pair (or pairs) that maximizes (or maximize) the value of the objective function $x\cos(y)$, with the added constraint that x lie in the interval $[-5, 5]$ (again, the actual maximum value of the expression does not matter). In this case, the solutions are the pairs of the form $(5, 2k\pi)$ and $(-5,(2k+1)\pi)$, where k ranges over all integers.

arg min and **arg max** are sometimes also written **argmin** and **argmax**, and stand for **argument of the minimum** and **argument of the maximum**.

History

Fermat and Lagrange found calculus-based formulae for identifying optima, while Newton and Gauss proposed iterative methods for moving towards an optimum.

The term "linear programming" for certain optimization cases was due to George B. Dantzig, although much of the theory had been introduced by Leonid Kantorovich in 1939. (*Programming* in this context does not refer to computer programming, but from the use of *program* by the United States military to refer to proposed training and logistics schedules, which were the problems Dantzig studied at that time.) Dantzig published the Simplex algorithm in 1947, and John von Neumann developed the theory of duality in the same year.

Other major researchers in mathematical optimization include the following:

- Aharon Ben-Tal
- Richard Bellman
- Roger Fletcher
- Ronald A. Howard
- Fritz John
- Narendra Karmarkar
- William Karush
- Leonid Khachiyan
- Bernard Koopman
- Harold Kuhn
- László Lovász
- Arkadi Nemirovski
- Yurii Nesterov
- Boris Polyak
- Lev Pontryagin
- James Renegar
- R. Tyrrell Rockafellar
- Cornelis Roos
- Naum Z. Shor
- Michael J. Todd
- Albert Tucker

Major subfields

- Convex programming studies the case when the objective function is convex (minimization) or concave (maximization) and the constraint set is convex. This can be viewed as a particular case of nonlinear programming or as generalization of linear or convex quadratic programming.

  - Linear programming (LP), a type of convex programming, studies the case in which the objective function f is linear and the constraints are specified using only linear equalities and inequalities. Such a constraint set is called a polyhedron or a polytope if it is bounded.
  - Second order cone programming (SOCP) is a convex program, and includes certain types of quadratic programs.
  - Semidefinite programming (SDP) is a subfield of convex optimization where the underlying variables are semidefinite matrices. It is a generalization of linear and convex quadratic programming.
  - Conic programming is a general form of convex programming. LP, SOCP and SDP can all be viewed as conic programs with the appropriate type of cone.
  - Geometric programming is a technique whereby objective and inequality constraints expressed as posynomials and equality constraints as monomials can be transformed into a convex program.
- Integer programming studies linear programs in which some or all variables are constrained to take on integer values. This is not convex, and in general much more difficult than regular linear programming.
- Quadratic programming allows the objective function to have quadratic terms, while the feasible set must be specified with linear equalities and inequalities. For specific forms of the quadratic term, this is a type of convex programming.
- Fractional programming studies optimization of ratios of two nonlinear functions. The special class of concave fractional programs can be transformed to a convex optimization problem.
- Nonlinear programming studies the general case in which the objective function or the constraints or both contain nonlinear parts. This may or may not be a convex program. In general, whether the program is convex affects the difficulty of solving it.
- Stochastic programming studies the case in which some of the constraints or parameters depend on random variables.
- Robust programming is, like stochastic programming, an attempt to capture uncertainty in the data underlying the optimization problem. Robust optimization targets to find solutions that are valid under all possible realizations of the uncertainties.
- Combinatorial optimization is concerned with problems where the set of feasible solutions is discrete or can be reduced to a discrete one.
- Stochastic optimization is used with random (noisy) function measurements or random inputs in the search process.
- Infinite-dimensional optimization studies the case when the set of feasible solutions is a subset of an infinite-dimensional space, such as a space of functions.

- Heuristics and metaheuristics make few or no assumptions about the problem being optimized. Usually, heuristics do not guarantee that any optimal solution need be found. On the other hand, heuristics are used to find approximate solutions for many complicated optimization problems.
- Constraint satisfaction studies the case in which the objective function *f* is constant (this is used in artificial intelligence, particularly in automated reasoning).
  - Constraint programming is a programming paradigm wherein relations between variables are stated in the form of constraints.
- Disjunctive programming is used where at least one constraint must be satisfied but not all. It is of particular use in scheduling.
- Space mapping is a concept for modeling and optimization of an engineering system to high-fidelity (fine) model accuracy exploiting a suitable physically meaningful coarse or surrogate model.

In a number of subfields, the techniques are designed primarily for optimization in dynamic contexts (that is, decision making over time):

- Calculus of variations seeks to optimize an action integral over some space to an extremum by varying a function of the coordinates.
- Optimal control theory is a generalization of the calculus of variations which introduces control policies.
- Dynamic programming studies the case in which the optimization strategy is based on splitting the problem into smaller subproblems. The equation that describes the relationship between these subproblems is called the Bellman equation.
- Mathematical programming with equilibrium constraints is where the constraints include variational inequalities or complementarities.

Multi-objective optimization

Adding more than one objective to an optimization problem adds complexity. For example, to optimize a structural design, one would desire a design that is both light and rigid. When two objectives conflict, a trade-off must be created. There may be one lightest design, one stiffest design, and an infinite number of designs that are some compromise of weight and rigidity. The set of trade-off designs that cannot be improved upon according to one criterion without hurting another criterion is known as the Pareto set. The curve created plotting weight against stiffness of the best designs is known as the Pareto frontier.

A design is judged to be "Pareto optimal" (equivalently, "Pareto efficient" or in the Pareto set) if it is not dominated by any other design: If it is worse than another design in some respects and no better in any respect, then it is dominated and is not Pareto optimal.

The choice among "Pareto optimal" solutions to determine the "favorite solution" is delegated to the decision maker. In other words, defining the problem as multi-objective optimization signals that some information is missing: desirable objectives are given but combinations of them are not rated relative to each other. In some cases, the missing information can be derived by interactive sessions with the decision maker.

Multi-objective optimization problems have been generalized further into vector optimization problems where the (partial) ordering is no longer given by the Pareto ordering.

Multi-modal optimization

Optimization problems are often multi-modal; that is, they possess multiple good solutions. They could all be globally good (same cost function value) or there could be a mix of globally good and locally good solutions. Obtaining all (or at least some of) the multiple solutions is the goal of a multi-modal optimizer.

Classical optimization techniques due to their iterative approach do not perform satisfactorily when they are used to obtain multiple solutions, since it is not guaranteed that different solutions will be obtained even with different starting points in multiple runs of the algorithm. Evolutionary algorithms, however, are a very popular approach to obtain multiple solutions in a multi-modal optimization task.

Classification of critical points and extrema

Feasibility problem

The **satisfiability problem**, also called the **feasibility problem**, is just the problem of finding any feasible solution at all without regard to objective value. This can be regarded as the special case of mathematical optimization where the objective value is the same for every solution, and thus any solution is optimal.

Many optimization algorithms need to start from a feasible point. One way to obtain such a point is to relax the feasibility conditions using a slack variable; with enough slack, any starting point is feasible. Then, minimize that slack variable until slack is null or negative.

Existence

The extreme value theorem of Karl Weierstrass states that a continuous real-valued function on a compact set attains its maximum and minimum value. More generally, a lower semi-continuous function on a compact set attains its minimum; an upper semi-continuous function on a compact set attains its maximum.

Necessary conditions for optimality

One of Fermat's theorems states that optima of unconstrained problems are found at stationary points, where the first derivative or the gradient of the objective function is zero (see first derivative test). More generally, they may be found at critical points, where the first derivative or gradient of the objective function is zero or is undefined, or on the boundary of the choice set. An equation (or set of equations) stating that the first derivative(s) equal(s) zero at an interior optimum is called a 'first-order condition' or a set of first-order conditions.

Optima of equality-constrained problems can be found by the Lagrange multiplier method. The optima of problems with equality and/or inequality constraints can be found using the 'Karush–Kuhn–Tucker conditions'.

Sufficient conditions for optimality

While the first derivative test identifies points that might be extrema, this test does not distinguish a point that is a minimum from one that is a maximum or one that is neither. When the objective function is twice differentiable, these cases can be distinguished by checking the second derivative or the matrix of second derivatives (called the Hessian matrix) in unconstrained problems, or the matrix of second derivatives of the objective function and the constraints called the bordered Hessian in constrained problems. The conditions that distinguish maxima, or minima, from other stationary points are called 'second-order conditions' (see 'Second derivative test'). If a candidate solution satisfies the first-order conditions, then satisfaction of the second-order conditions as well is sufficient to establish at least local optimality.

Sensitivity and continuity of optima

The envelope theorem describes how the value of an optimal solution changes when an underlying parameter changes. The process of computing this change is called comparative statics.

The maximum theorem of Claude Berge (1963) describes the continuity of an optimal solution as a function of underlying parameters.

Calculus of optimization

For unconstrained problems with twice-differentiable functions, some critical points can be found by finding the points where the gradient of the objective function is zero (that is, the stationary points). More generally, a zero subgradient certifies that a local minimum has been found for minimization problems with convex functions and other locally Lipschitz functions.

Further, critical points can be classified using the definiteness of the Hessian matrix: If the Hessian is *positive* definite at a critical point, then the point is a local minimum; if the Hessian matrix is negative definite, then the point is a local maximum; finally, if indefinite, then the point is some kind of saddle point.

Constrained problems can often be transformed into unconstrained problems with the help of Lagrange multipliers. Lagrangian relaxation can also provide approximate solutions to difficult constrained problems.

When the objective function is convex, then any local minimum will also be a global minimum. There exist efficient numerical techniques for minimizing convex functions, such as interior-point methods.

Computational optimization techniques

To solve problems, researchers may use algorithms that terminate in a finite number of steps, or iterative methods that converge to a solution (on some specified class of problems), or heuristics that may provide approximate solutions to some problems (although their iterates need not converge).

Optimization algorithms

- Simplex algorithm of George Dantzig, designed for linear programming.
- Extensions of the simplex algorithm, designed for quadratic programming and for linear-fractional programming.
- Variants of the simplex algorithm that are especially suited for network optimization.
- Combinatorial algorithms
- Quantum optimization algorithms

Iterative methods

The iterative methods used to solve problems of nonlinear programming differ according to whether they evaluate Hessians, gradients, or only function values. While evaluating Hessians (H) and gradients (G) improves the rate of convergence, for functions for which these quantities exist and vary sufficiently smoothly, such evaluations increase the computational complexity (or computational cost) of each iteration. In some cases, the computational complexity may be excessively high.

One major criterion for optimizers is just the number of required function evaluations as this often is already a large computational effort, usually much more effort than within the optimizer itself, which mainly has to operate over the N variables. The derivatives provide detailed information for such optimizers, but are even harder to calculate, e.g. approximating the gradient takes at least N+1 function evaluations. For approximations of the 2nd derivatives (collected in the Hessian matrix) the number of function evaluations is in the order of N^2. Newton's method requires the 2nd order derivates, so for each iteration the number of function calls is in the order of N^2, but for a simpler pure gradient optimizer it is only N. However, gradient optimizers need usually more iterations than Newton's algorithm. Which one is best with respect to the number of function calls depends on the problem itself.

- Methods that evaluate Hessians (or approximate Hessians, using finite differences):
  - Newton's method
  - Sequential quadratic programming: A Newton-based method for small-medium scale *constrained* problems. Some versions can handle large-dimensional problems.
  - Interior point methods: This is a large class of methods for constrained optimization. Some interior-point methods use only (sub)gradient information, and others of which require the evaluation of Hessians.
- Methods that evaluate gradients, or approximate gradients in some way (or even subgradients):
  - Coordinate descent methods: Algorithms which update a single coordinate in each iteration
  - Conjugate gradient methods: Iterative methods for large problems. (In theory, these methods terminate in a finite number of steps with quadratic objective functions, but this finite termination is not observed in practice on finite–precision computers.)
  - Gradient descent (alternatively, "steepest descent" or "steepest ascent"): A (slow) method of historical and theoretical interest, which has had renewed interest for finding approximate solutions of enormous problems.

  - Subgradient methods - An iterative method for large locally Lipschitz functions using generalized gradients. Following Boris T. Polyak, subgradient–projection methods are similar to conjugate–gradient methods.
  - Bundle method of descent: An iterative method for small–medium-sized problems with locally Lipschitz functions, particularly for convex minimization problems. (Similar to conjugate gradient methods)
  - Ellipsoid method: An iterative method for small problems with quasiconvex objective functions and of great theoretical interest, particularly in establishing the polynomial time complexity of some combinatorial optimization problems. It has similarities with Quasi-Newton methods.
  - Conditional gradient method (Frank–Wolfe) for approximate minimization of specially structured problems with linear constraints, especially with traffic networks. For general unconstrained problems, this method reduces to the gradient method, which is regarded as obsolete (for almost all problems).
  - Quasi-Newton methods: Iterative methods for medium-large problems (e.g. N<1000).
  - Simultaneous perturbation stochastic approximation (SPSA) method for stochastic optimization; uses random (efficient) gradient approximation.
- Methods that evaluate only function values: If a problem is continuously differentiable, then gradients can be approximated using finite differences, in which case a gradient-based method can be used.
  - Interpolation methods
  - Pattern search methods, which have better convergence properties than the Nelder–Mead heuristic (with simplices), which is listed below.

Global convergence

More generally, if the objective function is not a quadratic function, then many optimization methods use other methods to ensure that some subsequence of iterations converges to an optimal solution. The first and still popular method for ensuring convergence relies on line searches, which optimize a function along one dimension. A second and increasingly popular method for ensuring convergence uses trust regions. Both line searches and trust regions are used in modern methods of non-differentiable optimization. Usually a global optimizer is much slower than advanced local optimizers (such as BFGS), so often an efficient global optimizer can be constructed by starting the local optimizer from different starting points.

Heuristics

Besides (finitely terminating) algorithms and (convergent) iterative methods, there are heuristics. A heuristic is any algorithm which is not guaranteed (mathematically) to find the solution, but which is nevertheless useful in certain practical situations. List of some well-known heuristics:

- Memetic algorithm
- Differential evolution
- Evolutionary algorithms
- Dynamic relaxation
- Genetic algorithms
- Hill climbing with random restart
- Nelder-Mead simplicial heuristic: A popular heuristic for approximate minimization (without calling gradients)
- Particle swarm optimization
- Cuckoo search
- Gravitational search algorithm
- Artificial bee colony optimization
- Simulated annealing
- Stochastic tunneling
- Tabu search
- Reactive Search Optimization (RSO) implemented in LIONsolver

Applications

Mechanics

Problems in rigid body dynamics (in particular articulated rigid body dynamics) often require mathematical programming techniques, since you can view rigid body dynamics as attempting to solve an ordinary differential equation on a constraint manifold; the constraints are various nonlinear geometric constraints such as "these two points must always coincide", "this surface must not penetrate any other", or "this point must always lie somewhere on this curve". Also, the problem of computing contact forces can be done by solving a linear complementarity problem, which can also be viewed as a QP (quadratic programming) problem.

Many design problems can also be expressed as optimization programs. This application is called design optimization. One subset is the engineering optimization, and another recent and growing subset of this field is multidisciplinary design optimization, which, while useful in many problems, has in particular been applied to aerospace engineering problems.

This approach may be applied in cosmology and astrophysics,.

Economics and finance

Economics is closely enough linked to optimization of agents that an influential definition relatedly describes economics *qua* science as the "study of human behavior as a relationship between ends and scarce means" with alternative uses.[567] Modern optimization theory includes traditional optimization theory but also overlaps with game theory and the study of economic equilibria. The *Journal of Economic Literature* codes classify mathematical programming, optimization techniques, and related topics under JEL:C61-C63.

In microeconomics, the utility maximization problem and its dual problem, the expenditure minimization problem, are economic optimization problems. Insofar as they behave consistently, consumers are assumed to maximize their utility, while firms are usually assumed to maximize their profit. Also, agents are often modeled as being risk-averse, thereby preferring to avoid risk. Asset prices are also modeled using optimization theory, though the underlying mathematics relies on optimizing stochastic processes rather than on static optimization. International trade theory also uses optimization to explain trade patterns between nations. The optimization of portfolios is an example of multi-objective optimization in economics.

Since the 1970s, economists have modeled dynamic decisions over time using control theory. For example, dynamic search models are used to study labor-market behavior. A crucial distinction is between deterministic and stochastic models.[568] Macroeconomists build dynamic stochastic general equilibrium (DSGE) models that describe the dynamics of the whole economy as the result of the interdependent optimizing decisions of workers, consumers, investors, and governments.[569]

Electrical engineering

Some common applications of optimization techniques in electrical engineering include active filter design, stray field reduction in superconducting magnetic energy storage systems, space mapping design of microwave structures,[570] handset antennas,[571,572,573] electromagnetics-based design. Electromagnetically validated design optimization of microwave components and antennas has made extensive use of an appropriate physics-based or empirical surrogate model and space mapping methodologies since the discovery of space mapping in 1993.[574,575]

Civil engineering

Optimization has been widely used in civil engineering. The most common civil engineering problems that are solved by optimization are cut and fill of roads, life-cycle analysis of structures and infrastructures,[576] resource leveling[577] and schedule optimization.

Operations research

Another field that uses optimization techniques extensively is operations research. Operations research also uses stochastic modeling and simulation to support improved decision-making. Increasingly, operations research uses stochastic programming to model dynamic decisions that adapt to events; such problems can be solved with large-scale optimization and stochastic optimization methods.

Control engineering

Mathematical optimization is used in much modern controller design. High-level controllers such as model predictive control (MPC) or real-time optimization (RTO) employ mathematical optimization. These algorithms run online and repeatedly determine values for decision variables, such as choke openings in a process plant, by iteratively solving a mathematical optimization problem including constraints and a model of the system to be controlled.

Geophysics

Optimization techniques are regularly used in geophysical parameter estimation problems. Given a set of geophysical measurements, e.g. seismic recordings, it is common to solve for the physical properties and geometrical shapes of the underlying rocks and fluids.

Molecular modeling

Nonlinear optimization methods are widely used in conformational analysis.

Further reading

Comprehensive

Undergraduate level

- Bradley, S.; Hax, A.; Magnanti, T. (1977). *Applied mathematical programming*. Addison Wesley.
- Parkinson, A.; Balling, R.; Hedengren, J. (2013). *Optimization Methods for Engineering Design*[578]. Provo, UT: Brigham Young University.
- Rardin, Ronald L. (1997). *Optimization in operations research*. Prentice Hall. p. 919. ISBN 0-02-398415-5. <q>copyright: 1998</q>
- Snyman, J.A.; Wilke, D.N. (2018). *Practical Mathematical Optimization - Basic Optimization Theory and Gradient-Based Algorithms*[579]. Springer Optimization and Its Applications Vol. 133 (2 ed.). Springer International Publishing. pp. xxvi+372. ISBN 978-3-319-77585-2.
- Strang, Gilbert (1986). *Introduction to applied mathematics*[580]. Wellesley, MA: Wellesley-Cambridge Press (Strang's publishing company). pp. xii+758. ISBN 0-9614088-0-4. MR 0870634[581].

Graduate level

- Magnanti, Thomas L. (1989). "Twenty years of mathematical programming". In Cornet, Bernard; Tulkens, Henry. *Contributions to Operations Research and Economics: The twentieth anniversary of CORE (Papers from the symposium held in Louvain-la-Neuve, January 1987)*. Cambridge, MA: MIT Press. pp. 163–227. ISBN 0-262-03149-3. MR 1104662[582].
- Minoux, M. (1986). *Mathematical programming: Theory and algorithms*. Egon Balas foreword) (Translated by Steven Vajda from the (1983 Paris: Dunod) French ed.). Chichester: A Wiley-Interscience Publication. John Wiley & Sons, Ltd. pp. xxviii+489. ISBN 0-471-90170-9. MR 2571910[583]. (2008 Second ed., in French: *Programmation mathématique: Théorie et algorithmes*. Editions Tec & Doc, Paris, 2008. xxx+711 pp. ISBN 978-2-7430-1000-3. MR 868279[584].
- Nemhauser, G. L.; Rinnooy Kan, A. H. G.; Todd, M. J., eds. (1989). *Optimization*. Handbooks in Operations Research and Management Science. **1**. Amsterdam: North-Holland Publishing Co. pp. xiv+709. ISBN 0-444-87284-1. MR 1105099[585].
  - J. E. Dennis, Jr. and Robert B. Schnabel, A view of unconstrained optimization (pp. 1–72);
  - Donald Goldfarb and Michael J. Todd, Linear programming (pp. 73–170);

  - Philip E. Gill, Walter Murray, Michael A. Saunders, and Margaret H. Wright, Constrained nonlinear programming (pp. 171–210);
  - Ravindra K. Ahuja, Thomas L. Magnanti, and James B. Orlin, Network flows (pp. 211–369);
  - W. R. Pulleyblank, Polyhedral combinatorics (pp. 371–446);
  - George L. Nemhauser and Laurence A. Wolsey, Integer programming (pp. 447–527);
  - Claude Lemaréchal, Nondifferentiable optimization (pp. 529–572);
  - Roger J-B Wets, Stochastic programming (pp. 573–629);
  - A. H. G. Rinnooy Kan and G. T. Timmer, Global optimization (pp. 631–662);
  - P. L. Yu, Multiple criteria decision making: five basic concepts (pp. 663–699).
- Shapiro, Jeremy F. (1979). *Mathematical programming: Structures and algorithms*. New York: Wiley-Interscience [John Wiley & Sons]. pp. xvi+388. ISBN 0-471-77886-9. MR 0544669[586].
- Spall, J. C. (2003), *Introduction to Stochastic Search and Optimization: Estimation, Simulation, and Control*, Wiley, Hoboken, NJ.
- University, Edwin K. P. Chong, Colorado State University, Stanislaw H. Żak, Purdue (2013). *An introduction to optimization* (Fourth edition. ed.). Hoboken, New Jersey: John Wiley & Sons. ISBN 9781118279014.

Continuous optimization

- Roger Fletcher (2000). *Practical methods of optimization*. Wiley. ISBN 978-0-471-49463-8.
- Mordecai Avriel (2003). *Nonlinear Programming: Analysis and Methods*. Dover Publishing. ISBN 0-486-43227-0.
- P. E. Gill, W. Murray and M. H. Wright (1982). *Practical Optimization*. Emerald Publishing. ISBN 978-0-12-283952-8.
- Xin-She Yang (2010). *Engineering Optimization: An Introduction with Metaheuristic Applications*. Wiley. ISBN 978-0-470-58246-6.
- Bonnans, J. Frédéric; Gilbert, J. Charles; Lemaréchal, Claude; Sagastizábal, Claudia A. (2006). *Numerical optimization: Theoretical and practical aspects*[587]. Universitext (Second revised ed. of translation of 1997 French ed.). Berlin: Springer-Verlag. pp. xiv+490. doi: 10.1007/978-3-540-35447-5[588]. ISBN 3-540-35445-X. MR 2265882[589].
- Bonnans, J. Frédéric; Shapiro, Alexander (2000). *Perturbation analysis of optimization problems*. Springer Series in Operations Research. New York: Springer-Verlag. pp. xviii+601. ISBN 0-387-98705-3. MR 1756264[590].

- Boyd, Stephen P.; Vandenberghe, Lieven (2004). *Convex Optimization*[591] (pdf). Cambridge University Press. ISBN 978-0-521-83378-3. Retrieved October 15, 2011.
- Jorge Nocedal and Stephen J. Wright (2006). *Numerical Optimization*[592]. Springer. ISBN 0-387-30303-0.
- Ruszczyński, Andrzej (2006). *Nonlinear Optimization*. Princeton, NJ: Princeton University Press. pp. xii+454. ISBN 978-0-691-11915-1. MR 2199043[593].
- Robert J. Vanderbei (2013). *Linear Programming: Foundations and Extensions, 4th Edition*[594]. Springer. ISBN 978-1-4614-7629-0.

Combinatorial optimization

- R. K. Ahuja, Thomas L. Magnanti, and James B. Orlin (1993). *Network Flows: Theory, Algorithms, and Applications*. Prentice-Hall, Inc. ISBN 0-13-617549-X.
- William J. Cook, William H. Cunningham, William R. Pulleyblank, Alexander Schrijver; *Combinatorial Optimization*; John Wiley & Sons; 1 edition (November 12, 1997); ISBN 0-471-55894-X.
- Gondran, Michel; Minoux, Michel (1984). *Graphs and algorithms*. Wiley-Interscience Series in Discrete Mathematics (Translated by Steven Vajda from the second (*Collection de la Direction des Études et Recherches d'Électricité de France* [Collection of the Department of Studies and Research of Électricité de France], v. 37. Paris: Éditions Eyrolles 1985. xxviii+545 pp. MR 868083[595]) French ed.). Chichester: John Wiley & Sons, Ltd. pp. xix+650. ISBN 978-2-7430-1035-5. MR 2552933[596]. (Fourth ed. Collection EDF R&D. Paris: Editions Tec & Doc 2009. xxxii+784 pp. MR 745802[597].
- Eugene Lawler (2001). *Combinatorial Optimization: Networks and Matroids*. Dover. ISBN 0-486-41453-1.
- Lawler, E. L.; Lenstra, J. K.; Rinnooy Kan, A. H. G.; Shmoys, D. B. (1985), *The traveling salesman problem: A guided tour of combinatorial optimization*, John Wiley & Sons, ISBN 0-471-90413-9.
- Jon Lee; *A First Course in Combinatorial Optimization*[598]; Cambridge University Press; 2004; ISBN 0-521-01012-8.
- Christos H. Papadimitriou and Kenneth Steiglitz *Combinatorial Optimization : Algorithms and Complexity*; Dover Pubns; (paperback, Unabridged edition, July 1998) ISBN 0-486-40258-4.

Relaxation (extension method)

Methods to obtain suitable (in some sense) natural extensions of optimization problems that otherwise lack of existence or stability of solutions to obtain problems with guaranteed existence of solutions and their stability in some sense (typically under various perturbation of data) are in general called relaxation. Solutions of such extended (=relaxed) problems in some sense characterizes (at least certain features) of the original problems, e.g. as far as their optimizing sequences concerns. Relaxed problems may also possesses their own natural linear structure that may yield specific optimality conditions different from optimality conditions for the original problems.

- H. O. Fattorini: Infinite Dimensional Optimization and Control Theory. Cambridge Univ. Press, 1999.
- P. Pedregal: Parametrized Measures and Variational Principles. Birkhäuser, Basel, 1997
- T. Roubicek: "Relaxation in Optimization Theory and Variational Calculus"[599]. W. de Gruyter, Berlin, 1997. ISBN 3-11-014542-1.
- J. Warga: Optimal control of differential and functional equations. Academic Press, 1972.

Journals

- *Computational Optimization and Applications*[600]
- Journal of Computational *Optimization in Economics and Finance*[601]
- *Journal of Economic Dynamics and Control*[602]
- *SIAM Journal on Optimization* (SIOPT)[603] and Editorial Policy[604]
- *SIAM Journal on Control and Optimization* (SICON)[605] and Editorial Policy[606]

External links

Wikimedia Commons has media related to ***Mathematical optimization***.

- COIN-OR[607]—Computational Infrastructure for Operations Research
- Decision Tree for Optimization Software[608] Links to optimization source codes
- Global optimization[609]
- Mathematical Programming Glossary[610]
- Mathematical Programming Society[611]

- NEOS Guide[612] currently being replaced by the NEOS Wiki[613]Wikipedia:Link rot
- Optimization Online[614] A repository for optimization e-prints
- Optimization Related Links[615]
- Convex Optimization I[616] EE364a: Course from Stanford University
- Convex Optimization – Boyd and Vandenberghe[617] Book on Convex Optimization
- Book and Course[578] on Optimization Methods for Engineering Design
- Mathematical Optimization in Operations Research[618] from the Institute for Operations Research and the Management Sciences (INFORMS)

Automated reasoning

Automated reasoning is an area of computer science and mathematical logic dedicated to understanding different aspects of reasoning. The study of automated reasoning helps produce computer programs that allow computers to reason completely, or nearly completely, automatically. Although automated reasoning is considered a sub-field of artificial intelligence, it also has connections with theoretical computer science, and even philosophy.

The most developed subareas of automated reasoning are automated theorem proving (and the less automated but more pragmatic subfield of interactive theorem proving) and automated proof checking (viewed as guaranteed correct reasoning under fixed assumptions). Extensive work has also been done in reasoning by analogy induction and abduction.

Other important topics include reasoning under uncertainty and non-monotonic reasoning. An important part of the uncertainty field is that of argumentation, where further constraints of minimality and consistency are applied on top of the more standard automated deduction. John Pollock's OSCAR system[619] is an example of an automated argumentation system that is more specific than being just an automated theorem prover.

Tools and techniques of automated reasoning include the classical logics and calculi, fuzzy logic, Bayesian inference, reasoning with maximal entropy and a large number of less formal *ad hoc* techniques.

Early years

The development of formal logic played a big role in the field of automated reasoning, which itself led to the development of artificial intelligence. A formal proof is a proof in which every logical inference has been checked back to the fundamental axioms of mathematics. All the intermediate logical steps are supplied, without exception. No appeal is made to intuition, even if the translation from intuition to logic is routine. Thus, a formal proof is less intuitive, and less susceptible to logical errors.[620]

Some consider the Cornell Summer meeting of 1957, which brought together a large number of logicians and computer scientists, as the origin of automated reasoning, or automated deduction.[621] Others say that it began before that with the 1955 Logic Theorist program of Newell, Shaw and Simon, or with Martin Davis' 1954 implementation of Presburger's decision procedure (which proved that the sum of two even numbers is even).[622] Automated reasoning, although a significant and popular area of research, went through an "AI winter" in the eighties and early nineties. The field subsequently revived, however. For example, in 2005, Microsoft started using verification technology in many of their internal projects and is planning to include a logical specification and checking language in their 2012 version of Visual C.

Significant contributions

Principia Mathematica was a milestone work in formal logic written by Alfred North Whitehead and Bertrand Russell. Principia Mathematica - also meaning Principles of Mathematics - was written with a purpose to derive all or some of the mathematical expressions, in terms of symbolic logic. Principia Mathematica was initially published in three volumes in 1910, 1912 and 1913.[623]

Logic Theorist (LT) was the first ever program developed in 1956 by Allen Newell, Cliff Shaw and Herbert A. Simon to "mimic human reasoning" in proving theorems and was demonstrated on fifty-two theorems from chapter two of Principia Mathematica, proving thirty-eight of them.[624] In addition to proving the theorems, the program found a proof for one of the theorems that was more elegant than the one provided by Whitehead and Russell. After an unsuccessful attempt at publishing their results, Newell, Shaw, and Herbert reported in their publication in 1958, *The Next Advance in Operation Research*:

> *'There are now in the world machines that think, that learn and that create. Moreover, their ability to do these things is going to increase rapidly until (in a visible future) the range of problems they can handle will be co- extensive with the range to which the human mind has been applied. '*[625]

Examples of Formal Proofs

Year	Theorem	Proof System	Formalizer	Traditional Proof
1986	First Incompleteness	Boyer-Moore	Shankar	Gödel
1990	Quadratic Reciprocity	Boyer-Moore	Russinoff	Eisenstein
1996	Fundamental- of Calculus	HOL Light	Harrison	Henstock
2000	Fundamental- of Algebra	Mizar	Milewski	Brynski
2000	Fundamental- of Algebra	Coq	Geuvers et al.	Kneser
2004	Four Color	Coq	Gonthier	Robertson et al.
2004	Prime Number	Isabelle	Avigad et al.	Selberg-Erdős
2005	Jordan Curve	HOL Light	Hales	Thomassen
2005	Brouwer Fixed Point	HOL Light	Harrison	Kuhn
2006	Flyspeck 1	Isabelle	Bauer-Nipkow	Hales
2007	Cauchy Residue	HOL Light	Harrison	Classical
2008	Prime Number	HOL Light	Harrison	analytic proof
2012	Feit-Thompson	Coq	Gonthier et al.	Bender, Glauberman and Peterfalvi
2016	Boolean Pythagorean triples problem	Formalized as SAT	Heule et al.[626]	none

Proof systems

Boyer-Moore Theorem Prover (NQTHM)

The design of NQTHM was influenced by John McCarthy and Woody Bledsoe. Started in 1971 at Edinburgh, Scotland, this was a fully automatic theorem prover built using Pure Lisp. The main aspects of NQTHM were:

1. the use of Lisp as a working logic.
2. the reliance on a principle of definition for total recursive functions.
3. the extensive use of rewriting and "symbolic evaluation".
4. an induction heuristic based the failure of symbolic evaluation.[627]

HOL Light

Written in OCaml, HOL Light is designed to have a simple and clean logical foundation and an uncluttered implementation. It is essentially another proof assistant for classical higher order logic.[628]

Coq

Developed in France, Coq is another automated proof assistant, which can automatically extract executable programs from specifications, as either Objective CAML or Haskell source code. Properties, programs and proofs are formalized in the same language called the Calculus of Inductive Constructions (CIC).[629]

Applications

Automated reasoning has been most commonly used to build automated theorem provers. Oftentimes, however, theorem provers require some human guidance to be effective and so more generally qualify as proof assistants. In some cases such provers have come up with new approaches to proving a theorem. Logic Theorist is a good example of this. The program came up with a proof for one of the theorems in Principia Mathematica that was more efficient (requiring fewer steps) than the proof provided by Whitehead and Russell. Automated reasoning programs are being applied to solve a growing number of problems in formal logic, mathematics and computer science, logic programming, software and hardware verification, circuit design, and many others. The TPTP (Sutcliffe and Suttner 1998) is a library of such problems that is updated on a regular basis. There is also a competition among automated theorem provers held regularly at the CADE conference (Pelletier, Sutcliffe and Suttner 2002); the problems for the competition are selected from the TPTP library.[630]

External links

- International Workshop on the Implementation of Logics[631]
- Workshop Series on Empirically Successful Topics in Automated Reasoning[632]

Search algorithm

In computer science, a **search algorithm** is any algorithm which solves the search problem, namely, to retrieve information stored within some data structure, or calculated in the search space of a problem domain. Examples of such structures include but are not limited to a linked list, an array data structure, or a search tree. The appropriate search algorithm often depends on the data structure being searched, and may also include prior knowledge about the data. Searching also encompasses algorithms that query the data structure, such as the SQL SELECT command.[633][634]

Search algorithms can be classified based on their mechanism of searching. Linear search algorithms check every record for the one associated with a target key in a linear fashion.[635][4] Binary, or half interval searches, repeatedly target the center of the search structure and divide the search space in half. Comparison search algorithms improve on linear searching by successively eliminating records based on comparisons of the keys until the target record is found, and can be applied on data structures with a defined order.[636] Digital search algorithms work based on the properties of digits in data structures that use numerical keys.[637] Finally, hashing directly maps keys to records based on a hash function.[638] Searches outside a linear search require that the data be sorted in some way.

Search functions are also evaluated on the basis of their complexity, or maximum theoretical run time. Binary search functions, for example, have a maximum complexity of $O(\log n)$, or logarithmic time. This means that the maximum number of operations needed to find the search target is a logarithmic function of the size of the search space.

Classes

For virtual search spaces

Algorithms for searching virtual spaces are used in the constraint satisfaction problem, where the goal is to find a set of value assignments to certain variables that will satisfy specific mathematical equations and inequations / equalities. They are also used when the goal is to find a variable assignment that will maximize or minimize a certain function of those variables. Algorithms for these problems include the basic brute-force search (also called "naïve" or "uninformed" search), and a variety of heuristics that try to exploit partial knowledge about the structure of this space, such as linear relaxation, constraint generation, and constraint propagation.

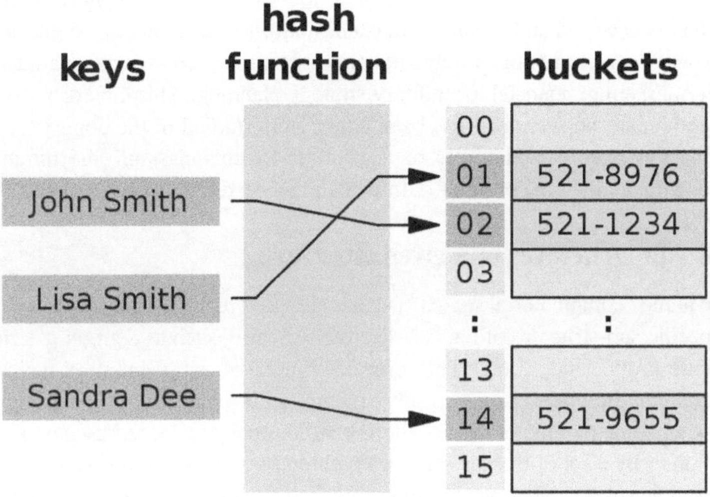

Figure 23: *Visual representation of a hash table, a data structure that allows for fast retrieval of information.*

An important subclass are the local search methods, that view the elements of the search space as the vertices of a graph, with edges defined by a set of heuristics applicable to the case; and scan the space by moving from item to item along the edges, for example according to the steepest descent or best-first criterion, or in a stochastic search. This category includes a great variety of general metaheuristic methods, such as simulated annealing, tabu search, A-teams, and genetic programming, that combine arbitrary heuristics in specific ways.

This class also includes various tree search algorithms, that view the elements as vertices of a tree, and traverse that tree in some special order. Examples of the latter include the exhaustive methods such as depth-first search and breadth-first search, as well as various heuristic-based search tree pruning methods such as backtracking and branch and bound. Unlike general metaheuristics, which at best work only in a probabilistic sense, many of these tree-search methods are guaranteed to find the exact or optimal solution, if given enough time. This is called "completeness".

Another important sub-class consists of algorithms for exploring the game tree of multiple-player games, such as chess or backgammon, whose nodes consist of all possible game situations that could result from the current situation.

The goal in these problems is to find the move that provides the best chance of a win, taking into account all possible moves of the opponent(s). Similar problems occur when humans or machines have to make successive decisions whose outcomes are not entirely under one's control, such as in robot guidance or in marketing, financial, or military strategy planning. This kind of problem — combinatorial search — has been extensively studied in the context of artificial intelligence. Examples of algorithms for this class are the minimax algorithm, alpha–beta pruning, * Informational search and the A* algorithm.

For sub-structures of a given structure

The name "combinatorial search" is generally used for algorithms that look for a specific sub-structure of a given discrete structure, such as a graph, a string, a finite group, and so on. The term combinatorial optimization is typically used when the goal is to find a sub-structure with a maximum (or minimum) value of some parameter. (Since the sub-structure is usually represented in the computer by a set of integer variables with constraints, these problems can be viewed as special cases of constraint satisfaction or discrete optimization; but they are usually formulated and solved in a more abstract setting where the internal representation is not explicitly mentioned.)

An important and extensively studied subclass are the graph algorithms, in particular graph traversal algorithms, for finding specific sub-structures in a given graph — such as subgraphs, paths, circuits, and so on. Examples include Dijkstra's algorithm, Kruskal's algorithm, the nearest neighbour algorithm, and Prim's algorithm.

Another important subclass of this category are the string searching algorithms, that search for patterns within strings. Two famous examples are the Boyer–Moore and Knuth–Morris–Pratt algorithms, and several algorithms based on the suffix tree data structure.

Search for the maximum of a function

In 1953, American statistician Jack Kiefer devised Fibonacci search which can be used to find the maximum of a unimodal function and has many other applications in computer science.

For quantum computers

There are also search methods designed for quantum computers, like Grover's algorithm, that are theoretically faster than linear or brute-force search even without the help of data structures or heuristics.

References

Bibliography

Books

- Knuth, Donald (1998). *Sorting and Searching*. The Art of Computer Programming. **3** (2nd ed.). Reading, MA: Addison-Wesley Professional.

Articles

- Beame, Paul; Fich, Faith E. (2002-08-01). "Optimal Bounds for the Predecessor Problem and Related Problems"[639]. *Journal of Computer and System Sciences*. **65** (1): 38–72. doi: 10.1006/jcss.2002.1822[640] ა .

External links

<templatestyles src="Template:Refbegin/styles.css" />

- Uninformed Search Project at the Wikiversity.
- Unsorted Data Searching Using Modulated Database[641].

Logic programming

Programming paradigms

- Action
- Agent-oriented
- Array-oriented
- Automata-based
- Concurrent computing
  - Relativistic programming
- Data-driven
- Declarative (contrast: Imperative)
  - Functional
    - Functional logic
    - Purely functional
  - Logic
    - Abductive logic
    - Answer set
    - Concurrent logic
    - Functional logic
    - Inductive logic
  - Constraint
    - Constraint logic
      - Concurrent constraint logic
  - Dataflow
    - Flow-based
    - Cell-oriented (spreadsheets)
    - Reactive
- Dynamic/scripting
- Event-driven
  - Service-oriented
  - Time-driven
- Function-level (contrast: Value-level)
  - Point-free style
    - Concatenative
- Generic
- Imperative (contrast: Declarative)
  - Procedural
  - Object-oriented
- Literate
- Language-oriented
  - Natural-language programming
  - Discipline-specific
  - Domain-specific
  - Grammar-oriented
  - Intentional
- Metaprogramming
  - Automatic
    - Inductive programming
  - Reflective
    - Attribute-oriented
  - Homoiconic
  - Macro
  - Template
- Non-structured (contrast: Structured)
  - Array

- Nondeterministic
- Parallel computing
  - Process-oriented
- Probabilistic
- Stack-based
- Structured (contrast: Non-structured)
  - Block-structured
  - Modular (contrast: Monolithic)
  - Object-oriented
    - Actor-based
    - Class-based
    - Concurrent
    - Prototype-based
    - By separation of concerns:
      - Aspect-oriented
      - Role-oriented
      - Subject-oriented
  - Recursive
- Value-level (contrast: Function-level)
- Quantum programming

- v
- t
- e[642]

Logic programming is a type of programming paradigm which is largely based on formal logic. Any program written in a logic programming language is a set of sentences in logical form, expressing facts and rules about some problem domain. Major logic programming language families include Prolog, Answer set programming (ASP) and Datalog. In all of these languages, rules are written in the form of *clauses*:

$$H :- B_1, \ldots, B_n.$$

and are read declaratively as logical implications:

$$H \text{ if } B_1 \text{ and } \ldots \text{ and } B_n.$$

H is called the *head* of the rule and $B_1, \ldots, B_n$ is called the *body*. Facts are rules that have no body, and are written in the simplified form:

```
H.
```

In the simplest case in which H, $B_1, \ldots, B_n$ are all atomic formulae, these clauses are called definite clauses or Horn clauses. However, there exist many extensions of this simple case, the most important one being the case in which conditions in the body of a clause can also be negations of atomic formulae. Logic programming languages that include this extension have the knowledge representation capabilities of a non-monotonic logic.

In ASP and Datalog, logic programs have only a declarative reading, and their execution is performed by means of a proof procedure or model generator whose behaviour is not meant to be under the control of the programmer. However, in the Prolog family of languages, logic programs also have a procedural interpretation as goal-reduction procedures:

to solve H, solve B_1, and ... and solve B_n.

Consider, for example, the following clause:

```
fallible(X) :- human(X).
```

based on an example used by Terry Winograd to illustrate the programming language Planner. As a clause in a logic program, it can be used both as a procedure to test whether X is `fallible` by testing whether X is `human`, and as a procedure to find an X that is `fallible` by finding an X that is `human`. Even facts have a procedural interpretation. For example, the clause:

```
human(socrates).
```

can be used both as a procedure to show that `socrates` is `human`, and as a procedure to find an X that is `human` by "assigning" `socrates` to X.

The declarative reading of logic programs can be used by a programmer to verify their correctness. Moreover, logic-based program transformation techniques can also be used to transform logic programs into logically equivalent programs that are more efficient. In the Prolog family of logic programming languages, the programmer can also use the known problem-solving behaviour of the execution mechanism to improve the efficiency of programs.

History

The use of mathematical logic to represent and execute computer programs is also a feature of the lambda calculus, developed by Alonzo Church in the 1930s. However, the first proposal to use the clausal form of logic for representing computer programs was made by Cordell Green.[643] This used an axiomatization of a subset of LISP, together with a representation of an input-output relation, to compute the relation by simulating the execution of the program in LISP. Foster and Elcock's Absys, on the other hand, employed a combination of equations and lambda calculus in an assertional programming language which places no constraints on the order in which operations are performed.[644]

Logic programming in its present form can be traced back to debates in the late 1960s and early 1970s about declarative versus procedural representations of knowledge in Artificial Intelligence. Advocates of declarative representations were notably working at Stanford, associated with John McCarthy, Bertram Raphael and Cordell Green, and in Edinburgh, with John Alan Robinson (an academic visitor from Syracuse University), Pat Hayes, and Robert Kowalski.

Advocates of procedural representations were mainly centered at MIT, under the leadership of Marvin Minsky and Seymour Papert.Wikipedia:Citation needed

Although it was based on the proof methods of logic, Planner, developed at MIT, was the first language to emerge within this proceduralist paradigm.[645] Planner featured pattern-directed invocation of procedural plans from goals (i.e. goal-reduction or backward chaining) and from assertions (i.e. forward chaining). The most influential implementation of Planner was the subset of Planner, called Micro-Planner, implemented by Gerry Sussman, Eugene Charniak and Terry Winograd. It was used to implement Winograd's natural-language understanding program SHRDLU, which was a landmark at that time. To cope with the very limited memory systems at the time, Planner used a backtracking control structure so that only one possible computation path had to be stored at a time. Planner gave rise to the programming languages QA-4, Popler, Conniver, QLISP, and the concurrent language Ether.Wikipedia:Citation needed

Hayes and Kowalski in Edinburgh tried to reconcile the logic-based declarative approach to knowledge representation with Planner's procedural approach. Hayes (1973) developed an equational language, Golux, in which different procedures could be obtained by altering the behavior of the theorem prover.[646] Kowalski, on the other hand, developed SLD resolution,[647] a variant of SL-resolution,[648] and showed how it treats implications as goal-reduction procedures. Kowalski collaborated with Colmerauer in Marseille, who developed these ideas in the design and implementation of the programming language Prolog.

The Association for Logic Programming was founded to promote Logic Programming in 1986.

Prolog gave rise to the programming languages ALF, Fril, Gödel, Mercury, Oz, Ciao, Visual Prolog, XSB, and λProlog, as well as a variety of concurrent logic programming languages,[649] constraint logic programming languages and datalog.Wikipedia:Citation needed

Concepts

Logic and control

Logic programming can be viewed as controlled deduction. An important concept in logic programming is the separation of programs into their logic component and their control component. With pure logic programming languages, the logic component alone determines the solutions produced. The control component can be varied to provide alternative ways of executing a logic program. This notion is captured by the slogan

Algorithm = Logic + Control

where "Logic" represents a logic program and "Control" represents different theorem-proving strategies.

Problem solving

In the simplified, propositional case in which a logic program and a top-level atomic goal contain no variables, backward reasoning determines an and-or tree, which constitutes the search space for solving the goal. The top-level goal is the root of the tree. Given any node in the tree and any clause whose head matches the node, there exists a set of child nodes corresponding to the sub-goals in the body of the clause. These child nodes are grouped together by an "and". The alternative sets of children corresponding to alternative ways of solving the node are grouped together by an "or".

Any search strategy can be used to search this space. Prolog uses a sequential, last-in-first-out, backtracking strategy, in which only one alternative and one sub-goal is considered at a time. Other search strategies, such as parallel search, intelligent backtracking, or best-first search to find an optimal solution, are also possible.

In the more general case, where sub-goals share variables, other strategies can be used, such as choosing the subgoal that is most highly instantiated or that is sufficiently instantiated so that only one procedure applies. Such strategies are used, for example, in concurrent logic programming.

Negation as failure

For most practical applications, as well as for applications that require non-monotonic reasoning in artificial intelligence, Horn clause logic programs need to be extended to normal logic programs, with negative conditions. A *clause* in a normal logic program has the form:

$$H \text{ :- } A_1, \ldots, A_n, \text{ not } B_1, \ldots, \text{ not } B_n.$$

and is read declaratively as a logical implication:

$$H \text{ if } A_1 \text{ and } \ldots \text{ and } A_n \text{ and not } B_1 \text{ and } \ldots \text{ and not } B_n.$$

where H and all the A_i and B_i are atomic formulas. The negation in the negative literals not B_i is commonly referred to as "negation as failure", because in most implementations, a negative condition not B_i is shown to hold by showing that the positive condition B_i fails to hold. For example:

```
canfly(X) :- bird(X), not abnormal(X).
abnormal(X) :-  wounded(X).
bird(john).
bird(mary).
wounded(john).
```

Given the goal of finding something that can fly:

```
:- canfly(X).
```

there are two candidate solutions, which solve the first subgoal bird(X), namely X = john and X = mary. The second subgoal not abnormal(john) of the first candidate solution fails, because wounded(john) succeeds and therefore abnormal(john) succeeds. However, The second subgoal not abnormal(mary) of the second candidate solution succeeds, because wounded(mary) fails and therefore abnormal(mary) fails. Therefore, X = mary is the only solution of the goal.

Micro-Planner had a construct, called "thnot", which when applied to an expression returns the value true if (and only if) the evaluation of the expression fails. An equivalent operator is normally built-in in modern Prolog's implementations. It is normally written as not(*Goal*) or \+ *Goal*, where *Goal* is some goal (proposition) to be proved by the program. This operator differs from negation in first-order logic: a negation such as \+ X == 1 fails when the variable X has been bound to the atom 1, but it succeeds in all other cases, including when X is unbound. This makes Prolog's reasoning non-monotonic: X = 1, \+ X == 1 always fails, while \+ X

`== 1, X = 1` can succeed, binding X to 1, depending on whether X was initially bound (note that standard Prolog executes goals in left-to-right order).

The logical status of negation as failure was unresolved until Keith Clark [1978] showed that, under certain natural conditions, it is a correct (and sometimes complete) implementation of classical negation with respect to the completion of the program. Completion amounts roughly to regarding the set of all the program clauses with the same predicate on the left hand side, say

```
H :- Body1.
 ...
H :- Bodyk.
```

as a definition of the predicate

```
H iff (Body1 or ... or Bodyk)
```

where "iff" means "if and only if". Writing the completion also requires explicit use of the equality predicate and the inclusion of a set of appropriate axioms for equality. However, the implementation of negation by failure needs only the if-halves of the definitions without the axioms of equality.

For example, the completion of the program above is:

```
canfly(X) iff bird(X), not abnormal(X).
abnormal(X) iff wounded(X).
 bird(X) iff X = john or X = mary.
 X = X.
 not john = mary.
 not mary = john.
```

The notion of completion is closely related to McCarthy's circumscription semantics for default reasoning, and to the closed world assumption.

As an alternative to the completion semantics, negation as failure can also be interpreted epistemically, as in the stable model semantics of answer set programming. In this interpretation not(B_i) means literally that B_i is not known or not believed. The epistemic interpretation has the advantage that it can be combined very simply with classical negation, as in "extended logic programming", to formalise such phrases as "the contrary can not be shown", where "contrary" is classical negation and "can not be shown" is the epistemic interpretation of negation as failure.

Knowledge representation

The fact that Horn clauses can be given a procedural interpretation and, vice versa, that goal-reduction procedures can be understood as Horn clauses + backward reasoning means that logic programs combine declarative and procedural representations of knowledge. The inclusion of negation as failure means that logic programming is a kind of non-monotonic logic.

Despite its simplicity compared with classical logic, this combination of Horn clauses and negation as failure has proved to be surprisingly expressive. For example, it provides a natural representation for the common-sense laws of cause and effect, as formalised by both the situation calculus and event calculus. It has also been shown to correspond quite naturally to the semi-formal language of legislation. In particular, Prakken and Sartor[650] credit the representation of the British Nationality Act as a logic program[651] with being "hugely influential for the development of computational representations of legislation, showing how logic programming enables intuitively appealing representations that can be directly deployed to generate automatic inferences".

Variants and extensions

Prolog

The programming language Prolog was developed in 1972 by Alain Colmerauer. It emerged from a collaboration between Colmerauer in Marseille and Robert Kowalski in Edinburgh. Colmerauer was working on natural language understanding, using logic to represent semantics and using resolution for question-answering. During the summer of 1971, Colmerauer and Kowalski discovered that the clausal form of logic could be used to represent formal grammars and that resolution theorem provers could be used for parsing. They observed that some theorem provers, like hyper-resolution, behave as bottom-up parsers and others, like SL-resolution (1971), behave as top-down parsers.

It was in the following summer of 1972, that Kowalski, again working with Colmerauer, developed the procedural interpretation of implications. This dual declarative/procedural interpretation later became formalised in the Prolog notation

$$H \text{ :- } B_1, \ldots, B_n.$$

which can be read (and used) both declaratively and procedurally. It also became clear that such clauses could be restricted to definite clauses or Horn clauses, where H, B_1, ..., B_n are all atomic predicate logic formulae, and that SL-resolution could be restricted (and generalised) to LUSH or SLD-resolution. Kowalski's procedural interpretation and LUSH were described in a 1973 memo, published in 1974.

Colmerauer, with Philippe Roussel, used this dual interpretation of clauses as the basis of Prolog, which was implemented in the summer and autumn of 1972. The first Prolog program, also written in 1972 and implemented in Marseille, was a French question-answering system. The use of Prolog as a practical programming language was given great momentum by the development of a compiler by David Warren in Edinburgh in 1977. Experiments demonstrated that Edinburgh Prolog could compete with the processing speed of other symbolic programming languages such as Lisp. Edinburgh Prolog became the *de facto* standard and strongly influenced the definition of ISO standard Prolog.

Abductive logic programming

Abductive logic programming is an extension of normal Logic Programming that allows some predicates, declared as abducible predicates, to be "open" or undefined. A clause in an abductive logic program has the form:

$$H :- B_1, \ldots, B_n, A_1, \ldots, A_n.$$

where H is an atomic formula that is not abducible, all the B_i are literals whose predicates are not abducible, and the A_i are atomic formulas whose predicates are abducible. The abducible predicates can be constrained by integrity constraints, which can have the form:

$$false :- B_1, \ldots, B_n.$$

where the B_i are arbitrary literals (defined or abducible, and atomic or negated). For example:

```
canfly(X) :- bird(X), normal(X).
false :-  normal(X), wounded(X).
bird(john).
bird(mary).
wounded(john).
```

where the predicate `normal` is abducible.

Problem solving is achieved by deriving hypotheses expressed in terms of the abducible predicates as solutions of problems to be solved. These problems can be either observations that need to be explained (as in classical abductive reasoning) or goals to be solved (as in normal logic programming). For example, the hypothesis `normal(mary)` explains the observation `canfly(mary)`. Moreover, the same hypothesis entails the only solution `X = mary` of the goal of finding something that can fly:

```
:- canfly(X).
```

Abductive logic programming has been used for fault diagnosis, planning, natural language processing and machine learning. It has also been used to interpret Negation as Failure as a form of abductive reasoning.

Metalogic programming

Because mathematical logic has a long tradition of distinguishing between object language and metalanguage, logic programming also allows metalevel programming. The simplest metalogic program is the so-called "vanilla" meta-interpreter:

```
    solve(true).
    solve((A,B)):- solve(A),solve(B).
    solve(A):- clause(A,B),solve(B).
```

where true represents an empty conjunction, and clause(A,B) means there is an object-level clause of the form A :- B.

Metalogic programming allows object-level and metalevel representations to be combined, as in natural language. It can also be used to implement any logic that is specified by means of inference rules. Metalogic is used in logic programming to implement metaprograms, which manipulate other programs, databases, knowledge bases or axiomatic theories as data.

Constraint logic programming

Constraint logic programming combines Horn clause logic programming with constraint solving. It extends Horn clauses by allowing some predicates, declared as constraints predicates, to occur as literals in the body of clauses. A constraint logic program is a set of clauses of the form:

$$H \ :- \ C_1, \ \ldots, \ C_n \ \Diamond \ B_1, \ \ldots, \ B_n.$$

where H and all the B_i are atomic formulas, and the C_i are constraints. Declaratively, such clauses are read as ordinary logical implications:

$$H \text{ if } C_1 \text{ and } \ldots \text{ and } C_n \text{ and } B_1 \text{ and } \ldots \text{ and } B_n.$$

However, whereas the predicates in the heads of clauses are defined by the constraint logic program, the predicates in the constraints are predefined by some domain-specific model-theoretic structure or theory.

Procedurally, subgoals whose predicates are defined by the program are solved by goal-reduction, as in ordinary logic programming, but constraints are checked for satisfiability by a domain-specific constraint-solver, which implements the semantics of the constraint predicates. An initial problem is solved by reducing it to a satisfiable conjunction of constraints.

The following constraint logic program represents a toy temporal database of `john's` history as a teacher:

```
teaches(john, hardware, T) :- 1990 ≤ T, T < 1999.
teaches(john, software, T) :- 1999 ≤ T, T < 2005.
teaches(john, logic, T) :- 2005 ≤ T, T ≤ 2012.
rank(john, instructor, T) :- 1990 ≤ T, T < 2010.
rank(john, professor, T) :- 2010 ≤ T, T < 2014.
```

Here $\leq$ and $<$ are constraint predicates, with their usual intended semantics. The following goal clause queries the database to find out when `john` both taught `logic` and was a `professor`:

```
:- teaches(john, logic, T), rank(john, pro-
fessor, T).
```

The solution is `2010 ≤ T, T ≤ 2012.`

Constraint logic programming has been used to solve problems in such fields as civil engineering, mechanical engineering, digital circuit verification, automated timetabling, air traffic control, and finance. It is closely related to abductive logic programming.

Concurrent logic programming

Concurrent logic programming integrates concepts of logic programming with concurrent programming. Its development was given a big impetus in the 1980s by its choice for the systems programming language of the Japanese Fifth Generation Project (FGCS).[652]

A concurrent logic program is a set of guarded Horn clauses of the form:

$$H \text{ :- } G_1, \ldots, G_n \mid B_1, \ldots, B_n.$$

The conjunction $G_1, \ldots, G_n$ is called the guard of the clause, and $\mid$ is the commitment operator. Declaratively, guarded Horn clauses are read as ordinary logical implications:

$$H \text{ if } G_1 \text{ and } \ldots \text{ and } G_n \text{ and } B_1 \text{ and } \ldots \text{ and } B_n.$$

However, procedurally, when there are several clauses whose heads H match a given goal, then all of the clauses are executed in parallel, checking whether their guards $G_1, \ldots, G_n$ hold. If the guards of more than one clause hold, then a committed choice is made to one of the clauses, and execution proceedes with the subgoals $B_1, \ldots, B_n$ of the chosen clause. These subgoals can also be executed in parallel. Thus concurrent logic programming implements a form of "don't care nondeterminism", rather than "don't know nondeterminism".

For example, the following concurrent logic program defines a predicate `shuffle(Left, Right, Merge)` , which can be used to shuffle two lists `Left` and `Right`, combining them into a single list `Merge` that preserves the ordering of the two lists `Left` and `Right`:

```
shuffle([], [], []).
shuffle(Left, Right, Merge) :-
    Left = [First | Rest] |
    Merge = [First | ShortMerge],
    shuffle(Rest, Right, ShortMerge).
shuffle(Left, Right, Merge) :-
    Right = [First | Rest] |
    Merge = [First | ShortMerge],
    shuffle(Left, Rest, ShortMerge).
```

Here, [] represents the empty list, and [`Head | Tail`] represents a list with first element `Head` followed by list `Tail`, as in Prolog. (Notice that the first occurrence of | in the second and third clauses is the list constructor, whereas the second occurrence of | is the commitment operator.) The program can be used, for example, to shuffle the lists [`ace, queen, king`] and [`1, 4, 2`] by invoking the goal clause:

```
shuffle([ace, queen, king], [1, 4, 2], Merge).
```

The program will non-deterministically generate a single solution, for example `Merge = [ace, queen, 1, king, 4, 2]`.

Arguably, concurrent logic programming is based on message passing and consequently is subject to the same indeterminacy as other concurrent message-passing systems, such as Actors (see Indeterminacy in concurrent computation). Carl Hewitt has argued that, concurrent logic programming is not based on logic in his sense that computational steps cannot be logically deduced. However, in concurrent logic programming, any result of a terminating computation is a logical consequence of the program, and any partial result of a partial computation is a logical consequence of the program and the residual goal (process network). Consequently, the indeterminacy of computations implies that not all logical consequences of the program can be deduced.Wikipedia:Neutral point of viewTalk:Logic programming#Hewitt

Concurrent constraint logic programming

Concurrent constraint logic programming combines concurrent logic programming and constraint logic programming, using constraints to control concurrency. A clause can contain a guard, which is a set of constraints that may block the applicability of the clause. When the guards of several clauses are satisfied, concurrent constraint logic programming makes a committed choice to the use of only one.

Inductive logic programming

Inductive logic programming is concerned with generalizing positive and negative examples in the context of background knowledge: machine learning of logic programs. Recent work in this area, combining logic programming, learning and probability, has given rise to the new field of statistical relational learning and probabilistic inductive logic programming.

Higher-order logic programming

Several researchers have extended logic programming with higher-order programming features derived from higher-order logic, such as predicate variables. Such languages include the Prolog extensions HiLog and λProlog.

Linear logic programming

Basing logic programming within linear logic has resulted in the design of logic programming languages that are considerably more expressive than those based on classical logic. Horn clause programs can only represent state change by the change in arguments to predicates. In linear logic programming, one can use the ambient linear logic to support state change. Some early designs of logic programming languages based on linear logic include LO [Andreoli & Pareschi, 1991], Lolli,[653] ACL,[654] and Forum [Miller, 1996]. Forum provides a goal-directed interpretation of all of linear logic.

Object-oriented logic programming

F-logic extends logic programming with objects and the frame syntax.

Logtalk extends the Prolog programming language with support for objects, protocols, and other OOP concepts. Highly portable, it supports most standard-compliant Prolog systems as backend compilers.

Transaction logic programming

Transaction logic is an extension of logic programming with a logical theory of state-modifying updates. It has both a model-theoretic semantics and a procedural one. An implementation of a subset of Transaction logic is available in the Flora-2 system. Other prototypes are also available.

References

General introductions

- Baral, C.; Gelfond, M. (1994). "Logic programming and knowledge representation"[655] (PDF). *The Journal of Logic Programming*. 19–20: 73–148. doi: 10.1016/0743-1066(94)90025-6[656].
- Robert Kowalski. **The Early Years of Logic Programming**[657] Kowalski, R. A. (1988). "The early years of logic programming"[657] (PDF). *Communications of the ACM*. **31**: 38–43. doi: 10.1145/35043.35046[658].
- Lloyd, J. W. (1987). *Foundations of Logic Programming. (2nd edition).* Springer-Verlag.

Other sources

- John McCarthy. **Programs with common sense** Symposium on Mechanization of Thought Processes. National Physical Laboratory. Teddington, England. 1958.
- D. Miller, G. Nadathur, F. Pfenning, A. Scedrov. **Uniform proofs as a foundation for logic programming**[659], Annals of Pure and Applied Logic, vol. 51, pp 125–157, 1991.
- Ehud Shapiro (Editor). **Concurrent Prolog** MIT Press. 1987.
- James Slagle. **Experiments with a Deductive Question-Answering Program** CACM. December 1965.

Further reading

- Carl Hewitt. **Procedural Embedding of Knowledge In Planner**[660] IJCAI 1971.
- Carl Hewitt. **The repeated demise of logic programming and why it will be reincarnated**[661]
- Evgeny Dantsin, Thomas Eiter, Georg Gottlob, Andrei Voronkov: Complexity and expressive power of logic programming[662]. ACM Comput. Surv. 33(3): 374–425 (2001)
- Ulf Nilsson and Jan Maluszynski, Logic, Programming and Prolog[663]

External links

Wikimedia Commons has media related to ***Logic programming***.

- *Logic Programming* Virtual Library entry[664]
- Bibliographies on Logic Programming[665]
- Association for Logic Programming (ALP)[666]
- Theory and Practice of Logic Programming[667] journal
- Logic programming in C++ with Castor[668]
- Logic programming in[669] Oz
- Prolog Development Center[670]
- Racklog: Logic Programming in Racket[671]

Bayesian network

Part of a series on Statistics

Bayesian statistics

Theory

- Admissible decision rule
- Bayesian efficiency
- Bayesian probability
- Probability interpretations
- Bayes' theorem
- Bayes factor
- Bayesian inference
- Bayesian network
- Prior
- Posterior
- Likelihood
- Conjugate prior
- Posterior predictive
- Hyperparameter
- Hyperprior
- Principle of indifference
- Principle of maximum entropy
- Empirical Bayes method
- Cromwell's rule

- Bernstein–von Mises theorem
- Schwarz criterion
- Credible interval
- Maximum a posteriori estimation
- Radical probabilism

Techniques

- Bayesian linear regression
- Bayesian estimator
- Approximate Bayesian computation

- **Statistics portal**

- v
- t
- e[672]

A **Bayesian network**, **Bayes network**, **belief network**, **Bayes(ian) model** or **probabilistic directed acyclic graphical model** is a probabilistic graphical model (a type of statistical model) that represents a set of variables and their conditional dependencies via a directed acyclic graph (DAG). For example, a Bayesian network could represent the probabilistic relationships between diseases and symptoms. Given symptoms, the network can be used to compute the probabilities of the presence of various diseases.

Formally, Bayesian networks are DAGs whose nodes represent variables in the Bayesian sense: they may be observable quantities, latent variables, unknown parameters or hypotheses. Edges represent conditional dependencies; nodes that are not connected (there is no path from one of the variables to the other in the Bayesian network) represent variables that are conditionally independent of each other. Each node is associated with a probability function that takes, as input, a particular set of values for the node's parent variables, and gives (as output) the probability (or probability distribution, if applicable) of the variable represented by the node. For example, if m parent nodes represent m Boolean variables then the probability function could be represented by a table of 2^m entries, one entry for each of the 2^m possible combinations of its parents being true or false. Similar ideas may be applied to undirected, and possibly cyclic, graphs; such as Markov networks.

Efficient algorithms exist that perform inference and learning in Bayesian networks. Bayesian networks that model sequences of variables (*e.g.* speech signals or protein sequences) are called dynamic Bayesian networks. Generalizations of Bayesian networks that can represent and solve decision problems under uncertainty are called influence diagrams (also known as Bayesian Decision Networks).

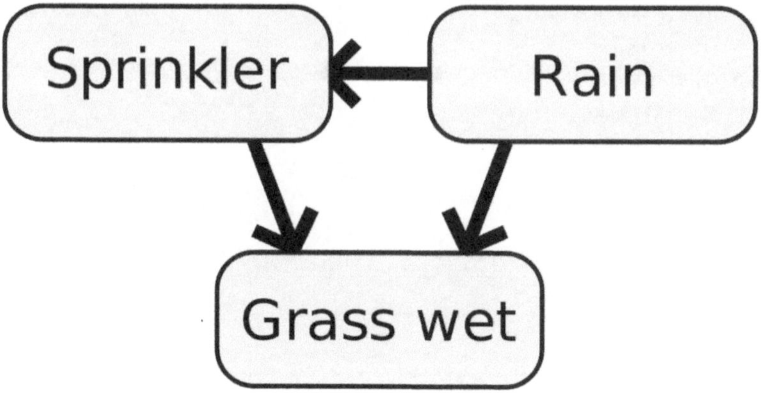

Figure 24: *A simple Bayesian network. Rain influences whether the sprinkler is activated, and both rain and the sprinkler influence whether the grass is wet.*

RAIN	SPRINKLER T	SPRINKLER F
F	0.4	0.6
T	0.01	0.99

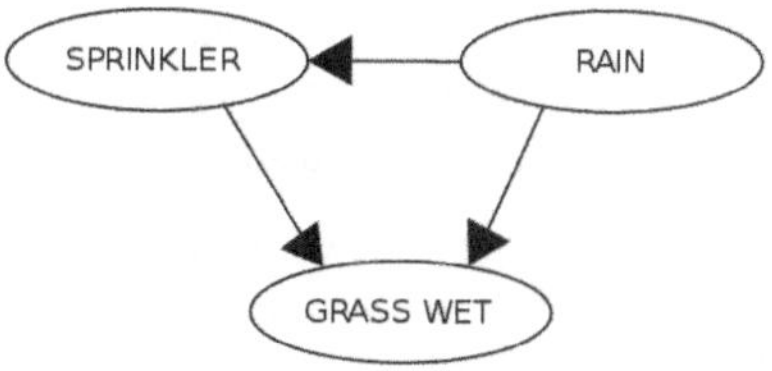

RAIN T	RAIN F
0.2	0.8

SPRINKLER	RAIN	GRASS WET T	GRASS WET F
F	F	0.0	1.0
F	T	0.8	0.2
T	F	0.9	0.1
T	T	0.99	0.01

Figure 25: *A simple Bayesian network with conditional probability tables*

Example

Suppose that there are two events which could cause grass to be wet: either the sprinkler is on or it's raining. Also, suppose that the rain has a direct effect on the use of the sprinkler (namely that when it rains, the sprinkler is usually not turned on). Then the situation can be modeled with a Bayesian network (shown to the right). All three variables have two possible values, T (for true) and F (for false).

The joint probability function is:

$$\Pr(G,S,R) = \Pr(G|S,R)\Pr(S|R)\Pr(R)$$

where the names of the variables have been abbreviated to G = *Grass wet (true/false)*, S = *Sprinkler turned on (true/false)*, and R = *Raining (true/false)*.

The model can answer questions like "What is the probability that it is raining, given the grass is wet?" by using the conditional probability formula and summing over all nuisance variables:

$$\Pr(R=T|G=T) = \frac{\Pr(G=T,R=T)}{\Pr(G=T)} = \frac{\sum_{S\in\{T,F\}}\Pr(G=T,S,R=T)}{\sum_{S,R\in\{T,F\}}\Pr(G=T,S,R)}$$

Using the expansion for the joint probability function $\Pr(G,S,R)$ and the conditional probabilities from the conditional probability tables (CPTs) stated in the diagram, one can evaluate each term in the sums in the numerator and denominator. For example,

$$\begin{aligned}\Pr(G=T,S=T,R=T) &= \Pr(G=T|S=T,R=T)\Pr(S=T|R=T)\Pr(R=T)\\ &= 0.99\times 0.01\times 0.2\\ &= 0.00198.\end{aligned}$$

Then the numerical results (subscripted by the associated variable values) are

$$\Pr(R=T|G=T) = \frac{0.00198_{TTT}+0.1584_{TFT}}{0.00198_{TTT}+0.288_{TTF}+0.1584_{TFT}+0.0_{TFF}} = \frac{891}{2491} \approx 35.77\%.$$

If, on the other hand, we wish to answer an interventional question: "What is the probability that it would rain, given that we wet the grass?" the answer would be governed by the post-intervention joint distribution function $\Pr(S,R|\text{do}(G=T)) = \Pr(S|R)P(R)$ obtained by removing the factor $\Pr(G|S,R)$ from the pre-intervention distribution. As expected, the probability of rain is unaffected by the action: $\Pr(R|\text{do}(G=T)) = \Pr(R)$.

If, moreover, we wish to predict the impact of turning the sprinkler on, we have

$$\Pr(R,G|\text{do}(S=T)) = \Pr(R)\Pr(G|R,S=T)$$

with the term $\Pr(S=T|R)$ removed, showing that the action has an effect on the grass but not on the rain.

These predictions may not be feasible when some of the variables are unobserved, as in most policy evaluation problems. The effect of the action $\text{do}(x)$ can still be predicted, however, whenever a criterion called "back-door" is satisfied. It states that, if a set Z of nodes can be observed that d-separates (or blocks) all back-door paths from X to Y then $\Pr(Y,Z|\text{do}(x)) = \Pr(Y,Z,X=x)/\Pr(X=x|Z)$. A back-door path is one that ends with an arrow into X. Sets that satisfy the back-door criterion are called "sufficient" or "admissible." For example, the set $Z = R$ is admissible for predicting the effect of $S = T$ on G, because R d-separate the (only) back-door path $S \leftarrow R \rightarrow G$. However, if S is not observed, there is no other set that d-separates this path and the effect of turning the sprinkler on ($S = T$) on the grass (G) cannot be predicted from passive observations. We then say that $P(G \mid do(S = T))$ is not "identified." This reflects the fact that, lacking interventional data, we cannot determine if the observed dependence between S and G is due to a causal connection or is spurious (apparent dependence arising from a common cause, R). (see Simpson's paradox)

To determine whether a causal relation is identified from an arbitrary Bayesian network with unobserved variables, one can use the three rules of "*do*-calculus" and test whether all *do* terms can be removed from the expression of that relation, thus confirming that the desired quantity is estimable from frequency data.

Using a Bayesian network can save considerable amounts of memory, if the dependencies in the joint distribution are sparse. For example, a naive way of storing the conditional probabilities of 10 two-valued variables as a table requires storage space for $2^{10} = 1024$ values. If the local distributions of no variable depends on more than three parent variables, the Bayesian network representation only needs to store at most $10 \cdot 2^3 = 80$ values.

One advantage of Bayesian networks is that it is intuitively easier for a human to understand (a sparse set of) direct dependencies and local distributions than complete joint distributions.

Inference and learning

There are three main inference tasks for Bayesian networks.

Inferring unobserved variables

Because a Bayesian network is a complete model for the variables and their relationships, it can be used to answer probabilistic queries about them. For example, the network can be used to find out updated knowledge of the state of a subset of variables when other variables (the *evidence* variables) are observed. This process of computing the *posterior* distribution of variables given evidence is called probabilistic inference. The posterior gives a universal sufficient statistic for detection applications, when one wants to choose values for the variable subset which minimize some expected loss function, for instance the probability of decision error. A Bayesian network can thus be considered a mechanism for automatically applying Bayes' theorem to complex problems.

The most common exact inference methods are: variable elimination, which eliminates (by integration or summation) the non-observed non-query variables one by one by distributing the sum over the product; clique tree propagation, which caches the computation so that many variables can be queried at one time and new evidence can be propagated quickly; and recursive conditioning and AND/OR search, which allow for a space–time tradeoff and match the efficiency of variable elimination when enough space is used. All of these methods have complexity that is exponential in the network's treewidth. The most common approximate inference algorithms are importance sampling, stochastic MCMC simulation, mini-bucket elimination, loopy belief propagation, generalized belief propagation, and variational methods.

Parameter learning

In order to fully specify the Bayesian network and thus fully represent the joint probability distribution, it is necessary to specify for each node *X* the probability distribution for *X* conditional upon *X*'s parents. The distribution of *X* conditional upon its parents may have any form. It is common to work with discrete or Gaussian distributions since that simplifies calculations. Sometimes only constraints on a distribution are known; one can then use the principle of maximum entropy to determine a single distribution, the one with the greatest entropy given the constraints. (Analogously, in the specific context of a dynamic Bayesian network, one commonly specifies the conditional distribution for the hidden state's temporal evolution to maximize the entropy rate of the implied stochastic process.)

Often these conditional distributions include parameters which are unknown and must be estimated from data, sometimes using the maximum likelihood approach. Direct maximization of the likelihood (or of the posterior probability) is often complex when there are unobserved variables. A classical approach to this problem is the expectation-maximization algorithm which alternates computing expected values of the unobserved variables conditional on observed data, with maximizing the complete likelihood (or posterior) assuming that previously computed expected values are correct. Under mild regularity conditions this process converges on maximum likelihood (or maximum posterior) values for parameters.

A more fully Bayesian approach to parameters is to treat parameters as additional unobserved variables and to compute a full posterior distribution over all nodes conditional upon observed data, then to integrate out the parameters. This approach can be expensive and lead to large dimension models, so in practice classical parameter-setting approaches are more common.

Structure learning

In the simplest case, a Bayesian network is specified by an expert and is then used to perform inference. In other applications the task of defining the network is too complex for humans. In this case the network structure and the parameters of the local distributions must be learned from data.

Automatically learning the graph structure of a Bayesian network (BN) is a challenge pursued within machine learning. The basic idea goes back to a recovery algorithm developed by Rebane and Pearl (1987) and rests on the distinction between the three possible types of adjacent triplets allowed in a directed acyclic graph (DAG):

1. $X \rightarrow Y \rightarrow Z$
2. $X \leftarrow Y \rightarrow Z$
3. $X \rightarrow Y \leftarrow Z$

Type 1 and type 2 represent the same dependencies (X and Z are independent given Y) and are, therefore, indistinguishable. Type 3, however, can be uniquely identified, since X and Z are marginally independent and all other pairs are dependent. Thus, while the *skeletons* (the graphs stripped of arrows) of these three triplets are identical, the directionality of the arrows is partially identifiable. The same distinction applies when X and Z have common parents, except that one must first condition on those parents. Algorithms have been developed to systematically determine the skeleton of the underlying graph and, then, orient all arrows whose directionality is dictated by the conditional independencies observed.

An alternative method of structural learning uses optimization based search. It requires a scoring function and a search strategy. A common scoring function is posterior probability of the structure given the training data, like the BIC or the BDeu. The time requirement of an exhaustive search returning a structure that maximizes the score is superexponential in the number of variables. A local search strategy makes incremental changes aimed at improving the score of the structure. A global search algorithm like Markov chain Monte Carlo can avoid getting trapped in local minima. Friedman et al. discuss using mutual information between variables and finding a structure that maximizes this. They do this by restricting the parent candidate set to *k* nodes and exhaustively searching therein.

A particularly fast method for exact BN learning is to cast the problem as an optimization problem, and solve it using integer programming. Acyclicity constraints are added to the integer program (IP) during solving in the form of cutting planes. Such method can handle problems with up to 100 variables.

In order to deal with problems with thousands of variables, it is necessary to use a different approach. One is to first sample one ordering, and then find the optimal BN structure with respect to that ordering. This implies working on the search space of the possible orderings, which is convenient as it is smaller than the space of network structures. Multiple orderings are then sampled and evaluated. This method has been proven to be the best available in literature when the number of variables is huge.

Another method consists of focusing on the sub-class of decomposable models, for which the MLE have a closed form. It is then possible to discover a consistent structure for hundreds of variables.

As previously noted, learning Bayesian networks with bounded treewidth is necessary to allow exact tractable inference, since the worst-case inference complexity is exponential in the treewidth k (under the exponential time hypothesis). Yet, being a global property of the graph, it considerably increases the difficulty of the learning process. In this context it is possible to use the concept of K-tree for effective learning.[673]

Statistical introduction

Given data x and parameter θ , a simple Bayesian analysis starts with a prior probability (*prior*) $p(\theta)$ and likelihood $p(x \mid \theta)$ to compute a posterior probability $p(\theta \mid x) \propto p(x \mid \theta)p(\theta)$.

Often the prior on θ depends in turn on other parameters φ that are not mentioned in the likelihood. So, the prior $p(\theta)$ must be replaced by a likelihood $p(\theta \mid \varphi)$, and a prior $p(\varphi)$ on the newly introduced parameters φ is required, resulting in a posterior probability

$$p(\theta, \varphi|x) \propto p(x|\theta)p(\theta|\varphi)p(\varphi).$$

This is the simplest example of a *hierarchical Bayes model*.Wikipedia:Please clarify

The process may be repeated; for example, the parameters φ may depend in turn on additional parameters ψ , which will require their own prior. Eventually the process must terminate, with priors that do not depend on any other unmentioned parameters.

Introductory examples

Suppose we have measured the quantities $x_1, \ldots, x_n$ each with normally distributed errors of known standard deviation σ ,

$$x_i \sim N(\theta_i, \sigma^2)$$

Suppose we are interested in estimating the θ_i . An approach would be to estimate the θ_i using a maximum likelihood approach; since the observations are independent, the likelihood factorizes and the maximum likelihood estimate is simply

$$\theta_i = x_i$$

However, if the quantities are related, so that for example we may think that the individual θ_i have themselves been drawn from an underlying distribution, then this relationship destroys the independence and suggests a more complex model, e.g.,

$$x_i \sim N(\theta_i, \sigma^2),$$

$$\theta_i \sim N(\varphi, \tau^2)$$

with improper priors $\varphi \sim$ flat, $\tau \sim$ flat $\in (0, \infty)$. When $n \geq 3$, this is an *identified model* (i.e. there exists a unique solution for the model's parameters), and the posterior distributions of the individual θ_i will tend to move, or *shrink* away from the maximum likelihood estimates towards their common mean. This *shrinkage* is a typical behavior in hierarchical Bayes models.

Restrictions on priors

Some care is needed when choosing priors in a hierarchical model, particularly on scale variables at higher levels of the hierarchy such as the variable τ in the example. The usual priors such as the Jeffreys prior often do not work, because the posterior distribution will be improper (not normalizable), and estimates made by minimizing the expected loss will be inadmissible.

Definitions and concepts

There are several equivalent definitions of a Bayesian network. For all the following, let $G = (V,E)$ be a directed acyclic graph (or DAG), and let $X = (X_v)$, $v \in V$ be a set of random variables indexed by V.

Factorization definition

X is a Bayesian network with respect to *G* if its joint probability density function (with respect to a product measure) can be written as a product of the individual density functions, conditional on their parent variables:[674]

$$p(x) = \prod_{v \in V} p\left(x_v \,\middle|\, x_{\operatorname{pa}(v)}\right)$$

where pa(*v*) is the set of parents of *v* (i.e. those vertices pointing directly to *v* via a single edge).

For any set of random variables, the probability of any member of a joint distribution can be calculated from conditional probabilities using the chain rule (given a topological ordering of *X*) as follows:[674]
$\mathrm{P}(X_1 = x_1, \ldots, X_n = x_n) = \prod_{v=1}^{n} \mathrm{P}\left(X_v = x_v \mid X_{v+1} = x_{v+1}, \ldots, X_n = x_n\right)$

Compare this with the definition above, which can be written as:

$\mathrm{P}(X_1 = x_1, \ldots, X_n = x_n) = \prod_{v=1}^{n} \mathrm{P}(X_v = x_v \mid X_j = x_j$ for each X_j which is a parent of X_v)

The difference between the two expressions is the conditional independence of the variables from any of their non-descendants, given the values of their parent variables.

Local Markov property

X is a Bayesian network with respect to *G* if it satisfies the *local Markov property*: each variable is conditionally independent of its non-descendants given its parent variables:[675]

$$X_v \perp\!\!\!\perp X_{V \setminus \operatorname{de}(v)} \mid X_{\operatorname{pa}(v)} \quad \text{for all } v \in V$$

where de(*v*) is the set of descendants and *V* \ de(*v*) is the set of non-descendants of *v*.

This can also be expressed in terms similar to the first definition, as

$\mathrm{P}(X_v = x_v \mid X_i = x_i$ for each X_i which is not a descendant of X_v) $= P(X_v = x_v \mid X_j = x_j$ for each X_j which is a parent of X_v)

Note that the set of parents is a subset of the set of non-descendants because the graph is acyclic.

Developing Bayesian networks

To develop a Bayesian network, we often first develop a DAG *G* such that we believe *X* satisfies the local Markov property with respect to *G*. Sometimes this is done by creating a causal DAG. We then ascertain the conditional probability distributions of each variable given its parents in *G*. In many cases, in particular in the case where the variables are discrete, if we define the joint distribution of *X* to be the product of these conditional distributions, then *X* is a Bayesian network with respect to *G*.

Markov blanket

The Markov blanket of a node is the set of nodes consisting of its parents, its children, and any other parents of its children. The Markov blanket renders the node independent of the rest of the network; the joint distribution of the variables in the Markov blanket of a node is sufficient knowledge for calculating the distribution of the node. *X* is a Bayesian network with respect to *G* if every node is conditionally independent of all other nodes in the network, given its Markov blanket.[675]

d-separation

This definition can be made more general by defining the "d"-separation of two nodes, where d stands for directional. Let *P* be a trail from node *u* to *v*. A trail is a loop-free, undirected (i.e. all edge directions are ignored) path between two nodes. Then *P* is said to be *d*-separated by a set of nodes *Z* if any of the following conditions holds:

1. *P* contains (but does not need to be entirely) a directed *chain*, $u \ldots \leftarrow m \leftarrow \ldots v$ or $u \ldots \rightarrow m \rightarrow \ldots v$, such that the middle node *m* is in *Z*,
2. *P* contains a *fork*, $u \ldots \leftarrow m \rightarrow \ldots v$, such that the middle node *m* is in *Z*, or
3. *P* contains an *inverted fork* (or *collider*), $u \ldots \rightarrow m \leftarrow \ldots v$, such that the middle node *m* is **not** in *Z* and no descendant of *m* is in *Z*.

The nodes *u* and *v* are said to be *d*-separated by *Z* if all trails between them are *d*-separated. If *u* and *v* are not d-separated, they are called d-connected.

X is a Bayesian network with respect to *G* if, for any two nodes *u*, *v*:

$$X_u \perp\!\!\!\perp X_v \mid X_Z$$

where *Z* is a set which *d*-separates *u* and *v*. (The Markov blanket is the minimal set of nodes which *d*-separates node *v* from all other nodes.)

Hierarchical models

The term *hierarchical model* is sometimes considered a particular type of Bayesian network, but has no formal definition. Sometimes the term is reserved for models with three or more levels of random variables; other times, it is reserved for models with latent variables. In general, however, any moderately complex Bayesian network is usually termed "hierarchical".

Causal networks

Although Bayesian networks are often used to represent causal relationships, this need not be the case: a directed edge from u to v does not require that X_v is causally dependent on X_u. This is demonstrated by the fact that Bayesian networks on the graphs:

$$a \rightarrow b \rightarrow c \qquad \text{and} \qquad a \leftarrow b \leftarrow c$$

are equivalent: that is they impose exactly the same conditional independence requirements.

A **causal network** is a Bayesian network with an explicit requirement that the relationships be causal. The additional semantics of the causal networks specify that if a node X is actively caused to be in a given state x (an action written as do($X = x$)), then the probability density function changes to the one of the network obtained by cutting the links from the parents of X to X, and setting X to the caused value x. Using these semantics, one can predict the impact of external interventions from data obtained prior to intervention.

Inference complexity and approximation algorithms

In 1990, while working at Stanford University on large bioinformatic applications, Greg Cooper proved that exact inference in Bayesian networks is NP-hard. This result prompted a surge in research on approximation algorithms with the aim of developing a tractable approximation to probabilistic inference. In 1993, Paul Dagum and Michael Luby proved two surprising results on the complexity of approximation of probabilistic inference in Bayesian networks. First, they proved that there is no tractable *deterministic algorithm* that can approximate probabilistic inference to within an absolute error $\varepsilon < 1/2$. Second, they proved that there is no tractable *randomized algorithm* that can approximate probabilistic inference to within an absolute error $\varepsilon < 1/2$ with confidence probability greater than $1/2$.

At about the same time, Dan Roth proved that exact inference in Bayesian networks is in fact #P-complete (and thus as hard as counting the number of satisfying assignments of a CNF formula) and that approximate inference within a factor $2^n 1-\varepsilon$ for every $\varepsilon > 0$, even for Bayesian networks with restricted architecture, is NP-hard.[676,677]

In practical terms, these complexity results suggested that while Bayesian networks were rich representations for AI and machine learning applications, their use in large real-world applications would need to be tempered by either topological structural constraints, such as naïve Bayes networks, or by restrictions on the conditional probabilities. The *bounded variance algorithm* was the first provable fast approximation algorithm to efficiently approximate probabilistic inference in Bayesian networks with guarantees on the error approximation. This powerful algorithm required the minor restriction on the conditional probabilities of the Bayesian network to be bounded away from zero and one by $1/p(n)$ where $p(n)$ was any polynomial on the number of nodes in the network n.

Applications

Bayesian networks are used for modelling beliefs in computational biology and bioinformatics (gene regulatory networks, protein structure, gene expression analysis, learning epistasis from GWAS data sets) medicine, forensics, biomonitoring, document classification, information retrieval, semantic search,[678] image processing, data fusion, decision support systems, engineering, sports betting, gaming, property market, geophysics and volcano monitoring, law, study design and risk analysis. There are texts applying Bayesian networks to bioinformatics and financial and marketing informatics.

Software

- Just another Gibbs sampler (JAGS) - Open-source alternative to WinBUGS. Uses Gibbs sampling.
- OpenBUGS - Open-source development of WinBUGS.
- SPSS Modeler - Commercial software that includes an implementation for Bayesian networks.
- Stan (software) - Stan is an open-source package for obtaining Bayesian inference using the No-U-Turn sampler, a variant of Hamiltonian Monte Carlo.
- WinBUGS - One of the first computational implementations of MCMC samplers. No longer maintained.

History

The term "Bayesian networks" was coined by Judea Pearl in 1985 to emphasize three aspects:

1. The often subjective nature of the input information.
2. The reliance on Bayes' conditioning as the basis for updating information.
3. The distinction between causal and evidential modes of reasoning, which underscores Thomas Bayes' posthumously published paper of 1763.

In the late 1980s Judea Pearl's text *Probabilistic Reasoning in Intelligent Systems* and Richard E. Neapolitan's text *Probabilistic Reasoning in Expert Systems* summarized the properties of Bayesian networks and established Bayesian networks as a field of study.

References

<templatestyles src="Template:Refbegin/styles.css" />

- Ben Gal I (2007). "Bayesian Networks". In Ruggeri F, Kennett RS, Faltin FW. *Encyclopedia of Statistics in Quality and Reliability*[679] (PDF). John Wiley & Sons. doi: 10.1002/9780470061572.eqr089[680]. ISBN 978-0-470-01861-3.
- Bertsch McGrayne S. *The Theory That Would not Die*. New Haven: Yale University Press.
- Borgelt C, Kruse R (March 2002). *Graphical Models: Methods for Data Analysis and Mining*[681]. Chichester, UK: Wiley. ISBN 0-470-84337-3.
- Borsuk ME (2008). "Ecological informatics: Bayesian networks". In Jørgensen, Sven Erik, Fath, Brian. *Encyclopedia of Ecology*. Elsevier. ISBN 978-0-444-52033-3.
- Castillo E, Gutiérrez JM, Hadi AS (1997). "Learning Bayesian Networks". *Expert Systems and Probabilistic Network Models*. Monographs in computer science. New York: Springer-Verlag. pp. 481–528. ISBN 0-387-94858-9.
- Comley JW, Dowe DL (June 2003). "General Bayesian networks and asymmetric languages"[682]. *Proceedings of the 2nd Hawaii International Conference on Statistics and Related Fields*. Hawaii.
- Comley JW, Dowe DL (2005). "Minimum Message Length and Generalized Bayesian Nets with Asymmetric Languages"[683]. In Grünwald PD, Myung IJ, Pitt MA. *Advances in Minimum Description Length: Theory and Applications*. Neural information processing series. Cambridge, Massachusetts: Bradford Books (MIT Press) (published April 2005). pp. 265–294. ISBN 0-262-07262-9. (This paper puts decision trees in internal nodes of Bayes networks using Minimum Message Length[684]

(MML). Camera-ready final version submitted on 15 October 2003. An earlier version is Comley and Dowe (2003)[682], .pdf[685].)

- Darwiche A (2009). *Modeling and Reasoning with Bayesian Networks*[686]. Cambridge University Press. ISBN 978-0521884389.
- Dowe, David L. (2010). MML, hybrid Bayesian network graphical models, statistical consistency, invariance and uniqueness[687], in Handbook of Philosophy of Science (Volume 7: Handbook of Philosophy of Statistics), Elsevier,[688] ISBN 978-0-444-51862-0, pp 901–982[687].
- Fenton N, Neil ME (November 2007). "Managing Risk in the Modern World: Applications of Bayesian Networks"[689] (PDF). *A Knowledge Transfer Report from the London Mathematical Society and the Knowledge Transfer Network for Industrial Mathematics*. London (England): London Mathematical Society.
- Fenton N, Neil ME (July 23, 2004). "Combining evidence in risk analysis using Bayesian Networks"[690] (PDF). *Safety Critical Systems Club Newsletter*. **13** (4). Newcastle upon Tyne, England. pp. 8–13. Archived from the original[691] (PDF) on 2007-09-27.
- Gelman A, Carlin JB, Stern HS, Rubin DB (2003). "Part II: Fundamentals of Bayesian Data Analysis: Ch.5 Hierarchical models"[692]. *Bayesian Data Analysis*[693]. CRC Press. pp. 120–. ISBN 978-1-58488-388-3.
- Heckerman, David (March 1, 1995). "Tutorial on Learning with Bayesian Networks"[694]. In Jordan, Michael Irwin. *Learning in Graphical Models*. Adaptive Computation and Machine Learning. Cambridge, Massachusetts: MIT Press (published 1998). pp. 301–354. ISBN 0-262-60032-3.

  Also appears as Heckerman, David (March 1997). "Bayesian Networks for Data Mining". *Data Mining and Knowledge Discovery*. **1** (1): 79–119. doi: 10.1023/A:1009730122752[695].

  An earlier version appears as Technical Report MSR-TR-95-06[696], Microsoft Research, March 1, 1995. The paper is about both parameter and structure learning in Bayesian networks.

- Jensen FV, Nielsen TD (June 6, 2007). *Bayesian Networks and Decision Graphs*. Information Science and Statistics series (2nd ed.). New York: Springer-Verlag. ISBN 978-0-387-68281-5.
- Karimi K, Hamilton HJ (2000). "Finding temporal relations: Causal bayesian networks vs. C4. 5"[697] (PDF). *Twelfth International Symposium on Methodologies for Intelligent Systems*.
- Korb KB, Nicholson AE (December 2010). *Bayesian Artificial Intelligence*. CRC Computer Science & Data Analysis (2nd ed.). Chapman & Hall (CRC Press). doi: 10.1007/s10044-004-0214-5[698]. ISBN 1-58488-387-1.

- Lunn D, Spiegelhalter D, Thomas A, Best N (November 2009). "The BUGS project: Evolution, critique and future directions". *Statistics in Medicine*. **28** (25): 3049–67. doi: 10.1002/sim.3680[699]. PMID 19630097[700].
- Neil M, Fenton N, Tailor M (August 2005). Greenberg, Michael R., ed. "Using Bayesian networks to model expected and unexpected operational losses"[701] (pdf). *Risk Analysis*. **25** (4): 963–72. doi: 10.1111/j.1539-6924.2005.00641.x[702]. PMID 16268944[703].
- Pearl J (September 1986). "Fusion, propagation, and structuring in belief networks". *Artificial Intelligence*. **29** (3): 241–288. doi: 10.1016/0004-3702(86)90072-X[704].
- Pearl J (1988). *Probabilistic Reasoning in Intelligent Systems: Networks of Plausible Inference*. Representation and Reasoning Series (2nd printing ed.). San Francisco, California: Morgan Kaufmann. ISBN 0-934613-73-7.
- Pearl J, Russell S (November 2002). "Bayesian Networks". In Arbib MA. *Handbook of Brain Theory and Neural Networks*. Cambridge, Massachusetts: Bradford Books (MIT Press). pp. 157–160. ISBN 0-262-01197-2.
- Russell, Stuart J.; Norvig, Peter (2003), *Artificial Intelligence: A Modern Approach*[705] (2nd ed.), Upper Saddle River, New Jersey: Prentice Hall, ISBN 0-13-790395-2.
- Zhang NL, Poole D (May 1994). "A simple approach to Bayesian network computations". *Proceedings of the Tenth Biennial Canadian Artificial Intelligence Conference (AI-94)*. Banff, Alberta: 171–178. This paper presents variable elimination for belief networks.

Further reading

- Conrady S, Jouffe L. *Bayesian Networks and BayesiaLab - A practical introduction for researchers*. Franklin, Tennessee: Bayesian USA. ISBN 978-0-9965333-0-0.
- Charniak E (Winter 1991). "Bayesian networks without tears"[706] (PDF). *AI Magazine*.
- Kruse R, Borgelt C, Klawonn F, Moewes C, Steinbrecher M, Held P (2013). *Computational Intelligence A Methodological Introduction*. London: Springer-Verlag. ISBN 978-1-4471-5012-1.
- Borgelt C, Steinbrecher M, Kruse R (2009). *Graphical Models - Representations for Learning, Reasoning and Data Mining* (Second ed.). Chichester: Wiley. ISBN 978-0-470-74956-2.

External links

- An Introduction to Bayesian Networks and their Contemporary Applications[707]
- On-line Tutorial on Bayesian nets and probability[708]
- Web-App to create Bayesian nets and run it with a Monte Carlo method[709]
- Continuous Time Bayesian Networks[710]
- Bayesian Networks: Explanation and Analogy[711]
- A live tutorial on learning Bayesian networks[712]
- A hierarchical Bayes Model for handling sample heterogeneity in classification problems[713], provides a classification model taking into consideration the uncertainty associated with measuring replicate samples.
- Hierarchical Naive Bayes Model for handling sample uncertainty[714], shows how to perform classification and learning with continuous and discrete variables with replicated measurements.

Hidden Markov model

<indicator name=”good-star”> ⊕ </indicator>

Machine learning and data mining

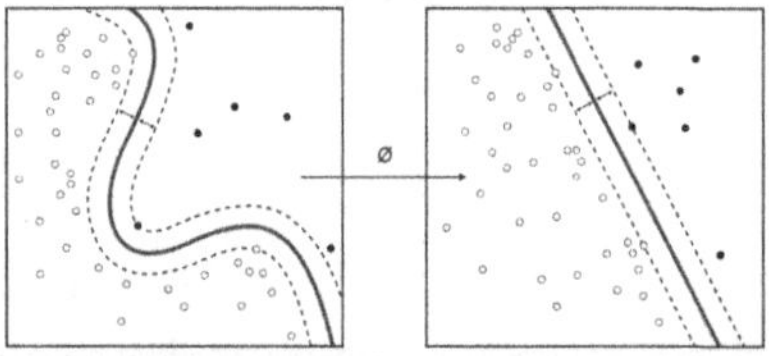

• Machine learning portal
• v • t • e[715]

Hidden Markov Model (**HMM**) is a statistical Markov model in which the system being modeled is assumed to be a Markov process with unobserved (i.e. *hidden*) states.

The hidden Markov model can be represented as the simplest dynamic Bayesian network. The mathematics behind the HMM were developed by L. E. Baum and coworkers. HMM is closely related to earlier work on the optimal nonlinear filtering problem by Ruslan L. Stratonovich, who was the first to describe the forward-backward procedure.

In simpler Markov models (like a Markov chain), the state is directly visible to the observer, and therefore the state transition probabilities are the only parameters, while in the hidden Markov model, the state is not directly visible, but the output (in the form of data or "token" in the following), dependent on the state, is visible. Each state has a probability distribution over the possible output tokens. Therefore, the sequence of tokens generated by an HMM gives some information about the sequence of states; this is also known as pattern theory, a topic of grammar induction.

The adjective *hidden* refers to the state sequence through which the model passes, not to the parameters of the model; the model is still referred to as a hidden Markov model even if these parameters are known exactly.

Hidden Markov models are especially known for their application in reinforcement learning and temporal pattern recognition such as speech, handwriting, gesture recognition,[716] part-of-speech tagging, musical score following,[717] partial discharges[718] and bioinformatics.

A hidden Markov model can be considered a generalization of a mixture model where the hidden variables (or latent variables), which control the mixture component to be selected for each observation, are related through a Markov process rather than independent of each other. Recently, hidden Markov models have been generalized to pairwise Markov models and triplet Markov models which allow consideration of more complex data structures[719] and the modeling of nonstationary data.[720,721]

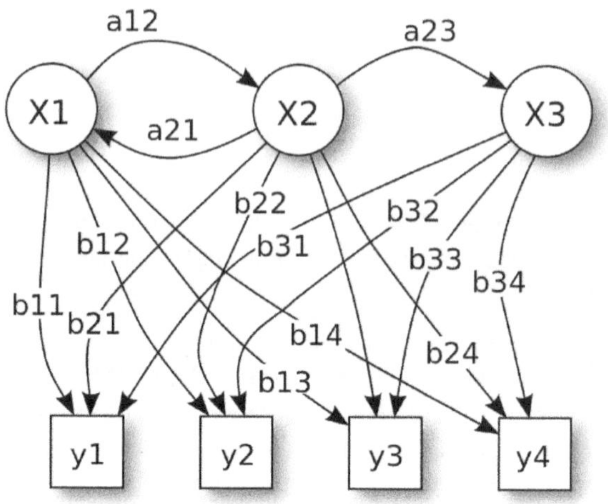

Figure 26: *Figure 1. Probabilistic parameters of a hidden Markov model (example)*
X — states
y — possible observations
a — state transition probabilities
b — output probabilities

Description in terms of urns

In its discrete form, a hidden Markov process can be visualized as a generalization of the Urn problem with replacement (where each item from the urn is returned to the original urn before the next step). Consider this example: in a room that is not visible to an observer there is a genie. The room contains urns X1, X2, X3, ... each of which contains a known mix of balls, each ball labeled y1, y2, y3, ... The genie chooses an urn in that room and randomly draws a ball from that urn. It then puts the ball onto a conveyor belt, where the observer can observe the sequence of the balls but not the sequence of urns from which they were drawn. The genie has some procedure to choose urns; the choice of the urn for the n-th ball depends only upon a random number and the choice of the urn for the $(n - 1)$-th ball. The choice of urn does not directly depend on the urns chosen before this single previous urn; therefore, this is called a Markov process. It can be described by the upper part of Figure 1.

The Markov process itself cannot be observed, only the sequence of labeled balls, thus this arrangement is called a "hidden Markov process". This is illustrated by the lower part of the diagram shown in Figure 1, where one can see that balls y1, y2, y3, y4 can be drawn at each state. Even if the observer knows the composition of the urns and has just observed a sequence of three balls, *e.g.* y1, y2 and y3 on the conveyor belt, the observer still cannot be *sure* which urn (*i.e.*, at which state) the genie has drawn the third ball from. However, the observer can work out other information, such as the likelihood that the third ball came from each of the urns.

Architecture

The diagram below shows the general architecture of an instantiated HMM. Each oval shape represents a random variable that can adopt any of a number of values. The random variable *x*(*t*) is the hidden state at time *t* (with the model from the above diagram, $x(t) \in \{ x_1, x_2, x_3 \}$). The random variable *y*(*t*) is the observation at time *t* (with $y(t) \in \{ y_1, y_2, y_3, y_4 \}$). The arrows in the diagram (often called a trellis diagram) denote conditional dependencies.

From the diagram, it is clear that the conditional probability distribution of the hidden variable *x*(*t*) at time *t*, given the values of the hidden variable *x* at all times, depends *only* on the value of the hidden variable *x*(*t* – 1); the values at time *t* – 2 and before have no influence. This is called the Markov property. Similarly, the value of the observed variable *y*(*t*) only depends on the value of the hidden variable *x*(*t*) (both at time *t*).

In the standard type of hidden Markov model considered here, the state space of the hidden variables is discrete, while the observations themselves can either be discrete (typically generated from a categorical distribution) or continuous (typically from a Gaussian distribution). The parameters of a hidden Markov model are of two types, *transition probabilities* and *emission probabilities* (also known as *output probabilities*). The transition probabilities control the way the hidden state at time *t* is chosen given the hidden state at time $t - 1$.

The hidden state space is assumed to consist of one of *N* possible values, modeled as a categorical distribution. (See the section below on extensions for other possibilities.) This means that for each of the *N* possible states that a hidden variable at time *t* can be in, there is a transition probability from this state to each of the *N* possible states of the hidden variable at time $t + 1$, for a total of N^2 transition probabilities. Note that the set of transition probabilities for transitions from any given state must sum to 1. Thus, the $N \times N$ matrix of transition probabilities is a Markov matrix. Because any one transition probability can be determined once the others are known, there are a total of $N(N - 1)$ transition parameters.

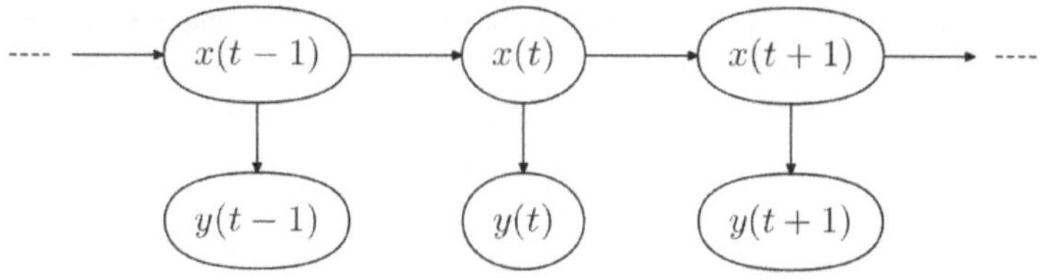

In addition, for each of the *N* possible states, there is a set of emission probabilities governing the distribution of the observed variable at a particular time given the state of the hidden variable at that time. The size of this set depends on the nature of the observed variable. For example, if the observed variable is discrete with *M* possible values, governed by a categorical distribution, there will be $M-1$ separate parameters, for a total of $N(M-1)$ emission parameters over all hidden states. On the other hand, if the observed variable is an M-dimensional vector distributed according to an arbitrary multivariate Gaussian distribution, there will be *M* parameters controlling the means and $\frac{M(M+1)}{2}$ parameters controlling the covariance matrix, for a total of $N\left(M+\frac{M(M+1)}{2}\right)=\frac{NM(M+3)}{2}=O(NM^2)$ emission parameters. (In such a case, unless the value of *M* is small, it may be more practical to restrict the nature of the covariances between individual elements of the observation vector, e.g. by assuming that the elements are independent of each other, or less restrictively, are independent of all but a fixed number of adjacent elements.)

Inference

Several inference problems are associated with hidden Markov models, as outlined below.

Probability of an observed sequence

The task is to compute in a best way, given the parameters of the model, the probability of a particular output sequence. This requires summation over all possible state sequences:

The probability of observing a sequence

$$Y=y(0),y(1),\dots,y(L-1)$$

of length *L* is given by

$$P(Y)=\sum_X P(Y\mid X)P(X),$$

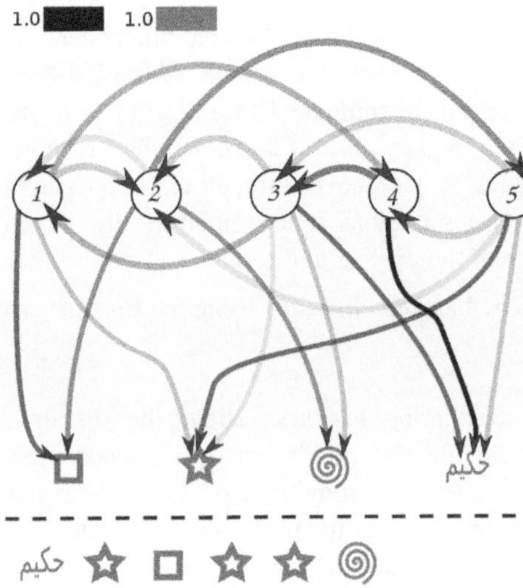

Figure 27: *The state transition and output probabilities of an HMM are indicated by the line opacity in the upper part of the diagram. Given that we have observed the output sequence in the lower part of the diagram, we may be interested in the most likely sequence of states that could have produced it. Based on the arrows that are present in the diagram, the following state sequences are candidates: 5 3 2 5 3 2 4 3 2 5 3 2 3 1 2 5 3 2 We can find the most likely sequence by evaluating the joint probability of both the state sequence and the observations for each case (simply by multiplying the probability values, which here correspond to the opacities of the arrows involved). In general, this type of problem (i.e. finding the most likely explanation for an observation sequence) can be solved efficiently using the Viterbi algorithm.*

where the sum runs over all possible hidden-node sequences

$$X = x(0), x(1), \ldots, x(L-1).$$

Applying the principle of dynamic programming, this problem, too, can be handled efficiently using the forward algorithm.

Probability of the latent variables

A number of related tasks ask about the probability of one or more of the latent variables, given the model's parameters and a sequence of observations $y(1), \ldots, y(t)$.

Filtering

The task is to compute, given the model's parameters and a sequence of observations, the distribution over hidden states of the last latent variable at the end of the sequence, i.e. to compute $P(x(t) \mid y(1), \dots, y(t))$. This task is normally used when the sequence of latent variables is thought of as the underlying states that a process moves through at a sequence of points of time, with corresponding observations at each point in time. Then, it is natural to ask about the state of the process at the end.

This problem can be handled efficiently using the forward algorithm.

Smoothing

This is similar to filtering but asks about the distribution of a latent variable somewhere in the middle of a sequence, i.e. to compute $P(x(k) \mid y(1), \dots, y(t))$ for some $k < t$. From the perspective described above, this can be thought of as the probability distribution over hidden states for a point in time *k* in the past, relative to time *t*.

The forward-backward algorithm is an efficient method for computing the smoothed values for all hidden state variables.

Most likely explanation

The task, unlike the previous two, asks about the joint probability of the *entire* sequence of hidden states that generated a particular sequence of observations (see illustration on the right). This task is generally applicable when HMM's are applied to different sorts of problems from those for which the tasks of filtering and smoothing are applicable. An example is part-of-speech tagging, where the hidden states represent the underlying parts of speech corresponding to an observed sequence of words. In this case, what is of interest is the entire sequence of parts of speech, rather than simply the part of speech for a single word, as filtering or smoothing would compute.

This task requires finding a maximum over all possible state sequences, and can be solved efficiently by the Viterbi algorithm.

Statistical significance

For some of the above problems, it may also be interesting to ask about statistical significance. What is the probability that a sequence drawn from some null distribution will have an HMM probability (in the case of the forward algorithm) or a maximum state sequence probability (in the case of the Viterbi algorithm) at least as large as that of a particular output sequence? When an HMM is used to evaluate the relevance of a hypothesis for a particular output sequence, the statistical significance indicates the false positive rate associated with failing to reject the hypothesis for the output sequence.

A concrete example

Consider two friends, Alice and Bob, who live far apart from each other and who talk together daily over the telephone about what they did that day. Bob is only interested in three activities: walking in the park, shopping, and cleaning his apartment. The choice of what to do is determined exclusively by the weather on a given day. Alice has no definite information about the weather, but she knows general trends. Based on what Bob tells her he did each day, Alice tries to guess what the weather must have been like.

Alice believes that the weather operates as a discrete Markov chain. There are two states, "Rainy" and "Sunny", but she cannot observe them directly, that is, they are *hidden* from her. On each day, there is a certain chance that Bob will perform one of the following activities, depending on the weather: "walk", "shop", or "clean". Since Bob tells Alice about his activities, those are the *observations*. The entire system is that of a hidden Markov model (HMM).

Alice knows the general weather trends in the area, and what Bob likes to do on average. In other words, the parameters of the HMM are known. They can be represented as follows in Python:

```
states = ('Rainy', 'Sunny')

observations = ('walk', 'shop', 'clean')

start_probability = {'Rainy': 0.6, 'Sunny': 0.4}

transition_probability = {
   'Rainy' : {'Rainy': 0.7, 'Sunny': 0.3},
   'Sunny' : {'Rainy': 0.4, 'Sunny': 0.6},
   }

emission_probability = {
   'Rainy' : {'walk': 0.1, 'shop': 0.4, 'clean': 0.5},
   'Sunny' : {'walk': 0.6, 'shop': 0.3, 'clean': 0.1},
   }
```

In this piece of code, `start_probability` represents Alice's belief about which state the HMM is in when Bob first calls her (all she knows is that it tends to be rainy on average). The particular probability distribution used here is not the equilibrium one, which is (given the transition probabilities) approximately `{'Rainy': 0.57, 'Sunny': 0.43}`. The `transition_probability` represents the change of the weather in the underlying Markov chain. In this example, there is only a 30% chance that tomorrow will be sunny if today is rainy. The `emission_probability`

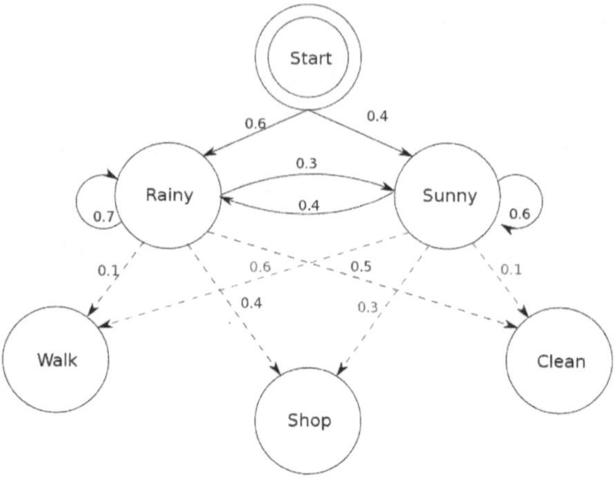

represents how likely Bob is to perform a certain activity on each day. If it is rainy, there is a 50% chance that he is cleaning his apartment; if it is sunny, there is a 60% chance that he is outside for a walk.

A similar example is further elaborated in the Viterbi algorithm page.

Learning

The parameter learning task in HMMs is to find, given an output sequence or a set of such sequences, the best set of state transition and emission probabilities. The task is usually to derive the maximum likelihood estimate of the parameters of the HMM given the set of output sequences. No tractable algorithm is known for solving this problem exactly, but a local maximum likelihood can be derived efficiently using the Baum–Welch algorithm or the Baldi–Chauvin algorithm. The Baum–Welch algorithm is a special case of the expectation-maximization algorithm. If the HMMs are used for time series prediction, more sophisticated Bayesian inference methods, like Markov chain Monte Carlo (MCMC) sampling are proven to be favorable over finding a single maximum likelihood model both in terms of accuracy and stability.[722] Since MCMC imposes significant computational burden, in cases where computational scalability is also of interest, one may alternatively resort to variational approximations to Bayesian inference, e.g. Indeed, approximate variational inference offers computational efficiency comparable to expectation-maximization, while yielding an accuracy profile only slightly inferior to exact MCMC-type Bayesian inference.

Mathematical description

General description

A basic hidden Markov model can be described as follows:

N	=	number of states
T	=	number of observations
$\theta_{i=1\ldots N}$	=	emission parameter for an observation associated with state i
$\phi_{i=1\ldots N,j=1\ldots N}$	=	probability of transition from state i to state j
$\boldsymbol{\phi}_{i=1\ldots N}$	=	N -dimensional vector, composed of $\phi_{i,j=1\ldots N}$; the i -th row of the matrix $\phi_{i=1\ldots N,j=1\ldots N}$ (sum of it is 1)
$x_{t=1\ldots T}$	=	(hidden) state at time t
$y_{t=1\ldots T}$	=	observation at time t
$F(y\|\theta)$	=	probability distribution of an observation, parametrized on θ
$x_{t=2\ldots T}$	~	$\text{Categorical}(\boldsymbol{\phi}_{x_{t-1}})$
$y_{t=1\ldots T}$	~	$F(\theta_{x_t})$

Note that, in the above model (and also the one below), the prior distribution of the initial state x_1 is not specified. Typical learning models correspond to assuming a discrete uniform distribution over possible states (i.e. no particular prior distribution is assumed).

In a Bayesian setting, all parameters are associated with random variables, as follows:

N, T	=	as above
$\theta_{i=1\ldots N}, \phi_{i=1\ldots N,j=1\ldots N}, \boldsymbol{\phi}_{i=1\ldots N}$	=	as above
$x_{t=1\ldots T}, y_{t=1\ldots T}, F(y\|\theta)$	=	as above
α	=	shared hyperparameter for emission parameters
β	=	shared hyperparameter for transition parameters
$H(\theta\|\alpha)$	=	prior probability distribution of emission parameters, parametrized on α
$\theta_{i=1\ldots N}$	~	$H(\alpha)$
$\boldsymbol{\phi}_{i=1\ldots N}$	~	$\text{Symmetric-Dirichlet}_N(\beta)$
$x_{t=2\ldots T}$	~	$\text{Categorical}(\boldsymbol{\phi}_{x_{t-1}})$
$y_{t=1\ldots T}$	~	$F(\theta_{x_t})$

These characterizations use F and H to describe arbitrary distributions over observations and parameters, respectively. Typically H will be the conjugate prior of F . The two most common choices of F are Gaussian and categorical; see below.

Compared with a simple mixture model

As mentioned above, the distribution of each observation in a hidden Markov model is a mixture density, with the states of the corresponding to mixture components. It is useful to compare the above characterizations for an HMM with the corresponding characterizations, of a mixture model, using the same notation.

A non-Bayesian mixture model:

N	$=$	number of mixture components
T	$=$	number of observations
$\theta_{i=1\dots N}$	$=$	parameter of distribution of observation associated with component i
$\phi_{i=1\dots N}$	$=$	mixture weight, i.e. prior probability of component i
$\boldsymbol{\phi}$	$=$	N -dimensional vector, composed of $\phi_{1\dots N}$; must sum to 1
$x_{t=1\dots T}$	$=$	component of observation t
$y_{t=1\dots T}$	$=$	observation t
$F(y\|\theta)$	$=$	probability distribution of an observation, parametrized on θ
$x_{t=1\dots T}$	$\sim$	$\operatorname{Categorical}(\boldsymbol{\phi})$
$y_{t=1\dots T}$	$\sim$	$F(\theta_{x_t})$

A Bayesian mixture model:

N, T	$=$	as above
$\theta_{i=1\dots N}, \phi_{i=1\dots N}, \boldsymbol{\phi}$	$=$	as above
$x_{t=1\dots T}, y_{t=1\dots T}, F(y\|\theta)$	$=$	as above
α	$=$	shared hyperparameter for component parameters
β	$=$	shared hyperparameter for mixture weights
$H(\theta\|\alpha)$	$=$	prior probability distribution of component parameters, parametrized on α
$\theta_{i=1\dots N}$	$\sim$	$H(\alpha)$
$\boldsymbol{\phi}$	$\sim$	$\operatorname{Symmetric-Dirichlet}_N(\beta)$
$x_{t=1\dots T}$	$\sim$	$\operatorname{Categorical}(\boldsymbol{\phi})$
$y_{t=1\dots T}$	$\sim$	$F(\theta_{x_t})$

Examples

The following mathematical descriptions are fully written out and explained, for ease of implementation.

A typical non-Bayesian HMM with Gaussian observations looks like this:

N	$=$	number of states
T	$=$	number of observations
$\phi_{i=1\ldots N,j=1\ldots N}$	$=$	probability of transition from state i to state j
$\boldsymbol{\phi}_{i=1\ldots N}$	$=$	N -dimensional vector, composed of $\phi_{i,1\ldots N}$; must sum to 1
$\mu_{i=1\ldots N}$	$=$	mean of observations associated with state i
$\sigma^2_{i=1\ldots N}$	$=$	variance of observations associated with state i
$x_{t=1\ldots T}$	$=$	state of observation at time t
$y_{t=1\ldots T}$	$=$	observation at time t
$x_{t=2\ldots T}$	$\sim$	$\text{Categorical}(\boldsymbol{\phi}_{x_{t-1}})$
$y_{t=1\ldots T}$	$\sim$	$\mathcal{N}(\mu_{x_t}, \sigma^2_{x_t})$

A typical Bayesian HMM with Gaussian observations looks like this:

N	$=$	number of states
T	$=$	number of observations
$\phi_{i=1\ldots N,j=1\ldots N}$	$=$	probability of transition from state i to state j
$\boldsymbol{\phi}_{i=1\ldots N}$	$=$	N -dimensional vector, composed of $\phi_{i,1\ldots N}$; must sum to 1
$\mu_{i=1\ldots N}$	$=$	mean of observations associated with state i
$\sigma^2_{i=1\ldots N}$	$=$	variance of observations associated with state i
$x_{t=1\ldots T}$	$=$	state of observation at time t
$y_{t=1\ldots T}$	$=$	observation at time t
β	$=$	concentration hyperparameter controlling the density of the transition matrix
μ_0, λ	$=$	shared hyperparameters of the means for each state
ν, σ_0^2	$=$	shared hyperparameters of the variances for each state
$\boldsymbol{\phi}_{i=1\ldots N}$	$\sim$	$\text{Symmetric-Dirichlet}_N(\beta)$
$x_{t=2\ldots T}$	$\sim$	$\text{Categorical}(\boldsymbol{\phi}_{x_{t-1}})$
$\mu_{i=1\ldots N}$	$\sim$	$\mathcal{N}(\mu_0, \lambda\sigma_i^2)$
$\sigma^2_{i=1\ldots N}$	$\sim$	$\text{Inverse-Gamma}(\nu, \sigma_0^2)$
$y_{t=1\ldots T}$	$\sim$	$\mathcal{N}(\mu_{x_t}, \sigma^2_{x_t})$

A typical non-Bayesian HMM with categorical observations looks like this:

N	=	number of states
T	=	number of observations
$\phi_{i=1\ldots N,j=1\ldots N}$	=	probability of transition from state i to state j
$\boldsymbol{\phi}_{i=1\ldots N}$	=	N -dimensional vector, composed of $\phi_{i,1\ldots N}$; must sum to 1
V	=	dimension of categorical observations, e.g. size of word vocabulary
$\theta_{i=1\ldots N,j=1\ldots V}$	=	probability for state i of observing the j th item
$\boldsymbol{\theta}_{i=1\ldots N}$	=	V -dimensional vector, composed of $\theta_{i,1\ldots V}$; must sum to 1
$x_{t=1\ldots T}$	=	state of observation at time t
$y_{t=1\ldots T}$	=	observation at time t
$x_{t=2\ldots T}$	$\sim$	$\text{Categorical}(\boldsymbol{\phi}_{x_{t-1}})$
$y_{t=1\ldots T}$	$\sim$	$\text{Categorical}(\boldsymbol{\theta}_{x_t})$

A typical Bayesian HMM with categorical observations looks like this:

N	=	number of states
T	=	number of observations
$\phi_{i=1\ldots N,j=1\ldots N}$	=	probability of transition from state i to state j
$\boldsymbol{\phi}_{i=1\ldots N}$	=	N -dimensional vector, composed of $\phi_{i,1\ldots N}$; must sum to 1
V	=	dimension of categorical observations, e.g. size of word vocabulary
$\theta_{i=1\ldots N,j=1\ldots V}$	=	probability for state i of observing the j th item
$\boldsymbol{\theta}_{i=1\ldots N}$	=	V -dimensional vector, composed of $\theta_{i,1\ldots V}$; must sum to 1
$x_{t=1\ldots T}$	=	state of observation at time t
$y_{t=1\ldots T}$	=	observation at time t
α	=	shared concentration hyperparameter of $\boldsymbol{\theta}$ for each state
β	=	concentration hyperparameter controlling the density of the transition matrix
$\boldsymbol{\phi}_{i=1\ldots N}$	$\sim$	$\text{Symmetric-Dirichlet}_N(\beta)$
$\boldsymbol{\theta}_{1\ldots V}$	$\sim$	$\text{Symmetric-Dirichlet}_V(\alpha)$
$x_{t=2\ldots T}$	$\sim$	$\text{Categorical}(\boldsymbol{\phi}_{x_{t-1}})$
$y_{t=1\ldots T}$	$\sim$	$\text{Categorical}(\boldsymbol{\theta}_{x_t})$

Note that in the above Bayesian characterizations, β (a concentration parameter) controls the density of the transition matrix. That is, with a high value of β (significantly above 1), the probabilities controlling the transition out of a particular state will all be similar, meaning there will be a significant probability of transitioning to any of the other states. In other words, the path followed by the Markov chain of hidden states will be highly random. With a low value of β (significantly below 1), only a small number of the possible transitions out of a given state will have significant probability, meaning that the path followed by the hidden states will be somewhat predictable.

A two-level Bayesian HMM

An alternative for the above two Bayesian examples would be to add another level of prior parameters for the transition matrix. That is, replace the lines

β	=	concentration hyperparameter controlling the density of the transition matrix
$\phi_{i=1\dots N}$	~	$\operatorname{Symmetric-Dirichlet}_N(\beta)$

with the following:

γ	=	concentration hyperparameter controlling how many states are intrinsically likely
β	=	concentration hyperparameter controlling the density of the transition matrix
$\boldsymbol{\eta}$	=	N -dimensional vector of probabilities, specifying the intrinsic probability of a given state
$\boldsymbol{\eta}$	~	$\operatorname{Symmetric-Dirichlet}_N(\gamma)$
$\phi_{i=1\dots N}$	~	$\operatorname{Dirichlet}_N(\beta N\boldsymbol{\eta})$

What this means is the following:

1. $\boldsymbol{\eta}$ is a probability distribution over states, specifying which states are inherently likely. The greater the probability of a given state in this vector, the more likely is a transition to that state (regardless of the starting state).
2. γ controls the density of $\boldsymbol{\eta}$. Values significantly above 1 cause a dense vector where all states will have similar prior probabilities. Values significantly below 1 cause a sparse vector where only a few states are inherently likely (have prior probabilities significantly above 0).
3. β controls the density of the transition matrix, or more specifically, the density of the *N* different probability vectors $\phi_{i=1\dots N}$ specifying the probability of transitions out of state *i* to any other state.

Imagine that the value of β is significantly above 1. Then the different ϕ vectors will be dense, i.e. the probability mass will be spread out fairly evenly over all states. However, to the extent that this mass is unevenly spread, $\boldsymbol{\eta}$ controls which states are likely to get more mass than others.

Now, imagine instead that β is significantly below 1. This will make the ϕ vectors sparse, i.e. almost all the probability mass is distributed over a small number of states, and for the rest, a transition to that state will be very unlikely. Notice that there are different ϕ vectors for each starting state, and so even if all the vectors are sparse, different vectors may distribute the mass to different ending states. However, for all of the vectors, η controls which ending states are likely to get mass assigned to them. For example, if β is 0.1, then each ϕ will be sparse and, for any given starting state i, the set of states $\mathbf{J}_i$ to which transitions are likely to occur will be very small, typically having only one or two members. Now, if the probabilities in η are all the same (or equivalently, one of the above models without η is used), then for different i, there will be different states in the corresponding $\mathbf{J}_i$, so that all states are equally likely to occur in any given $\mathbf{J}_i$. On the other hand, if the values in η are unbalanced, so that one state has a much higher probability than others, almost all $\mathbf{J}_i$ will contain this state; hence, regardless of the starting state, transitions will nearly always occur to this given state.

Hence, a two-level model such as just described allows independent control over (1) the overall density of the transition matrix, and (2) the density of states to which transitions are likely (i.e. the density of the prior distribution of states in any particular hidden variable x_i). In both cases this is done while still assuming ignorance over which particular states are more likely than others. If it is desired to inject this information into the model, the probability vector η can be directly specified; or, if there is less certainty about these relative probabilities, a non-symmetric Dirichlet distribution can be used as the prior distribution over η . That is, instead of using a symmetric Dirichlet distribution with the single parameter γ (or equivalently, a general Dirichlet with a vector all of whose values are equal to γ), use a general Dirichlet with values that are variously greater or less than γ , according to which state is more or less preferred.

Poisson hidden Markov model

Poisson hidden Markov models (PHMM) are special cases of hidden Markov models where a Poisson process has a rate which varies in association with changes between the different states of a Markov model.[723] PHMMs are not necessarily Markovian processes themselves because the underlying Markov chain or Markov process cannot be observed and only the Poisson signal is observed.

Applications

HMMs can be applied in many fields where the goal is to recover a data sequence that is not immediately observable (but other data that depend on the sequence are). Applications include:

- Computational finance
- Single-molecule kinetic analysis

- Cryptanalysis
- Speech recognition, including Siri
- Speech synthesis
- Part-of-speech tagging
- Document separation in scanning solutions
- Machine translation
- Partial discharge
- Gene prediction
- Handwriting recognition
- Alignment of bio-sequences
- Time series analysis
- Activity recognition
- Protein folding
- Sequence classification
- Metamorphic virus detection
- DNA motif discovery
- ChromHMM[724]

History

The forward and backward recursions used in HMM as well as computations of marginal smoothing probabilities were first described by Ruslan L. Stratonovich in 1960 (pages 160—162) and in the late 1950s in his papers in Russian. The Hidden Markov Models were later described in a series of statistical papers by Leonard E. Baum and other authors in the second half of the 1960s. One of the first applications of HMMs was speech recognition, starting in the mid-1970s.

In the second half of the 1980s, HMMs began to be applied to the analysis of biological sequences, in particular DNA. Since then, they have become ubiquitous in the field of bioinformatics.

Types

Hidden Markov models can model complex Markov processes where the states emit the observations according to some probability distribution. One such example is the Gaussian distribution, in such a Hidden Markov Model the states output are represented by a Gaussian distribution.

Moreover, it could represent even more complex behavior when the output of the states is represented as mixture of two or more Gaussians, in which case the probability of generating an observation is the product of the probability of first selecting one of the Gaussians and the probability of generating that observation from that Gaussian. In cases of modeled data exhibiting artifacts such as outliers and skewness, one may resort to finite mixtures of heavier-tailed elliptical distributions, such as the multivariate Student's-t distribution, or appropriate non-elliptical distributions, such as the multivariate Normal Inverse-Gaussian.[725]

Extensions

In the hidden Markov models considered above, the state space of the hidden variables is discrete, while the observations themselves can either be discrete (typically generated from a categorical distribution) or continuous (typically from a Gaussian distribution). Hidden Markov models can also be generalized to allow continuous state spaces. Examples of such models are those where the Markov process over hidden variables is a linear dynamical system, with a linear relationship among related variables and where all hidden and observed variables follow a Gaussian distribution. In simple cases, such as the linear dynamical system just mentioned, exact inference is tractable (in this case, using the Kalman filter); however, in general, exact inference in HMMs with continuous latent variables is infeasible, and approximate methods must be used, such as the extended Kalman filter or the particle filter.

Hidden Markov models are generative models, in which the joint distribution of observations and hidden states, or equivalently both the prior distribution of hidden states (the *transition probabilities*) and conditional distribution of observations given states (the *emission probabilities*), is modeled. The above algorithms implicitly assume a uniform prior distribution over the transition probabilities. However, it is also possible to create hidden Markov models with other types of prior distributions. An obvious candidate, given the categorical distribution of the transition probabilities, is the Dirichlet distribution, which is the conjugate prior distribution of the categorical distribution. Typically, a symmetric Dirichlet distribution is chosen, reflecting ignorance about which states are inherently more likely than others. The single parameter of this distribution (termed the *concentration parameter*) controls the relative density or sparseness of the resulting transition matrix. A choice of 1 yields a uniform distribution. Values greater than 1 produce a dense matrix, in which the transition probabilities between pairs of states are likely to be nearly equal. Values less than 1 result in a sparse matrix in which, for each given source state, only a small number of destination states have non-negligible transition probabilities. It is also possible to use a two-level prior Dirichlet distribution, in which one Dirichlet distribution (the upper distribution) governs the parameters of another Dirichlet distribution (the lower distribution), which in turn governs the transition probabilities. The upper distribution governs the overall distribution of states, determining how likely each state is to occur; its concentration parameter determines the density or sparseness of states. Such a two-level prior distribution, where both concentration parameters are set to produce sparse distributions, might be useful for example in unsupervised part-of-speech tagging, where some parts of speech occur much more commonly than others; learning algorithms that assume a uniform prior distribution generally perform poorly on this task. The parameters of models of this sort, with non-uniform prior distributions, can be learned using Gibbs sampling or extended versions of the expectation-maximization algorithm.

An extension of the previously described hidden Markov models with Dirichlet priors uses a Dirichlet process in place of a Dirichlet distribution. This type of model allows for an unknown and potentially infinite number of states. It is common to use a two-level Dirichlet process, similar to the previously described model with two levels of Dirichlet distributions. Such a model is called a *hierarchical Dirichlet process hidden Markov model*, or *HDP-HMM* for short. It was originally described under the name "Infinite Hidden Markov Model"[726] and was further formalized in[727].

A different type of extension uses a discriminative model in place of the generative model of standard HMMs. This type of model directly models the conditional distribution of the hidden states given the observations, rather than modeling the joint distribution. An example of this model is the so-called *maximum entropy Markov model* (MEMM), which models the conditional distribution of the states using logistic regression (also known as a "maximum entropy model"). The advantage of this type of model is that arbitrary features (i.e. functions) of the observations can be modeled, allowing domain-specific knowledge of the problem at hand to be injected into the model. Models of this sort are not limited to modeling direct dependencies between a hidden state and its associated observation; rather, features of nearby observations, of combinations of the associated observation and nearby observations, or in fact of arbitrary observations at any distance from a given hidden state can be included in the process used to determine the value of a hidden state. Furthermore, there is no need for these features to be statistically independent of each other, as would be the case if such features were used in a generative model. Finally, arbitrary features over pairs of adjacent hidden states can be used rather than simple transition probabilities. The disadvantages of such models are: (1) The types of prior distributions that can be placed on hidden states are severely limited; (2) It is not possible to predict the probability of seeing an arbitrary observation. This second limitation is often not an issue in practice, since many common usages of HMM's do not require such predictive probabilities.

A variant of the previously described discriminative model is the linear-chain conditional random field. This uses an undirected graphical model (aka Markov random field) rather than the directed graphical models of MEMM's and similar models. The advantage of this type of model is that it does not suffer from the so-called *label bias* problem of MEMM's, and thus may make more accurate predictions. The disadvantage is that training can be slower than for MEMM's.

Yet another variant is the *factorial hidden Markov model*, which allows for a single observation to be conditioned on the corresponding hidden variables of a set of K independent Markov chains, rather than a single Markov chain. It is equivalent to a single HMM, with N^K states (assuming there are N states for each chain), and therefore, learning in such a model is difficult: for a sequence of length T , a straightforward Viterbi algorithm has complexity $O(N^{2K}\,T)$. To find an exact solution, a junction tree algorithm could be used, but it results in an $O(N^{K+1}\,K\,T)$ complexity. In practice, approximate techniques, such as variational approaches, could be used.

All of the above models can be extended to allow for more distant dependencies among hidden states, e.g. allowing for a given state to be dependent on the previous two or three states rather than a single previous state; i.e. the transition probabilities are extended to encompass sets of three or four adjacent states (or in general K adjacent states). The disadvantage of such models is that dynamic-programming algorithms for training them have an $O(N^K\,T)$ running time, for K adjacent states and T total observations (i.e. a length- T Markov chain).

Another recent extension is the *triplet Markov model*, in which an auxiliary underlying process is added to model some data specificities. Many variants of this model have been proposed. One should also mention the interesting link that has been established between the *theory of evidence* and the *triplet Markov models* and which allows to fuse data in Markovian context and to model nonstationary data. Note that alternative multi-stream data fusion strategies have also been proposed in the recent literature, e.g.[728]

Finally, a different rationale towards addressing the problem of modeling nonstationary data by means of hidden Markov models was suggested in 2012. It consists in employing a small recurrent neural network (RNN), specifically a reservoir network,[729] to capture the evolution of the temporal dynamics in the observed data. This information, encoded in the form of a high-dimensional vector, is used as a conditioning variable of the HMM state transition probabilities. Under such a setup, we eventually obtain a nonstationary HMM the transition probabilities of which evolve over time in a manner that is inferred from the data itself, as opposed to some unrealistic ad-hoc model of temporal evolution.

External links

Wikimedia Commons has media related to ***Hidden Markov Model***.

Concepts

- Teif, V. B.; Rippe, K. (2010). "Statistical–mechanical lattice models for protein–DNA binding in chromatin"[730]. *J. Phys.: Condens. Matter.* **22** (41): 414105. arXiv: 1004.5514[731]. Bibcode: 2010JPCM...22O4105T[732]. doi: 10.1088/0953-8984/22/41/414105[733].
- A Revealing Introduction to Hidden Markov Models[734] by Mark Stamp, San Jose State University.
- Fitting HMM's with expectation-maximization – complete derivation[735]
- A step-by-step tutorial on HMMs[736] *(University of Leeds)*
- Hidden Markov Models[737] *(an exposition using basic mathematics)*
- Hidden Markov Models[738] *(by Narada Warakagoda)*
- Hidden Markov Models: Fundamentals and Applications Part 1[739], Part 2[740] *(by V. Petrushin)*
- Lecture on a Spreadsheet by Jason Eisner, Video[741] and interactive spreadsheet[742]

Software

- Hidden Markov Model (HMM) Toolbox for Matlab[743] *(by Kevin Murphy)*
- Hidden Markov Model Toolkit (HTK)[744] *(a portable toolkit for building and manipulating hidden Markov models)*
- Hidden Markov Model R-Package[745] to set up, apply and make inference with discrete time and discrete space Hidden Markov Models
- zipHMMlib[746] *(a library for general (discrete) hidden Markov models, exploiting repetitions in the input sequence to greatly speed up the forward algorithm. Implementation of the posterior decoding algorithm and the Viterbi algorithm are also provided.)*
- GHMM Library[747] *(home page of the GHMM Library project)*
- Jahmm Java Library[748] *(general-purpose Java library)*
- HMM and other statistical programs[749] *(Implementation in C by Tapas Kanungo)*
- The hmm package[750] A Haskell[751] library for working with Hidden Markov Models.
- GT2K[752] Georgia Tech Gesture Toolkit (referred to as GT2K)
- Hidden Markov Models -online calculator for HMM – Viterbi path and probabilities. Examples with perl source code.[753]

- A discrete Hidden Markov Model class, based on OpenCV.[754]
- depmixS4[755] R-Package (Hidden Markov Models of GLMs and Other Distributions in S4)
- MLPACK contains a C++ implementation of HMMs
- Hidden Markov Models Java Library[756] contains basic HMMs abstractions in Java 8
- SFIHMM[757] high-speed C code for the estimation of Hidden Markov Models, Viterbi Path Reconstruction, and the generation of simulated data from HMMs.

Kalman filter

Kalman filtering, also known as **linear quadratic estimation** (**LQE**), is an algorithm that uses a series of measurements observed over time, containing statistical noise and other inaccuracies, and produces estimates of unknown variables that tend to be more accurate than those based on a single measurement alone, by estimating a joint probability distribution over the variables for each timeframe. The filter is named after Rudolf E. Kálmán, one of the primary developers of its theory.

The Kalman filter has numerous applications in technology. A common application is for guidance, navigation, and control of vehicles, particularly aircraft and spacecraft. Furthermore, the Kalman filter is a widely applied concept in time series analysis used in fields such as signal processing and econometrics. Kalman filters also are one of the main topics in the field of robotic motion planning and control, and they are sometimes included in trajectory optimization. The Kalman filter also works for modeling the central nervous system's control of movement. Due to the time delay between issuing motor commands and receiving sensory feedback, use of the Kalman filter supports a realistic model for making estimates of the current state of the motor system and issuing updated commands.

The algorithm works in a two-step process. In the prediction step, the Kalman filter produces estimates of the current state variables, along with their uncertainties. Once the outcome of the next measurement (necessarily corrupted with some amount of error, including random noise) is observed, these estimates are updated using a weighted average, with more weight being given to estimates with higher certainty. The algorithm is recursive. It can run in real time, using only the present input measurements and the previously calculated state and its uncertainty matrix; no additional past information is required.

Using a Kalman filter does not assume that the errors are Gaussian. However, the filter yields the exact conditional probability estimate in the special case that all errors are Gaussian.

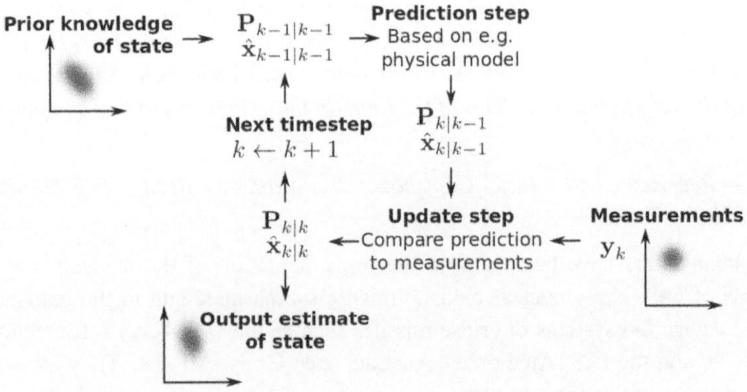

Figure 28: *The Kalman filter keeps track of the estimated state of the system and the variance or uncertainty of the estimate. The estimate is updated using a state transition model and measurements.* $\hat{x}_{k|k-1}$ *denotes the estimate of the system's state at time step k before the k-th measurement* y_k *has been taken into account;* $P_{k|k-1}$ *is the corresponding uncertainty.*

Extensions and generalizations to the method have also been developed, such as the extended Kalman filter and the unscented Kalman filter which work on nonlinear systems. The underlying model is similar to a hidden Markov model except that the state space of the latent variables is continuous and all latent and observed variables have Gaussian distributions.

History

The filter is named after Hungarian émigré Rudolf E. Kálmán, although Thorvald Nicolai Thiele and Peter Swerling developed a similar algorithm earlier. Richard S. Bucy of the University of Southern California contributed to the theory, leading to it sometimes being called the Kalman–Bucy filter. Stanley F. Schmidt is generally credited with developing the first implementation of a Kalman filter. He realized that the filter could be divided into two distinct parts, with one part for time periods between sensor outputs and another part for incorporating measurements.[758] It was during a visit by Kálmán to the NASA Ames Research Center that Schmidt saw the applicability of Kálmán's ideas to the nonlinear problem of trajectory estimation for the Apollo program leading to its incorporation in the Apollo navigation computer. This Kalman filter was first described and partially developed in technical papers by Swerling (1958), Kalman (1960) and Kalman and Bucy (1961).

The Apollo computer used 2k of magnetic core RAM and 36k wire rope [...]. The CPU was built from ICs [...]. Clock speed was under 100 kHz [...]. The fact that the MIT engineers were able to pack such good software (one of the very first applications of the Kalman filter) into such a tiny computer is truly remarkable.

—Interview with Jack Crenshaw, by Matthew Reed, TRS-80.org (2009)[759]

Kalman filters have been vital in the implementation of the navigation systems of U.S. Navy nuclear ballistic missile submarines, and in the guidance and navigation systems of cruise missiles such as the U.S. Navy's Tomahawk missile and the U.S. Air Force's Air Launched Cruise Missile. They are also used in the guidance and navigation systems of reusable launch vehicles and the attitude control and navigation systems of spacecraft which dock at the International Space Station.

This digital filter is sometimes called the *Stratonovich–Kalman–Bucy filter* because it is a special case of a more general, non-linear filter developed somewhat earlier by the Soviet mathematician Ruslan Stratonovich.[760,761,762,763] In fact, some of the special case linear filter's equations appeared in these papers by Stratonovich that were published before summer 1960, when Kalman met with Stratonovich during a conference in Moscow.Wikipedia:Citation needed

Overview of the calculation

The Kalman filter uses a system's dynamics model (e.g., physical laws of motion), known control inputs to that system, and multiple sequential measurements (such as from sensors) to form an estimate of the system's varying quantities (its state) that is better than the estimate obtained by using only one measurement alone. As such, it is a common sensor fusion and data fusion algorithm.

Noisy sensor data, approximations in the equations that describe the system evolution, and external factors that are not accounted for all place limits on how well it is possible to determine the system's state. The Kalman filter deals effectively with the uncertainty due to noisy sensor data and to some extent also with random external factors. The Kalman filter produces an estimate of the state of the system as an average of the system's predicted state and of the new measurement using a weighted average. The purpose of the weights is that values with better (i.e., smaller) estimated uncertainty are "trusted" more. The weights are calculated from the covariance, a measure of the estimated uncertainty of the prediction of the system's state. The result of the weighted average is a new state estimate that lies between the predicted and measured state, and has a better estimated uncertainty than either alone. This process is repeated at every time step, with the new estimate and its covariance informing the prediction used in the following iteration. This means that the Kalman filter works recursively and requires only the last "best guess", rather than the entire history, of a system's state to calculate a new state.

The relative certainty of the measurements and current state estimate is an important consideration, and it is common to discuss the response of the filter in terms of the Kalman filter's *gain*. The Kalman gain is the relative weight given to the measurements and current state estimate, and can be "tuned" to achieve particular performance. With a high gain, the filter places more weight on the most recent measurements, and thus follows them more responsively. With a low gain, the filter follows the model predictions more closely. At the extremes, a high gain close to one will result in a more jumpy estimated trajectory, while low gain close to zero will smooth out noise but decrease the responsiveness.

When performing the actual calculations for the filter (as discussed below), the state estimate and covariances are coded into matrices to handle the multiple dimensions involved in a single set of calculations. This allows for a representation of linear relationships between different state variables (such as position, velocity, and acceleration) in any of the transition models or covariances.

Example application

As an example application, consider the problem of determining the precise location of a truck. The truck can be equipped with a GPS unit that provides an estimate of the position within a few meters. The GPS estimate is likely to be noisy; readings 'jump around' rapidly, though remaining within a few meters of the real position. In addition, since the truck is expected to follow the laws of physics, its position can also be estimated by integrating its velocity over time, determined by keeping track of wheel revolutions and the angle of the steering wheel. This is a technique known as dead reckoning. Typically, the dead reckoning will provide a very smooth estimate of the truck's position, but it will drift over time as small errors accumulate.

In this example, the Kalman filter can be thought of as operating in two distinct phases: predict and update. In the prediction phase, the truck's old position will be modified according to the physical laws of motion (the dynamic or "state transition" model). Not only will a new position estimate be calculated, but a new covariance will be calculated as well. Perhaps the covariance is proportional to the speed of the truck because we are more uncertain about the accuracy of the dead reckoning position estimate at high speeds but very certain about the position estimate when moving slowly. Next, in the update phase, a measurement of the truck's position is taken from the GPS unit. Along with this measurement comes some amount of uncertainty, and its covariance relative to that of the prediction from the previous phase determines how much the new measurement will affect the updated prediction. Ideally, as the dead reckoning estimates tend to drift away from the real position, the GPS measurement should pull the position estimate back towards the real position but not disturb it to the point of becoming rapidly jumping and noisy.

Technical description and context

The Kalman filter is an efficient recursive filter that estimates the internal state of a linear dynamic system from a series of noisy measurements. It is used in a wide range of engineering and econometric applications from radar and computer vision to estimation of structural macroeconomic models, and is an important topic in control theory and control systems engineering. Together with the linear-quadratic regulator (LQR), the Kalman filter solves the linear–quadratic–Gaussian control problem (LQG). The Kalman filter, the linear-quadratic regulator and the linear–quadratic–Gaussian controller are solutions to what arguably are the most fundamental problems in control theory.

In most applications, the internal state is much larger (more degrees of freedom) than the few "observable" parameters which are measured. However, by combining a series of measurements, the Kalman filter can estimate the entire internal state.

In Dempster–Shafer theory, each state equation or observation is considered a special case of a linear belief function and the Kalman filter is a special case of combining linear belief functions on a join-tree or Markov tree. Additional approaches include belief filters which use Bayes or evidential updates to the state equations.

A wide variety of Kalman filters have now been developed, from Kalman's original formulation, now called the "simple" Kalman filter, the Kalman–Bucy filter, Schmidt's "extended" filter, the information filter, and a variety of "square-root" filters that were developed by Bierman, Thornton and many others. Perhaps the most commonly used type of very simple Kalman filter is the phase-locked loop, which is now ubiquitous in radios, especially frequency modulation (FM) radios, television sets, satellite communications receivers, outer space communications systems, and nearly any other electronic communications equipment.

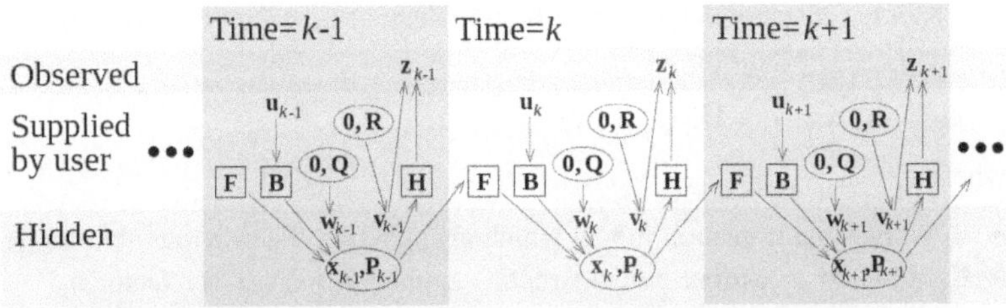

Figure 29: *Model underlying the Kalman filter. Squares represent matrices. Ellipses represent multivariate normal distributions (with the mean and covariance matrix enclosed). Unenclosed values are vectors. In the simple case, the various matrices are constant with time, and thus the subscripts are dropped, but the Kalman filter allows any of them to change each time step.*

Underlying dynamical system model

The Kalman filters are based on linear dynamical systems discretized in the time domain. They are modeled on a Markov chain built on linear operators perturbed by errors that may include Gaussian noise. The state of the system is represented as a vector of real numbers. At each discrete time increment, a linear operator is applied to the state to generate the new state, with some noise mixed in, and optionally some information from the controls on the system if they are known. Then, another linear operator mixed with more noise generates the observed outputs from the true ("hidden") state. The Kalman filter may be regarded as analogous to the hidden Markov model, with the key difference that the hidden state variables take values in a continuous space (as opposed to a discrete state space as in the hidden Markov model). There is a strong analogy between the equations of the Kalman Filter and those of the hidden Markov model. A review of this and other models is given in Roweis and Ghahramani (1999), and Hamilton (1994), Chapter 13.[764]

In order to use the Kalman filter to estimate the internal state of a process given only a sequence of noisy observations, one must model the process in accordance with the framework of the Kalman filter. This means specifying the following matrices:

- $\mathbf{F}_k$, the state-transition model;
- $\mathbf{H}_k$, the observation model;
- $\mathbf{Q}_k$, the covariance of the process noise;
- $\mathbf{R}_k$, the covariance of the observation noise;
- and sometimes $\mathbf{B}_k$, the control-input model, for each time-step, k, as described below.

The Kalman filter model assumes the true state at time k is evolved from the state at $(k - 1)$ according to

$$\mathbf{x}_k = \mathbf{F}_k \mathbf{x}_{k-1} + \mathbf{B}_k \mathbf{u}_k + \mathbf{w}_k$$

where

- $\mathbf{F}_k$ is the state transition model which is applied to the previous state $\mathbf{x}_{k-1}$;
- $\mathbf{B}_k$ is the control-input model which is applied to the control vector $\mathbf{u}_k$;
- $\mathbf{w}_k$ is the process noise which is assumed to be drawn from a zero mean multivariate normal distribution, $\mathcal{N}$, with covariance, $\mathbf{Q}_k$: $\mathbf{w}_k \sim \mathcal{N}(0, \mathbf{Q}_k)$.

At time k an observation (or measurement) $\mathbf{z}_k$ of the true state $\mathbf{x}_k$ is made according to

$$\mathbf{z}_k = \mathbf{H}_k \mathbf{x}_k + \mathbf{v}_k$$

where

- $\mathbf{H}_k$ is the observation model which maps the true state space into the observed space and
- $\mathbf{v}_k$ is the observation noise which is assumed to be zero mean Gaussian white noise with covariance $\mathbf{R}_k$: $\mathbf{v}_k \sim \mathcal{N}(0, \mathbf{R}_k)$.

The initial state, and the noise vectors at each step $\{\mathbf{x}_0, \mathbf{w}_1, ..., \mathbf{w}_k, \mathbf{v}_1 ... \mathbf{v}_k\}$ are all assumed to be mutually independent.

Many real dynamical systems do not exactly fit this model. In fact, unmodeled dynamics can seriously degrade the filter performance, even when it was supposed to work with unknown stochastic signals as inputs. The reason for this is that the effect of unmodeled dynamics depends on the input, and, therefore, can bring the estimation algorithm to instability (it diverges). On the other hand, independent white noise signals will not make the algorithm diverge. The problem of distinguishing between measurement noise and unmodeled dynamics is a difficult one and is treated in control theory under the framework of robust control.

Details

The Kalman filter is a recursive estimator. This means that only the estimated state from the previous time step and the current measurement are needed to compute the estimate for the current state. In contrast to batch estimation techniques, no history of observations and/or estimates is required. In what follows, the notation $\hat{\mathbf{x}}_{n|m}$ represents the estimate of $\mathbf{x}$ at time n given observations up to and including at time $m \le n$.

The state of the filter is represented by two variables:

- $\hat{\mathbf{x}}_{k|k}$, the *posteriori* state estimate at time k given observations up to and including at time k;
- $\mathbf{P}_{k|k}$, the *posteriori* error covariance matrix (a measure of the estimated accuracy of the state estimate).

The Kalman filter can be written as a single equation, however it is most often conceptualized as two distinct phases: "Predict" and "Update". The predict phase uses the state estimate from the previous timestep to produce an estimate of the state at the current timestep. This predicted state estimate is also known as the *a priori* state estimate because, although it is an estimate of the state at the current timestep, it does not include observation information from the current timestep. In the update phase, the current *a priori* prediction is combined with current observation information to refine the state estimate. This improved estimate is termed the *a posteriori* state estimate.

Typically, the two phases alternate, with the prediction advancing the state until the next scheduled observation, and the update incorporating the observation. However, this is not necessary; if an observation is unavailable for some reason, the update may be skipped and multiple prediction steps performed. Likewise, if multiple independent observations are available at the same time, multiple update steps may be performed (typically with different observation matrices $\mathbf{H}_k$).[765]

Predict

Predicted (*a priori*) state estimate	$\hat{\mathbf{x}}_{k\mid k-1} = \mathbf{F}_k\hat{\mathbf{x}}_{k-1\mid k-1} + \mathbf{B}_k\mathbf{u}_k$
Predicted (*a priori*) error covariance	$\mathbf{P}_{k\mid k-1} = \mathbf{F}_k\mathbf{P}_{k-1\mid k-1}\mathbf{F}_k^{\mathrm{T}} + \mathbf{Q}_k$

Update

Innovation or measurement pre-fit residual	$\tilde{\mathbf{y}}_k = \mathbf{z}_k - \mathbf{H}_k\hat{\mathbf{x}}_{k\mid k-1}$
Innovation (or pre-fit residual) covariance	$\mathbf{S}_k = \mathbf{R}_k + \mathbf{H}_k\mathbf{P}_{k\mid k-1}\mathbf{H}_k^{\mathrm{T}}$
Optimal Kalman gain	$\mathbf{K}_k = \mathbf{P}_{k\mid k-1}\mathbf{H}_k^{\mathrm{T}}\mathbf{S}_k^{-1}$
Updated (*a posteriori*) state estimate	$\hat{\mathbf{x}}_{k\mid k} = \hat{\mathbf{x}}_{k\mid k-1} + \mathbf{K}_k\tilde{\mathbf{y}}_k$
Updated (*a posteriori*) estimate covariance	$\mathbf{P}_{k\mid k} = (\mathbf{I} - \mathbf{K}_k\mathbf{H}_k)\mathbf{P}_{k\mid k-1}(\mathbf{I} - \mathbf{K}_k\mathbf{H}_k)^T + \mathbf{K}_k\mathbf{R}_k\mathbf{K}_k^T$
Measurement post-fit residual	$\tilde{\mathbf{y}}_{k\mid k} = \mathbf{z}_k - \mathbf{H}_k\hat{\mathbf{x}}_{k\mid k}$

The formula for the updated (*a posteriori*) estimate covariance above is valid for any gain $\mathbf{K}_k$ and is sometimes called the **Joseph form**. For the optimal Kalman gain the formula further simplifies to $\mathbf{P}_{k|k} = (\mathbf{I} - \mathbf{K}_k\mathbf{H}_k)\mathbf{P}_{k|k-1}$, in which form it is most widely used in applications. However, one must keep in mind, that it is valid only for the optimal gain that minimizes the residual error. Proof of the formulae is found in the *derivations* section.

Invariants

If the model is accurate, and the values for $\hat{\mathbf{x}}_{0|0}$ and $\mathbf{P}_{0|0}$ accurately reflect the distribution of the initial state values, then the following invariants are preserved:

$$\begin{aligned} \mathrm{E}[\mathbf{x}_k - \hat{\mathbf{x}}_{k|k}] = \mathrm{E}[\mathbf{x}_k - \hat{\mathbf{x}}_{k|k-1}] = 0 \\ \mathrm{E}[\tilde{\mathbf{y}}_k] = 0 \end{aligned}$$

where $\mathrm{E}[\xi]$ is the expected value of ξ. That is, all estimates have a mean error of zero.

Also:

$$\begin{aligned} \mathbf{P}_{k|k} &= \mathrm{cov}(\mathbf{x}_k - \hat{\mathbf{x}}_{k|k}) \\ \mathbf{P}_{k|k-1} &= \mathrm{cov}(\mathbf{x}_k - \hat{\mathbf{x}}_{k|k-1}) \\ \mathbf{S}_k &= \mathrm{cov}(\tilde{\mathbf{y}}_k) \end{aligned}$$

so covariance matrices accurately reflect the covariance of estimates.

Estimation of the noise covariances $\mathbf{Q}_k$ and $\mathbf{R}_k$

Practical implementation of the Kalman Filter is often difficult due to the difficulty of getting a good estimate of the noise covariance matrices $\mathbf{Q}_k$ and $\mathbf{R}_k$. Extensive research has been done in this field to estimate these covariances from data. One practical approach to do this is the *autocovariance least-squares (ALS)* technique that uses the time-lagged autocovariances of routine operating data to estimate the covariances. The GNU Octave and Matlab code used to calculate the noise covariance matrices using the ALS technique is available online under the GNU General Public License.

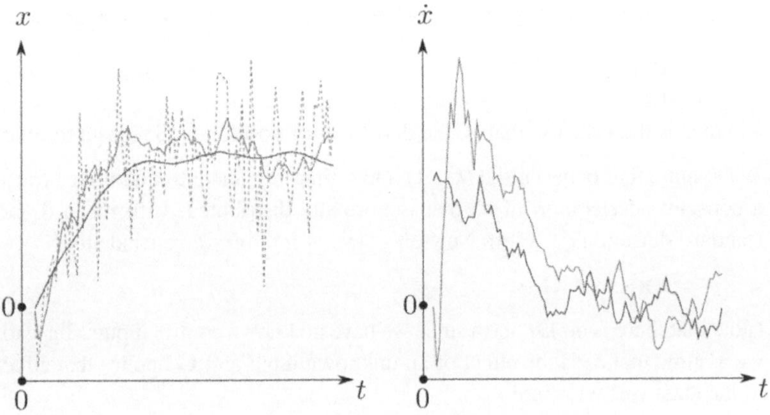

Figure 30: *Truth; filtered process; observations.*

Optimality and performance

It follows from theory that the Kalman filter is the optimal linear filter in cases where a) the model perfectly matches the real system, b) the entering noise is white (uncorrelated) and c) the covariances of the noise are exactly known. Several methods for the noise covariance estimation have been proposed during past decades, including ALS, mentioned in the section above. After the covariances are estimated, it is useful to evaluate the performance of the filter; i.e., whether it is possible to improve the state estimation quality. If the Kalman filter works optimally, the innovation sequence (the output prediction error) is a white noise, therefore the whiteness property of the innovations measures filter performance. Several different methods can be used for this purpose.[766] If the noise terms are non-Gaussian distributed, methods for assessing performance of the filter estimate, which use probability inequalities or large-sample theory, are known in the literature.

Example application, technical

Consider a truck on frictionless, straight rails. Initially, the truck is stationary at position 0, but it is buffeted this way and that by random uncontrolled forces. We measure the position of the truck every Δt seconds, but these measurements are imprecise; we want to maintain a model of where the truck is and what its velocity is. We show here how we derive the model from which we create our Kalman filter.

Since $\mathbf{F}, \mathbf{H}, \mathbf{R}, \mathbf{Q}$ are constant, their time indices are dropped.

The position and velocity of the truck are described by the linear state space

$$\mathbf{x}_k = \begin{bmatrix} x \\ \dot{x} \end{bmatrix}$$

where $\dot{x}$ is the velocity, that is, the derivative of position with respect to time.

We assume that between the (k – 1) and k timestep uncontrolled forces cause a constant acceleration of a_k that is normally distributed, with mean 0 and standard deviation σ_a. From Newton's laws of motion we conclude that

$$\mathbf{x}_k = \mathbf{F}\mathbf{x}_{k-1} + \mathbf{G}a_k$$

(note that there is no $\mathbf{B}u$ term since we have no known control inputs. Instead, we assume that a_k is the effect of an unknown input and $\mathbf{G}$ applies that effect to the state vector) where

$$\mathbf{F} = \begin{bmatrix} 1 & \Delta t \\ 0 & 1 \end{bmatrix}$$

$$\mathbf{G} = \begin{bmatrix} \frac{1}{2}\Delta t^2 \\ \Delta t \end{bmatrix}$$

so that

$$\mathbf{x}_k = \mathbf{F}\mathbf{x}_{k-1} + \mathbf{w}_k$$

where

$$\mathbf{w}_k \sim N(0, \mathbf{Q})$$

$$\mathbf{Q} = \mathbf{G}\mathbf{G}^{\mathrm{T}}\sigma_a^2 = \begin{bmatrix} \frac{1}{4}\Delta t^4 & \frac{1}{2}\Delta t^3 \\ \frac{1}{2}\Delta t^3 & \Delta t^2 \end{bmatrix}\sigma_a^2.$$

Please note that the matrix $\mathbf{Q}$ is not full rank (it is of rank one if $\Delta t \neq 0$). Hence, the distribution $N(0, \mathbf{Q})$ is not absolutely continuous and has no probability density function. Another way to express this, avoiding explicit degenerate distributions is given by

$$\mathbf{w}_k \sim \mathbf{G} \cdot N\left(0, \sigma_a^2\right).$$

At each time step, a noisy measurement of the true position of the truck is made. Let us suppose the measurement noise v_k is also normally distributed, with mean 0 and standard deviation σ_z.

$$\mathbf{z}_k = \mathbf{H}\mathbf{x}_k + \mathbf{v}_k$$

where

$$\mathbf{H} = \begin{bmatrix} 1 & 0 \end{bmatrix}$$

and

$$\mathbf{R} = \mathrm{E}\left[\mathbf{v}_k\mathbf{v}_k^{\mathrm{T}}\right] = \left[\sigma_z^2\right]$$

We know the initial starting state of the truck with perfect precision, so we initialize

$$\hat{\mathbf{x}}_{0|0} = \begin{bmatrix} 0 \\ 0 \end{bmatrix}$$

and to tell the filter that we know the exact position and velocity, we give it a zero covariance matrix:

$$\mathbf{P}_{0|0} = \begin{bmatrix} 0 & 0 \\ 0 & 0 \end{bmatrix}$$

If the initial position and velocity are not known perfectly, the covariance matrix should be initialized with suitable variances on its diagonal:

$$\mathbf{P}_{0|0} = \begin{bmatrix} \sigma_x^2 & 0 \\ 0 & \sigma_{\dot{x}}^2 \end{bmatrix}$$

The filter will then prefer the information from the first measurements over the information already in the model.

Derivations

Deriving the *posteriori* estimate covariance matrix

Starting with our invariant on the error covariance $\mathbf{P}_{k\mid k}$ as above

$$\mathbf{P}_{k|k} = \mathrm{cov}\left(\mathbf{x}_k - \hat{\mathbf{x}}_{k|k}\right)$$

substitute in the definition of $\hat{\mathbf{x}}_{k|k}$

$$\mathbf{P}_{k|k} = \mathrm{cov}\left[\mathbf{x}_k - \left(\hat{\mathbf{x}}_{k|k-1} + \mathbf{K}_k\tilde{\mathbf{y}}_k\right)\right]$$

and substitute $\tilde{\mathbf{y}}_k$

$$\mathbf{P}_{k|k} = \mathrm{cov}\left(\mathbf{x}_k - \left[\hat{\mathbf{x}}_{k|k-1} + \mathbf{K}_k\left(\mathbf{z}_k - \mathbf{H}_k\hat{\mathbf{x}}_{k|k-1}\right)\right]\right)$$

and $\mathbf{z}_k$

$$\mathbf{P}_{k|k} = \mathrm{cov}\left(\mathbf{x}_k - \left[\hat{\mathbf{x}}_{k|k-1} + \mathbf{K}_k\left(\mathbf{H}_k\mathbf{x}_k + \mathbf{v}_k - \mathbf{H}_k\hat{\mathbf{x}}_{k|k-1}\right)\right]\right)$$

and by collecting the error vectors we get

$$\mathbf{P}_{k|k} = \mathrm{cov}\left[(\mathbf{I} - \mathbf{K}_k\mathbf{H}_k)(\mathbf{x}_k - \hat{\mathbf{x}}_{k|k-1}) - \mathbf{K}_k\mathbf{v}_k\right]$$

Since the measurement error $\mathbf{v}_k$ is uncorrelated with the other terms, this becomes

$$\mathbf{P}_{k|k} = \mathrm{cov}\left[(\mathbf{I} - \mathbf{K}_k\mathbf{H}_k)\left(\mathbf{x}_k - \hat{\mathbf{x}}_{k|k-1}\right)\right] + \mathrm{cov}\left[\mathbf{K}_k\mathbf{v}_k\right]$$

by the properties of vector covariance this becomes

$$\mathbf{P}_{k|k} = (\mathbf{I} - \mathbf{K}_k\mathbf{H}_k)\mathrm{cov}(\mathbf{x}_k - \hat{\mathbf{x}}_{k|k-1})\left(\mathbf{I} - \mathbf{K}_k\mathbf{H}_k\right)^{\mathrm{T}} + \mathbf{K}_k\mathrm{cov}\left(\mathbf{v}_k\right)\mathbf{K}_k^{\mathrm{T}}$$

which, using our invariant on $\mathbf{P}_{k\mid k-1}$ and the definition of $\mathbf{R}_k$ becomes

$$\mathbf{P}_{k|k} = (\mathbf{I} - \mathbf{K}_k\mathbf{H}_k)\mathbf{P}_{k|k-1}(\mathbf{I} - \mathbf{K}_k\mathbf{H}_k)^{\mathrm{T}} + \mathbf{K}_k\mathbf{R}_k\mathbf{K}_k^{\mathrm{T}}$$

This formula (sometimes known as the **Joseph form** of the covariance update equation) is valid for any value of $\mathbf{K}_k$. It turns out that if $\mathbf{K}_k$ is the optimal Kalman gain, this can be simplified further as shown below.

Kalman gain derivation

The Kalman filter is a minimum mean-square error estimator. The error in the *a posteriori* state estimation is

$$\mathbf{x}_k - \hat{\mathbf{x}}_{k|k}$$

We seek to minimize the expected value of the square of the magnitude of this vector, $\mathrm{E}\left[\left\|\mathbf{x}_k - \hat{\mathbf{x}}_{k|k}\right\|^2\right]$. This is equivalent to minimizing the trace of the *a posteriori* estimate covariance matrix $\mathbf{P}_{k|k}$. By expanding out the terms in the equation above and collecting, we get:

$$\begin{aligned}\mathbf{P}_{k|k} &= \mathbf{P}_{k|k-1} - \mathbf{K}_k\mathbf{H}_k\mathbf{P}_{k|k-1} - \mathbf{P}_{k|k-1}\mathbf{H}_k^{\mathrm{T}}\mathbf{K}_k^{\mathrm{T}} + \mathbf{K}_k(\mathbf{H}_k\mathbf{P}_{k|k-1}\mathbf{H}_k^{\mathrm{T}} + \mathbf{R}_k)\mathbf{K}_k^{\mathrm{T}} \\ &= \mathbf{P}_{k|k-1} - \mathbf{K}_k\mathbf{H}_k\mathbf{P}_{k|k-1} - \mathbf{P}_{k|k-1}\mathbf{H}_k^{\mathrm{T}}\mathbf{K}_k^{\mathrm{T}} + \mathbf{K}_k\mathbf{S}_k\mathbf{K}_k^{\mathrm{T}}\end{aligned}$$

The trace is minimized when its matrix derivative with respect to the gain matrix is zero. Using the gradient matrix rules and the symmetry of the matrices involved we find that

$$\frac{\partial\, \mathrm{tr}(\mathbf{P}_{k|k})}{\partial\, \mathbf{K}_k} = -2(\mathbf{H}_k\mathbf{P}_{k|k-1})^{\mathrm{T}} + 2\mathbf{K}_k\mathbf{S}_k = 0.$$

Solving this for $\mathbf{K}_k$ yields the Kalman gain:

$$\begin{aligned}\mathbf{K}_k\mathbf{S}_k &= (\mathbf{H}_k\mathbf{P}_{k|k-1})^{\mathrm{T}} = \mathbf{P}_{k|k-1}\mathbf{H}_k^{\mathrm{T}} \\ \mathbf{K}_k &= \mathbf{P}_{k|k-1}\mathbf{H}_k^{\mathrm{T}}\mathbf{S}_k^{-1}\end{aligned}$$

This gain, which is known as the *optimal Kalman gain*, is the one that yields MMSE estimates when used.

Simplification of the *posteriori* error covariance formula

The formula used to calculate the *a posteriori* error covariance can be simplified when the Kalman gain equals the optimal value derived above. Multiplying both sides of our Kalman gain formula on the right by $\mathbf{S}_k\mathbf{K}_k^T$, it follows that

$$\mathbf{K}_k\mathbf{S}_k\mathbf{K}_k^{\mathrm{T}} = \mathbf{P}_{k|k-1}\mathbf{H}_k^{\mathrm{T}}\mathbf{K}_k^{\mathrm{T}}$$

Referring back to our expanded formula for the *a posteriori* error covariance,

$$\mathbf{P}_{k|k} = \mathbf{P}_{k|k-1} - \mathbf{K}_k\mathbf{H}_k\mathbf{P}_{k|k-1} - \mathbf{P}_{k|k-1}\mathbf{H}_k^{\mathrm{T}}\mathbf{K}_k^{\mathrm{T}} + \mathbf{K}_k\mathbf{S}_k\mathbf{K}_k^{\mathrm{T}}$$

we find the last two terms cancel out, giving

$$\mathbf{P}_{k|k} = \mathbf{P}_{k|k-1} - \mathbf{K}_k\mathbf{H}_k\mathbf{P}_{k|k-1} = (\mathbf{I} - \mathbf{K}_k\mathbf{H}_k)\mathbf{P}_{k|k-1}.$$

This formula is computationally cheaper and thus nearly always used in practice, but is only correct for the optimal gain. If arithmetic precision is unusually low causing problems with numerical stability, or if a non-optimal Kalman gain is deliberately used, this simplification cannot be applied; the *a posteriori* error covariance formula as derived above (Joseph form) must be used.

Sensitivity analysis

The Kalman filtering equations provide an estimate of the state $\hat{\mathbf{x}}_{k|k}$ and its error covariance $\mathbf{P}_{k|k}$ recursively. The estimate and its quality depend on the system parameters and the noise statistics fed as inputs to the estimator. This section analyzes the effect of uncertainties in the statistical inputs to the filter. In the absence of reliable statistics or the true values of noise covariance matrices $\mathbf{Q}_k$ and $\mathbf{R}_k$, the expression

$$\mathbf{P}_{k|k} = (\mathbf{I} - \mathbf{K}_k\mathbf{H}_k)\mathbf{P}_{k|k-1}(\mathbf{I} - \mathbf{K}_k\mathbf{H}_k)^{\mathrm{T}} + \mathbf{K}_k\mathbf{R}_k\mathbf{K}_k^{\mathrm{T}}$$

no longer provides the actual error covariance. In other words, $\mathbf{P}_{k|k} \neq E\left[\left(\mathbf{x}_k - \hat{\mathbf{x}}_{k|k}\right)\left(\mathbf{x}_k - \hat{\mathbf{x}}_{k|k}\right)^{\mathrm{T}}\right]$. In most real-time applications, the covariance matrices that are used in designing the Kalman filter are different from the actual (true) noise covariances matrices.Wikipedia:Citation needed This sensitivity analysis describes the behavior of the estimation error covariance when the noise covariances as well as the system matrices $\mathbf{F}_k$ and $\mathbf{H}_k$ that are fed as inputs to the filter are incorrect. Thus, the sensitivity analysis describes the robustness (or sensitivity) of the estimator to misspecified statistical and parametric inputs to the estimator.

This discussion is limited to the error sensitivity analysis for the case of statistical uncertainties. Here the actual noise covariances are denoted by $\mathbf{Q}_k^a$ and $\mathbf{R}_k^a$ respectively, whereas the design values used in the estimator are $\mathbf{Q}_k$ and $\mathbf{R}_k$ respectively. The actual error covariance is denoted by $\mathbf{P}_{k|k}^a$ and $\mathbf{P}_{k|k}$

as computed by the Kalman filter is referred to as the Riccati variable. When $\mathbf{Q}_k \equiv \mathbf{Q}_k^a$ and $\mathbf{R}_k \equiv \mathbf{R}_k^a$, this means that $\mathbf{P}_{k|k} = \mathbf{P}_{k|k}^a$. While computing the actual error covariance using $\mathbf{P}_{k|k}^a = E\left[\left(\mathbf{x}_k - \hat{\mathbf{x}}_{k|k}\right)\left(\mathbf{x}_k - \hat{\mathbf{x}}_{k|k}\right)^{\mathrm{T}}\right]$, substituting for $\widehat{\mathbf{x}}_{k|k}$ and using the fact that $E\left[\mathbf{w}_k\mathbf{w}_k^{\mathrm{T}}\right] = \mathbf{Q}_k^a$ and $E\left[\mathbf{v}_k\mathbf{v}_k^{\mathrm{T}}\right] = \mathbf{R}_k^a$, results in the following recursive equations for $\mathbf{P}_{k|k}^a$:

$$\mathbf{P}_{k|k-1}^a = \mathbf{F}_k\mathbf{P}_{k-1|k-1}^a\mathbf{F}_k^{\mathrm{T}} + \mathbf{Q}_k^a$$

and

$$\mathbf{P}_{k|k}^a = \left(\mathbf{I} - \mathbf{K}_k\mathbf{H}_k\right)\mathbf{P}_{k|k-1}^a\left(\mathbf{I} - \mathbf{K}_k\mathbf{H}_k\right)^{\mathrm{T}} + \mathbf{K}_k\mathbf{R}_k^a\mathbf{K}_k^{\mathrm{T}}$$

While computing $\mathbf{P}_{k|k}$, by design the filter implicitly assumes that $E\left[\mathbf{w}_k\mathbf{w}_k^{\mathrm{T}}\right] = \mathbf{Q}_k$ and $E\left[\mathbf{v}_k\mathbf{v}_k^{\mathrm{T}}\right] = \mathbf{R}_k$. Note that the recursive expressions for $\mathbf{P}_{k|k}^a$ and $\mathbf{P}_{k|k}$ are identical except for the presence of $\mathbf{Q}_k^a$ and $\mathbf{R}_k^a$ in place of the design values $\mathbf{Q}_k$ and $\mathbf{R}_k$ respectively. Researches have been done to analyze Kalman filter system's robustness.[767]

Square root form

One problem with the Kalman filter is its numerical stability. If the process noise covariance $\mathbf{Q}_k$ is small, round-off error often causes a small positive eigenvalue to be computed as a negative number. This renders the numerical representation of the state covariance matrix **P** indefinite, while its true form is positive-definite.

Positive definite matrices have the property that they have a triangular matrix square root $\mathbf{P} = \mathbf{S}\cdot\mathbf{S}^{\mathrm{T}}$. This can be computed efficiently using the Cholesky factorization algorithm, but more importantly, if the covariance is kept in this form, it can never have a negative diagonal or become asymmetric. An equivalent form, which avoids many of the square root operations required by the matrix square root yet preserves the desirable numerical properties, is the U-D decomposition form, $\mathbf{P} = \mathbf{U}\cdot\mathbf{D}\cdot\mathbf{U}^{\mathrm{T}}$, where **U** is a unit triangular matrix (with unit diagonal), and **D** is a diagonal matrix.

Between the two, the U-D factorization uses the same amount of storage, and somewhat less computation, and is the most commonly used square root form. (Early literature on the relative efficiency is somewhat misleading, as it assumed that square roots were much more time-consuming than divisions,[:69] while on 21-st century computers they are only slightly more expensive.)

Efficient algorithms for the Kalman prediction and update steps in the square root form were developed by G. J. Bierman and C. L. Thornton.

The $\mathbf{L}\cdot\mathbf{D}\cdot\mathbf{L}^{\mathrm{T}}$ decomposition of the innovation covariance matrix $\mathbf{S}_k$ is the basis for another type of numerically efficient and robust square root filter. The

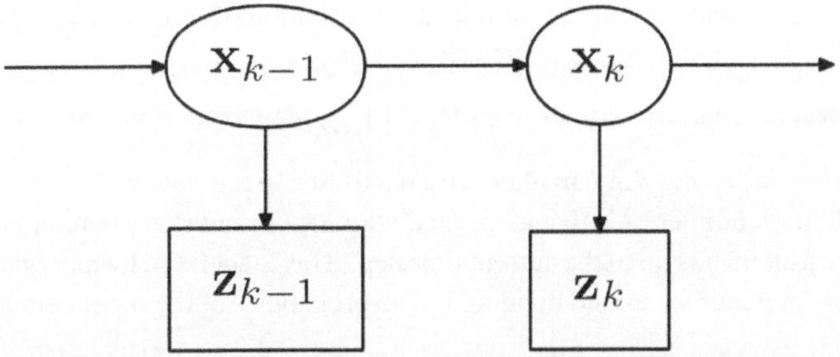

algorithm starts with the LU decomposition as implemented in the Linear Algebra PACKage (LAPACK). These results are further factored into the $\mathbf{L}\cdot\mathbf{D}\cdot\mathbf{L}^T$ structure with methods given by Golub and Van Loan (algorithm 4.1.2) for a symmetric nonsingular matrix. Any singular covariance matrix is pivoted so that the first diagonal partition is nonsingular and well-conditioned. The pivoting algorithm must retain any portion of the innovation covariance matrix directly corresponding to observed state-variables $\mathbf{H}_k \cdot \mathbf{x}_{k|k-1}$ that are associated with auxiliary observations in $\mathbf{y}_k$. The $\mathbf{l}\cdot\mathbf{d}\cdot\mathbf{l}^t$ square-root filter requires orthogonalization of the observation vector. This may be done with the inverse square-root of the covariance matrix for the auxiliary variables using Method 2 in Higham (2002, p. 263).

Relationship to recursive Bayesian estimation

The Kalman filter can be presented as one of the simplest dynamic Bayesian networks. The Kalman filter calculates estimates of the true values of states recursively over time using incoming measurements and a mathematical process model. Similarly, recursive Bayesian estimation calculates estimates of an unknown probability density function (PDF) recursively over time using incoming measurements and a mathematical process model.

In recursive Bayesian estimation, the true state is assumed to be an unobserved Markov process, and the measurements are the observed states of a hidden Markov model (HMM).

because of the Markov assumption, the true state is conditionally independent of all earlier states given the immediately previous state.

$$p(\mathbf{x}_k \mid \mathbf{x}_0, \dots, \mathbf{x}_{k-1}) = p(\mathbf{x}_k \mid \mathbf{x}_{k-1})$$

Similarly, the measurement at the k-th timestep is dependent only upon the current state and is conditionally independent of all other states given the current state.

$$p(\mathbf{z}_k \mid \mathbf{x}_0, \dots, \mathbf{x}_k) = p(\mathbf{z}_k \mid \mathbf{x}_k)$$

Using these assumptions the probability distribution over all states of the hidden Markov model can be written simply as:

$$p(\mathbf{x}_0, \ldots, \mathbf{x}_k, \mathbf{z}_1, \ldots, \mathbf{z}_k) = p(\mathbf{x}_0) \prod_{i=1}^{k} p(\mathbf{z}_i \mid \mathbf{x}_i) p(\mathbf{x}_i \mid \mathbf{x}_{i-1})$$

However, when the Kalman filter is used to estimate the state **x**, the probability distribution of interest is that associated with the current states conditioned on the measurements up to the current timestep. This is achieved by marginalizing out the previous states and dividing by the probability of the measurement set.

This leads to the *predict* and *update* steps of the Kalman filter written probabilistically. The probability distribution associated with the predicted state is the sum (integral) of the products of the probability distribution associated with the transition from the (k – 1)-th timestep to the k-th and the probability distribution associated with the previous state, over all possible Z_{k-1} .

$$p(\mathbf{x}_k \mid \mathbf{Z}_{k-1}) = \int p(\mathbf{x}_k \mid \mathbf{x}_{k-1}) p(\mathbf{x}_{k-1} \mid \mathbf{Z}_{k-1}) \, d\mathbf{x}_{k-1}$$

The measurement set up to time t is

$$\mathbf{Z}_t = \{\mathbf{z}_1, \ldots, \mathbf{z}_t\}$$

The probability distribution of the update is proportional to the product of the measurement likelihood and the predicted state.

$$p(\mathbf{x}_k \mid \mathbf{Z}_k) = \frac{p(\mathbf{z}_k \mid \mathbf{x}_k) p(\mathbf{x}_k \mid \mathbf{Z}_{k-1})}{p(\mathbf{z}_k \mid \mathbf{Z}_{k-1})}$$

The denominator

$$p(\mathbf{z}_k \mid \mathbf{Z}_{k-1}) = \int p(\mathbf{z}_k \mid \mathbf{x}_k) p(\mathbf{x}_k \mid \mathbf{Z}_{k-1}) \, d\mathbf{x}_k$$

is a normalization term.

The remaining probability density functions are

$$\begin{aligned} p(\mathbf{x}_k \mid \mathbf{x}_{k-1}) &= \mathcal{N}(\mathbf{F}_k \mathbf{x}_{k-1}, \mathbf{Q}_k) \\ p(\mathbf{z}_k \mid \mathbf{x}_k) &= \mathcal{N}(\mathbf{H}_k \mathbf{x}_k, \mathbf{R}_k) \\ p(\mathbf{x}_{k-1} \mid \mathbf{Z}_{k-1}) &= \mathcal{N}(\hat{\mathbf{x}}_{k-1}, \mathbf{P}_{k-1}) \end{aligned}$$

Note that the PDF at the previous timestep is inductively assumed to be the estimated state and covariance. This is justified because, as an optimal estimator, the Kalman filter makes best use of the measurements, therefore the PDF for $\mathbf{x}_k$ given the measurements $\mathbf{Z}_k$ is the Kalman filter estimate.

Marginal likelihood

Related to the recursive Bayesian interpretation described above, the Kalman filter can be viewed as a generative model, i.e., a process for *generating* a stream of random observations $\mathbf{z} = (\mathbf{z}_0, \mathbf{z}_1, \mathbf{z}_2, \ldots)$. Specifically, the process is

1. Sample a hidden state $\mathbf{x}_0$ from the Gaussian prior distribution $p(\mathbf{x}_0) = \mathcal{N}(\hat{\mathbf{x}}_{0|0}, \mathbf{P}_{0|0})$.
2. Sample an observation $\mathbf{z}_0$ from the observation model $p(\mathbf{z}_0 \mid \mathbf{x}_0) = \mathcal{N}(\mathbf{H}_0\mathbf{x}_0, \mathbf{R}_0)$.
3. For $k = 1, 2, 3, \ldots$, do
   1. Sample the next hidden state $\mathbf{x}_k$ from the transition model $p(\mathbf{x}_k \mid \mathbf{x}_{k-1}) = \mathcal{N}(\mathbf{F}_k\mathbf{x}_{k-1} + \mathbf{B}_k\mathbf{u}_k, \mathbf{Q}_k)$.
   2. Sample an observation $\mathbf{z}_k$ from the observation model $p(\mathbf{z}_k \mid \mathbf{x}_k) = \mathcal{N}(\mathbf{H}_k\mathbf{x}_k, \mathbf{R}_k)$.

Note that this process has identical structure to the hidden Markov model, except that the discrete state and observations are replaced with continuous variables sampled from Gaussian distributions.

In some applications, it is useful to compute the *probability* that a Kalman filter with a given set of parameters (prior distribution, transition and observation models, and control inputs) would generate a particular observed signal. This probability is known as the marginal likelihood because it integrates over ("marginalizes out") the values of the hidden state variables, so it can be computed using only the observed signal. The marginal likelihood can be useful to evaluate different parameter choices, or to compare the Kalman filter against other models using Bayesian model comparison.

It is straightforward to compute the marginal likelihood as a side effect of the recursive filtering computation. By the chain rule, the likelihood can be factored as the product of the probability of each observation given previous observations,

$$p(\mathbf{z}) = \prod_{k=0}^{T} p(\mathbf{z}_k \mid \mathbf{z}_{k-1}, \ldots, \mathbf{z}_0),$$

and because the Kalman filter describes a Markov process, all relevant information from previous observations is contained in the current state estimate $\hat{\mathbf{x}}_{k|k-1}, \mathbf{P}_{k|k-1}$. Thus the marginal likelihood is given by

$$\begin{aligned}
p(\mathbf{z}) &= \prod_{k=0}^{T} \int p(\mathbf{z}_k \mid \mathbf{x}_k) p(\mathbf{x}_k \mid \mathbf{z}_{k-1}, \ldots, \mathbf{z}_0) d\mathbf{x}_k \\
&= \prod_{k=0}^{T} \int \mathcal{N}(\mathbf{z}_k; \mathbf{H}_k \mathbf{x}_k, \mathbf{R}_k) \mathcal{N}(\mathbf{x}_k; \hat{\mathbf{x}}_{k|k-1}, \mathbf{P}_{k|k-1}) d\mathbf{x}_k \\
&= \prod_{k=0}^{T} \mathcal{N}(\mathbf{z}_k; \mathbf{H}_k \hat{\mathbf{x}}_{k|k-1}, \mathbf{R}_k + \mathbf{H}_k \mathbf{P}_{k|k-1} \mathbf{H}_k^T) \\
&= \prod_{k=0}^{T} \mathcal{N}(\mathbf{z}_k; \mathbf{H}_k \hat{\mathbf{x}}_{k|k-1}, \mathbf{S}_k),
\end{aligned}$$

i.e., a product of Gaussian densities, each corresponding to the density of one observation $\mathbf{z}_k$ under the current filtering distribution $\mathbf{H}_k \hat{\mathbf{x}}_{k|k-1}, \mathbf{S}_k$. This can easily be computed as a simple recursive update; however, to avoid numeric underflow, in a practical implementation it is usually desirable to compute the *log* marginal likelihood $\ell = \log p(\mathbf{z})$ instead. Adopting the convention $\ell^{(-1)} = 0$, this can be done via the recursive update rule

$$\ell^{(k)} = \ell^{(k-1)} - \frac{1}{2} \left(\tilde{\mathbf{y}}_k^T \mathbf{S}_k^{-1} \tilde{\mathbf{y}}_k + \log |\mathbf{S}_k| + d_y \log 2\pi \right),$$

where d_y is the dimension of the measurement vector.

An important application where such a (log) likelihood of the observations (given the filter parameters) is used is multi-target tracking. For example, consider an object tracking scenario where a stream of observations is the input, however, it is unknown how many objects are in the scene (or, the number of objects is known but is greater than one). In such a scenario, it can be unknown apriori which observations/measurements were generated by which object. A multiple hypothesis tracker (MHT) typically will form different track association hypotheses, where each hypothesis can be viewed as a Kalman filter (in the linear Gaussian case) with a specific set of parameters associated with the hypothesized object. Thus, it is important to compute the likelihood of the observations for the different hypotheses under consideration, such that the most-likely one can be found.

Information filter

In the information filter, or inverse covariance filter, the estimated covariance and estimated state are replaced by the information matrix and information vector respectively. These are defined as:

$$\begin{aligned}
\mathbf{Y}_{k|k} &= \mathbf{P}_{k|k}^{-1} \\
\hat{\mathbf{y}}_{k|k} &= \mathbf{P}_{k|k}^{-1} \hat{\mathbf{x}}_{k|k}
\end{aligned}$$

Similarly the predicted covariance and state have equivalent information forms, defined as:

$$\mathbf{Y}_{k|k-1} = \mathbf{P}_{k|k-1}^{-1}$$
$$\hat{\mathbf{y}}_{k|k-1} = \mathbf{P}_{k|k-1}^{-1}\hat{\mathbf{x}}_{k|k-1}$$

as have the measurement covariance and measurement vector, which are defined as:

$$\mathbf{I}_k = \mathbf{H}_k^{\mathrm{T}}\mathbf{R}_k^{-1}\mathbf{H}_k$$
$$\mathbf{i}_k = \mathbf{H}_k^{\mathrm{T}}\mathbf{R}_k^{-1}\mathbf{z}_k$$

The information update now becomes a trivial sum.

$$\mathbf{Y}_{k|k} = \mathbf{Y}_{k|k-1} + \mathbf{I}_k$$
$$\hat{\mathbf{y}}_{k|k} = \hat{\mathbf{y}}_{k|k-1} + \mathbf{i}_k$$

The main advantage of the information filter is that *N* measurements can be filtered at each timestep simply by summing their information matrices and vectors.

$$\mathbf{Y}_{k|k} = \mathbf{Y}_{k|k-1} + \sum_{j=1}^{N} \mathbf{I}_{k,j}$$
$$\hat{\mathbf{y}}_{k|k} = \hat{\mathbf{y}}_{k|k-1} + \sum_{j=1}^{N} \mathbf{i}_{k,j}$$

To predict the information filter the information matrix and vector can be converted back to their state space equivalents, or alternatively the information space prediction can be used.

$$\mathbf{M}_k = \left[\mathbf{F}_k^{-1}\right]^{\mathrm{T}} \mathbf{Y}_{k-1|k-1}\mathbf{F}_k^{-1}$$
$$\mathbf{C}_k = \mathbf{M}_k\left[\mathbf{M}_k + \mathbf{Q}_k^{-1}\right]^{-1}$$
$$\mathbf{L}_k = \mathbf{I} - \mathbf{C}_k$$
$$\mathbf{Y}_{k|k-1} = \mathbf{L}_k\mathbf{M}_k\mathbf{L}_k^{\mathrm{T}} + \mathbf{C}_k\mathbf{Q}_k^{-1}\mathbf{C}_k^{\mathrm{T}}$$
$$\hat{\mathbf{y}}_{k|k-1} = \mathbf{L}_k\left[\mathbf{F}_k^{-1}\right]^{\mathrm{T}}\hat{\mathbf{y}}_{k-1|k-1}$$

Note that if *F* and *Q* are time invariant these values can be cached. Note also that *F* and *Q* need to be invertible.

Fixed-lag smoother

The optimal fixed-lag smoother provides the optimal estimate of $\hat{\mathbf{x}}_{k-N|k}$ for a given fixed-lag N using the measurements from $\mathbf{z}_1$ to $\mathbf{z}_k$. It can be derived using the previous theory via an augmented state, and the main equation of the filter is the following:

$$\begin{bmatrix} \hat{\mathbf{x}}_{t|t} \\ \hat{\mathbf{x}}_{t-1|t} \\ \vdots \\ \hat{\mathbf{x}}_{t-N+1|t} \end{bmatrix} = \begin{bmatrix} \mathbf{I} \\ 0 \\ \vdots \\ 0 \end{bmatrix} \hat{\mathbf{x}}_{t|t-1} + \begin{bmatrix} 0 & \dots & 0 \\ \mathbf{I} & 0 & \vdots \\ \vdots & \ddots & \vdots \\ 0 & \dots & \mathbf{I} \end{bmatrix} \begin{bmatrix} \hat{\mathbf{x}}_{t-1|t-1} \\ \hat{\mathbf{x}}_{t-2|t-1} \\ \vdots \\ \hat{\mathbf{x}}_{t-N+1|t-1} \end{bmatrix} + \begin{bmatrix} \mathbf{K}^{(0)} \\ \mathbf{K}^{(1)} \\ \vdots \\ \mathbf{K}^{(N-1)} \end{bmatrix} \mathbf{y}_{t|t-1}$$

where:

- $\hat{\mathbf{x}}_{t|t-1}$ is estimated via a standard Kalman filter;
- $\mathbf{y}_{t|t-1} = \mathbf{z}_t - \mathbf{H}\hat{\mathbf{x}}_{t|t-1}$ is the innovation produced considering the estimate of the standard Kalman filter;
- the various $\hat{\mathbf{x}}_{t-i|t}$ with $i = 1, \dots, N-1$ are new variables; i.e., they do not appear in the standard Kalman filter;
- the gains are computed via the following scheme:

  $$\mathbf{K}^{(i)} = \mathbf{P}^{(i)}\mathbf{H}^{\mathrm{T}}\left[\mathbf{H}\mathbf{P}\mathbf{H}^{\mathrm{T}} + \mathbf{R}\right]^{-1}$$

  and

  $$\mathbf{P}^{(i)} = \mathbf{P}\left[(\mathbf{F} - \mathbf{K}\mathbf{H})^{\mathrm{T}}\right]^{i}$$

  where $\mathbf{P}$ and $\mathbf{K}$ are the prediction error covariance and the gains of the standard Kalman filter (i.e., $\mathbf{P}_{t|t-1}$).

If the estimation error covariance is defined so that

$$\mathbf{P}_i := E\left[\left(\mathbf{x}_{t-i} - \hat{\mathbf{x}}_{t-i|t}\right)^* \left(\mathbf{x}_{t-i} - \hat{\mathbf{x}}_{t-i|t}\right) \mid z_1 \dots z_t\right],$$

then we have that the improvement on the estimation of $\mathbf{x}_{t-i}$ is given by:

$$\mathbf{P} - \mathbf{P}_i = \sum_{j=0}^{i}\left[\mathbf{P}^{(j)}\mathbf{H}^{\mathrm{T}}\left(\mathbf{H}\mathbf{P}\mathbf{H}^{\mathrm{T}} + \mathbf{R}\right)^{-1}\mathbf{H}\left(\mathbf{P}^{(i)}\right)^{\mathrm{T}}\right]$$

Fixed-interval smoothers

The optimal fixed-interval smoother provides the optimal estimate of $\hat{\mathbf{x}}_{k|n}$ ($k < n$) using the measurements from a fixed interval $\mathbf{z}_1$ to $\mathbf{z}_n$. This is also called "Kalman Smoothing". There are several smoothing algorithms in common use.

Rauch–Tung–Striebel

The Rauch–Tung–Striebel (RTS) smoother is an efficient two-pass algorithm for fixed interval smoothing.

The forward pass is the same as the regular Kalman filter algorithm. These *filtered* a-priori and a-posteriori state estimates $\hat{\mathbf{x}}_{k|k-1}$, $\hat{\mathbf{x}}_{k|k}$ and covariances $\mathbf{P}_{k|k-1}$, $\mathbf{P}_{k|k}$ are saved for use in the backwards pass.

In the backwards pass, we compute the *smoothed* state estimates $\hat{\mathbf{x}}_{k|n}$ and covariances $\mathbf{P}_{k|n}$. We start at the last time step and proceed backwards in time using the following recursive equations:

$$\hat{\mathbf{x}}_{k|n} = \hat{\mathbf{x}}_{k|k} + \mathbf{C}_k(\hat{\mathbf{x}}_{k+1|n} - \hat{\mathbf{x}}_{k+1|k})$$

$$\mathbf{P}_{k|n} = \mathbf{P}_{k|k} + \mathbf{C}_k(\mathbf{P}_{k+1|n} - \mathbf{P}_{k+1|k})\mathbf{C}_k^{\mathrm{T}}$$

where

$$\mathbf{C}_k = \mathbf{P}_{k|k}\mathbf{F}_{k+1}^{\mathrm{T}}\mathbf{P}_{k+1|k}^{-1}.$$

Note that $\mathbf{x}_{k|k}$ is the a-posteriori state estimate of timestep k and $\mathbf{x}_{k+1|k}$ is the a-priori state estimate of timestep $k+1$. The same notation applies to the covariance.

Modified Bryson–Frazier smoother

An alternative to the RTS algorithm is the modified Bryson–Frazier (MBF) fixed interval smoother developed by Bierman. This also uses a backward pass that processes data saved from the Kalman filter forward pass. The equations for the backward pass involve the recursive computation of data which are used at each observation time to compute the smoothed state and covariance.

The recursive equations are

$$\begin{aligned}
\tilde{\Lambda}_k &= \mathbf{H}_k^{\mathrm{T}}\mathbf{S}_k^{-1}\mathbf{H}_k + \hat{\mathbf{C}}_k^{\mathrm{T}}\hat{\Lambda}_k\hat{\mathbf{C}}_k \\
\hat{\Lambda}_{k-1} &= \mathbf{F}_k^{\mathrm{T}}\tilde{\Lambda}_k\mathbf{F}_k \\
\hat{\Lambda}_n &= 0 \\
\tilde{\lambda}_k &= -\mathbf{H}_k^{\mathrm{T}}\mathbf{S}_k^{-1}\mathbf{y}_k + \hat{\mathbf{C}}_k^{\mathrm{T}}\hat{\lambda}_k \\
\hat{\lambda}_{k-1} &= \mathbf{F}_k^{\mathrm{T}}\tilde{\lambda}_k \\
\hat{\lambda}_n &= 0
\end{aligned}$$

where $\mathbf{S}_k$ is the residual covariance and $\hat{\mathbf{C}}_k = \mathbf{I} - \mathbf{K}_k\mathbf{H}_k$. The smoothed state and covariance can then be found by substitution in the equations

$$\mathbf{P}_{k|n} = \mathbf{P}_{k|k} - \mathbf{P}_{k|k}\hat{\Lambda}_k\mathbf{P}_{k|k}$$

$$\mathbf{x}_{k|n} = \mathbf{x}_{k|k} - \mathbf{P}_{k|k}\hat{\lambda}_k$$

or

$$\mathbf{P}_{k|n} = \mathbf{P}_{k|k-1} - \mathbf{P}_{k|k-1}\tilde{\Lambda}_k\mathbf{P}_{k|k-1}$$
$$\mathbf{x}_{k|n} = \mathbf{x}_{k|k-1} - \mathbf{P}_{k|k-1}\tilde{\lambda}_k.$$

An important advantage of the MBF is that it does not require finding the inverse of the covariance matrix.

Minimum-variance smoother

The minimum-variance smoother can attain the best-possible error performance, provided that the models are linear, their parameters and the noise statistics are known precisely. This smoother is a time-varying state-space generalization of the optimal non-causal Wiener filter.

The smoother calculations are done in two passes. The forward calculations involve a one-step-ahead predictor and are given by

$$\hat{\mathbf{x}}_{k+1|k} = (\mathbf{F}_k - \mathbf{K}_k\mathbf{H}_k)\hat{\mathbf{x}}_{k|k-1} + \mathbf{K}_k\mathbf{z}_k$$
$$\alpha_k = -\mathbf{S}_k^{-\frac{1}{2}}\mathbf{H}_k\hat{\mathbf{x}}_{k|k-1} + \mathbf{S}_k^{-\frac{1}{2}}\mathbf{z}_k$$

The above system is known as the inverse Wiener-Hopf factor. The backward recursion is the adjoint of the above forward system. The result of the backward pass β_k may be calculated by operating the forward equations on the time-reversed α_k and time reversing the result. In the case of output estimation, the smoothed estimate is given by

$$\hat{\mathbf{y}}_{k|N} = \mathbf{z}_k - \mathbf{R}_k\beta_k$$

Taking the causal part of this minimum-variance smoother yields

$$\hat{\mathbf{y}}_{k|k} = \mathbf{z}_k - \mathbf{R}_k\mathbf{S}_k^{-\frac{1}{2}}\alpha_k$$

which is identical to the minimum-variance Kalman filter. The above solutions minimize the variance of the output estimation error. Note that the Rauch–Tung–Striebel smoother derivation assumes that the underlying distributions are Gaussian, whereas the minimum-variance solutions do not. Optimal smoothers for state estimation and input estimation can be constructed similarly.

A continuous-time version of the above smoother is described in.

Expectation-maximization algorithms may be employed to calculate approximate maximum likelihood estimates of unknown state-space parameters within minimum-variance filters and smoothers. Often uncertainties remain within problem assumptions. A smoother that accommodates uncertainties can be designed by adding a positive definite term to the Riccati equation.

In cases where the models are nonlinear, step-wise linearizations may be within the minimum-variance filter and smoother recursions (extended Kalman filtering).

Frequency-weighted Kalman filters

Pioneering research on the perception of sounds at different frequencies was conducted by Fletcher and Munson in the 1930s. Their work led to a standard way of weighting measured sound levels within investigations of industrial noise and hearing loss. Frequency weightings have since been used within filter and controller designs to manage performance within bands of interest.

Typically, a frequency shaping function is used to weight the average power of the error spectral density in a specified frequency band. Let $\mathbf{y} - \hat{\mathbf{y}}$ denote the output estimation error exhibited by a conventional Kalman filter. Also, let $\mathbf{W}$ denote a causal frequency weighting transfer function. The optimum solution which minimizes the variance of $\mathbf{W}\left(\mathbf{y} - \hat{\mathbf{y}}\right)$ arises by simply constructing $\mathbf{W}^{-1}\hat{\mathbf{y}}$.

The design of $\mathbf{W}$ remains an open question. One way of proceeding is to identify a system which generates the estimation error and setting $\mathbf{W}$ equal to the inverse of that system. This procedure may be iterated to obtain mean-square error improvement at the cost of increased filter order. The same technique can be applied to smoothers.

Non-linear filters

The basic Kalman filter is limited to a linear assumption. More complex systems, however, can be nonlinear. The non-linearity can be associated either with the process model or with the observation model or with both.

Extended Kalman filter

In the extended Kalman filter (EKF), the state transition and observation models need not be linear functions of the state but may instead be non-linear functions. These functions are of differentiable type.

$$\mathbf{x}_k = f(\mathbf{x}_{k-1}, \mathbf{u}_k) + \mathbf{w}_k$$

$$\mathbf{z}_k = h(\mathbf{x}_k) + \mathbf{v}_k$$

The function f can be used to compute the predicted state from the previous estimate and similarly the function h can be used to compute the predicted measurement from the predicted state. However, f and h cannot be applied to the covariance directly. Instead a matrix of partial derivatives (the Jacobian) is computed.

At each timestep the Jacobian is evaluated with current predicted states. These matrices can be used in the Kalman filter equations. This process essentially linearizes the non-linear function around the current estimate.

Unscented Kalman filter

When the state transition and observation models—that is, the predict and update functions f and h —are highly non-linear, the extended Kalman filter can give particularly poor performance. This is because the covariance is propagated through linearization of the underlying non-linear model. The unscented Kalman filter (UKF) uses a deterministic sampling technique known as the unscented transform (UT) to pick a minimal set of sample points (called sigma points) around the mean. The sigma points are then propagated through the non-linear functions, from which a new mean and covariance estimate are then formed. The resulting filter depends on how the transformed statistics of the UT are calculated and which set of sigma points are used—it should be remarked that is always possible to construct new UKFs in a consistent way. For certain systems, the resulting UKF filter more accurately estimates the true mean and covariance. This can be verified with Monte Carlo sampling or Taylor series expansion of the posterior statistics. In addition, this technique removes the requirement to explicitly calculate Jacobians, which for complex functions can be a difficult task in itself (i.e., requiring complicated derivatives if done analytically or being computationally costly if done numerically), if not impossible (if those functions are not differentiable).

Predict

As with the EKF, the UKF prediction can be used independently from the UKF update, in combination with a linear (or indeed EKF) update, or vice versa.

The estimated state and covariance are augmented with the mean and covariance of the process noise.

$$\mathbf{x}^a_{k-1|k-1} = \begin{bmatrix}\hat{\mathbf{x}}^{\mathrm{T}}_{k-1|k-1} & E\left[\mathbf{w}^{\mathrm{T}}_k\right]\end{bmatrix}^{\mathrm{T}}$$

$$\mathbf{P}^a_{k-1|k-1} = \begin{bmatrix}\mathbf{P}_{k-1|k-1} & 0 \\ 0 & \mathbf{Q}_k\end{bmatrix}$$

A set of $2L + 1$ sigma points is derived from the augmented state and covariance where L is the dimension of the augmented state.

$$\chi^0_{k-1|k-1} = \mathbf{x}^a_{k-1|k-1}$$

$$\chi^i_{k-1|k-1} = \mathbf{x}^a_{k-1|k-1} + \left(\sqrt{(L+\lambda)\mathbf{P}^a_{k-1|k-1}}\right)_i, \qquad i = 1,\ldots,L$$

$$\chi^i_{k-1|k-1} = \mathbf{x}^a_{k-1|k-1} - \left(\sqrt{(L+\lambda)\mathbf{P}^a_{k-1|k-1}}\right)_{i-L}, \qquad i = L+1,\ldots,2L$$

where

$$\left(\sqrt{(L+\lambda)\mathbf{P}^a_{k-1|k-1}}\right)_i$$

is the ith column of the matrix square root of

$$(L+\lambda)\mathbf{P}^a_{k-1|k-1}$$

using the definition: square root $\mathbf{A}$ of matrix $\mathbf{B}$ satisfies

$$\mathbf{B} \triangleq \mathbf{A}\mathbf{A}.$$

The matrix square root should be calculated using numerically efficient and stable methods such as the Cholesky decomposition.

The sigma points are propagated through the transition function f.

$$\chi^i_{k|k-1} = f\left(\chi^i_{k-1|k-1}\right) \quad i = 0,\ldots,2L$$

where $f : R^L \to R^{|\mathbf{x}|}$. The weighted sigma points are recombined to produce the predicted state and covariance.

$$\hat{\mathbf{x}}_{k|k-1} = \sum_{i=0}^{2L} W^i_s \chi^i_{k|k-1}$$

$$\mathbf{P}_{k|k-1} = \sum_{i=0}^{2L} W^i_c \left[\chi^i_{k|k-1} - \hat{\mathbf{x}}_{k|k-1}\right]\left[\chi^i_{k|k-1} - \hat{\mathbf{x}}_{k|k-1}\right]^{\mathrm{T}}$$

where the weights for the state and covariance are given by:

$$W^0_s = \frac{\lambda}{L+\lambda}$$

$$W^0_c = \frac{\lambda}{L+\lambda} + (1-\alpha^2+\beta)$$

$$W^i_s = W^i_c = \frac{1}{2(L+\lambda)}$$

$$\lambda = \alpha^2(L+\kappa) - L$$

α and κ control the spread of the sigma points. β is related to the distribution of x .

Appropriate values depend on the problem at hand, but a typical recommendation is $\alpha = 10^{-3}$, $\kappa = 0$ and $\beta = 2$. However, a larger value of α (e.g., $\alpha = 1$) may be beneficial in order to better capture the spread of the distribution and possible non-linearities, as discussed in , where a detailed inspection of how to choose sigma points can be found. If the true distribution of x is Gaussian, $\beta = 2$ is optimal.

Update

The predicted state and covariance are augmented as before, except now with the mean and covariance of the measurement noise.

$$\mathbf{x}^a_{k|k-1} = \left[\hat{\mathbf{x}}^{\mathrm{T}}_{k|k-1} \quad E\left[\mathbf{v}^{\mathrm{T}}_k\right]\right]^{\mathrm{T}}$$

$$\mathbf{P}^a_{k|k-1} = \begin{bmatrix} \mathbf{P}_{k|k-1} & 0 \\ 0 & \mathbf{R}_k \end{bmatrix}$$

As before, a set of $2L + 1$ sigma points is derived from the augmented state and covariance where L is the dimension of the augmented state.

$$\chi^0_{k|k-1} = \mathbf{x}^a_{k|k-1}$$

$$\chi^i_{k|k-1} = \mathbf{x}^a_{k|k-1} + \left(\sqrt{(L+\lambda)\mathbf{P}^a_{k|k-1}}\right)_i, \qquad i = 1, \ldots, L$$

$$\chi^i_{k|k-1} = \mathbf{x}^a_{k|k-1} - \left(\sqrt{(L+\lambda)\mathbf{P}^a_{k|k-1}}\right)_{i-L}, \qquad i = L+1, \ldots, 2L$$

Alternatively if the UKF prediction has been used the sigma points themselves can be augmented along the following lines

$$\chi_{k|k-1} := \left[\chi^{\mathrm{T}}_{k|k-1} \quad E\left[\mathbf{v}^{\mathrm{T}}_k\right]\right]^{\mathrm{T}} \pm \sqrt{(L+\lambda)\mathbf{R}^a_k}$$

where

$$\mathbf{R}^a_k = \begin{bmatrix} 0 & 0 \\ 0 & \mathbf{R}_k \end{bmatrix}$$

The sigma points are projected through the observation function h.

$$\gamma^i_k = h(\chi^i_{k|k-1}) \quad i = 0, \ldots, 2L$$

The weighted sigma points are recombined to produce the predicted measurement and predicted measurement covariance.

$$\hat{\mathbf{z}}_k = \sum_{i=0}^{2L} W^i_s \gamma^i_k$$

$$\mathbf{P}_{z_k z_k} = \sum_{i=0}^{2L} W^i_c \left[\gamma^i_k - \hat{\mathbf{z}}_k\right] \left[\gamma^i_k - \hat{\mathbf{z}}_k\right]^{\mathrm{T}}$$

The state-measurement cross-covariance matrix,

$$\mathbf{P}_{x_k z_k} = \sum_{i=0}^{2L} W^i_c \left[\chi^i_{k|k-1} - \hat{\mathbf{x}}_{k|k-1}\right]\left[\gamma^i_k - \hat{\mathbf{z}}_k\right]^{\mathrm{T}}$$

is used to compute the UKF Kalman gain.

$$K_k = \mathbf{P}_{x_k z_k} \mathbf{P}^{-1}_{z_k z_k}$$

As with the Kalman filter, the updated state is the predicted state plus the innovation weighted by the Kalman gain,

$$\hat{\mathbf{x}}_{k|k} = \hat{\mathbf{x}}_{k|k-1} + K_k(\mathbf{z}_k - \hat{\mathbf{z}}_k)$$

And the updated covariance is the predicted covariance, minus the predicted measurement covariance, weighted by the Kalman gain.

$$\mathbf{P}_{k|k} = \mathbf{P}_{k|k-1} - K_k \mathbf{P}_{z_k z_k} K_k^{\mathrm{T}}$$

Kalman–Bucy filter

The Kalman–Bucy filter (named after Richard Snowden Bucy) is a continuous time version of the Kalman filter.[768,769]

It is based on the state space model

$$\frac{d}{dt}\mathbf{x}(t) = \mathbf{F}(t)\mathbf{x}(t) + \mathbf{B}(t)\mathbf{u}(t) + \mathbf{w}(t)$$

$$\mathbf{z}(t) = \mathbf{H}(t)\mathbf{x}(t) + \mathbf{v}(t)$$

where $\mathbf{Q}(t)$ and $\mathbf{R}(t)$ represent the intensities (or, more accurately: the Power Spectral Density - PSD - matrices) of the two white noise terms $\mathbf{w}(t)$ and $\mathbf{v}(t)$, respectively.

The filter consists of two differential equations, one for the state estimate and one for the covariance:

$$\frac{d}{dt}\hat{\mathbf{x}}(t) = \mathbf{F}(t)\hat{\mathbf{x}}(t) + \mathbf{B}(t)\mathbf{u}(t) + \mathbf{K}(t)\left(\mathbf{z}(t) - \mathbf{H}(t)\hat{\mathbf{x}}(t)\right)$$

$$\frac{d}{dt}\mathbf{P}(t) = \mathbf{F}(t)\mathbf{P}(t) + \mathbf{P}(t)\mathbf{F}^{T}(t) + \mathbf{Q}(t) - \mathbf{K}(t)\mathbf{R}(t)\mathbf{K}^{\mathrm{T}}(t)$$

where the Kalman gain is given by

$$\mathbf{K}(t) = \mathbf{P}(t)\mathbf{H}^{\mathrm{T}}(t)\mathbf{R}^{-1}(t)$$

Note that in this expression for $\mathbf{K}(t)$ the covariance of the observation noise $\mathbf{R}(t)$ represents at the same time the covariance of the prediction error (or *innovation*) $\tilde{\mathbf{y}}(t) = \mathbf{z}(t) - \mathbf{H}(t)\hat{\mathbf{x}}(t)$; these covariances are equal only in the case of continuous time.

The distinction between the prediction and update steps of discrete-time Kalman filtering does not exist in continuous time.

The second differential equation, for the covariance, is an example of a Riccati equation.

Hybrid Kalman filter

Most physical systems are represented as continuous-time models while discrete-time measurements are frequently taken for state estimation via a digital processor. Therefore, the system model and measurement model are given by

$$\dot{\mathbf{x}}(t) = \mathbf{F}(t)\mathbf{x}(t) + \mathbf{B}(t)\mathbf{u}(t) + \mathbf{w}(t), \quad \mathbf{w}(t) \sim N\left(\mathbf{0}, \mathbf{Q}(t)\right)$$
$$\mathbf{z}_k = \mathbf{H}_k\mathbf{x}_k + \mathbf{v}_k, \quad \mathbf{v}_k \sim N(\mathbf{0}, \mathbf{R}_k)$$

where

$$\mathbf{x}_k = \mathbf{x}(t_k).$$

Initialize

$$\hat{\mathbf{x}}_{0|0} = E\left[\mathbf{x}(t_0)\right], \mathbf{P}_{0|0} = \text{Var}\left[\mathbf{x}(t_0)\right]$$

Predict

$$\dot{\hat{\mathbf{x}}}(t) = \mathbf{F}(t)\hat{\mathbf{x}}(t) + \mathbf{B}(t)\mathbf{u}(t), \text{ with } \hat{\mathbf{x}}(t_{k-1}) = \hat{\mathbf{x}}_{k-1|k-1}$$
$$\Rightarrow \hat{\mathbf{x}}_{k|k-1} = \hat{\mathbf{x}}(t_k)$$
$$\dot{\mathbf{P}}(t) = \mathbf{F}(t)\mathbf{P}(t) + \mathbf{P}(t)\mathbf{F}(t)^{\mathrm{T}} + \mathbf{Q}(t), \text{ with } \mathbf{P}(t_{k-1}) = \mathbf{P}_{k-1|k-1}$$
$$\Rightarrow \mathbf{P}_{k|k-1} = \mathbf{P}(t_k)$$

The prediction equations are derived from those of continuous-time Kalman filter without update from measurements, i.e., $\mathbf{K}(t) = 0$. The predicted state and covariance are calculated respectively by solving a set of differential equations with the initial value equal to the estimate at the previous step.

Update

$$\mathbf{K}_k = \mathbf{P}_{k|k-1}\mathbf{H}_k^{\mathrm{T}}\left(\mathbf{H}_k\mathbf{P}_{k|k-1}\mathbf{H}_k^{\mathrm{T}} + \mathbf{R}_k\right)^{-1}$$
$$\hat{\mathbf{x}}_{k|k} = \hat{\mathbf{x}}_{k|k-1} + \mathbf{K}_k(\mathbf{z}_k - \mathbf{H}_k\hat{\mathbf{x}}_{k|k-1})$$
$$\mathbf{P}_{k|k} = (\mathbf{I} - \mathbf{K}_k\mathbf{H}_k)\mathbf{P}_{k|k-1}$$

The update equations are identical to those of the discrete-time Kalman filter.

Variants for the recovery of sparse signals

The traditional Kalman filter has also been employed for the recovery of sparse, possibly dynamic, signals from noisy observations. Recent works utilize notions from the theory of compressed sensing/sampling, such as the restricted isometry property and related probabilistic recovery arguments, for sequentially estimating the sparse state in intrinsically low-dimensional systems.

Applications

- Attitude and heading reference systems
- Autopilot
- Battery state of charge (SoC) estimation
- Brain-computer interface
- Chaotic signals
- Tracking and vertex fitting of charged particles in particle detectors
- Tracking of objects in computer vision
- Dynamic positioning
- Economics, in particular macroeconomics, time series analysis, and econometrics
- Inertial guidance system
- Nuclear medicine – single photon emission computed tomography image restoration
- Orbit Determination
- Power system state estimation
- Radar tracker
- Satellite navigation systems
- Seismology
- Sensorless control of AC motor variable-frequency drives
- Simultaneous localization and mapping
- Speech enhancement
- Visual odometry
- Weather forecasting
- Navigation system
- 3D modeling
- Structural health monitoring
- Human sensorimotor processing

Further reading

<templatestyles src="Template:Refbegin/styles.css" />

- Einicke, G.A. (2012). *Smoothing, Filtering and Prediction: Estimating the Past, Present and Future*[770]. Rijeka, Croatia: Intech. ISBN 978-953-307-752-9.
- Jinya Su; Baibing Li; Wen-Hua Chen (2015). "On existence, optimality and asymptotic stability of the Kalman filter with partially observed inputs". *Automatica*. **53**: 149–154. doi: 10.1016/j.automatica.2014.12.044[771].
- Gelb, A. (1974). *Applied Optimal Estimation*. MIT Press.

- Kalman, R.E. (1960). "A new approach to linear filtering and prediction problems"[772] (PDF). *Journal of Basic Engineering*. **82** (1): 35–45. doi: 10.1115/1.3662552[773]. Archived from the original[774] (PDF) on 2008-05-29. Retrieved 2008-05-03.
- Kalman, R.E.; Bucy, R.S. (1961). "New Results in Linear Filtering and Prediction Theory"[775].
- Harvey, A.C. (1990). *Forecasting, Structural Time Series Models and the Kalman Filter*. Cambridge University Press.
- Roweis, S.; Ghahramani, Z. (1999). "A Unifying Review of Linear Gaussian Models". *Neural Computation*. **11** (2): 305–345. doi: 10.1162/089976699300016674[776]. PMID 9950734[777].
- Simon, D. (2006). *Optimal State Estimation: Kalman, H Infinity, and Nonlinear Approaches*[778]. Wiley-Interscience.
- Stengel, R.F. (1994). *Optimal Control and Estimation*[779]. Dover Publications. ISBN 0-486-68200-5.
- Warwick, K. (1987). "Optimal observers for ARMA models"[780] (PDF). *International Journal of Control*. **46** (5): 1493–1503. doi: 10.1080/00207178708933989[781]. Retrieved 2008-05-03.
- Bierman, G.J. (1977). *Factorization Methods for Discrete Sequential Estimation. Mathematics in Science and Engineering*. **128**. Mineola, N.Y.: Dover Publications. ISBN 978-0-486-44981-4.
- Bozic, S.M. (1994). *Digital and Kalman filtering*. Butterworth–Heinemann.
- Haykin, S. (2002). *Adaptive Filter Theory*. Prentice Hall.
- Liu, W.; Principe, J.C. and Haykin, S. (2010). *Kernel Adaptive Filtering: A Comprehensive Introduction*. John Wiley.
- Manolakis, D.G. (1999). *Statistical and Adaptive signal processing*. Artech House.
- Welch, Greg; Bishop, Gary (1997). "SCAAT: incremental tracking with incomplete information"[782] (PDF). *SIGGRAPH '97 Proceedings of the 24th annual conference on Computer graphics and interactive techniques*. ACM Press/Addison-Wesley Publishing Co. pp. 333–344. doi: 10.1145/258734.258876[783]. ISBN 0-89791-896-7.
- Jazwinski, Andrew H. (1970). *Stochastic Processes and Filtering*. Mathematics in Science and Engineering. New York: Academic Press. p. 376. ISBN 0-12-381550-9.
- Maybeck, Peter S. (1979). "Chapter 1". *Stochastic Models, Estimation, and Control*[784] (PDF). Mathematics in Science and Engineering. 141-1. New York: Academic Press. ISBN 0-12-480701-1.
- Moriya, N. (2011). *Primer to Kalman Filtering: A Physicist Perspective*. New York: Nova Science Publishers, Inc. ISBN 978-1-61668-311-5.
- Dunik, J.; Simandl M.; Straka O. (2009). "Methods for Estimating State

and Measurement Noise Covariance Matrices: Aspects and Comparison". *15th IFAC Symposium on System Identification, 2009*[785]. 15th IFAC Symposium on System Identification, 2009. France. pp. 372–377. doi: 10.3182/20090706-3-FR-2004.00061[786]. ISBN 978-3-902661-47-0.
- Chui, Charles K.; Chen, Guanrong (2009). *Kalman Filtering with Real-Time Applications*. Springer Series in Information Sciences. **17** (4th ed.). New York: Springer. p. 229. ISBN 978-3-540-87848-3.
- Spivey, Ben; Hedengren, J. D. and Edgar, T. F. (2010). "Constrained Nonlinear Estimation for Industrial Process Fouling"[787]. *Industrial & Engineering Chemistry Research*. **49** (17): 7824–7831. doi: 10.1021/ie9018116[788].
- Thomas Kailath; Ali H. Sayed; Babak Hassibi (2000). *Linear Estimation*. NJ: Prentice–Hall. ISBN 978-0-13-022464-4.
- Ali H. Sayed (2008). *Adaptive Filters*. NJ: Wiley. ISBN 978-0-470-25388-5.

External links

- A New Approach to Linear Filtering and Prediction Problems[789], by R. E. Kalman, 1960
- Kalman and Bayesian Filters in Python[790]. Open source Kalman filtering textbook.
- How a Kalman filter works, in pictures[791]. Illuminates the Kalman filter with pictures and colors
- *Kalman–Bucy Filter*[792], a derivation of the Kalman–Bucy Filter
- MIT Video Lecture on the Kalman filter[793] on YouTube
- An Introduction to the Kalman Filter[794], SIGGRAPH 2001 Course, Greg Welch and Gary Bishop
- Kalman Filter[795] webpage, with lots of links
- "Kalman filters used in Weather models"[796] (PDF). *SIAM News*. **36** (8). October 2003.
- Haseltine, Eric L.; Rawlings, James B. (2005). "Critical Evaluation of Extended Kalman Filtering and Moving-Horizon Estimation". *Industrial & Engineering Chemistry Research*. **44** (8): 2451. doi: 10.1021/ie0343081[797].
- Gerald J. Bierman's Estimation Subroutine Library[798]: Corresponds to the code in the research monograph "Factorization Methods for Discrete Sequential Estimation" originally published by Academic Press in 1977. Republished by Dover.
- Matlab Toolbox implementing parts of Gerald J. Bierman's Estimation Subroutine Library[799]: UD / UDU' and LD / LDL' factorization with associated time and measurement updates making up the Kalman filter.

- Matlab Toolbox of Kalman Filtering applied to Simultaneous Localization and Mapping[800]: Vehicle moving in 1D, 2D and 3D
- The Kalman Filter in Reproducing Kernel Hilbert Spaces[801] A comprehensive introduction.
- Matlab code to estimate Cox–Ingersoll–Ross interest rate model with Kalman Filter[802]: Corresponds to the paper "estimating and testing exponential-affine term structure models by kalman filter" published by Review of Quantitative Finance and Accounting in 1999.
- Online demo of the Kalman Filter[803]. Demonstration of Kalman Filter (and other data assimilation methods) using twin experiments.
- Botella, Guillermo; Martín h., José Antonio; Santos, Matilde; Meyer-Baese, Uwe (2011). "FPGA-Based Multimodal Embedded Sensor System Integrating Low- and Mid-Level Vision". *Sensors*. **11** (12): 1251–1259. doi: 10.3390/s110808164[804].
- Examples and how-to on using Kalman Filters with MATLAB[805] A Tutorial on Filtering and Estimation
- Explaining Filtering (Estimation) in One Hour, Ten Minutes, One Minute, and One Sentence[806] by Yu-Chi Ho

Particle filter

Particle filters or **Sequential Monte Carlo** (SMC) methods are a set of genetic, Monte Carlo algorithms used to solve filtering problems arising in signal processing and Bayesian statistical inference. The **filtering problem** consists of estimating the internal states in dynamical systems when partial observations are made, and random perturbations are present in the sensors as well as in the dynamical system. The objective is to compute the posterior distributions of the states of some Markov process, given some noisy and partial observations. The term "particle filters" was first coined in 1996 by Del Moral in reference to mean field interacting particle methods used in fluid mechanics since the beginning of the 1960s. The terminology "sequential Monte Carlo" was proposed by Liu and Chen in 1998.

Particle filtering uses a genetic mutation-selection sampling approach, with a set of particles (also called samples) to represent the posterior distribution of some stochastic process given noisy and/or partial observations. The state-space model can be nonlinear and the initial state and noise distributions can take any form required. Particle filter techniques provide a well-established methodology for generating samples from the required distribution without requiring assumptions about the state-space model or the state distributions. However, these methods do not perform well when applied to very high-dimensional systems.

Particle filters implement the **prediction-updating** transitions of the filtering equation directly by using a genetic type **mutation-selection** particle algorithm. The samples from the distribution are represented by a set of particles; each particle has a likelihood weight assigned to it that represents the probability of that particle being sampled from the probability density function. Weight disparity leading to weight collapse is a common issue encountered in these filtering algorithms; however it can be mitigated by including a resampling step before the weights become too uneven. Several adaptive resampling criteria can be used, including the variance of the weights and the relative entropy with respect to the uniform distribution. In the resampling step, the particles with negligible weights are replaced by new particles in the proximity of the particles with higher weights.

From the statistical and probabilistic point of view, particle filters can be interpreted as mean field particle interpretations of Feynman-Kac probability measures. These particle integration techniques were developed in molecular chemistry and computational physics by Theodore E. Harris and Herman Kahn in 1951, Marshall N. Rosenbluth and Arianna W. Rosenbluth in 1955 and more recently by Jack H. Hetherington in 1984. In computational physics, these Feynman-Kac type path particle integration methods are also used in Quantum Monte Carlo, and more specifically Diffusion Monte Carlo methods. Feynman-Kac interacting particle methods are also strongly related to mutation-selection genetic algorithms currently used in evolutionary computing to solve complex optimization problems.

The particle filter methodology is used to solve Hidden Markov Model (HMM) and nonlinear filtering problems. With the notable exception of linear-Gaussian signal-observation models (Kalman filter) or wider classes of models (Benes filter) Mireille Chaleyat-Maurel and Dominique Michel proved in 1984 that the sequence of posterior distributions of the random states of the signal given the observations (a.k.a. optimal filter) have no finitely recursive recursion. Various numerical methods based on fixed grid approximations, Markov Chain Monte Carlo techniques (MCMC), conventional linearization, extended Kalman filters, or determining the best linear system (in expect cost-error sense) have never really coped with large scale systems, unstable processes or when the nonlinearities are not sufficiently smooth.

Particle filters and Feynman-Kac particle methodologies find application in signal and image processing, Bayesian inference, machine learning, risk analysis and rare event sampling, engineering and robotics, artificial intelligence, bioinformatics, phylogenetics, computational science, Economics and mathematical finance, molecular chemistry, computational physics, pharmacokinetic and other fields.

History

Heuristic like algorithms

From the statistical and probabilistic viewpoint, particle filters belong to the class of branching/genetic type algorithms, and mean field type interacting particle methodologies. The interpretation of these particle methods depends on the scientific discipline. In Evolutionary Computing, mean field genetic type particle methodologies are often used as a heuristic and natural search algorithms (a.k.a. Metaheuristic). In computational physics and molecular chemistry they are used to solve Feynman-Kac path integration problems, or they compute Boltzmann-Gibbs measures, top eigenvalues and ground states of Schrödinger operators. In Biology and Genetics they also represent the evolution of a population of individuals or genes in some environment.

The origins of mean field type evolutionary computational techniques can be traced to 1950 and 1954 with the seminal work of Alan Turing on genetic type mutation-selection learning machines and the articles by Nils Aall Barricelli at the Institute for Advanced Study in Princeton, New Jersey. The first trace of particle filters in statistical methodology dates back to the mid-50's; the 'Poor Man's Monte Carlo', that was proposed by Hammersley et al., in 1954, contained hints of the genetic type particle filtering methods used today. In 1963, Nils Aall Barricelli simulated a genetic type algorithm to mimic the ability of individuals to play a simple game. In evolutionary computing literature, genetic type mutation-selection algorithms became popular through the seminal work of John Holland in the early 1970s, and particularly his book published in 1975.

In Biology and Genetics, the Australian geneticist Alex Fraser also published in 1957 a series of papers on the genetic type simulation of artificial selection of organisms. The computer simulation of evolution by biologists became more common in the early 1960s, and the methods were described in books by Fraser and Burnell (1970) and Crosby (1973). Fraser's simulations included all of the essential elements of modern mutation-selection genetic particle algorithms.

From the mathematical viewpoint, the conditional distribution of the random states of a signal given some partial and noisy observations is described by a Feynman-Kac probability on the random trajectories of the signal weighted by a sequence of likelihood potential functions. Quantum Monte Carlo, and more specifically Diffusion Monte Carlo methods can also be interpreted as a mean field genetic type particle approximation of Feynman-Kac path integrals. The origins of Quantum Monte Carlo methods are often attributed to Enrico Fermi and Robert Richtmyer who developed in 1948 a mean field particle interpretation of neutron-chain reactions, but the first heuristic-like and genetic type particle algorithm (a.k.a. Resampled or Reconfiguration Monte Carlo methods) for estimating ground state energies of quantum systems (in reduced matrix models) is due to Jack H. Hetherington in 1984. We also quote an earlier seminal works of Theodore E. Harris and Herman Kahn in particle physics, published in 1951, using mean field but heuristic-like genetic methods for estimating particle transmission energies. In molecular chemistry, the use of genetic heuristic-like particle methodologies (a.k.a. pruning and enrichment strategies) can be traced back to 1955 with the seminal work of Marshall. N. Rosenbluth and Arianna. W. Rosenbluth.

The use of genetic particle algorithms in advanced signal processing and Bayesian inference is more recent. It was in 1993, that Gordon et al., published in their seminal work the first application of genetic type algorithm in Bayesian statistical inference. The authors named their algorithm 'the bootstrap filter', and demonstrated that compared to other filtering methods, their bootstrap algorithm does not require any assumption about that state-space or the noise of the system. We also quote another pioneering article in this field of Genshiro Kitagawa on a related "Monte Carlo filter", and the ones by Pierre Del Moral and Himilcon Carvalho, Pierre Del Moral, André Monin and Gérard Salut on particle filters published in the mid-1990s. Particle filters were also developed in signal processing in the early 1989-1992 by P. Del Moral, J.C. Noyer, G. Rigal, and G. Salut in the LAAS-CNRS in a series of restricted and classified research reports with STCAN (Service Technique des Constructions et Armes Navales), the IT company DIGILOG, and the LAAS-CNRS[807] (the Laboratory for Analysis and Architecture of Systems) on RADAR/SONAR and GPS signal processing problems.[808,809,810,811,812,813]

Mathematical foundations

From 1950 to 1996, all the publications on particle filters, genetic algorithms, including the pruning and resample Monte Carlo methods introduced in computational physics and molecular chemistry, present natural and heuristic-like algorithms applied to different situations without a single proof of their consistency, nor a discussion on the bias of the estimates and on genealogical and ancestral tree based algorithms.

The mathematical foundations and the first rigorous analysis of these particle algorithms are due to Pierre Del Moral in 1996. The article also contains a proof of the unbiased properties of a particle approximations of likelihood functions and unnormalized conditional probability measures. The unbiased particle estimator of the likelihood functions presented in this article is used today in Bayesian statistical inference.

Branching type particle methodologies with varying population sizes were also developed toward the end of the 1990s by Dan Crisan, Jessica Gaines and

Terry Lyons, and by Dan Crisan, Pierre Del Moral and Terry Lyons. Further developments in this field were developed in 2000 by P. Del Moral, A. Guionnet and L. Miclo. The first central limit theorems are due to Pierre Del Moral and Alice Guionnet in 1999 and Pierre Del Moral and Laurent Miclo in 2000. The first uniform convergence results with respect to the time parameter for particle filters were developed in the end of the 1990s by Pierre Del Moral and Alice Guionnet. The first rigorous analysis of genealogical tree based particle filter smoothers is due to P. Del Moral and L. Miclo in 2001

The theory on Feynman-Kac particle methodologies and related particle filters algorithms has been developed in 2000 and 2004 in the books. These abstract probabilistic models encapsulate genetic type algorithms, particle and bootstrap filters, interacting Kalman filters (a.k.a. Rao–Blackwellized particle filter), importance sampling and resampling style particle filter techniques, including genealogical tree based and particle backward methodologies for solving filtering and smoothing problems. Other classes of particle filtering methodologies includes genealogical tree based models, backward Markov particle models, adaptive mean field particle models, island type particle models, and particle Markov chain Monte Carlo methodologies.

The filtering problem

Objective

The objective of a particle filter is to estimate the posterior density of the state variables given the observation variables. The particle filter is designed for a hidden Markov Model, where the system consists of hidden and observable variables. The observable variables (observation process) are related to the hidden variables (state-process) by some functional form that is known. Similarly the dynamical system describing the evolution of the state variables is also known probabilistically.

A generic particle filter estimates the posterior distribution of the hidden states using the observation measurement process. Consider a state-space shown in the diagram below.

$$\begin{array}{cccccccccc} X_0 & \rightarrow & X_1 & \rightarrow & X_2 & \rightarrow & X_3 & \rightarrow & \cdots & \text{signal} \\ \downarrow & & \downarrow & & \downarrow & & \downarrow & & \cdots & \\ Y_0 & & Y_1 & & Y_2 & & Y_3 & & \cdots & \text{observation} \end{array}$$

The filtering problem is to estimate **sequentially** the values of the hidden states X_k , given the values of the observation process $Y_0, \cdots, Y_k$, at any time step k.

All Bayesian estimates of X_k follow from the posterior density $p(x_k \mid y_0, y_1, \ldots, y_k)$. The particle filter methodology provides an approximation of these conditional probabilities using the empirical measure associated with a genetic type particle algorithm. In contrast, the MCMC or importance sampling approach would model the full posterior $p(x_0,x_1,\ldots,x_k \mid y_0,y_1,\ldots,y_k)$.

The Signal-Observation Model

Particle methods often assume X_k and the observations Y_k can be modeled in this form:

- $X_0, X_1, \cdots$ is a Markov process on $\mathbb{R}^{d_x}$ (for some $d_x \geqslant 1$) that evolves according to the transition probability density $p(x_k|x_{k-1})$. This model is also often written in a synthetic way as
  $X_k|X_{k-1} = x_k \sim p(x_k|x_{k-1})$
  with an initial probability density $p(x_0)$.
- The observations $Y_0, Y_1, \cdots$ take values in some state space on $\mathbb{R}^{d_y}$ (for some $d_y \geqslant 1$) are conditionally independent provided that $X_0, X_1, \cdots$ are known. In other words, each Y_k only depends on X_k . In addition, we assume conditional distribution for Y_k given $X_k = x_k$ are absolutely continuous, and in a synthetic way we have
  $Y_k|X_k = y_k \sim p(y_k|x_k)$

An example of system with these properties is:

$$X_k = g(X_{k-1}) + W_{k-1}$$
$$Y_k = h(X_k) + V_k$$

where both W_k and V_k are mutually independent sequences with known probability density functions and g and h are known functions. These two equations can be viewed as state space equations and look similar to the state space equations for the Kalman filter. If the functions g and h in the above example are linear, and if both W_k and V_k are Gaussian, the Kalman filter finds the exact Bayesian filtering distribution. If not, Kalman filter based methods are a first-order approximation (EKF) or a second-order approximation (UKF in general, but if probability distribution is Gaussian a third-order approximation is possible).

The assumption that the initial distribution and the transitions of the Markov chain are absolutely continuous with respect to the Lebesgue measure can be relaxed. To design a particle filter we simply need to assume that we can sample the transitions $X_{k-1} \to X_k$ of the Markov chain X_k, and to compute the likelihood function $x_k \mapsto p(y_k|x_k)$ (see for instance the genetic selection mutation description of the particle filter given below). The absolutely continuous assumption on the Markov transitions of X_k are only used to derive in an informal (and rather abusive) way different formulae between posterior distributions using the Bayes' rule for conditional densities.

Approximate Bayesian Computation models

In some important problems, the conditional distribution of the observations given the random states of the signal may fail to have a density or may be impossible or too complex to compute. In this situation, we need to resort to an additional level of approximation. One strategy is to replace the signal X_k by the Markov chain $\mathcal{X}_k = (X_k, Y_k)$ and to introduce a virtual observation of the form

$$\mathcal{Y}_k = Y_k + \epsilon \mathcal{V}_k \quad \text{for some parameter} \quad \epsilon \in [0,1]$$

for some sequence of independent sequences with known probability density functions. The central idea is to observe that

$$\text{Law}\left(X_k | \mathcal{Y}_0 = y_0, \cdots, \mathcal{Y}_k = y_k\right) \approx_{\epsilon \downarrow 0} \text{Law}\left(X_k | Y_0 = y_0, \cdots, Y_k = y_k\right)$$

The particle filter associated with the Markov process $\mathcal{X}_k = (X_k, Y_k)$ given the partial observations $\mathcal{Y}_0 = y_0, \cdots, \mathcal{Y}_k = y_k$, is defined in terms of particles evolving in $\mathbb{R}^{d_x + d_y}$ with a likelihood function given with some obvious abusive notation by $p(\mathcal{Y}_k | \mathcal{X}_k)$. These probabilistic techniques are closely related to Approximate Bayesian Computation (ABC). In the context of particle filters, these ABC particle filtering techniques were introduced in 1998 by P. Del Moral, J. Jacod and P. Protter. They were further developed by P. Del Moral, A. Doucet and A. Jasra.

The nonlinear filtering equation

The Bayes' rule for conditional probability gives:

$$p(x_0, \cdots, x_k | y_0, \cdots, y_k) = \frac{p(y_0, \cdots, y_k | x_0, \cdots, x_k) p(x_0, \cdots, x_k)}{p(y_0, \cdots, y_k)}$$

where

$$p(y_0, \cdots, y_k) = \int p(y_0, \cdots, y_k | x_0, \cdots, x_k) p(x_0, \cdots, x_k) dx_0 \cdots dx_k$$

$$p(y_0, \cdots, y_k | x_0, \cdots, x_k) = \prod_{l=0}^{k} p(y_l | x_l)$$

$$p(x_0, \cdots, x_k) = p_0(x_0) \prod_{l=1}^{k} p(x_l | x_{l-1})$$

Particle filters are also an approximation, but with enough particles they can be much more accurate. The nonlinear filtering equation is given by the recursion

$$p(x_k|y_0,\cdots,y_{k-1}) \stackrel{\text{updating}}{\longrightarrow} p(x_k|y_0,\cdots,y_k) = \frac{p(y_k|x_k)p(x_k|y_0,\cdots,y_{k-1})}{\int p(y_k|x'_k)p(x'_k|y_0,\cdots,y_{k-1})dx'_k}$$
$$\stackrel{\text{prediction}}{\longrightarrow} p(x_{k+1}|y_0,\cdots,y_k) = \int p(x_{k+1}|x_k)p(x_k|y_0,\cdots,y_k)dx_k \quad \textbf{(Eq. 1)}$$

with the convention $p(x_0|y_0,\cdots,y_{-1}) = p(x_0)$ for $k = 0$. The nonlinear filtering problem consists of computing sequentially these sequence of conditional distributions.

Feynman-Kac formulation

We fix a time horizon n and a sequence of observations $Y_0 = y_0,\cdots,Y_n = y_n$, and for each $k = 0, ..., n$ we set:

$$G_k(x_k) = p(y_k|x_k).$$

In this notation, for any bounded function F on the set of trajectories of X_k from the origin $k = 0$ up to time $k = n$, we have the Feynman-Kac formula

$$\int F(x_0,\cdots,x_n)p(x_0,\cdots,x_n|y_0,\cdots,y_n)dx_0\cdots dx_n = \frac{\int F(x_0,\cdots,x_n)\left\{\prod_{k=0}^{n} p(y_k|x_k)\right\}p(x_0,\cdots,x_n)dx_0\cdots dx_n}{\int\left\{\prod_{k=0}^{n} p(y_k|x_k)\right\}p(x_0,\cdots,x_n)dx_0\cdots dx_n}$$
$$= \frac{E\left(F(X_0,\cdots,X_n)\prod_{k=0}^{n} G_k(X_k)\right)}{E\left(\prod_{k=0}^{n} G_k(X_k)\right)}$$

These Feynman-Kac path integration models arise in a variety of scientific disciplines, including in computational physics, biology, information theory and computer sciences. Their interpretations depend on the application domain. For instance, if we choose the indicator function $G_n(x_n) = 1_A(x_n)$ of some subset of the state space, they represent the conditional distribution of a Markov chain given it stays in a given tube; that is, we have:

$$E\left(F(X_0,\cdots,X_n)|X_0 \in A,\cdots,X_n \in A\right) = \frac{E\left(F(X_0,\cdots,X_n)\prod_{k=0}^{n} G_k(X_k)\right)}{E\left(\prod_{k=0}^{n} G_k(X_k)\right)}$$

and

$$P\left(X_0 \in A,\cdots,X_n \in A\right) = E\left(\prod_{k=0}^{n} G_k(X_k)\right)$$

as soon as the normalizing constant is strictly positive.

Particle filters

A Genetic type particle algorithm

Initially we start with N independent random variables $\left(\xi_0^i\right)_{1\leqslant i\leqslant N}$ with common probability density $p(x_0)$. The genetic algorithm selection-mutation transitions

$$\xi_k := \left(\xi_k^i\right)_{1\leqslant i\leqslant N} \stackrel{\text{selection}}{\longrightarrow} \widehat{\xi}_k := \left(\widehat{\xi_k^i}\right)_{1\leqslant i\leqslant N} \stackrel{\text{mutation}}{\longrightarrow} \xi_{k+1} := \left(\xi_{k+1}^i\right)_{1\leqslant i\leqslant N}$$

mimic/approximate the updating-prediction transitions of the optimal filter evolution (**Eq. 1**):

- **During the selection-updating transition** we sample N (conditionally) independent random variables $\widehat{\xi}_k := \left(\widehat{\xi_k^i}\right)_{1\leqslant i\leqslant N}$ with common (conditional) distribution

$$\sum_{i=1}^{N} \frac{p(y_k|\xi_k^i)}{\sum_{j=1}^{N} p(y_k|\xi_k^j)} \delta_{\xi_k^i}(dx_k)$$

- **During the mutation-prediction transition,** from each selected particle $\widehat{\xi_k^i}$ we sample independently a transition

$$\widehat{\xi_k^i} \longrightarrow \xi_{k+1}^i \sim p(x_{k+1}|\widehat{\xi_k^i}), \qquad i = 1, \cdots, N.$$

In the above displayed formulae $p(y_k|\xi_k^i)$ stands for the likelihood function $x_k \mapsto p(y_k|x_k)$ evaluated at $x_k = \xi_k^i$, and $p(x_{k+1}|\widehat{\xi_k^i})$ stands for the conditional density $p(x_{k+1}|x_k)$ evaluated at $x_k = \widehat{\xi_k^i}$.

At each time k, we have the particle approximations

$$\widehat{p}(dx_k|y_0, \cdots, y_k) := \frac{1}{N}\sum_{i=1}^{N} \delta_{\widehat{\xi_k^i}}(dx_k) \approx_{N\uparrow\infty} p(dx_k|y_0, \cdots, y_k) \approx_{N\uparrow\infty} \sum_{i=1}^{N} \frac{p(y_k|\xi_k^i)}{\sum_{i=1}^{N} p(y_k|\xi_k^j)} \delta_{\xi_k^i}(dx_k)$$

and

$$\widehat{p}(dx_k|y_0, \cdots, y_{k-1}) := \frac{1}{N}\sum_{i=1}^{N} \delta_{\xi_k^i}(dx_k) \approx_{N\uparrow\infty} p(dx_k|y_0, \cdots, y_{k-1})$$

A detailed proof of these convergence results can be found in, see also the more recent developments provided in the books. In Genetic algorithms and Evolutionary computing community, the mutation-selection Markov chain described above is often called the genetic algorithm with proportional selection. Several branching variants, including with random population sizes have also been proposed in the articles.

Monte Carlo principles

Particle methods, like all sampling-based approaches (e.g., MCMC), generate a set of samples that approximate the filtering density

$$p(x_k|y_0, \cdots, y_k).$$

For example, we may have N samples from the approximate posterior distribution of X_k , where the samples are labeled with superscripts as

$$\widehat{\xi}^1_k, \cdots, \widehat{\xi}^N_k.$$

Then, expectations with respect to the filtering distribution are approximated by

$$\int f(x_k)p(x_k|y_0, \cdots, y_k)\, dx_k \approx_{N\uparrow\infty} \frac{1}{N}\sum_{i=1}^N f\left(\widehat{\xi}^i_k\right) = \int f(x_k)\widehat{p}(dx_k|y_0, \cdots, y_k) \quad \textbf{(Eq. 2)}$$

with

$$\widehat{p}(dx_k|y_0, \cdots, y_k) = \frac{1}{N}\sum_{i=1}^N \delta_{\widehat{\xi}^i_k}(dx_k)$$

where δ_a stands for the **Dirac measure** at a given state a. The function f, in the usual way for Monte Carlo, can give all the moments etc. of the distribution up to some degree of approximation. When the approximation equation (**Eq. 2**) is satisfied for any bounded function f we write

$$p(dx_k|y_0, \cdots, y_k) := p(x_k|y_0, \cdots, y_k)dx_k \approx_{N\uparrow\infty} \widehat{p}(dx_k|y_0, \cdots, y_k) = \frac{1}{N}\sum_{i=1}^N \delta_{\widehat{\xi}^i_k}(dx_k)$$

Particle filters can be interpreted as a genetic type particle algorithm evolving with mutation and selection transitions. We can keep track of the ancestral lines

$$\left(\widehat{\xi}^i_{0,k}, \widehat{\xi}^i_{1,k}, \cdots, \widehat{\xi}^i_{k-1,k}, \widehat{\xi}^i_{k,k}\right)$$

of the particles $i = 1, \cdots, N$. The random states $\widehat{\xi}^i_{l,k}$, with the lower indices l=0,...,k, stands for the ancestor of the individual $\widehat{\xi}^i_{k,k} = \widehat{\xi}^i_k$ at level l=0,...,k. In this situation, we have the approximation formula

$$\int F(x_0,\cdots,x_k)p(x_0,\cdots,x_k|y_0,\cdots,y_k)\,dx_0\cdots dx_k \approx_{N\uparrow\infty} \frac{1}{N}\sum_{i=1}^{N} F\left(\widehat{\xi}^i_{0,k},\widehat{\xi}^i_{1,k},\cdots,\widehat{\xi}^i_{k,k}\right)$$
$$= \int F(x_0,\cdots,x_k)\widehat{p}(d(x_0,\cdots,x_k)|y_0,\cdots,y_k) \quad \textbf{(Eq. 3)}$$

with the empirical measure

$$\widehat{p}(d(x_0,\cdots,x_k)|y_0,\cdots,y_k) := \tfrac{1}{N}\sum_{i=1}^{N}\delta_{\left(\widehat{\xi}^i_{0,k},\widehat{\xi}^i_{1,k},\cdots,\widehat{\xi}^i_{k,k}\right)}(d(x_0,\cdots,x_k))$$

Here F stands for any founded function on the path space of the signal. In a more synthetic form (**Eq. 3**) is equivalent to

$$\begin{aligned} p(d(x_0,\cdots,x_k)|y_0,\cdots,y_k) &:= p(x_0,\cdots,x_k|y_0,\cdots,y_k)\,dx_0\cdots dx_k \\ &\approx_{N\uparrow\infty} \widehat{p}(d(x_0,\cdots,x_k)|y_0,\cdots,y_k) \\ &:= \frac{1}{N}\sum_{i=1}^{N}\delta_{\left(\widehat{\xi}^i_{0,k},\cdots,\widehat{\xi}^i_{k,k}\right)}(d(x_0,\cdots,x_k)) \end{aligned}$$

Particle filters can be interpreted in many different ways. From the probabilistic point of view they coincide with a mean field particle interpretation of the nonlinear filtering equation. The updating-prediction transitions of the optimal filter evolution can also be interpreted as the classical genetic type selection-mutation transitions of individuals. The sequential importance resampling technique provides another interpretation of the filtering transitions coupling importance sampling with the bootstrap resampling step. Last, but not least, particle filters can be seen as an acceptance-rejection methodology equipped with a recycling mechanism.

Mean field particle simulation

The general probabilistic principle

The nonlinear filtering evolution can be interpreted as a dynamical system in the set of probability measures of the following form $\eta_{n+1} = \Phi_{n+1}(\eta_n)$ where Φ_{n+1} stands for some mapping from the set of probability distribution into itself. For instance, the evolution of the one-step optimal predictor $\eta_n(dx_n) = p(x_n|y_0,\cdots,y_{n-1})dx_n$

satisfies a nonlinear evolution starting with the probability distribution $\eta_0(dx_0) = p(x_0)dx_0$. One of the simplest ways to approximate these probability measures is to start with N independent random variables $\left(\xi^i_0\right)_{1\leqslant i\leqslant N}$ with common probability distribution $\eta_0(dx_0) = p(x_0)dx_0$. Suppose we have defined a sequence of N random variables $\left(\xi^i_n\right)_{1\leqslant i\leqslant N}$ such that

$$\frac{1}{N}\sum_{i=1}^{N}\delta_{\xi_n^i}(dx_n) \approx_{N\uparrow\infty} \eta_n(dx_n)$$

At the next step we sample N (conditionally) independent random variables $\xi_{n+1} := \left(\xi_{n+1}^i\right)_{1\leqslant i\leqslant N}$ with common law .

$$\Phi_{n+1}\left(\frac{1}{N}\sum_{i=1}^{N}\delta_{\xi_n^i}\right) \approx_{N\uparrow\infty} \Phi_{n+1}(\eta_n) = \eta_{n+1}$$

A particle interpretation of the filtering equation

We illustrate this mean field particle principle in the context of the evolution of the one step optimal predictors

$p(x_k\|y_0,\cdots,y_{k-1})dx_k \to p(x_{k+1}\|y_0,\cdots,y_k) = \int p(x_{k+1}\|x'_k)\frac{p(y_k\|x'_k)p(x'_k\|y_0,\cdots,y_{k-1})dx'_k}{\int p(y_k\|x''_k)p(x''_k\|y_0,\cdots,y_{k-1})dx''_k}$	**(Eq. 4)**

For $k = 0$ we use the convention $p(x_0|y_0,\cdots,y_{-1}) := p(x_0)$.

By the law of large numbers, we have

$$\widehat{p}(dx_0) = \frac{1}{N}\sum_{i=1}^{N}\delta_{\xi_0^i}(dx_0) \approx_{N\uparrow\infty} p(x_0)dx_0$$

in the sense that

$$\int f(x_0)\widehat{p}(dx_0) = \frac{1}{N}\sum_{i=1}^{N} f(\xi_0^i) \approx_{N\uparrow\infty} \int f(x_0)p(dx_0)dx_0$$

for any bounded function f . We further assume that we have constructed a sequence of particles $\left(\xi_k^i\right)_{1\leqslant i\leqslant N}$ at some rank k such that

$$\widehat{p}(dx_k|y_0,\cdots,y_{k-1}) := \tfrac{1}{N}\textstyle\sum_{i=1}^{N}\delta_{\xi_k^i}(dx_k) \approx_{N\uparrow\infty} p(x_k \mid y_0,\cdots,y_{k-1})dx_k$$

in the sense that for any bounded function f we have

$$\textstyle\int f(x_k)\widehat{p}(dx_k|y_0,\cdots,y_{k-1}) = \frac{1}{N}\sum_{i=1}^{N} f(\xi_k^i) \approx_{N\uparrow\infty} \int f(x_k)p(dx_k|y_0,\cdots,y_{k-1})dx_k$$

In this situation, replacing $p(x_k|y_0,\cdots,y_{k-1})dx_k$ by the empirical measure $\widehat{p}(dx_k|y_0,\cdots,y_{k-1})$ in the evolution equation of the one-step optimal filter stated in (**Eq. 4**) we find that

$$p(x_{k+1}|y_0,\cdots,y_k) \approx_{N\uparrow\infty} \int p(x_{k+1}|x'_k)\frac{p(y_k|x'_k)\widehat{p}(dx'_k|y_0,\cdots,y_{k-1})}{\int p(y_k|x''_k)\widehat{p}(dx''_k|y_0,\cdots,y_{k-1})}$$

Notice that the right hand side in the above formula is a weighted probability mixture

$$\int p(x_{k+1}|x'_k)\frac{p(y_k|x'_k)\widehat{p}(dx'_k|y_0,\cdots,y_{k-1})}{\int p(y_k|x''_k)\widehat{p}(dx''_k|y_0,\cdots,y_{k-1})} = \sum_{i=1}^{N}\frac{p(y_k|\xi_k^i)}{\sum_{i=1}^{N} p(y_k|\xi_k^j)}p(x_{k+1}|\xi_k^i) =: \widehat{q}(x_{k+1}|y_0,\cdots,y_k)$$

where $p(y_k|\xi_k^i)$ stands for the density $p(y_k|x_k)$ evaluated at $x_k = \xi_k^i$, and $p(x_{k+1}|\xi_k^i)$ stands for the density $p(x_{k+1}|x_k)$ evaluated at $x_k = \xi_k^i$ for $i = 1, \cdots, N$.

Then, we sample N independent random variable $\left(\xi_{k+1}^i\right)_{1 \leqslant i \leqslant N}$ with common probability density $\widehat{q}(x_{k+1}|y_0, \cdots, y_k)$ so that

$$\widehat{p}(dx_{k+1}|y_0, \cdots, y_k) := \frac{1}{N}\sum_{i=1}^N \delta_{\xi_{k+1}^i}(dx_{k+1}) \approx_{N\uparrow\infty} \widehat{q}(x_{k+1}|y_0, \cdots, y_k)dx_{k+1} \approx_{N\uparrow\infty} p(x_{k+1}|y_0, \cdots, y_k)dx_{k+1}$$

Iterating this procedure, we design a Markov chain such that

$$\widehat{p}(dx_k|y_0, \cdots, y_{k-1}) := \frac{1}{N}\sum_{i=1}^N \delta_{\xi_k^i}(dx_k) \approx_{N\uparrow\infty} p(dx_k|y_0, \cdots, y_{k-1}) := p(x_k|y_0, \cdots, y_{k-1})dx_k$$

Notice that the optimal filter is approximated at each time step k using the Bayes' formulae

$$p(dx_k|y_0, \cdots, y_k) \approx_{N\uparrow\infty} \frac{p(y_k|x_k)\widehat{p}(dx_k|y_0, \cdots, y_{k-1})}{\int p(y_k|x_k')\widehat{p}(dx_k'|y_0, \cdots, y_{k-1})} = \sum_{i=1}^N \frac{p(y_k|\xi_k^i)}{\sum_{j=1}^N p(y_k|\xi_k^j)} \delta_{\xi_k^i}(dx_k)$$

The terminology "mean field approximation" comes from the fact that we replace at each time step the probability measure $p(dx_k|y_0, \cdots, y_{k-1})$ by the empirical approximation $\widehat{p}(dx_k|y_0, \cdots, y_{k-1})$. The mean field particle approximation of the filtering problem is far from being unique. Several strategies are developed in the books.

Some convergence results

The analysis of the convergence of particle filters was started in 1996 and in 2000 in the book and the series of articles. More recent developments can be found in the books, When the filtering equation is stable (in the sense that it corrects any erroneous initial condition), the bias and the variance of the particle particle estimates

$$I_k(f) := \int f(x_k)p(dx_k|y_0, \cdots, y_{k-1}) \approx_{N\uparrow\infty} \widehat{I}_k(f) := \int f(x_k)\widehat{p}(dx_k|y_0, \cdots, y_{k-1})$$

are controlled by the non asymptotic uniform estimates

$$\sup_{k\geqslant 0}\left|E\left(\widehat{I}_k(f)\right) - I_k(f)\right| \leqslant \frac{c_1}{N}$$

$$\sup_{k\geqslant 0} E\left(\left[\widehat{I}_k(f) - I_k(f)\right]^2\right) \leqslant \frac{c_2}{N}$$

for any function f bounded by 1, and for some finite constants c_1, c_2. In addition, for any $x \geqslant 0$:

$$\mathbf{P}\left(\left|\widehat{I}_k(f) - I_k(f)\right| \leqslant c_1\frac{x}{N} + c_2\sqrt{\frac{x}{N}} \wedge \sup_{0\leqslant k\leqslant n}\left|\widehat{I}_k(f) - I_k(f)\right| \leqslant c\sqrt{\frac{x\log(n)}{N}}\right) > 1 - e^{-x}$$

for some finite constants c_1, c_2 related to the asymptotic bias and variance of the particle estimate, and some finite constant c. The same results are satisfied if we replace the one step optimal predictor by the optimal filter approximation.

Genealogical trees and Unbiasedness properties

Genealogical tree based particle smoothing

Tracing back in time the ancestral lines

$$\left(\widehat{\xi}^i_{0,k}, \widehat{\xi}^i_{1,k}, \cdots, \widehat{\xi}^i_{k-1,k}, \widehat{\xi}^i_{k,k}\right), \quad \left(\xi^i_{0,k}, \xi^i_{1,k}, \cdots, \xi^i_{k-1,k}, \xi_{k,k}\right)$$

of the individuals $\widehat{\xi}^i_k \left(= \widehat{\xi}^i_{k,k}\right)$ and $\xi^i_k \left(= \xi^i_{k,k}\right)$ at every time step k, we also have the particle approximations

$$\begin{aligned}
\widehat{p}(d(x_0, \cdots, x_k)|y_0, \cdots, y_k) &:= \frac{1}{N}\sum_{i=1}^{N} \delta_{\left(\widehat{\xi}^i_{0,k}, \cdots, \widehat{\xi}^i_{0,k}\right)}(d(x_0, \cdots, x_k)) \\
&\approx_{N\uparrow\infty} p(d(x_0, \cdots, x_k)|y_0, \cdots, y_k) \\
&\approx_{N\uparrow\infty} \sum_{i=1}^{N} \frac{p(y_k|\xi^i_{k,k})}{\sum_{j=1}^{N} p(y_k|\xi^j_{k,k})} \delta_{\left(\xi^i_{0,k}, \cdots, \xi^i_{0,k}\right)}(d(x_0, \cdots, x_k))
\end{aligned}$$

$$\begin{aligned}
\widehat{p}(d(x_0, \cdots, x_k)|y_0, \cdots, y_{k-1}) &:= \frac{1}{N}\sum_{i=1}^{N} \delta_{\left(\xi^i_{0,k}, \cdots, \xi^i_{k,k}\right)}(d(x_0, \cdots, x_k)) \\
&\approx_{N\uparrow\infty} p(d(x_0, \cdots, x_k)|y_0, \cdots, y_{k-1}) \\
&:= p(x_0, \cdots, x_k|y_0, \cdots, y_{k-1})dx_0, \cdots, dx_k
\end{aligned}$$

These empirical approximations are equivalent to the particle integral approximations

$$\begin{aligned}
\int F(x_0, \cdots, x_n)\widehat{p}(d(x_0, \cdots, x_k)|y_0, \cdots, y_k) &:= \frac{1}{N}\sum_{i=1}^{N} F\left(\widehat{\xi}^i_{0,k}, \cdots, \widehat{\xi}^i_{0,k}\right) \\
&\approx_{N\uparrow\infty} \int F(x_0, \cdots, x_n)p(d(x_0, \cdots, x_k)|y_0, \cdots, y_k) \\
&\approx_{N\uparrow\infty} \sum_{i=1}^{N} \frac{p(y_k|\xi^i_{k,k})}{\sum_{j=1}^{N} p(y_k|\xi^j_{k,k})} F\left(\xi^i_{0,k}, \cdots, \xi^i_{k,k}\right)
\end{aligned}$$

$$\begin{aligned}
\int F(x_0, \cdots, x_n)\widehat{p}(d(x_0, \cdots, x_k)|y_0, \cdots, y_{k-1}) &:= \frac{1}{N}\sum_{i=1}^{N} F\left(\xi^i_{0,k}, \cdots, \xi^i_{k,k}\right) \\
&\approx_{N\uparrow\infty} \int F(x_0, \cdots, x_n)p(d(x_0, \cdots, x_k)|y_0, \cdots, y_{k-1})
\end{aligned}$$

for any bounded function F on the random trajectories of the signal. As shown in the evolution of the genealogical tree coincides with a mean field particle interpretation of the evolution equations associated with the posterior densities of the signal trajectories. For more details on these path space models, we refer to the books.

Unbiased particle estimates of likelihood functions

We use the product formula

$$p(y_0, \cdots, y_n) = \prod_{k=0}^{n} p(y_k | y_0, \cdots, y_{k-1})$$

with

$$p(y_k | y_0, \cdots, y_{k-1}) = \int p(y_k | x_k) p(dx_k | y_0, \cdots, y_{k-1})$$

and the conventions $p(y_0 | y_0, \cdots, y_{-1}) = p(y_0)$ and $p(x_0 | y_0, \cdots, y_{-1}) = p(x_0)$, for $k = 0$. Replacing $p(x_k | y_0, \cdots, y_{k-1}) dx_k$ by the empirical approximation

$$\widehat{p}(dx_k | y_0, \cdots, y_{k-1}) := \frac{1}{N} \sum_{i=1}^{N} \delta_{\xi_k^i}(dx_k) \approx_{N \uparrow \infty} p(dx_k | y_0, \cdots, y_{k-1})$$

in the above displayed formula, we design the following unbiased particle approximation of the likelihood function

$$p(y_0, \cdots, y_n) \approx_{N \uparrow \infty} \widehat{p}(y_0, \cdots, y_n) = \prod_{k=0}^{n} \widehat{p}(y_k | y_0, \cdots, y_{k-1})$$

with

$$\widehat{p}(y_k | y_0, \cdots, y_{k-1}) = \int p(y_k | x_k) \widehat{p}(dx_k | y_0, \cdots, y_{k-1}) = \frac{1}{N} \sum_{i=1}^{N} p(y_k | \xi_k^i)$$

where $p(y_k | \xi_k^i)$ stands for the density $p(y_k | x_k)$ evaluated at $x_k = \xi_k^i$. The design of this particle estimate and the unbiasedness property has been proved in 1996 in the article. Refined variance estimates can be found in and.

Backward particle smoothers

Using Bayes' rule, we have the formula

$$p(x_0, \cdots, x_n | y_0, \cdots, y_{n-1}) = p(x_n | y_0, \cdots, y_{n-1}) p(x_{n-1} | x_n, y_0, \cdots, y_{n-1}) \cdots p(x_1 | x_2, y_0, y_1) p(x_0 | x_1, y_0)$$

Notice that

$$p(x_{k-1} | x_k, (y_0, \cdots, y_{k-1})) \propto p(x_k | x_{k-1}) p(x_{k-1} | (y_0, \cdots, y_{k-1}))$$
$$p(x_{k-1} | (y_0, \cdots, y_{k-1}) \propto p(y_{k-1} | x_{k-1}) p(x_{k-1} | (y_0, \cdots, y_{k-2})$$

This implies that

$$p(x_{k-1} | x_k, (y_0, \cdots, y_{k-1})) = \frac{p(y_{k-1} | x_{k-1}) p(x_k | x_{k-1}) p(x_{k-1} | y_0, \cdots, y_{k-2})}{\int p(y_{k-1} | x'_{k-1}) p(x_k | x'_{k-1}) p(x'_{k-1} | y_0, \cdots, y_{k-2}) dx'_{k-1}}$$

Replacing the one-step optimal predictors $p(x_{k-1} | (y_0, \cdots, y_{k-2})) dx_{k-1}$ by the particle empirical measures

$$\widehat{p}(dx_{k-1} | (y_0, \cdots, y_{k-2})) = \frac{1}{N} \sum_{i=1}^{N} \delta_{\xi_{k-1}^i}(dx_{k-1}) \left(\approx_{N \uparrow \infty} p(dx_{k-1} | (y_0, \cdots, y_{k-2})) := p(x_{k-1} | (y_0, \cdots, y_{k-2})) dx_{k-1}\right)$$

we find that

$$\begin{aligned} p(dx_{k-1}|x_k,(y_0,\cdots,y_{k-1})) &\approx_{N\uparrow\infty} \widehat{p}(dx_{k-1}|x_k,(y_0,\cdots,y_{k-1})) \\ &:= \frac{p(y_{k-1}|x_{k-1})p(x_k|x_{k-1})\widehat{p}(dx_{k-1}|y_0,\cdots,y_{k-2})}{\int p(y_{k-1}|x'_{k-1})\,p(x_k|x'_{k-1})\widehat{p}(dx'_{k-1}|y_0,\cdots,y_{k-2})} \\ &= \sum_{i=1}^{N} \frac{p(y_{k-1}|\xi^i_{k-1})p(x_k|\xi^i_{k-1})}{\sum_{j=1}^{N} p(y_{k-1}|\xi^j_{k-1})p(x_k|\xi^j_{k-1})}\delta_{\xi^i_{k-1}}(dx_{k-1}) \end{aligned}$$

We conclude that

$$p(d(x_0,\cdots,x_n)|(y_0,\cdots,y_{n-1})) \approx_{N\uparrow\infty} \widehat{p}_{backward}(d(x_0,\cdots,x_n)|(y_0,\cdots,y_{n-1}))$$

with the backward particle approximation

$$\widehat{p}_{backward}(d(x_0,\cdots,x_n)|(y_0,\cdots,y_{n-1})) = \widehat{p}(dx_n|(y_0,\cdots,y_{n-1}))\widehat{p}(dx_{n-1}|x_n,(y_0,\cdots,y_{n-1}))\cdots\widehat{p}(dx_1|x_2,(y_0,y_1))\widehat{p}(dx_0|x_1,y_0)$$

The probability measure

$$\widehat{p}_{backward}(d(x_0,\cdots,x_n)|(y_0,\cdots,y_{n-1}))$$

is the probability of the random paths of a Markov chain $\left(\mathbb{X}^\flat_{k,n}\right)_{0\leqslant k\leqslant n}$ running backward in time from time k=n to time k=0, and evolving at each time step k in the state space associated with the population of particles $\xi^i_k, i=1,\cdots,N$.

- Initially (at time k=n) the chain $\mathbb{X}^\flat_{n,n}$ chooses randomly a state with the distribution

$$\widehat{p}(dx_n|(y_0,\cdots,y_{n-1})) = \frac{1}{N}\sum_{i=1}^{N}\delta_{\xi^i_n}(dx_n)$$

- From time k to the time (k-1), the chain starting at some state $\mathbb{X}^\flat_{k,n}=\xi^i_k$ for some $i=1,\cdots,N$ at time k moves at time (k-1) to a random state $\mathbb{X}^\flat_{k-1,n}$ chosen with the discrete weighted probability

$$\widehat{p}(dx_{k-1}|\xi^i_k,(y_0,\cdots,y_{k-1})) = \sum_{j=1}^{N} \frac{p(y_{k-1}|\xi^j_{k-1})p(\xi^i_k|\xi^j_{k-1})}{\sum_{l=1}^{N} p(y_{k-1}|\xi^l_{k-1})p(\xi^i_k|\xi^l_{k-1})}\,\delta_{\xi^j_{k-1}}(dx_{k-1})$$

In the above displayed formula, $\widehat{p}(dx_{k-1}|\xi^i_k,(y_0,\cdots,y_{k-1}))$ stands for the conditional distribution $\widehat{p}(dx_{k-1}|x_k,(y_0,\cdots,y_{k-1}))$ evaluated at $x_k=\xi^i_k$. In the same vein, $p(y_{k-1}|\xi^j_{k-1})$ and $p(\xi^i_k|\xi^j_{k-1})$ stand for the conditional densities $p(y_{k-1}|x_{k-1})$ and $p(x_k|x_{k-1})$ evaluated at $x_k=\xi^i_k$ and $x_{k-1}=\xi^j_{k-1}$. These models allows to reduce integration with respect to the densities $p((x_0,\cdots,x_n)|(y_0,\cdots,y_{n-1}))$ in terms of matrix operations with respect to the Markov transitions of the chain described above. For instance, for any function f_k we have the particle estimates

$$\begin{aligned}
&\int p(d(x_0,\cdots,x_n)|(y_0,\cdots,y_{n-1}))f_k(x_k)\\
&\approx_{N\uparrow\infty} \int \widehat{p}_{backward}(d(x_0,\cdots,x_n)|(y_0,\cdots,y_{n-1}))f_k(x_k)\\
&= \int \widehat{p}(dx_n|(y_0,\cdots,y_{n-1}))\widehat{p}(dx_{n-1}|x_n,(y_0,\cdots,y_{n-1}))\cdots\widehat{p}(dx_k|x_{k+1},(y_0,\cdots,y_k))f_k(x_k)\\
&= \underbrace{\left[\tfrac{1}{N},\cdots,\tfrac{1}{N}\right]}_{N \text{ times}} \mathbb{M}_{n-1}\cdots\mathbb{M}_k \begin{bmatrix} f_k(\xi_k^1)\\ \vdots \\ f_k(\xi_k^N)\end{bmatrix}
\end{aligned}$$

where

$$\mathbb{M}_k = (\mathbb{M}_k(i,j))_{1\leqslant i,j\leqslant N}: \qquad \mathbb{M}_k(i,j) = \frac{p(\xi_k^i|\xi_{k-1}^j)\,p(y_{k-1}|\xi_{k-1}^j)}{\sum_{l=1}^N p(\xi_k^i|\xi_{k-1}^l)p(y_{k-1}|\xi_{k-1}^l)}$$

This also shows that if

$$\overline{F}(x_0,\cdots,x_n) := \frac{1}{n+1}\sum_{k=0}^n f_k(x_k)$$

then

$$\begin{aligned}
\int \overline{F}(x_0,\cdots,x_n)p(d(x_0,\cdots,x_n)|(y_0,\cdots,y_{n-1})) &\approx_{N\uparrow\infty} \int \overline{F}(x_0,\cdots,x_n)\widehat{p}_{backward}(d(x_0,\cdots,x_n)|(y_0,\cdots,y_{n-1}))\\
&= \frac{1}{n+1}\sum_{k=0}^n \underbrace{\left[\tfrac{1}{N},\cdots,\tfrac{1}{N}\right]}_{N \text{ times}} \mathbb{M}_{n-1}\mathbb{M}_{n-2}\cdots\mathbb{M}_k \begin{bmatrix} f_k(\xi_k^1)\\ \vdots \\ f_k(\xi_k^N)\end{bmatrix}
\end{aligned}$$

Some convergence results

We shall assume that filtering equation is stable, in the sense that it corrects any erroneous initial condition.

In this situation, the **particle approximations of the likelihood functions** are unbiased and the relative variance is controlled by

$$E\left(\widehat{p}(y_0,\cdots,y_n)\right) = p(y_0,\cdots,y_n), \qquad E\left(\left[\frac{\widehat{p}(y_0,\cdots,y_n)}{p(y_0,\cdots,y_n)}-1\right]^2\right) \leqslant \frac{cn}{N},$$

for some finite constant c. In addition, for any $x \geqslant 0$:

$$\mathbf{P}\left(\left|\tfrac{1}{n}\log\widehat{p}(y_0,\cdots,y_n) - \tfrac{1}{n}\log p(y_0,\cdots,y_n)\right| \leqslant c_1\tfrac{x}{N} + c_2\sqrt{\tfrac{x}{N}}\right) > 1-e^{-x}$$

for some finite constants c_1, c_2 related to the asymptotic bias and variance of the particle estimate, and for some finite constant c.

The bias and the variance of **the particle particle estimates based on the ancestral lines of the genealogical trees**

$$
\begin{aligned}
I_k^{path}(F) &:= \int F(x_0, \cdots, x_k) p(d(x_0, \cdots, x_k) | y_0, \cdots, y_{k-1}) \\
&\approx_{N \uparrow \infty} \widehat{I}_k^{path}(F) \\
&:= \int F(x_0, \cdots, x_k) \widehat{p}(d(x_0, \cdots, x_k) | y_0, \cdots, y_{k-1}) \\
&= \frac{1}{N} \sum_{i=1}^{N} F\left(\xi_{0,k}^i, \cdots, \xi_{k,k}^i\right)
\end{aligned}
$$

are controlled by the non asymptotic uniform estimates

$$
\left| E\left(\widehat{I}_k^{path}(F)\right) - I_k^{path}(F) \right| \leqslant \tfrac{c_1 k}{N}, \qquad E\left(\left[\widehat{I}_k^{path}(F) - I_k^{path}(F)\right]^2\right) \leqslant \tfrac{c_2 k}{N},
$$

for any function F bounded by 1, and for some finite constants c_1, c_2. In addition, for any $x \geqslant 0$:

$$
\mathbf{P}\left(\left| \widehat{I}_k^{path}(F) - I_k^{path}(F) \right| \leqslant c_1 \tfrac{kx}{N} + c_2 \sqrt{\tfrac{kx}{N}} \wedge \sup_{0 \leqslant k \leqslant n} \left| \widehat{I}_k^{path}(F) - I_k^{path}(F) \right| \leqslant c \sqrt{\tfrac{xn \log(n)}{N}} \right) > 1 - e^{-x}
$$

for some finite constants c_1, c_2 related to the asymptotic bias and variance of the particle estimate, and for some finite constant c. The same type of bias and variance estimates hold for the backward particle smoothers. For additive functionals of the form

$$
\overline{F}(x_0, \cdots, x_n) := \frac{1}{n+1} \sum_{0 \leqslant k \leqslant n} f_k(x_k)
$$

with

$$
I_n^{path}(\overline{F}) \approx_{N \uparrow \infty} I_n^{\flat, path}(\overline{F}) := \int \overline{F}(x_0, \cdots, x_n) \widehat{p}_{backward}(d(x_0, \cdots, x_n) | (y_0, \cdots, y_{n-1}))
$$

with functions f_k bounded by 1, we have

$$
\sup_{n \geqslant 0} \left| E\left(\widehat{I}_n^{\flat, path}(\overline{F})\right) - I_n^{path}(\overline{F}) \right| \leqslant \frac{c_1}{N}
$$

and

$$
E\left(\left[\widehat{I}_n^{\flat, path}(F) - I_n^{path}(F)\right]^2\right) \leqslant \frac{c_2}{nN} + \frac{c_3}{N^2}
$$

for some finite constants c_1, c_2, c_3. More refined estimates including exponentially small probability of errors are developed in.

Sequential Importance Resampling (SIR)

The bootstrap filter

Sequential importance Resampling (SIR), the original bootstrap filtering algorithm (Gordon et al. 1993), is also a very commonly used filtering algorithm, which approximates the filtering probability density $p(x_k|y_0, \cdots, y_k)$ by a weighted set of *N* samples

$$\left\{\left(w_k^{(i)}, x_k^{(i)}\right) : i \in \{1, \cdots, N\}\right\}.$$

The *importance weights* $w_k^{(i)}$ are approximations to the relative posterior probabilities (or densities) of the samples such that

$$\sum_{i=1}^{N} w_k^{(i)} = 1.$$

SIS is a sequential (i.e., recursive) version of importance sampling. As in importance sampling, the expectation of a function *f* can be approximated as a weighted average

$$\int f(x_k) p(x_k|y_0, \ldots, y_k) dx_k \approx \sum_{i=1}^{N} w_k^{(i)} f(x_k^{(i)}).$$

For a finite set of samples, the algorithm performance is dependent on the choice of the *proposal distribution*

$$\pi(x_k|x_{0:k-1}, y_{0:k}).$$

The "*optimal*" *proposal distribution* is given as the *target distribution*

$$\pi(x_k|x_{0:k-1}, y_{0:k}) = p(x_k|x_{k-1}, y_k) = \frac{p(y_k|x_k)}{\int p(y_k|x_k) p(x_k|x_{k-1}) dx_k} p(x_k|x_{k-1}).$$

This particular choice of proposal transition has been proposed by P. Del Moral in 1996 and 1998. When it is difficult to sample transitions according to the distribution $p(x_k|x_{k-1}, y_k)$ one natural strategy is to use the following particle approximation

$$\begin{aligned}\frac{p(y_k|x_k)}{\int p(y_k|x_k) p(x_k|x_{k-1}) dx_k} p(x_k|x_{k-1}) dx_k &\simeq_{N \uparrow \infty} \frac{p(y_k|x_k)}{\int p(y_k|x_k) \widehat{p}(dx_k|x_{k-1})} \widehat{p}(dx_k|x_{k-1}) \\ &= \sum_{i=1}^{N} \frac{p(y_k|X_k^i(x_{k-1}))}{\sum_{j=1}^{N} p(y_k|X_k^j(x_{k-1}))} \delta_{X_k^i(x_{k-1})}(dx_k)\end{aligned}$$

with the empirical approximation

$$\widehat{p}(dx_k|x_{k-1}) = \frac{1}{N} \sum_{i=1}^{N} \delta_{X_k^i(x_{k-1})}(dx_k) \simeq_{N \uparrow \infty} p(x_k|x_{k-1}) dx_k$$

associated with *N* (or any other large number of samples) independent random samples $X_k^i(x_{k-1}), i = 1, \cdots, N$ with the conditional distribution of the random state X_k given $X_{k-1} = x_{k-1}$. The consistency of the resulting particle filter of this approximation and other extensions are developed in. In the above display δ_a stands for the **Dirac measure** at a given state a.

However, the transition prior probability distribution is often used as importance function, since it is easier to draw particles (or samples) and perform subsequent importance weight calculations:

$$\pi(x_k|x_{0:k-1}, y_{0:k}) = p(x_k|x_{k-1}).$$

Sequential Importance Resampling (SIR) filters with transition prior probability distribution as importance function are commonly known as bootstrap filter and condensation algorithm.

Resampling is used to avoid the problem of degeneracy of the algorithm, that is, avoiding the situation that all but one of the importance weights are close to zero. The performance of the algorithm can be also affected by proper choice of resampling method. The *stratified sampling* proposed by Kitagawa (1996) is optimal in terms of variance.

A single step of sequential importance resampling is as follows:

1) For $i = 1, \cdots, N$ draw samples from the *proposal distribution*

$$x_k^{(i)} \sim \pi(x_k|x_{0:k-1}^{(i)}, y_{0:k})$$

2) For $i = 1, \cdots, N$ update the importance weights up to a normalizing constant:

$$\hat{w}_k^{(i)} = w_{k-1}^{(i)} \frac{p(y_k|x_k^{(i)})p(x_k^{(i)}|x_{k-1}^{(i)})}{\pi(x_k^{(i)}|x_{0:k-1}^{(i)}, y_{0:k})}.$$

Note that when we use the transition prior probability distribution as the importance function,

$$\pi(x_k^{(i)}|x_{0:k-1}^{(i)}, y_{0:k}) = p(x_k^{(i)}|x_{k-1}^{(i)}),$$

this simplifies to the following :

$$\hat{w}_k^{(i)} = w_{k-1}^{(i)} p(y_k|x_k^{(i)}),$$

3) For $i = 1, \cdots, N$ compute the normalized importance weights:

$$w_k^{(i)} = \frac{\hat{w}_k^{(i)}}{\sum_{j=1}^N \hat{w}_k^{(j)}}$$

4) Compute an estimate of the effective number of particles as

$$\hat{N}_{\mathit{eff}} = \frac{1}{\sum_{i=1}^N \left(w_k^{(i)}\right)^2}$$

This criterion reflects the variance of the weights, other criteria can be found in the article, including their rigorous analysis and central limit theorems.

5) If the effective number of particles is less than a given threshold $\hat{N}_{eff} < N_{thr}$, then perform resampling:

a) Draw N particles from the current particle set with probabilities proportional to their weights. Replace the current particle set with this new one.

b) For $i = 1, \cdots, N$ set $w_k^{(N)} = 1/N$.

The term *Sampling Importance Resampling* is also sometimes used when referring to SIR filters.

Sequential importance sampling (SIS)

- Is the same as sequential importance resampling, but without the resampling stage.

"direct version" algorithm

The "direct version" algorithm Wikipedia:Citation needed is rather simple (compared to other particle filtering algorithms) and it uses composition and rejection. To generate a single sample *x* at *k* from $p_{x_k|y_{1:k}}(x|y_{1:k})$:

1) Set n=0 (This will count the number of particles generated so far)

2) Uniformly choose an index i from the range $\{1, ..., N\}$

3) Generate a test $\hat{x}$ from the distribution $p(x_k|x_{k-1})$ with $x_{k-1} = x_{k-1|k-1}^{(i)}$

4) Generate the probability of $\hat{y}$ using $\hat{x}$ from $p(y_k|x_k)$, with $x_k = \hat{x}$ where y_k is the measured value

5) Generate another uniform u from $[0, m_k]$ where $m_k = \sup_{x_k} p(y_k|x_k)$

6) Compare u and $p\left(\hat{y}\right)$

6a) If u is larger then repeat from step 2

6b) If u is smaller then save $\hat{x}$ as $x_{k|k}^{(i)}$ and increment n

7) If n == N then quit

The goal is to generate P "particles" at *k* using only the particles from $k-1$. This requires that a Markov equation can be written (and computed) to generate a x_k based only upon x_{k-1} . This algorithm uses composition of the P particles from $k-1$ to generate a particle at *k* and repeats (steps 2–6) until P particles are generated at *k*.

This can be more easily visualized if *x* is viewed as a two-dimensional array. One dimension is *k* and the other dimensions is the particle number. For example, $x(k,i)$ would be the ith particle at k and can also be written $x_k^{(i)}$ (as done above in the algorithm). Step 3 generates a *potential* x_k based on a randomly chosen particle ($x_{k-1}^{(i)}$) at time $k-1$ and rejects or accepts it in step 6. In other words, the x_k values are generated using the previously generated x_{k-1} .

Other particle filters

- Exponential Natural Particle Filter
- Auxiliary particle filter
- Regularized auxiliary particle filter
- Gaussian particle filter
- Unscented particle filter
- Gauss–Hermite particle filter
- Cost Reference particle filter
- Hierarchical/Scalable particle filter
- Rao–Blackwellized particle filter
- Rejection-sampling based optimal particle filter
- Feynman-Kac and mean field particle methodologies
- Particle Markov-Chain Monte-Carlo

Bibliography

- Del Moral, Pierre (1996). "Non Linear Filtering: Interacting Particle Solution"[814] (PDF). *Markov Processes and Related Fields*. **2** (4): 555–580.
- Del Moral, Pierre (2004). *Feynman-Kac formulae. Genealogical and interacting particle approximations*[815]. Springer. p. 575. <q>Series: Probability and Applications</q>
- Del Moral, Pierre (2013). *Mean field simulation for Monte Carlo integration*[816] .[816] Chapman & Hall/CRC Press. p. 626. <q>Monographs on Statistics & Applied Probability</q>
- Cappe, O.; Moulines, E.; Ryden, T. (2005). *Inference in Hidden Markov Models*. Springer.
- Liu, J.S.,; Chen, R. (1998). "Sequential Monte Carlo methods for dynamic systems"[817] (PDF). *Journal of the American Statistical Association*. Taylor & Francis Group. **93** (443): 1032–1044. doi: 10.1080/01621459.1998.10473765[818].

- Liu, J.S. (2001). *Monte Carlo strategies in Scientific Computing*. Springer.
- Kong, A.; Liu, J.S.; Wong, W.H. (1994). "Sequential imputations and Bayesian missing data problems"[819] (PDF). *Journal of the American Statistical Association*. Taylor & Francis Group. **89** (425): 278–288. doi: 10.1080/01621459.1994.10476469[820].
- Liu, J.S.; Chen, R. (1995). "Blind deconvolution via sequential imputations"[821] (PDF). *Journal of the American Statistical Association*. Taylor & Francis Group. **90** (430): 567–576. doi: 10.2307/2291068[822].
- Ristic, B.; Arulampalam, S.; Gordon, N. (2004). *Beyond the Kalman Filter: Particle Filters for Tracking Applications*. Artech House.
- Doucet, A.; Johansen, A.M. (December 2008). "A tutorial on particle filtering and smoothing: fifteen years later"[823] (PDF). *Technical report*. Department of Statistics, University of British Columbia.
- Doucet, A.; Godsill, S.; Andrieu, C. (2000). "On sequential Monte Carlo sampling methods for Bayesian filtering"[824]. *Statistics and Computing*. **10** (3): 197–208. doi: 10.1023/A:1008935410038[825].
- Arulampalam, M.S.; Maskell, S.; Gordon, N.; Clapp, T. (2002). "A tutorial on particle filters for online nonlinear/non-Gaussian Bayesian tracking"[826]. *IEEE Transactions on Signal Processing*. **50** (2): 174–188. Bibcode: 2002ITSP...50..174A[827]. doi: 10.1109/78.978374[828].
- Cappe, O.; Godsill, S.; Moulines, E. (2007). "An overview of existing methods and recent advances in sequential Monte Carlo". *Proceedings of the IEEE*. **95** (5): 899–924. doi: 10.1109/JPROC.2007.893250[829].
- Kitagawa, G. (1996). "Monte carlo filter and smoother for non-Gaussian nonlinear state space models". *Journal of Computational and Graphical Statistics*. **5** (1): 1–25. doi: 10.2307/1390750[830]. JSTOR 1390750[831].
- Kotecha, J.H.; Djuric, P. (2003). "Gaussian Particle filtering". *IEEE Transactions Signal Processing*. **51** (10).
- Haug, A.J. (2005). "A Tutorial on Bayesian Estimation and Tracking Techniques Applicable to Nonlinear and Non-Gaussian Processes"[832] (PDF). *The MITRE Corporation, USA, Tech. Rep., Feb*. Retrieved 2008-05-06.
- Pitt, M.K.; Shephard, N. (1999). "Filtering Via Simulation: Auxiliary Particle Filters"[833]. *Journal of the American Statistical Association*. Journal of the American Statistical Association, Vol. 94, No. 446. **94** (446): 590–591. doi: 10.2307/2670179[834]. JSTOR 2670179[835]. Retrieved 2008-05-06.
- Gordon, N. J.; Salmond, D. J.; Smith, A. F. M. (1993). "Novel approach to nonlinear/non-Gaussian Bayesian state estimation"[836]. *IEE Proceedings F on Radar and Signal Processing*. **140** (2): 107–113. doi: 10.1049/ip-f-2.1993.0015[837]. Retrieved 2009-09-19.

- Chen, Z. (2003). "Bayesian Filtering: From Kalman Filters to Particle Filters, and Beyond". CiteSeerX 10.1.1.107.7415[838] ª.
- Vaswani, N.; Rathi, Y.; Yezzi, A.; Tannenbaum, A. (2007). "Tracking deforming objects using particle filtering for geometric active contours". *IEEE Trans. on Pattern Analysis and Machine Intelligence*. **29** (8): 1470–1475. doi: 10.1109/tpami.2007.1081[839].

External links

<templatestyles src="Template:Refbegin/styles.css" />

- Feynman–Kac models and interacting particle algorithms (a.k.a. Particle Filtering)[840] Theoretical aspects and a list of application domains of particle filters
- Sequential Monte Carlo Methods (Particle Filtering)[841] homepage on University of Cambridge
- Dieter Fox's MCL Animations[842]
- Rob Hess' free software[843]
- SMCTC: A Template Class for Implementing SMC algorithms in C++[844]
- Java applet on particle filtering[845]
- vSMC : Vectorized Sequential Monte Carlo[846]
- Particle filter explained in the context of self driving car[847]

Decision theory

Decision theory (or the **theory of choice**) is the study of the reasoning underlying an agent's choices.[848] Decision theory can be broken into three branches: normative decision theory, which gives advice on how to make the best decisions, given a set of uncertain beliefs and a set of values; descriptive decision theory, which analyzes how existing, possibly irrational agents actually make decisions; and prescriptive decision theory, which tries to guide or give procedures on how or what we should do in order to make best decisions in line with the normative theory.

Closely related to the field of game theory, decision theory is concerned with the choices of individual agents whereas game theory is concerned with interactions of agents whose decisions affect each other. Decision theory is an interdisciplinary topic, studied by economists, statisticians, psychologists, biologists, political and other social scientists, philosophers,[849] and computer scientists.

Empirical applications of this rich theory are usually done with the help of statistical and econometric methods, especially via the so-called choice models, such as probit and logit models. Estimation of such models is usually done via parametric, semi-parametric and non-parametric maximum likelihood methods.

Normative and descriptive

Normative decision theory is concerned with identifying the best decision to make, modelling an ideal decision maker who is able to compute with perfect accuracy and is fully rational. The practical application of this prescriptive approach (how people *ought to* make decisions) is called decision analysis, and is aimed at finding tools, methodologies and software (decision support systems) to help people make better decisions.

In contrast, positive or descriptive decision theory is concerned with describing observed behaviors under the assumption that the decision-making agents are behaving under some consistent rules. These rules may, for instance, have a procedural framework (e.g. Amos Tversky's elimination by aspects model) or an axiomatic framework, reconciling the Von Neumann-Morgenstern axioms with behavioral violations of the expected utility hypothesis, or they may explicitly give a functional form for time-inconsistent utility functions (e.g. Laibson's quasi-hyperbolic discounting).

The prescriptions or predictions about behaviour that positive decision theory produces allow for further tests of the kind of decision-making that occurs in practice. There is a thriving dialogue with experimental economics, which uses laboratory and field experiments to evaluate and inform theory. In recent decades, there has also been increasing interest in what is sometimes called "behavioral decision theory" and this has contributed to a re-evaluation of what rational decision-making requires.[850]

What kinds of decisions need a theory?

Choice under uncertainty

The area of choice under uncertainty represents the heart of decision theory. Known from the 17th century (Blaise Pascal invoked it in his famous wager, which is contained in his *Pensées*, published in 1670), the idea of expected value is that, when faced with a number of actions, each of which could give rise to more than one possible outcome with different probabilities, the rational procedure is to identify all possible outcomes, determine their values (positive or negative) and the probabilities that will result from each course of action, and multiply the two to give an "expected value", or the average expectation for an outcome; the action to be chosen should be the one that gives rise to the highest total expected value. In 1738, Daniel Bernoulli published an influential paper entitled *Exposition of a New Theory on the Measurement of Risk*, in which he uses the St. Petersburg paradox to show that expected value theory must be normatively wrong. He gives an example in which a Dutch merchant is trying to decide whether to insure a cargo being sent from Amsterdam to St Petersburg in winter. In his solution, he defines a utility function and computes expected utility rather than expected financial value (see for a review).

In the 20th century, interest was reignited by Abraham Wald's 1939 paper pointing out that the two central procedures of sampling-distribution-based statistical-theory, namely hypothesis testing and parameter estimation, are special cases of the general decision problem. Wald's paper renewed and synthesized many concepts of statistical theory, including loss functions, risk functions, admissible decision rules, antecedent distributions, Bayesian procedures, and minimax procedures. The phrase "decision theory" itself was used in 1950 by E. L. Lehmann.

The revival of subjective probability theory, from the work of Frank Ramsey, Bruno de Finetti, Leonard Savage and others, extended the scope of expected utility theory to situations where subjective probabilities can be used. At the time, von Neumann and Morgenstern theory of expected utility proved that expected utility maximization followed from basic postulates about rational behavior.

The work of Maurice Allais and Daniel Ellsberg showed that human behavior has systematic and sometimes important departures from expected-utility maximization. The prospect theory of Daniel Kahneman and Amos Tversky renewed the empirical study of economic behavior with less emphasis on rationality presuppositions. Kahneman and Tversky found three regularities – in actual human decision-making, "losses loom larger than gains"; persons focus more on *changes* in their utility-states than they focus on absolute utilities; and the estimation of subjective probabilities is severely biased by anchoring.

Intertemporal choice

Intertemporal choice is concerned with the kind of choice where different actions lead to outcomes that are realised at different points in time. If someone received a windfall of several thousand dollars, they could spend it on an expensive holiday, giving them immediate pleasure, or they could invest it in a pension scheme, giving them an income at some time in the future. What is the optimal thing to do? The answer depends partly on factors such as the expected rates of interest and inflation, the person's life expectancy, and their confidence in the pensions industry. However even with all those factors taken into account, human behavior again deviates greatly from the predictions of prescriptive decision theory, leading to alternative models in which, for example, objective interest rates are replaced by subjective discount rates.

Interaction of decision makers

Some decisions are difficult because of the need to take into account how other people in the situation will respond to the decision that is taken. The analysis of such social decisions is more often treated under the label of game theory, rather than decision theory, though it involves the same mathematical methods. From the standpoint of game theory most of the problems treated in decision theory are one-player games (or the one player is viewed as playing against an impersonal background situation). In the emerging socio-cognitive engineering, the research is especially focused on the different types of distributed decision-making in human organizations, in normal and abnormal/emergency/crisis situations.[851]

Complex decisions

Other areas of decision theory are concerned with decisions that are difficult simply because of their complexity, or the complexity of the organization that has to make them. Individuals making decisions may be limited in resources or are boundedly rational (have finite time or intelligence); in such cases the issue, more than the deviation between real and optimal behaviour, is the difficulty of determining the optimal behaviour in the first place. One example is the model of economic growth and resource usage developed by the Club of Rome to help politicians make real-life decisions in complex situationsWikipedia:Citation needed. Decisions are also affected by whether options are framed together or separately; this is known as the distinction bias. In 2011, Dwayne Rosenburgh explored and showed how decision theory can be applied to complex decisions that arise in areas such as wireless communications.[852]

Heuristics

The heuristic approach to decision-making makes decisions based on routine thinking, which, while quicker than step-by-step processing, opens the risk of introducing inaccuracies, mistakes and fallacies, which may be easily disproved in a step-by-step process of thinking. One example of common and incorrect thought process is the gambler's fallacy, or believing that a random event is affected by previous random events (truth is, there is a fifty percent chance of a coin landing on heads even after a long sequence of tails). Another example is that decision-makers may be biased towards preferring moderate alternatives to extreme ones; the "Compromise Effect" operates under a mindset driven by the belief that the most moderate option, amid extremes, carries the most benefits from each extreme.

Alternatives

A highly controversial issue is whether one can replace the use of probability in decision theory by other alternatives.

Probability theory

Advocates for the use of probability theory point to:

- the work of Richard Threlkeld Cox for justification of the probability axioms,
- the Dutch book paradoxes of Bruno de Finetti as illustrative of the theoretical difficulties that can arise from departures from the probability axioms, and
- the complete class theorems, which show that all admissible decision rules are equivalent to the Bayesian decision rule for some utility function and some prior distribution (or for the limit of a sequence of prior distributions). Thus, for every decision rule, either the rule may be reformulated as a Bayesian procedure (or a limit of a sequence of such), or there is a rule that is sometimes better and never worse.

Alternatives to probability theory

The proponents of fuzzy logic, possibility theory, quantum cognition, Dempster–Shafer theory, and info-gap decision theory maintain that probability is only one of many alternatives and point to many examples where non-standard alternatives have been implemented with apparent success; notably, probabilistic decision theory is sensitive to assumptions about the probabilities of various events, while non-probabilistic rules such as minimax are robust, in that they do not make such assumptions.

Ludic fallacy

A general criticism of decision theory based on a fixed universe of possibilities is that it considers the "known unknowns", not the "unknown unknowns"Wikipedia:Citation needed: it focuses on expected variations, not on unforeseen events, which some argue (as in black swan theory) have outsized impact and must be considered – significant events may be "outside model". This line of argument, called the ludic fallacy, is that there are inevitable imperfections in modeling the real world by particular models, and that unquestioning reliance on models blinds one to their limits.

Further reading

<templatestyles src="Template:Refbegin/styles.css" />

- Akerlof, George A.; Yellen, Janet L. (May 1987). "Rational Models of Irrational Behavior". **77** (2): 137–142.
- Anand, Paul (1993). *Foundations of Rational Choice Under Risk*. Oxford: Oxford University Press. ISBN 0-19-823303-5. (*an overview of the philosophical foundations of key mathematical axioms in subjective expected utility theory – mainly normative*)
- Arthur, W. Brian (May 1991). "Designing Economic Agents that Act like Human Agents: A Behavioral Approach to Bounded Rationality". *The American Economic Review*. **81** (2): 353–9.
- Berger, James O. (1985). *Statistical decision theory and Bayesian Analysis* (2nd ed.). New York: Springer-Verlag. ISBN 0-387-96098-8. MR 0804611[853].
- Bernardo, José M.; Smith, Adrian F. M. (1994). *Bayesian Theory*. Wiley. ISBN 0-471-92416-4. MR 1274699[854].
- Clemen, Robert; Reilly, Terence (2014). *Making Hard Decisions with DecisionTools: An Introduction to Decision Analysis* (3rd ed.). Stamford CT: Cengage. ISBN 0-538-79757-6. *(covers normative decision theory)*
- De Groot, Morris, *Optimal Statistical Decisions*. Wiley Classics Library. 2004. (Originally published 1970.) ISBN 0-471-68029-X.
- Goodwin, Paul; Wright, George (2004). *Decision Analysis for Management Judgment* (3rd ed.). Chichester: Wiley. ISBN 0-470-86108-8. *(covers both normative and descriptive theory)*
- Hansson, Sven Ove. "Decision Theory: A Brief Introduction"[855] (PDF). Archived from the original[856] (PDF) on July 5, 2006.
- Khemani, Karan, Ignorance is Bliss: A study on how and why humans depend on recognition heuristics in social relationships, the equity markets and the brand market-place, thereby making successful decisions, 2005.
- Leach, Patrick (2006). *Why Can't You Just Give Me the Number? An Executive's Guide to Using Probabilistic Thinking to Manage Risk and to Make Better Decisions*. Probabilistic. ISBN 0-9647938-5-7. A rational presentation of probabilistic analysis.
- Miller L (1985). "Cognitive risk-taking after frontal or temporal lobectomy—I. The synthesis of fragmented visual information". *Neuropsychologia*. **23** (3): 359–69. doi: 10.1016/0028-3932(85)90022-3[857]. PMID 4022303[858].
- Miller L, Milner B (1985). "Cognitive risk-taking after frontal or temporal lobectomy—II. The synthesis of phonemic and semantic information". *Neuropsychologia*. **23** (3): 371–9. doi: 10.1016/0028-3932(85)90023-5[859]. PMID 4022304[860].

- North, D.W. (1968). "A tutorial introduction to decision theory". *IEEE Transactions on Systems Science and Cybernetics*. **4** (3): 200–210. doi: 10.1109/TSSC.1968.300114[861]. Reprinted in Shafer & Pearl. *(also about normative decision theory)*
- Peterson, Martin (2009). *An Introduction to Decision Theory*. Cambridge University Press. ISBN 978-0-521-71654-3.
- Raiffa, Howard (1997). *Decision Analysis: Introductory Lectures on Choices Under Uncertainty*. McGraw Hill. ISBN 0-07-052579-X.
- Robert, Christian (2007). *The Bayesian Choice* (2nd ed.). New York: Springer. doi: 10.1007/0-387-71599-1[862]. ISBN 0-387-95231-4. MR 1835885[863].
- Shafer, Glenn; Pearl, Judea, eds. (1990). *Readings in uncertain reasoning*. San Mateo, CA: Morgan Kaufmann.
- Smith, J.Q. (1988). *Decision Analysis: A Bayesian Approach*. Chapman and Hall. ISBN 0-412-27520-1.
- Charles Sanders Peirce and Joseph Jastrow (1885). "On Small Differences in Sensation"[864]. *Memoirs of the National Academy of Sciences*. **3**: 73–83. http://psychclassics.yorku.ca/Peirce/small-diffs.htm
- Ramsey, Frank Plumpton; "Truth and Probability" (PDF[865]), Chapter VII in *The Foundations of Mathematics and other Logical Essays* (1931).
- de Finetti, Bruno (September 1989). "Probabilism: A Critical Essay on the Theory of Probability and on the Value of Science". *Erkenntnis*. **31**. (translation of 1931 article)
- de Finetti, Bruno (1937). "La Prévision: ses lois logiques, ses sources subjectives". *Annales de l'Institut Henri Poincaré*.

  de Finetti, Bruno. "Foresight: its Logical Laws, Its Subjective Sources," (translation of the 1937 article[866] in French) in H. E. Kyburg and H. E. Smokler (eds), *Studies in Subjective Probability,* New York: Wiley, 1964.
- de Finetti, Bruno. *Theory of Probability*, (translation by AFM Smith of 1970 book) 2 volumes, New York: Wiley, 1974-5.
- Donald Davidson, Patrick Suppes and Sidney Siegel (1957). *Decision-Making: An Experimental Approach*. Stanford University Press.
- Pfanzagl, J (1967). "Morgenstern". In Martin Shubik. *Essays in Mathematical Economics In Honor of Oskar Morgenstern*. Princeton University Press. pp. 237–251.
- Pfanzagl, J. in cooperation with V. Baumann and H. Huber (1968). "Events, Utility and Subjective Probability". *Theory of Measurement*. Wiley. pp. 195–220.
- Morgenstern, Oskar (1976). "Some Reflections on Utility". In Andrew Schotter. *Selected Economic Writings of Oskar Morgenstern*. New York University Press. pp. 65–70. ISBN 0-8147-7771-6.

- Non-Robust Models in Statistics by Lev B. Klebanov, Svetlozat T. Rachev and Frank J. Fabozzi, Nova Scientific Publishers, Inc. New York, 2009.

Utility

Part of a series on
Utilitarianism
Politics portal
• v • t • e[867]

Within economics the concept of **utility** is used to model worth or value, but its usage has evolved significantly over time. The term was introduced initially as a measure of pleasure or satisfaction within the theory of utilitarianism by moral philosophers such as Jeremy Bentham and John Stuart Mill. But the term has been adapted and reapplied within neoclassical economics, which dominates modern economic theory, as a **utility function** which is a mathematical mechanism for representing a consumer's preference ordering over a choice set. As such, it is devoid of its original interpretation as a measurement of the pleasure or satisfaction obtained by the consumer from that choice.

Utility function

Consider a set of alternatives facing an individual, and over which the individual has a preference ordering. A **utility function** is able to represent those preferences if it is possible to assign a real number to each alternative, in such a way that *alternative a* is assigned a number greater than *alternative b* if, and only if, the individual prefers *alternative a* to *alternative b*. In this situation an individual that selects the most preferred alternative available is necessarily also selecting the alternative that maximises the associated utility function.

Gérard Debreu precisely defined the conditions required for a preference ordering to be representable by a utility function. For a finite set of alternatives these require only that the preference ordering is complete (so the individual is able to determine which of any two alternatives is preferred, or that they are equally preferred), and that the preference order is transitive.

Applications

Utility is usually applied by economists in such constructs as the indifference curve, which plot the combination of commodities that an individual or a society would accept to maintain a given level of satisfaction. Utility and indifference curves are used by economists to understand the underpinnings of demand curves, which are half of the supply and demand analysis that is used to analyze the workings of goods markets.

Individual utility and social utility can be construed as the value of a utility function and a social welfare function respectively. When coupled with production or commodity constraints, under some assumptions these functions can be used to analyze Pareto efficiency, such as illustrated by Edgeworth boxes in contract curves. Such efficiency is a central concept in welfare economics.

In finance, utility is applied to generate an individual's price for an asset called the indifference price. Utility functions are also related to risk measures, with the most common example being the entropic risk measure.

Revealed preference

It was recognized that utility could not be measured or observed directly, so instead economists devised a way to infer underlying relative utilities from observed choice. These 'revealed preferences', as they were named by Paul Samuelson, were revealed e.g. in people's willingness to pay:

> *Utility is taken to be correlative to Desire or Want. It has been already argued that desires cannot be measured directly, but only indirectly, by the outward phenomena to which they give rise: and that in those cases with which economics is chiefly concerned the measure is found in the price which a person is willing to pay for the fulfillment or satisfaction of his desire.*:78

Functions

There has been some controversy over the question whether the utility of a commodity can be measured or not. At one time, it was assumed that the consumer was able to say exactly how much utility he got from the commodity. The economists who made this assumption belonged to the 'cardinalist school' of economics. Today **utility functions**, expressing utility as a function of the amounts of the various goods consumed, are treated as either *cardinal* or *ordinal*, depending on whether they are or are not interpreted as providing more information than simply the rank ordering of preferences over bundles of goods, such as information on the strength of preferences.

Cardinal

When cardinal utility is used, the magnitude of utility differences is treated as an ethically or behaviorally significant quantity. For example, suppose a cup of orange juice has utility of 120 utils, a cup of tea has a utility of 80 utils, and a cup of water has a utility of 40 utils. With cardinal utility, it can be concluded that the cup of orange juice is better than the cup of tea by exactly the same amount by which the cup of tea is better than the cup of water. Formally speaking, this means that if one has a cup of tea, she would be willing to take any bet with a probability, p, greater than .5 of getting a cup of juice, with a risk of getting a cup of water equal to 1-p. One cannot conclude, however, that the cup of tea is two thirds as good as the cup of juice, because this conclusion would depend not only on magnitudes of utility differences, but also on the "zero" of utility. For example, if the "zero" of utility was located at -40, then a cup of orange juice would be 160 utils more than zero, a cup of tea 120 utils more than zero.

Neoclassical economics has largely retreated from using cardinal utility functions as the basis of economic behavior. A notable exception is in the context of analyzing choice under conditions of risk (see below).

Sometimes cardinal utility is used to aggregate utilities across persons, to create a social welfare function.

Ordinal

When ordinal utilities are used, differences in utils (values taken on by the utility function) are treated as ethically or behaviorally meaningless: the utility index encodes a full behavioral ordering between members of a choice set, but tells nothing about the related *strength of preferences*. In the above example, it would only be possible to say that juice is preferred to tea to water, but no more.

Ordinal utility functions are unique up to increasing monotone (or monotonic) transformations. For example, if a function $u(x)$ is taken as ordinal, it is equivalent to the function $u(x)^3$, because taking the 3rd power is an increasing monotone transformation (or monotonic transformation). This means that the ordinal preference induced by these functions is the same (although they are two different functions). In contrast, cardinal utilities are unique only up to increasing linear transformations, so if $u(x)$ is taken as cardinal, it is not equivalent to $u(x)^3$.

Preferences

Although preferences are the conventional foundation of microeconomics, it is often convenient to represent preferences with a utility function and analyze human behavior indirectly with utility functions. Let *X* be the **consumption set**, the set of all mutually-exclusive baskets the consumer could conceivably consume. The consumer's **utility function** $u\colon X \to \mathbb{R}$ ranks each package in the consumption set. If the consumer strictly prefers *x* to *y* or is indifferent between them, then $u(x) \geq u(y)$.

For example, suppose a consumer's consumption set is *X* = {nothing, 1 apple,1 orange, 1 apple and 1 orange, 2 apples, 2 oranges}, and its utility function is *u*(nothing) = 0, *u*(1 apple) = 1, *u*(1 orange) = 2, *u*(1 apple and 1 orange) = 4, *u*(2 apples) = 2 and *u*(2 oranges) = 3. Then this consumer prefers 1 orange to 1 apple, but prefers one of each to 2 oranges.

In micro-economic models, there are usually a finite set of L commodities, and a consumer may consume an arbitrary amount of each commodity. This gives a consumption set of $\mathbb{R}^L_+$, and each package $x \in \mathbb{R}^L_+$ is a vector containing the amounts of each commodity. In the previous example, we might say there are two commodities: apples and oranges. If we say apples is the first commodity, and oranges the second, then the consumption set $X = \mathbb{R}^2_+$ and *u*(0, 0) = 0, *u*(1, 0) = 1, *u*(0, 1) = 2, *u*(1, 1) = 4, *u*(2, 0) = 2, *u*(0, 2) = 3 as before. Note that for *u* to be a utility function on *X*, it must be defined for every package in *X*.

A utility function $u\colon X \to \mathbb{R}$ **represents** a preference relation $\preceq$ on X iff for every $x, y \in X$, $u(x) \leq u(y)$ implies $x \preceq y$. If u represents $\preceq$, then this implies $\preceq$ is complete and transitive, and hence rational.

Revealed preferences in finance

In financial applications, e.g. portfolio optimization, an investor chooses financial portfolio which maximizes his/her own utility function, or, equivalently, minimizes his/her risk measure. For example, modern portfolio theory selects variance as a measure of risk; other popular theories are expected utility theory[868] and prospect theory.[869] To determine specific utility function for any given investor, one could design a questionnaire procedure with questions in the form: How much would you pay for $x\%$ chance of getting y? Revealed preference theory suggests a more direct approach: observe a portfolio $X*$ which an investor currently holds, and then find a utility function/risk measure such that $X*$ becomes an optimal portfolio.[870]

Examples

In order to simplify calculations, various alternative assumptions have been made concerning details of human preferences, and these imply various alternative utility functions such as:

- CES (*constant elasticity of substitution*, or *isoelastic*) utility
- Isoelastic utility
- Exponential utility
- Quasilinear utility
- Homothetic preferences
- Stone–Geary utility function
- Gorman polar form
  - Greenwood–Hercowitz–Huffman preferences
  - King–Plosser–Rebelo preferences
- Hyperbolic absolute risk aversion

Most utility functions used in modeling or theory are **well-behaved.** They are usually monotonic and quasi-concave. However, it is possible for preferences not to be representable by a utility function. An example is lexicographic preferences which are not continuous and cannot be represented by a continuous utility function.

Expected

The expected utility theory deals with the analysis of choices among **risky** projects with multiple (possibly multidimensional) outcomes.

The St. Petersburg paradox was first proposed by Nicholas Bernoulli in 1713 and solved by Daniel Bernoulli in 1738. D. Bernoulli argued that the paradox could be resolved if decision-makers displayed risk aversion and argued for a logarithmic cardinal utility function. (Analyses of international survey data in the 21st century have shown that insofar as utility represents happiness, as in utilitarianism, it is indeed proportional to log income.)

The first important use of the expected utility theory was that of John von Neumann and Oskar Morgenstern, who used the assumption of expected utility maximization in their formulation of game theory.

von Neumann–Morgenstern

Von Neumann and Morgenstern addressed situations in which the outcomes of choices are not known with certainty, but have probabilities attached to them.

A notation for a *lottery* is as follows: if options A and B have probability p and $1 - p$ in the lottery, we write it as a linear combination:

$$L = pA + (1 - p)B$$

More generally, for a lottery with many possible options:

$$L = \sum_i p_i A_i,$$

where $\sum_i p_i = 1$.

By making some reasonable assumptions about the way choices behave, von Neumann and Morgenstern showed that if an agent can choose between the lotteries, then this agent has a utility function such that the desirability of an arbitrary lottery can be calculated as a linear combination of the utilities of its parts, with the weights being their probabilities of occurring.

This is called the *expected utility theorem.* The required assumptions are four axioms about the properties of the agent's preference relation over 'simple lotteries', which are lotteries with just two options. Writing $B \preceq A$ to mean 'A is weakly preferred to B' ('A is preferred at least as much as B'), the axioms are:

1. completeness: For any two simple lotteries L and M , either $L \preceq M$ or $M \preceq L$ (or both, in which case they are viewed as equally desirable).
2. transitivity: for any three lotteries L, M, N , if $L \preceq M$ and $M \preceq N$, then $L \preceq N$.
3. convexity/continuity (Archimedean property): If $L \preceq M \preceq N$, then there is a p between 0 and 1 such that the lottery $pL + (1 - p)N$ is equally desirable as M .
4. independence: for any three lotteries L, M, N and any probability p, $L \preceq M$ if and only if $pL + (1 - p)N \preceq pM + (1 - p)N$. Intuitively, if the lottery formed by the probabilistic combination of L and N is no more preferable than the lottery formed by the same probabilistic combination of M and N, then and only then $L \preceq M$.

Axioms 3 and 4 enable us to decide about the relative utilities of two assets or lotteries.

In more formal language: A von Neumann–Morgenstern utility function is a function from choices to the real numbers:

$$u \colon X \to \mathbb{R}$$

which assigns a real number to every outcome in a way that captures the agent's preferences over simple lotteries. Under the four assumptions mentioned above, the agent will prefer a lottery L_2 to a lottery L_1 if and only if, for the utility function characterizing that agent, the expected utility of L_2 is greater than the expected utility of L_1 :

$$L_1 \preceq L_2 \text{ iff } u(L_1) \leq u(L_2).$$

Of all the axioms, independence is the most often discarded. A variety of generalized expected utility theories have arisen, most of which drop or relax the independence axiom.

As probability of success

Castagnoli and LiCalzi and Bordley and LiCalzi (2000) provided another interpretation for Von Neumann and Morgenstern's theory. Specifically for any utility function, there exists a hypothetical reference lottery with the expected utility of an arbitrary lottery being its probability of performing no worse than the reference lottery. Suppose success is defined as getting an outcome no worse than the outcome of the reference lottery. Then this mathematical equivalence means that maximizing expected utility is equivalent to maximizing the probability of success. In many contexts, this makes the concept of utility easier to justify and to apply. For example, a firm's utility might be the probability of meeting uncertain future customer expectations.[871,872,873]

Indirect

An indirect utility function gives the optimal attainable value of a given utility function, which depends on the prices of the goods and the income or wealth level that the individual possesses.

Money

One use of the indirect utility concept is the notion of the utility of money. The (indirect) utility function for money is a nonlinear function that is bounded and asymmetric about the origin. The utility function is concave in the positive region, reflecting the phenomenon of diminishing marginal utility. The boundedness reflects the fact that beyond a certain point money ceases being useful at all, as the size of any economy at any point in time is itself bounded. The asymmetry about the origin reflects the fact that gaining and losing money can have radically different implications both for individuals and businesses. The non-linearity of the utility function for money has profound implications in decision making processes: in situations where outcomes of choices influence utility through gains or losses of money, which are the norm in most business settings, the optimal choice for a given decision depends on the possible outcomes of all other decisions in the same time-period.

Discussion and criticism

Cambridge economist Joan Robinson famously criticized utility for being a circular concept: "Utility is the quality in commodities that makes individuals want to buy them, and the fact that individuals want to buy commodities shows that they have utility":48 Robinson also pointed out that because the theory assumes that preferences are fixed this means that utility is not a testable assumption. This is so because if we take changes in peoples' behavior in relation to a change in prices or a change in the underlying budget constraint we can never be sure to what extent the change in behavior was due to the change in price or budget constraint and how much was due to a change in preferences. This criticism is similar to that of the philosopher Hans Albert who argued that the ceteris paribus conditions on which the marginalist theory of demand rested rendered the theory itself an empty tautology and completely closed to experimental testing. In essence, demand and supply curve (theoretical line of quantity of a product which would have been offered or requested for given price) is purely ontological and could never been demonstrated empirically.

Another criticism comes from the assertion that neither cardinal nor ordinal utility is empirically observable in the real world. In the case of cardinal utility it is impossible to measure the level of satisfaction "quantitatively" when someone consumes or purchases an apple. In case of ordinal utility, it is impossible to determine what choices were made when someone purchases, for example, an orange. Any act would involve preference over a vast set of choices (such as apple, orange juice, other vegetable, vitamin C tablets, exercise, not purchasing, etc.).

Other questions of what arguments ought to enter into a utility function are difficult to answer, yet seem necessary to understanding utility. Whether people gain utility from coherence of wants, beliefs or a sense of duty is key to understanding their behavior in the utility organon. Likewise, choosing between alternatives is itself a process of determining what to consider as alternatives, a question of choice within uncertainty.

An evolutionary psychology perspective is that utility may be better viewed as due to preferences that maximized evolutionary fitness in the ancestral environment but not necessarily in the current one.[874]

Further reading

- Anand, Paul (1993). *Foundations of Rational Choice Under Risk*. Oxford: Oxford University Press. ISBN 0-19-823303-5.
- Fishburn, Peter C. (1970). *Utility Theory for Decision Making*. Huntington, NY: Robert E. Krieger. ISBN 0-88275-736-9.

- Georgescu-Roegen, Nicholas (Aug 1936). "The Pure Theory of Consumer's Behavior". *Quarterly Journal of Economics*. **50** (4): 545–593. doi: 10.2307/1891094[875]. JSTOR 1891094[876].
- Gilboa, Itzhak (2009). *Theory of Decision under Uncertainty*. Cambridge: Cambridge University Press. ISBN 978-0-521-74123-1.
- Kreps, David M. (1988). *Notes on the Theory of Choice*. Boulder, CO: West-view Press. ISBN 0-8133-7553-3.
- Nash, John F. (1950). "The Bargaining Problem". *Econometrica*. **18** (2): 155–162. doi: 10.2307/1907266[877]. JSTOR 1907266[878].
- Neumann, John von & Morgenstern, Oskar (1944). *Theory of Games and Economic Behavior*. Princeton, NJ: Princeton University Press.
- Nicholson, Walter (1978). *Micro-economic Theory* (Second ed.). Hinsdale: Dryden Press. pp. 53–87. ISBN 0-03-020831-9.
- Plous, S. (1993). *The Psychology of Judgement and Decision Making*. New York: McGraw-Hill. ISBN 0-07-050477-6.

External links

Wikimedia Commons has media related to ***Utility (decision theory)***.

- Definition of Utility by Investopedia[879]
- Anatomy of Cobb-Douglas Type Utility Functions in 3D[880]
- Anatomy of CES Type Utility Functions in 3D[881]
- Simpler Definition with example from Investopedia[882]
- Maximization of Originality - redefinition of classic utility[883]
- Utility Model of [[Marketing[884]]] - Form[885], Place[886], Time[887], Possession[888] and perhaps also Task[889]

Statistical classification

Machine learning and data mining

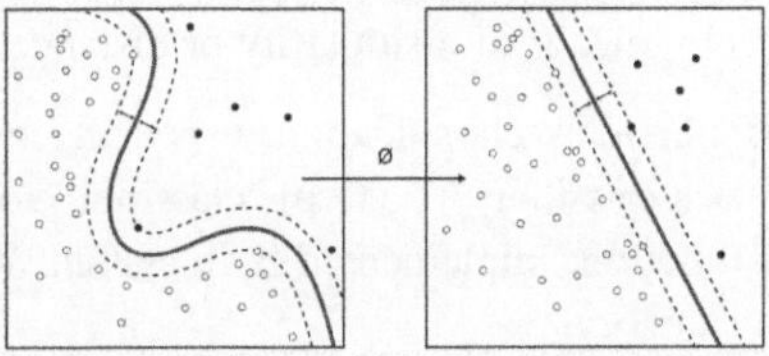

- Machine learning portal

- v
- t
- e[890]

In machine learning and statistics, **classification** is the problem of identifying to which of a set of categories (sub-populations) a new observation belongs, on the basis of a training set of data containing observations (or instances) whose category membership is known. Examples are assigning a given email to the "spam" or "non-spam" class, and assigning a diagnosis to a given patient based on observed characteristics of the patient (gender, blood pressure, presence or absence of certain symptoms, etc.). Classification is an example of pattern recognition.

In the terminology of machine learning, classification is considered an instance of supervised learning, i.e. learning where a training set of correctly identified observations is available. The corresponding unsupervised procedure is known as clustering, and involves grouping data into categories based on some measure of inherent similarity or distance.

Often, the individual observations are analyzed into a set of quantifiable properties, known variously as explanatory variables or *features*. These properties may variously be categorical (e.g. "A", "B", "AB" or "O", for blood type), ordinal (e.g. "large", "medium" or "small"), integer-valued (e.g. the number of occurrences of a particular word in an email) or real-valued (e.g. a measurement of blood pressure). Other classifiers work by comparing observations to previous observations by means of a similarity or distance function.

An algorithm that implements classification, especially in a concrete implementation, is known as a **classifier**. The term "classifier" sometimes also refers to the mathematical function, implemented by a classification algorithm, that maps input data to a category.

Terminology across fields is quite varied. In statistics, where classification is often done with logistic regression or a similar procedure, the properties of observations are termed explanatory variables (or independent variables, regressors, etc.), and the categories to be predicted are known as outcomes, which are considered to be possible values of the dependent variable. In machine learning, the observations are often known as *instances*, the explanatory variables are termed *features* (grouped into a feature vector), and the possible categories to be predicted are *classes*. Other fields may use different terminology: e.g. in community ecology, the term "classification" normally refers to cluster analysis, i.e. a type of unsupervised learning, rather than the supervised learning described in this article.

Relation to other problems

Classification and clustering are examples of the more general problem of pattern recognition, which is the assignment of some sort of output value to a given input value. Other examples are regression, which assigns a real-valued output to each input; sequence labeling, which assigns a class to each member of a sequence of values (for example, part of speech tagging, which assigns a part of speech to each word in an input sentence); parsing, which assigns a parse tree to an input sentence, describing the syntactic structure of the sentence; etc.

A common subclass of classification is probabilistic classification. Algorithms of this nature use statistical inference to find the best class for a given instance. Unlike other algorithms, which simply output a "best" class, probabilistic algorithms output a probability of the instance being a member of each of the possible classes. The best class is normally then selected as the one with the highest probability. However, such an algorithm has numerous advantages over non-probabilistic classifiers:

- It can output a confidence value associated with its choice (in general, a classifier that can do this is known as a *confidence-weighted classifier*).
- Correspondingly, it can *abstain* when its confidence of choosing any particular output is too low.
- Because of the probabilities which are generated, probabilistic classifiers can be more effectively incorporated into larger machine-learning tasks, in a way that partially or completely avoids the problem of *error propagation*.

Frequentist procedures

Early work on statistical classification was undertaken by Fisher,[891,892] in the context of two-group problems, leading to Fisher's linear discriminant function as the rule for assigning a group to a new observation.[893] This early work assumed that data-values within each of the two groups had a multivariate normal distribution. The extension of this same context to more than two-groups has also been considered with a restriction imposed that the classification rule should be linear.[894] Later work for the multivariate normal distribution allowed the classifier to be nonlinear:[895] several classification rules can be derived based on slight different adjustments of the Mahalanobis distance, with a new observation being assigned to the group whose centre has the lowest adjusted distance from the observation.

Bayesian procedures

Unlike frequentist procedures, Bayesian classification procedures provide a natural way of taking into account any available information about the relative sizes of the sub-populations associated with the different groups within the overall population.[896] Bayesian procedures tend to be computationally expensive and, in the days before Markov chain Monte Carlo computations were developed, approximations for Bayesian clustering rules were devised.[897]

Some Bayesian procedures involve the calculation of group membership probabilities: these can be viewed as providing a more and more informative outcome of a data analysis than a simple attribution of a single group-label to each new observation.

Binary and multiclass classification

Classification can be thought of as two separate problems – binary classification and multiclass classification. In binary classification, a better understood task, only two classes are involved, whereas multiclass classification involves assigning an object to one of several classes.[898] Since many classification methods have been developed specifically for binary classification, multiclass classification often requires the combined use of multiple binary classifiers.

Feature vectors

Most algorithms describe an individual instance whose category is to be predicted using a feature vector of individual, measurable properties of the instance. Each property is termed a feature, also known in statistics as an explanatory variable (or independent variable, although features may or may not be statistically independent). Features may variously be binary (e.g. "on" or "off"); categorical (e.g. "A", "B", "AB" or "O", for blood type); ordinal (e.g. "large", "medium" or "small"); integer-valued (e.g. the number of occurrences of a particular word in an email); or real-valued (e.g. a measurement of blood pressure). If the instance is an image, the feature values might correspond to the pixels of an image; if the instance is a piece of text, the feature values might be occurrence frequencies of different words. Some algorithms work only in terms of discrete data and require that real-valued or integer-valued data be *discretized* into groups (e.g. less than 5, between 5 and 10, or greater than 10)

Linear classifiers

A large number of algorithms for classification can be phrased in terms of a linear function that assigns a score to each possible category *k* by combining the feature vector of an instance with a vector of weights, using a dot product. The predicted category is the one with the highest score. This type of score function is known as a linear predictor function and has the following general form:

$$\operatorname{score}(\mathbf{X}_i, k) = \boldsymbol{\beta}_k \cdot \mathbf{X}_i,$$

where $\mathbf{X}_i$ is the feature vector for instance i, $\boldsymbol{\beta}_k$ is the vector of weights corresponding to category k, and score($\mathbf{X}_i$, k) is the score associated with assigning instance i to category k. In discrete choice theory, where instances represent people and categories represent choices, the score is considered the utility associated with person i choosing category k.

Algorithms with this basic setup are known as linear classifiers. What distinguishes them is the procedure for determining (training) the optimal weights/coefficients and the way that the score is interpreted.

Examples of such algorithms are

- Logistic regression and Multinomial logistic regression
- Probit regression
- The perceptron algorithm
- Support vector machines
- Linear discriminant analysis.

Algorithms

Examples of classification algorithms include:

- Linear classifiers
  - Fisher's linear discriminant
  - Logistic regression
  - Naive Bayes classifier
  - Perceptron
- Support vector machines
  - Least squares support vector machines
- Quadratic classifiers
- Kernel estimation
  - k-nearest neighbor
- Boosting (meta-algorithm)
- Decision trees
  - Random forests
- Neural networks
- Learning vector quantization

Evaluation

Classifier performance depends greatly on the characteristics of the data to be classified. There is no single classifier that works best on all given problems (a phenomenon that may be explained by the no-free-lunch theorem). Various empirical tests have been performed to compare classifier performance and to find the characteristics of data that determine classifier performance. Determining a suitable classifier for a given problem is however still more an art than a science.

The measures precision and recall are popular metrics used to evaluate the quality of a classification system. More recently, receiver operating characteristic (ROC) curves have been used to evaluate the tradeoff between true- and false-positive rates of classification algorithms.

As a performance metric, the uncertainty coefficient has the advantage over simple accuracy in that it is not affected by the relative sizes of the different classes. Further, it will not penalize an algorithm for simply *rearranging* the classes.

Application domains

Classification has many applications. In some of these it is employed as a data mining procedure, while in others more detailed statistical modeling is undertaken.

- Computer vision
  - Medical imaging and medical image analysis
  - Optical character recognition
  - Video tracking
- Drug discovery and development
  - Toxicogenomics
  - Quantitative structure-activity relationship
- Geostatistics
- Speech recognition
- Handwriting recognition
- Biometric identification
- Biological classification
- Statistical natural language processing
- Document classification
- Internet search engines
- Credit scoring
- Pattern recognition
- Micro-array classification

Artificial neural network

Machine learning and data mining

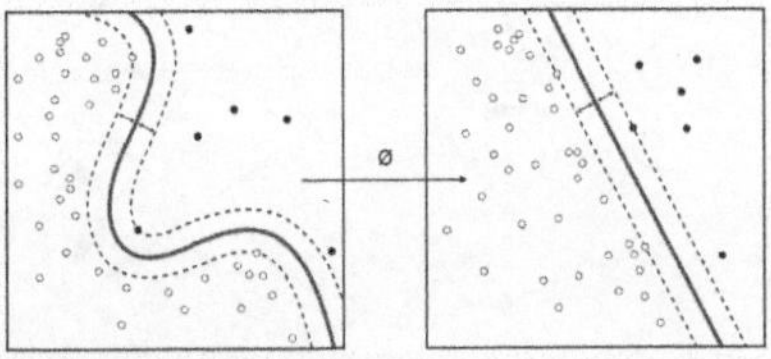

- Machine learning portal

- v
- t
- e[899]

Artificial neural networks (**ANNs**) or **connectionist systems** are computing systems vaguely inspired by the biological neural networks that constitute animal brains. Such systems "learn" to perform tasks by considering examples, generally without being programmed with any task-specific rules. For example, in image recognition, they might learn to identify images that contain cats by analyzing example images that have been manually labeled as "cat" or "no cat" and using the results to identify cats in other images. They do this without any prior knowledge about cats, e.g., that they have fur, tails, whiskers and cat-like faces. Instead, they automatically generate identifying characteristics from the learning material that they process.

An ANN is based on a collection of connected units or nodes called artificial neurons which loosely model the neurons in a biological brain. Each connection, like the synapses in a biological brain, can transmit a signal from one

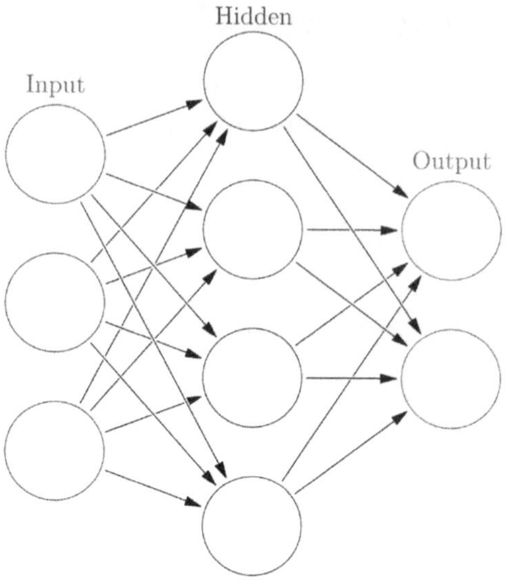

Figure 31: *An artificial neural network is an interconnected group of nodes, akin to the vast network of neurons in a brain. Here, each circular node represents an artificial neuron and an arrow represents a connection from the output of one artificial neuron to the input of another.*

artificial neuron to another. An artificial neuron that receives a signal can process it and then signal additional artificial neurons connected to it.

In common ANN implementations, the signal at a connection between artificial neurons is a real number, and the output of each artificial neuron is computed by some non-linear function of the sum of its inputs. The connections between artificial neurons are called 'edges'. Artificial neurons and edges typically have a weight that adjusts as learning proceeds. The weight increases or decreases the strength of the signal at a connection. Artificial neurons may have a threshold such that the signal is only sent if the aggregate signal crosses that threshold. Typically, artificial neurons are aggregated into layers. Different layers may perform different kinds of transformations on their inputs. Signals travel from the first layer (the input layer), to the last layer (the output layer), possibly after traversing the layers multiple times.

The original goal of the ANN approach was to solve problems in the same way that a human brain would. However, over time, attention moved to performing specific tasks, leading to deviations from biology. ANNs have been used on a variety of tasks, including computer vision, speech recognition, machine translation, social network filtering, playing board and video games and medical diagnosis.

History

Warren McCulloch and Walter Pitts (1943) created a computational model for neural networks based on mathematics and algorithms called threshold logic. This model paved the way for neural network research to split into two approaches. One approach focused on biological processes in the brain while the other focused on the application of neural networks to artificial intelligence. This work led to work on nerve networks and their link to finite automata.

Hebbian learning

In the late 1940s, D.O. Hebb created a learning hypothesis based on the mechanism of neural plasticity that became known as Hebbian learning. Hebbian learning is unsupervised learning. This evolved into models for long term potentiation. Researchers started applying these ideas to computational models in 1948 with Turing's B-type machines.

Farley and Clark (1954) first used computational machines, then called "calculators", to simulate a Hebbian network. Other neural network computational machines were created by Rochester, Holland, Habit and Duda (1956).

Rosenblatt (1958) created the perceptron, an algorithm for pattern recognition. With mathematical notation, Rosenblatt described circuitry not in the basic perceptron, such as the exclusive-or circuit that could not be processed by neural networks at the time.

In 1959, a biological model proposed by Nobel laureates Hubel and Wiesel was based on their discovery of two types of cells in the primary visual cortex: simple cells and complex cells.

The first functional networks with many layers were published by Ivakhnenko and Lapa in 1965, becoming the Group Method of Data Handling.

Neural network research stagnated after machine learning research by Minsky and Papert (1969), who discovered two key issues with the computational machines that processed neural networks. The first was that basic perceptrons were incapable of processing the exclusive-or circuit. The second was that computers didn't have enough processing power to effectively handle the work required by large neural networks. Neural network research slowed until computers achieved far greater processing power.

Much of artificial intelligence had focused on high-level (symbolic) models that are processed by using algorithms, characterized for example by expert systems with knowledge embodied in *if-then* rules, until in the late 1980s research expanded to low-level (sub-symbolic) machine learning, characterized by knowledge embodied in the parameters of a cognitive model.Wikipedia:Citation needed

Backpropagation

A key trigger for renewed interest in neural networks and learning was Werbos's (1975) backpropagation algorithm that effectively solved the exclusive-or problem and more generally accelerated the training of multi-layer networks. Backpropagation distributed the error term back up through the layers, by modifying the weights at each node.

In the mid-1980s, parallel distributed processing became popular under the name connectionism. Rumelhart and McClelland (1986) described the use of connectionism to simulate neural processes.

Support vector machines and other, much simpler methods such as linear classifiers gradually overtook neural networks in machine learning popularity. However, using neural networks transformed some domains, such as the prediction of protein structures.

In 1992, max-pooling was introduced to help with least shift invariance and tolerance to deformation to aid in 3D object recognition.[900,901,902] In 2010, Backpropagation training through max-pooling was accelerated by GPUs and shown to perform better than other pooling variants.[903]

The vanishing gradient problem affects many-layered feedforward networks that used backpropagation and also recurrent neural networks (RNNs).[904] As errors propagate from layer to layer, they shrink exponentially with the number of layers, impeding the tuning of neuron weights that is based on those errors, particularly affecting deep networks.

To overcome this problem, Schmidhuber adopted a multi-level hierarchy of networks (1992) pre-trained one level at a time by unsupervised learning and fine-tuned by backpropagation.[905] Behnke (2003) relied only on the sign of the gradient (Rprop) on problems such as image reconstruction and face localization.

Hinton et al. (2006) proposed learning a high-level representation using successive layers of binary or real-valued latent variables with a restricted Boltzmann machine to model each layer. Once sufficiently many layers have been learned, the deep architecture may be used as a generative model by reproducing the data when sampling down the model (an "ancestral pass") from the top level feature activations. In 2012, Ng and Dean created a network that learned to recognize higher-level concepts, such as cats, only from watching unlabeled images taken from YouTube videos.

Earlier challenges in training deep neural networks were successfully addressed with methods such as unsupervised pre-training, while available computing power increased through the use of GPUs and distributed computing. Neural networks were deployed on a large scale, particularly in image and visual recognition problems. This became known as "deep learning".Wikipedia:Citation needed

Hardware-based designs

Computational devices were created in CMOS, for both biophysical simulation and neuromorphic computing. Nanodevices for very large scale principal components analyses and convolution may create a new class of neural computing because they are fundamentally analog rather than digital (even though the first implementations may use digital devices). Ciresan and colleagues (2010) in Schmidhuber's group showed that despite the vanishing gradient problem, GPUs makes back-propagation feasible for many-layered feedforward neural networks.

Contests

Between 2009 and 2012, recurrent neural networks and deep feedforward neural networks developed in Schmidhuber's research group won eight international competitions in pattern recognition and machine learning.[906] For example, the bi-directional and multi-dimensional long short-term memory (LSTM)[907] of Graves et al. won three competitions in connected handwriting recognition at the 2009 International Conference on Document Analysis and Recognition (ICDAR), without any prior knowledge about the three languages to be learned.

Ciresan and colleagues won pattern recognition contests, including the IJCNN 2011 Traffic Sign Recognition Competition, the ISBI 2012 Segmentation of Neuronal Structures in Electron Microscopy Stacks challenge and others. Their neural networks were the first pattern recognizers to achieve human-competitive or even superhuman performance on benchmarks such as traffic sign recognition (IJCNN 2012), or the MNIST handwritten digits problem.

Researchers demonstrated (2010) that deep neural networks interfaced to a hidden Markov model with context-dependent states that define the neural network output layer can drastically reduce errors in large-vocabulary speech recognition tasks such as voice search.

GPU-based implementations of this approach won many pattern recognition contests, including the IJCNN 2011 Traffic Sign Recognition Competition, the ISBI 2012 Segmentation of neuronal structures in EM stacks challenge, the ImageNet Competition and others.

Deep, highly nonlinear neural architectures similar to the neocognitron and the "standard architecture of vision", inspired by simple and complex cells, were pre-trained by unsupervised methods by Hinton. A team from his lab won a 2012 contest sponsored by Merck to design software to help find molecules that might identify new drugs.

Convolutional networks

As of 2011, the state of the art in deep learning feedforward networks alternated between convolutional layers and max-pooling layers, topped by several fully or sparsely connected layers followed by a final classification layer. Learning is usually done without unsupervised pre-training.

Such supervised deep learning methods were the first to achieve human-competitive performance on certain tasks.

ANNs were able to guarantee shift invariance to deal with small and large natural objects in large cluttered scenes, only when invariance extended beyond shift, to all ANN-learned concepts, such as location, type (object class label), scale, lighting and others. This was realized in Developmental Networks (DNs)[908] whose embodiments are Where-What Networks, WWN-1 (2008)[909] through WWN-7 (2013).[910]

Models

An *artificial neural network* is a network of simple elements called *artificial neurons*, which receive input, change their internal state (*activation*) according to that input, and produce output depending on the input and activation. The *network* forms by connecting the output of certain neurons to the input of other neurons forming a directed, weighted graph. The weights as well as the functions that compute the activation can be modified by a process called *learning* which is governed by a *learning rule*.

Components of an artificial neural network

Neurons

A neuron with label j receiving an input $p_j(t)$ from predecessor neurons consists of the following components:

- an *activation* $a_j(t)$, depending on a discrete time parameter,
- possibly a *threshold* θ_j , which stays fixed unless changed by a learning function,
- an *activation function* f that computes the new activation at a given time $t+1$ from $a_j(t)$, θ_j and the net input $p_j(t)$ giving rise to the relation

  $$a_j(t+1) = f(a_j(t), p_j(t), \theta_j),$$

- and an *output function* f_{out} computing the output from the activation

  $$o_j(t) = f_{out}(a_j(t)).$$

Often the output function is simply the Identity function.

An *input neuron* has no predecessor but serves as input interface for the whole network. Similarly an *output neuron* has no successor and thus serves as output interface of the whole network.

Connections and weights

The *network* consists of connections, each connection transferring the output of a neuron i to the input of a neuron j . In this sense i is the predecessor of j and j is the successor of i . Each connection is assigned a weight w_{ij} .

Propagation function

The *propagation function* computes the *input* $p_j(t)$ to the neuron j from the outputs $o_i(t)$ of predecessor neurons and typically has the form

$$p_j(t) = \sum_i o_i(t) w_{ij}.$$

Learning rule

The *learning rule* is a rule or an algorithm which modifies the parameters of the neural network, in order for a given input to the network to produce a favored output. This *learning* process typically amounts to modifying the weights and thresholds of the variables within the network.

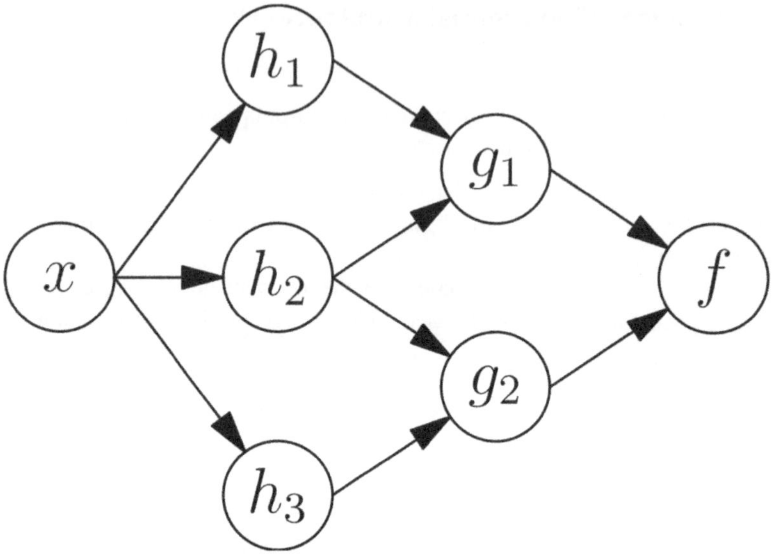

Figure 32: *ANN dependency graph*

Neural networks as functions

Neural network models can be viewed as simple mathematical models defining a function $f : X \to Y$ or a distribution over X or both X and Y . Sometimes models are intimately associated with a particular learning rule. A common use of the phrase "ANN model" is really the definition of a *class* of such functions (where members of the class are obtained by varying parameters, connection weights, or specifics of the architecture such as the number of neurons or their connectivity).

Mathematically, a neuron's network function $f(x)$ is defined as a composition of other functions $g_i(x)$, that can further be decomposed into other functions. This can be conveniently represented as a network structure, with arrows depicting the dependencies between functions. A widely used type of composition is the *nonlinear weighted sum*, where $f(x) = K\left(\sum_i w_i g_i(x)\right)$, where K (commonly referred to as the activation function) is some predefined function, such as the hyperbolic tangent or sigmoid function or softmax function or rectifier function. The important characteristic of the activation function is that it provides a smooth transition as input values change, i.e. a small change in input produces a small change in output. The following refers to a collection of functions g_i as a vector $g = (g_1, g_2, \ldots, g_n)$.

This figure depicts such a decomposition of f , with dependencies between variables indicated by arrows. These can be interpreted in two ways.

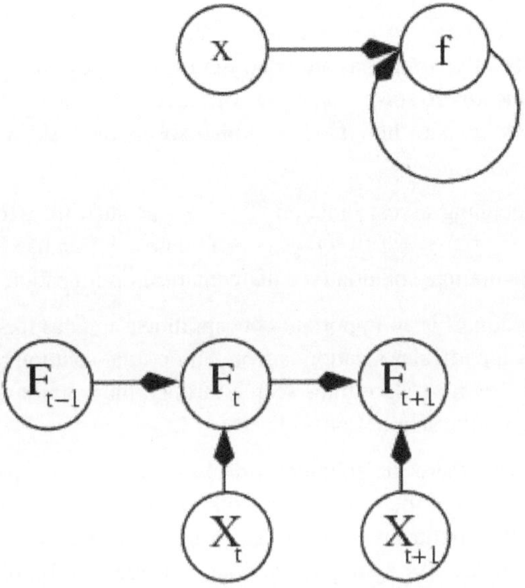

Figure 33: *Two separate depictions of the recurrent ANN dependency graph*

The first view is the functional view: the input x is transformed into a 3-dimensional vector h , which is then transformed into a 2-dimensional vector g , which is finally transformed into f . This view is most commonly encountered in the context of optimization.

The second view is the probabilistic view: the random variable $F = f(G)$ depends upon the random variable $G = g(H)$, which depends upon $H = h(X)$, which depends upon the random variable X . This view is most commonly encountered in the context of graphical models.

The two views are largely equivalent. In either case, for this particular architecture, the components of individual layers are independent of each other (e.g., the components of g are independent of each other given their input h). This naturally enables a degree of parallelism in the implementation.

Networks such as the previous one are commonly called feedforward, because their graph is a directed acyclic graph. Networks with cycles are commonly called recurrent. Such networks are commonly depicted in the manner shown at the top of the figure, where f is shown as being dependent upon itself. However, an implied temporal dependence is not shown.

Learning

The possibility of learning has attracted the most interest in neural networks. Given a specific *task* to solve, and a class of functions F , learning means using a set of observations to find $f^* \in F$ which solves the task in some optimal sense.

This entails defining a cost function $C : F \to \mathbb{R}$ such that, for the optimal solution f^* , $C(f^*) \leq C(f) \, \forall f \in F$ – i.e., no solution has a cost less than the cost of the optimal solution (see mathematical optimization).

The cost function C is an important concept in learning, as it is a measure of how far away a particular solution is from an optimal solution to the problem to be solved. Learning algorithms search through the solution space to find a function that has the smallest possible cost.

For applications where the solution is data dependent, the cost must necessarily be a function of the observations, otherwise the model would not relate to the data. It is frequently defined as a statistic to which only approximations can be made. As a simple example, consider the problem of finding the model f , which minimizes $C = E\left[(f(x) - y)^2\right]$, for data pairs (x, y) drawn from some distribution $\mathcal{D}$. In practical situations we would only have N samples from $\mathcal{D}$ and thus, for the above example, we would only minimize $\hat{C} = \frac{1}{N}\sum_{i=1}^{N}(f(x_i) - y_i)^2$. Thus, the cost is minimized over a sample of the data rather than the entire distribution.

When $N \to \infty$ some form of online machine learning must be used, where the cost is reduced as each new example is seen. While online machine learning is often used when $\mathcal{D}$ is fixed, it is most useful in the case where the distribution changes slowly over time. In neural network methods, some form of online machine learning is frequently used for finite datasets.

Choosing a cost function

While it is possible to define an ad hoc cost function, frequently a particular cost (function) is used, either because it has desirable properties (such as convexity) or because it arises naturally from a particular formulation of the problem (e.g., in a probabilistic formulation the posterior probability of the model can be used as an inverse cost). Ultimately, the cost function depends on the task.

Backpropagation

A DNN can be discriminatively trained with the standard backpropagation algorithm. Backpropagation is a method to calculate the gradient of the loss function (produces the cost associated with a given state) with respect to the weights in an ANN.

The basics of continuous backpropagation[911] were derived in the context of control theory by Kelley in 1960 and by Bryson in 1961,[912] using principles of dynamic programming. In 1962, Dreyfus published a simpler derivation based only on the chain rule. Bryson and Ho described it as a multi-stage dynamic system optimization method in 1969. In 1970, Linnainmaa finally published the general method for automatic differentiation (AD) of discrete connected networks of nested differentiable functions.[913] This corresponds to the modern version of backpropagation which is efficient even when the networks are sparse. In 1973, Dreyfus used backpropagation to adapt parameters of controllers in proportion to error gradients. In 1974, Werbos mentioned the possibility of applying this principle to ANNs,[914] and in 1982, he applied Linnainmaa's AD method to neural networks in the way that is widely used today. In 1986, Rumelhart, Hinton and Williams noted that this method can generate useful internal representations of incoming data in hidden layers of neural networks. In 1993, Wan was the first to win an international pattern recognition contest through backpropagation.[915]

The weight updates of backpropagation can be done via stochastic gradient descent using the following equation:

$$w_{ij}(t+1) = w_{ij}(t) + \eta\frac{\partial C}{\partial w_{ij}} + \xi(t)$$

where, η is the learning rate, C is the cost (loss) function and $\xi(t)$ a stochastic term. The choice of the cost function depends on factors such as the learning type (supervised, unsupervised, reinforcement, etc.) and the activation function. For example, when performing supervised learning on a multiclass classification problem, common choices for the activation function and cost function are the softmax function and cross entropy function, respectively. The softmax function is defined as $p_j = \frac{\exp(x_j)}{\sum_k \exp(x_k)}$ where p_j represents the class probability (output of the unit j) and x_j and x_k represent the total input to units j and k of the same level respectively. Cross entropy is defined as $C = -\sum_j d_j \log(p_j)$ where d_j represents the target probability for output unit j and p_j is the probability output for j after applying the activation function.

These can be used to output object bounding boxes in the form of a binary mask. They are also used for multi-scale regression to increase localization precision. DNN-based regression can learn features that capture geometric information in addition to serving as a good classifier. They remove the requirement to explicitly model parts and their relations. This helps to broaden the variety of objects that can be learned. The model consists of multiple layers, each of which has a rectified linear unit as its activation function for non-linear transformation. Some layers are convolutional, while others are fully connected. Every convolutional layer has an additional max pooling. The network is trained to minimize L^2 error for predicting the mask ranging over the entire training set containing bounding boxes represented as masks.

Alternatives to backpropagation include Extreme Learning Machines, "No-prop" networks, training without backtracking, "weightless" networks,[916] and non-connectionist neural networks.

Learning paradigms

The three major learning paradigms each correspond to a particular learning task. These are supervised learning, unsupervised learning and reinforcement learning.

Supervised learning

Supervised learning uses a set of example pairs $(x, y), x \in X, y \in Y$ and the aim is to find a function $f : X \to Y$ in the allowed class of functions that matches the examples. In other words, we wish to infer the mapping implied by the data; the cost function is related to the mismatch between our mapping and the data and it implicitly contains prior knowledge about the problem domain.

A commonly used cost is the mean-squared error, which tries to minimize the average squared error between the network's output, $f(x)$, and the target value y over all the example pairs. Minimizing this cost using gradient descent for the class of neural networks called multilayer perceptrons (MLP), produces the backpropagation algorithm for training neural networks.

Tasks that fall within the paradigm of supervised learning are pattern recognition (also known as classification) and regression (also known as function approximation). The supervised learning paradigm is also applicable to sequential data (e.g., for hand writing, speech and gesture recognition). This can be thought of as learning with a "teacher", in the form of a function that provides continuous feedback on the quality of solutions obtained thus far.

Unsupervised learning

In unsupervised learning, some data x is given and the cost function to be minimized, that can be any function of the data x and the network's output, f .

The cost function is dependent on the task (the model domain) and any *a priori* assumptions (the implicit properties of the model, its parameters and the observed variables).

As a trivial example, consider the model $f(x) = a$ where a is a constant and the cost $C = E[(x - f(x))^2]$. Minimizing this cost produces a value of a that is equal to the mean of the data. The cost function can be much more complicated. Its form depends on the application: for example, in compression it could be related to the mutual information between x and $f(x)$, whereas in statistical modeling, it could be related to the posterior probability of the model given the data (note that in both of those examples those quantities would be maximized rather than minimized).

Tasks that fall within the paradigm of unsupervised learning are in general estimation problems; the applications include clustering, the estimation of statistical distributions, compression and filtering.

Reinforcement learning

In reinforcement learning, data x are usually not given, but generated by an agent's interactions with the environment. At each point in time t , the agent performs an action y_t and the environment generates an observation x_t and an instantaneous cost c_t , according to some (usually unknown) dynamics. The aim is to discover a policy for selecting actions that minimizes some measure of a long-term cost, e.g., the expected cumulative cost. The environment's dynamics and the long-term cost for each policy are usually unknown, but can be estimated.

More formally the environment is modeled as a Markov decision process (MDP) with states $s_1, ..., s_n \in S$ and actions $a_1, ..., a_m \in A$ with the following probability distributions: the instantaneous cost distribution $P(c_t|s_t)$, the observation distribution $P(x_t|s_t)$ and the transition $P(s_{t+1}|s_t, a_t)$, while a policy is defined as the conditional distribution over actions given the observations. Taken together, the two then define a Markov chain (MC). The aim is to discover the policy (i.e., the MC) that minimizes the cost.

ANNs are frequently used in reinforcement learning as part of the overall algorithm. Dynamic programming was coupled with ANNs (giving neurodynamic programming) by Bertsekas and Tsitsiklis and applied to multi-dimensional nonlinear problems such as those involved in vehicle routing, natural resources management or medicine because of the ability of ANNs to mitigate losses of accuracy even when reducing the discretization grid density for numerically approximating the solution of the original control problems.

Tasks that fall within the paradigm of reinforcement learning are control problems, games and other sequential decision making tasks.

Convergent recursive learning algorithm

This is a learning method specially designed for cerebellar model articulation controller (CMAC) neural networks. In 2004 a recursive least squares algorithm was introduced to train CMAC neural network online.[917] This algorithm can converge in one step and update all weights in one step with any new input data. Initially, this algorithm had computational complexity of $O(N^3)$. Based on QR decomposition, this recursive learning algorithm was simplified to be $O(N)$.[918]

Learning algorithms

Training a neural network model essentially means selecting one model from the set of allowed models (or, in a Bayesian framework, determining a distribution over the set of allowed models) that minimizes the cost. Numerous algorithms are available for training neural network models; most of them can be viewed as a straightforward application of optimization theory and statistical estimation.

Most employ some form of gradient descent, using backpropagation to compute the actual gradients. This is done by simply taking the derivative of the cost function with respect to the network parameters and then changing those parameters in a gradient-related direction. Backpropagation training algorithms fall into three categories:

- steepest descent (with variable learning rate and momentum, resilient backpropagation);
- quasi-Newton (Broyden-Fletcher-Goldfarb-Shanno, one step secant);
- Levenberg-Marquardt and conjugate gradient (Fletcher-Reeves update, Polak-Ribiére update, Powell-Beale restart, scaled conjugate gradient).

Evolutionary methods, gene expression programming, simulated annealing, expectation-maximization, non-parametric methods and particle swarm optimization are other methods for training neural networks.

Variants

Group method of data handling

The Group Method of Data Handling (GMDH) features fully automatic structural and parametric model optimization. The node activation functions are Kolmogorov-Gabor polynomials that permit additions and multiplications. It used a deep feedforward multilayer perceptron with eight layers. It is a supervised learning network that grows layer by layer, where each layer is trained by regression analysis. Useless items are detected using a validation set, and pruned through regularization. The size and depth of the resulting network depends on the task.

Convolutional neural networks

A convolutional neural network (CNN) is a class of deep, feed-forward networks, composed of one or more convolutional layers with fully connected layers (matching those in typical ANNs) on top. It uses tied weights and pooling layers. In particular, max-pooling is often structured via Fukushima's convolutional architecture. This architecture allows CNNs to take advantage of the 2D structure of input data.

CNNs are suitable for processing visual and other two-dimensional data.[919,920] They have shown superior results in both image and speech applications. They can be trained with standard backpropagation. CNNs are easier to train than other regular, deep, feed-forward neural networks and have many fewer parameters to estimate. Examples of applications in computer vision include DeepDream and robot navigation.

Long short-term memory

Long short-term memory (LSTM) networks are RNNs that avoid the vanishing gradient problem. LSTM is normally augmented by recurrent gates called forget gates. LSTM networks prevent backpropagated errors from vanishing or exploding. Instead errors can flow backwards through unlimited numbers of virtual layers in space-unfolded LSTM. That is, LSTM can learn "very deep learning" tasks that require memories of events that happened thousands or even millions of discrete time steps ago. Problem-specific LSTM-like topologies can be evolved. LSTM can handle long delays and signals that have a mix of low and high frequency components.

Stacks of LSTM RNNs trained by Connectionist Temporal Classification (CTC) can find an RNN weight matrix that maximizes the probability of the label sequences in a training set, given the corresponding input sequences. CTC achieves both alignment and recognition.

In 2003, LSTM started to become competitive with traditional speech recognizers. In 2007, the combination with CTC achieved first good results on speech data. In 2009, a CTC-trained LSTM was the first RNN to win pattern recognition contests, when it won several competitions in connected handwriting recognition. In 2014, Baidu used CTC-trained RNNs to break the Switchboard Hub5'00 speech recognition benchmark, without traditional speech processing methods. LSTM also improved large-vocabulary speech recognition, text-to-speech synthesis, for Google Android, and photo-real talking heads. In 2015, Google's speech recognition experienced a 49% improvement through CTC-trained LSTM.

LSTM became popular in Natural Language Processing. Unlike previous models based on HMMs and similar concepts, LSTM can learn to recognise context-sensitive languages. LSTM improved machine translation, language modeling and multilingual language processing. LSTM combined with CNNs improved automatic image captioning.

Deep reservoir computing

Deep Reservoir Computing and Deep Echo State Networks (deepESNs) provide a framework for efficiently trained models for hierarchical processing of temporal data, while enabling the investigation of the inherent role of RNN layered composition.Wikipedia:Please clarify

Deep belief networks

A deep belief network (DBN) is a probabilistic, generative model made up of multiple layers of hidden units. It can be considered a composition of simple learning modules that make up each layer.

A DBN can be used to generatively pre-train a DNN by using the learned DBN weights as the initial DNN weights. Backpropagation or other discriminative algorithms can then tune these weights. This is particularly helpful when training data are limited, because poorly initialized weights can significantly hinder model performance. These pre-trained weights are in a region of the weight space that is closer to the optimal weights than were they randomly chosen. This allows for both improved modeling and faster convergence of the fine-tuning phase.

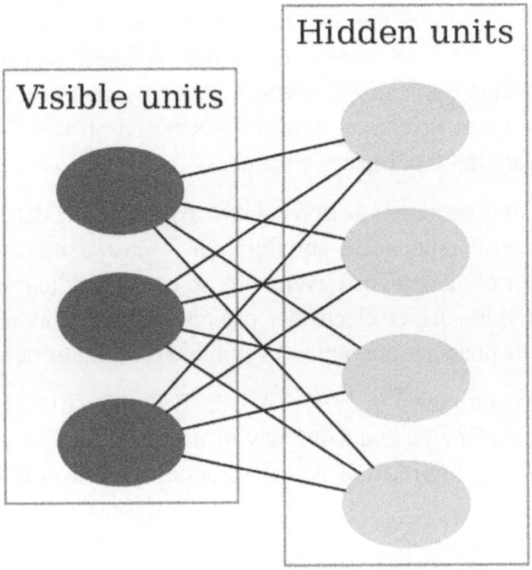

Figure 34: *A restricted Boltzmann machine (RBM) with fully connected visible and hidden units. Note there are no hidden-hidden or visible-visible connections.*

Large memory storage and retrieval neural networks

Large memory storage and retrieval neural networks (LAMSTAR) are fast deep learning neural networks of many layers that can use many filters simultaneously. These filters may be nonlinear, stochastic, logic, non-stationary, or even non-analytical. They are biologically motivated and learn continuously.

A LAMSTAR neural network may serve as a dynamic neural network in spatial or time domains or both. Its speed is provided by Hebbian link-weights[921] that integrate the various and usually different filters (preprocessing functions) into its many layers and to dynamically rank the significance of the various layers and functions relative to a given learning task. This grossly imitates biological learning which integrates various preprocessors (cochlea, retina, *etc.*) and cortexes (auditory, visual, *etc.*) and their various regions. Its deep learning capability is further enhanced by using inhibition, correlation and its ability to cope with incomplete data, or "lost" neurons or layers even amidst a task. It is fully transparent due to its link weights. The link-weights allow dynamic determination of innovation and redundancy, and facilitate the ranking of layers, of filters or of individual neurons relative to a task.

LAMSTAR has been applied to many domains, including medical and financial predictions, adaptive filtering of noisy speech in unknown noise, still-image recognition,[922] video image recognition, software security and adaptive control of non-linear systems. LAMSTAR had a much faster learning speed and somewhat lower error rate than a CNN based on ReLU-function filters and max pooling, in 20 comparative studies.

These applications demonstrate delving into aspects of the data that are hidden from shallow learning networks and the human senses, such as in the cases of predicting onset of sleep apnea events, of an electrocardiogram of a fetus as recorded from skin-surface electrodes placed on the mother's abdomen early in pregnancy, of financial prediction or in blind filtering of noisy speech.

LAMSTAR was proposed in 1996 (A U.S. Patent 5,920,852 A[923]) and was further developed Graupe and Kordylewski from 1997–2002. A modified version, known as LAMSTAR 2, was developed by Schneider and Graupe in 2008.

Stacked (de-noising) auto-encoders

The auto encoder idea is motivated by the concept of a *good* representation. For example, for a classifier, a good representation can be defined as one that yields a better-performing classifier.

An *encoder* is a deterministic mapping f_θ that transforms an input vector $\boldsymbol{x}$ into hidden representation $\boldsymbol{y}$, where $\theta = \{\boldsymbol{W}, b\}$, $\boldsymbol{W}$ is the weight matrix and **b** is an offset vector (bias). A *decoder* maps back the hidden representation **y** to the reconstructed input z via g_θ . The whole process of auto encoding is to compare this reconstructed input to the original and try to minimize the error to make the reconstructed value as close as possible to the original.

In *stacked denoising auto encoders*, the partially corrupted output is cleaned (de-noised). This idea was introduced in 2010 by Vincent et al. with a specific approach to *good* representation, a *good representation* is one that can be obtained robustly from a corrupted input and that will be useful for recovering the corresponding clean input. Implicit in this definition are the following ideas:

- The higher level representations are relatively stable and robust to input corruption;
- It is necessary to extract features that are useful for representation of the input distribution.

The algorithm starts by a stochastic mapping of $\boldsymbol{x}$ to $\tilde{\boldsymbol{x}}$ through $q_D(\tilde{\boldsymbol{x}}|\boldsymbol{x})$, this is the corrupting step. Then the corrupted input $\tilde{\boldsymbol{x}}$ passes through a basic auto-encoder process and is mapped to a hidden representation $\boldsymbol{y} = f_\theta(\tilde{\boldsymbol{x}}) = s(\boldsymbol{W}\tilde{\boldsymbol{x}} + b)$. From this hidden representation, we can reconstruct $\boldsymbol{z} = g_\theta(\boldsymbol{y})$. In the last stage, a minimization algorithm runs in order to have *z* as close as possible to uncorrupted input $\boldsymbol{x}$. The reconstruction error $L_H(\boldsymbol{x}, \boldsymbol{z})$ might be either the cross-entropy loss with an affine-sigmoid decoder, or the squared error loss with an affine decoder.

In order to make a deep architecture, auto encoders stack. Once the encoding function f_θ of the first denoising auto encoder is learned and used to uncorrupt the input (corrupted input), the second level can be trained.

Once the stacked auto encoder is trained, its output can be used as the input to a supervised learning algorithm such as support vector machine classifier or a multi-class logistic regression.

Deep stacking networks

A deep stacking network (DSN) (deep convex network) is based on a hierarchy of blocks of simplified neural network modules. It was introduced in 2011 by Deng and Dong. It formulates the learning as a convex optimization problem with a closed-form solution, emphasizing the mechanism's similarity to stacked generalization. Each DSN block is a simple module that is easy to train by itself in a supervised fashion without backpropagation for the entire blocks.

Each block consists of a simplified multi-layer perceptron (MLP) with a single hidden layer. The hidden layer ***h*** has logistic sigmoidal units, and the output layer has linear units. Connections between these layers are represented by weight matrix ***U;*** input-to-hidden-layer connections have weight matrix ***W***. Target vectors ***t*** form the columns of matrix ***T***, and the input data vectors ***x*** form the columns of matrix ***X***. The matrix of hidden units is $\boldsymbol{H} = \sigma(\boldsymbol{W}^T\boldsymbol{X})$. Modules are trained in order, so lower-layer weights ***W*** are known at each stage. The function performs the element-wise logistic sigmoid operation. Each block estimates the same final label class *y*, and its estimate is concatenated with original input ***X*** to form the expanded input for the next block. Thus, the input to the first block contains the original data only, while downstream blocks' input adds the output of preceding blocks. Then learning the upper-layer weight matrix ***U*** given other weights in the network can be formulated as a convex optimization problem:

$$\min_{U^T} f = ||\boldsymbol{U}^T\boldsymbol{H} - \boldsymbol{T}||_F^2,$$

which has a closed-form solution.

Unlike other deep architectures, such as DBNs, the goal is not to discover the transformed feature representation. The structure of the hierarchy of this kind of architecture makes parallel learning straightforward, as a batch-mode optimization problem. In purely discriminative tasks, DSNs perform better than conventional DBNs.

Tensor deep stacking networks

This architecture is a DSN extension. It offers two important improvements: it uses higher-order information from covariance statistics, and it transforms the non-convex problem of a lower-layer to a convex sub-problem of an upper-layer. TDSNs use covariance statistics in a bilinear mapping from each of two distinct sets of hidden units in the same layer to predictions, via a third-order tensor.

While parallelization and scalability are not considered seriously in conventional DNNs, all learning for DSNs and TDSNs is done in batch mode, to allow parallelization. Parallelization allows scaling the design to larger (deeper) architectures and data sets.

The basic architecture is suitable for diverse tasks such as classification and regression.

Spike-and-slab RBMs

The need for deep learning with real-valued inputs, as in Gaussian restricted Boltzmann machines, led to the *spike-and-slab* RBM (*ss*RBM), which models continuous-valued inputs with strictly binary latent variables. Similar to basic RBMs and its variants, a spike-and-slab RBM is a bipartite graph, while like GRBMs, the visible units (input) are real-valued. The difference is in the hidden layer, where each hidden unit has a binary spike variable and a real-valued slab variable. A spike is a discrete probability mass at zero, while a slab is a density over continuous domain; their mixture forms a prior.

An extension of ssRBM called µ-ssRBM provides extra modeling capacity using additional terms in the energy function. One of these terms enables the model to form a conditional distribution of the spike variables by marginalizing out the slab variables given an observation.

Compound hierarchical-deep models

Compound hierarchical-deep models compose deep networks with non-parametric Bayesian models. Features can be learned using deep architectures such as DBNs, DBMs, deep auto encoders, convolutional variants, ssRBMs, deep coding networks, DBNs with sparse feature learning, RNNs, conditional DBNs, de-noising auto encoders. This provides a better representation, allowing faster learning and more accurate classification with high-dimensional data. However, these architectures are poor at learning novel classes with few examples, because all network units are involved in representing the input (a **distributed representation**) and must be adjusted together (high degree of freedom). Limiting the degree of freedom reduces the number of parameters to learn, facilitating learning of new classes from few examples. *Hierarchical Bayesian (HB)* models allow learning from few examples, for example for computer vision, statistics and cognitive science.

Compound HD architectures aim to integrate characteristics of both HB and deep networks. The compound HDP-DBM architecture is a *hierarchical Dirichlet process (HDP)* as a hierarchical model, incorporated with DBM architecture. It is a full generative model, generalized from abstract concepts flowing through the layers of the model, which is able to synthesize new examples in novel classes that look "reasonably" natural. All the levels are learned jointly by maximizing a joint log-probability score.

In a DBM with three hidden layers, the probability of a visible input $\boldsymbol{\nu}$ is:

$$p(\boldsymbol{\nu}, \psi) = \frac{1}{Z} \sum_{h} e^{\sum_{ij} W_{ij}^{(1)} \nu_i h_j^1 + \sum_{jl} W_{jl}^{(2)} h_j^1 h_l^2 + \sum_{lm} W_{lm}^{(3)} h_l^2 h_m^3},$$

where $\boldsymbol{h} = \{\boldsymbol{h}^{(1)}, \boldsymbol{h}^{(2)}, \boldsymbol{h}^{(3)}\}$ is the set of hidden units, and $\psi = \{\boldsymbol{W}^{(1)}, \boldsymbol{W}^{(2)}, \boldsymbol{W}^{(3)}\}$ are the model parameters, representing visible-hidden and hidden-hidden symmetric interaction terms.

A learned DBM model is an undirected model that defines the joint distribution $P(\nu, h^1, h^2, h^3)$. One way to express what has been learned is the conditional model $P(\nu, h^1, h^2|h^3)$ and a prior term $P(h^3)$.

Here $P(\nu, h^1, h^2|h^3)$ represents a conditional DBM model, which can be viewed as a two-layer DBM but with bias terms given by the states of h^3 :

$$P(\nu, h^1, h^2|h^3) = \frac{1}{Z(\psi, h^3)} e^{\sum_{ij} W_{ij}^{(1)} \nu_i h_j^1 + \sum_{jl} W_{jl}^{(2)} h_j^1 h_l^2 + \sum_{lm} W_{lm}^{(3)} h_l^2 h_m^3}.$$

Deep predictive coding networks

A deep predictive coding network (DPCN) is a predictive coding scheme that uses top-down information to empirically adjust the priors needed for a bottom-up inference procedure by means of a deep, locally connected, generative model. This works by extracting sparse features from time-varying observations using a linear dynamical model. Then, a pooling strategy is used to learn invariant feature representations. These units compose to form a deep architecture and are trained by greedy layer-wise unsupervised learning. The layers constitute a kind of Markov chain such that the states at any layer depend only on the preceding and succeeding layers.

DPCNs predict the representation of the layer, by using a top-down approach using the information in upper layer and temporal dependencies from previous states.

DPCNs can be extended to form a convolutional network.

Networks with separate memory structures

Integrating external memory with ANNs dates to early research in distributed representations and Kohonen's self-organizing maps. For example, in sparse distributed memory or hierarchical temporal memory, the patterns encoded by neural networks are used as addresses for content-addressable memory, with "neurons" essentially serving as address encoders and decoders. However, the early controllers of such memories were not differentiable.

LSTM-related differentiable memory structures

Apart from long short-term memory (LSTM), other approaches also added differentiable memory to recurrent functions. For example:

- Differentiable push and pop actions for alternative memory networks called neural stack machines[924]
- Memory networks where the control network's external differentiable storage is in the fast weights of another network
- LSTM forget gates
- Self-referential RNNs with special output units for addressing and rapidly manipulating the RNN's own weights in differentiable fashion (internal storage)
- Learning to transduce with unbounded memory[925]

Neural Turing machines

Neural Turing machines[926] couple LSTM networks to external memory resources, with which they can interact by attentional processes. The combined system is analogous to a Turing machine but is differentiable end-to-end, allowing it to be efficiently trained by gradient descent. Preliminary results demonstrate that neural Turing machines can infer simple algorithms such as copying, sorting and associative recall from input and output examples.

Differentiable neural computers (DNC) are an NTM extension. They outperformed Neural turing machines, long short-term memory systems and memory networks on sequence-processing tasks.

Semantic hashing

Approaches that represent previous experiences directly and use a similar experience to form a local model are often called nearest neighbour or k-nearest neighbors methods. Deep learning is useful in semantic hashing[927] where a deep graphical model the word-count vectors obtained from a large set of documents.Wikipedia:Please clarify Documents are mapped to memory addresses in such a way that semantically similar documents are located at nearby addresses. Documents similar to a query document can then be found by accessing all the addresses that differ by only a few bits from the address of the query document. Unlike sparse distributed memory that operates on 1000-bit addresses, semantic hashing works on 32 or 64-bit addresses found in a conventional computer architecture.

Memory networks

Memory networks[928,929] are another extension to neural networks incorporating long-term memory. The long-term memory can be read and written to, with the goal of using it for prediction. These models have been applied in the context of question answering (QA) where the long-term memory effectively acts as a (dynamic) knowledge base and the output is a textual response.[930]

Pointer networks

Deep neural networks can be potentially improved by deepening and parameter reduction, while maintaining trainability. While training extremely deep (e.g., 1 million layers) neural networks might not be practical, CPU-like architectures such as pointer networks[931] and neural random-access machines[932] overcome this limitation by using external random-access memory and other components that typically belong to a computer architecture such as registers, ALU and pointers. Such systems operate on probability distribution vectors stored in memory cells and registers. Thus, the model is fully differentiable and trains end-to-end. The key characteristic of these models is that their depth, the size of their short-term memory, and the number of parameters can be altered independently – unlike models like LSTM, whose number of parameters grows quadratically with memory size.

Encoder–decoder networks

Encoder–decoder frameworks are based on neural networks that map highly structured input to highly structured output. The approach arose in the context of machine translation, where the input and output are written sentences in two natural languages. In that work, an LSTM RNN or CNN was used as an encoder to summarize a source sentence, and the summary was decoded using a conditional RNN language model to produce the translation.[933] These systems share building blocks: gated RNNs and CNNs and trained attention mechanisms.

Multilayer kernel machine

Multilayer kernel machines (MKM) are a way of learning highly nonlinear functions by iterative application of weakly nonlinear kernels. They use the kernel principal component analysis (KPCA), as a method for the unsupervised greedy layer-wise pre-training step of deep learning.

Layer $l+1$ learns the representation of the previous layer l , extracting the n_l principal component (PC) of the projection layer l output in the feature domain induced by the kernel. For the sake of dimensionality reduction of the updated representation in each layer, a supervised strategy selects the best informative features among features extracted by KPCA. The process is:

- rank the n_l features according to their mutual information with the class labels;
- for different values of *K* and $m_l \in \{1, \dots, n_l\}$, compute the classification error rate of a *K-nearest neighbor (K-NN)* classifier using only the m_l most informative features on a validation set;
- the value of m_l with which the classifier has reached the lowest error rate determines the number of features to retain.

Some drawbacks accompany the KPCA method as the building cells of an MKM.

A more straightforward way to use kernel machines for deep learning was developed for spoken language understanding. The main idea is to use a kernel machine to approximate a shallow neural net with an infinite number of hidden units, then use stacking to splice the output of the kernel machine and the raw input in building the next, higher level of the kernel machine. The number of levels in the deep convex network is a hyper-parameter of the overall system, to be determined by cross validation.

Neural architecture search

Neural architecture search (NAS) uses machine learning to automate the design of ANNs. Various approaches to NAS have designed networks that compare well with hand-designed systems. The basic search algorithm is to propose a candidate model, evaluate it against a dataset and use the results as feedback to teach the NAS network.

Use

Using ANNs requires an understanding of their characteristics.

- Choice of model: This depends on the data representation and the application. Overly complex models slow learning.
- Learning algorithm: Numerous trade-offs exist between learning algorithms. Almost any algorithm will work well with the correct hyperparameters for training on a particular data set. However, selecting and tuning an algorithm for training on unseen data requires significant experimentation.
- Robustness: If the model, cost function and learning algorithm are selected appropriately, the resulting ANN can become robust.

ANN capabilities fall within the following broad categories:Wikipedia:Citation needed

- Function approximation, or regression analysis, including time series prediction, fitness approximation and modeling.
- Classification, including pattern and sequence recognition, novelty detection and sequential decision making.
- Data processing, including filtering, clustering, blind source separation and compression.
- Robotics, including directing manipulators and prostheses.
- Control, including computer numerical control.

Applications

Because of their ability to reproduce and model nonlinear processes, ANNs have found many applications in a wide range of disciplines.

Application areas include system identification and control (vehicle control, trajectory prediction, process control, natural resource management), quantum chemistry, game-playing and decision making (backgammon, chess, poker), pattern recognition (radar systems, face identification, signal classification, object recognition and more), sequence recognition (gesture, speech, handwritten and printed text recognition), medical diagnosis, finance (e.g. automated trading systems), data mining, visualization, machine translation, social network filtering and e-mail spam filtering.

ANNs have been used to diagnose cancers, including lung cancer, prostate cancer, colorectal cancer and to distinguish highly invasive cancer cell lines from less invasive lines using only cell shape information.

ANNs have been used to accelerate reliability analysis of infrastructures subject to natural disasters.

ANNs have also been used for building black-box models in geoscience: hydrology, ocean modelling and coastal engineering, and geomorphology, are just few examples of this kind.

Types of models

Many types of models are used, defined at different levels of abstraction and modeling different aspects of neural systems. They range from models of the short-term behavior of individual neurons, models of how the dynamics of neural circuitry arise from interactions between individual neurons and finally to models of how behavior can arise from abstract neural modules that represent complete subsystems. These include models of the long-term, and short-term plasticity, of neural systems and their relations to learning and memory from the individual neuron to the system level.

Theoretical properties

Computational power

The multilayer perceptron is a universal function approximator, as proven by the universal approximation theorem. However, the proof is not constructive regarding the number of neurons required, the network topology, the weights and the learning parameters.

A specific recurrent architecture with rational valued weights (as opposed to full precision real number-valued weights) has the full power of a universal Turing machine, using a finite number of neurons and standard linear connections. Further, the use of irrational values for weights results in a machine with super-Turing power.

Capacity

Models' "capacity" property roughly corresponds to their ability to model any given function. It is related to the amount of information that can be stored in the network and to the notion of complexity.Wikipedia:Citation needed

Convergence

Models may not consistently converge on a single solution, firstly because many local minima may exist, depending on the cost function and the model. Secondly, the optimization method used might not guarantee to converge when it begins far from any local minimum. Thirdly, for sufficiently large data or parameters, some methods become impractical. However, for CMAC neural network, a recursive least squares algorithm was introduced to train it, and this algorithm can be guaranteed to converge in one step.

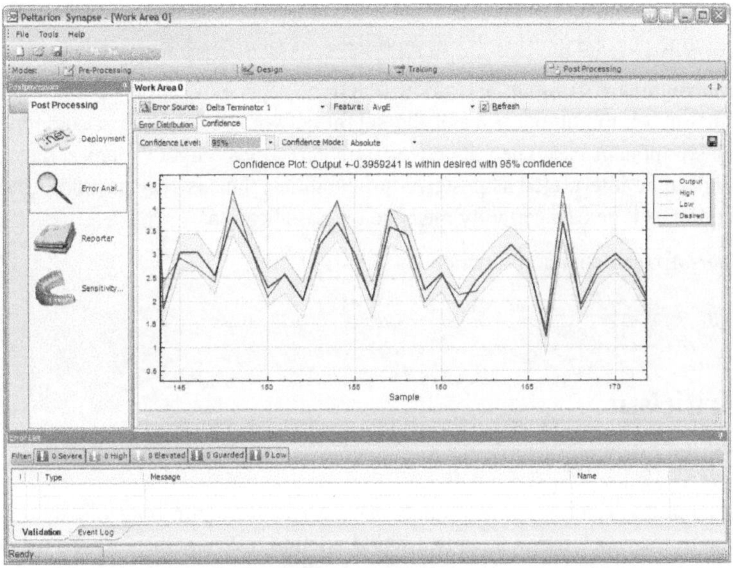

Figure 35: *Confidence analysis of a neural network*

Generalization and statistics

Applications whose goal is to create a system that generalizes well to unseen examples, face the possibility of over-training. This arises in convoluted or over-specified systems when the capacity of the network significantly exceeds the needed free parameters. Two approaches address over-training. The first is to use cross-validation and similar techniques to check for the presence of over-training and optimally select hyperparameters to minimize the generalization error. The second is to use some form of *regularization*. This concept emerges in a probabilistic (Bayesian) framework, where regularization can be performed by selecting a larger prior probability over simpler models; but also in statistical learning theory, where the goal is to minimize over two quantities: the 'empirical risk' and the 'structural risk', which roughly corresponds to the error over the training set and the predicted error in unseen data due to overfitting.

Supervised neural networks that use a mean squared error (MSE) cost function can use formal statistical methods to determine the confidence of the trained model. The MSE on a validation set can be used as an estimate for variance. This value can then be used to calculate the confidence interval of the output of the network, assuming a normal distribution. A confidence analysis made this way is statistically valid as long as the output probability distribution stays the same and the network is not modified.

By assigning a softmax activation function, a generalization of the logistic function, on the output layer of the neural network (or a softmax component in a component-based neural network) for categorical target variables, the outputs can be interpreted as posterior probabilities. This is very useful in classification as it gives a certainty measure on classifications.

The softmax activation function is:

$$y_i = \frac{e^{x_i}}{\sum_{j=1}^{c} e^{x_j}}$$

Criticism

Training issues

A common criticism of neural networks, particularly in robotics, is that they require too much training for real-world operation.Wikipedia:Citation needed Potential solutions include randomly shuffling training examples, by using a numerical optimization algorithm that does not take too large steps when changing the network connections following an example and by grouping examples in so-called mini-batches. Improving the training efficiency and convergence capability has always been an ongoing research area for neural network. For example, by introducing a recursive least squares algorithm for CMAC neural network, the training process only takes one step to converge.

Theoretical issues

No neural network has solved computationally difficult problems such as the n-Queens problem, the travelling salesman problem, or the problem of factoring large integers.

A fundamental objection is that they do not reflect how real neurons function. Back propagation is a critical part of most artificial neural networks, although no such mechanism exists in biological neural networks. How information is coded by real neurons is not known. Sensor neurons fire action potentials more frequently with sensor activation and muscle cells pull more strongly when their associated motor neurons receive action potentials more frequently. Other than the case of relaying information from a sensor neuron to a motor neuron, almost nothing of the principles of how information is handled by biological neural networks is known.

The motivation behind ANNs is not necessarily to strictly replicate neural function, but to use biological neural networks as an inspiration. A central claim of ANNs is therefore that it embodies some new and powerful general principle for processing information. Unfortunately, these general principles are ill-defined. It is often claimed that they are emergent from the network itself. This allows simple statistical association (the basic function of artificial neural networks) to be described as learning or recognition. Alexander Dewdney commented that, as a result, artificial neural networks have a "something-for-nothing quality, one that imparts a peculiar aura of laziness and a distinct lack of curiosity about just how good these computing systems are. No human hand (or mind) intervenes; solutions are found as if by magic; and no one, it seems, has learned anything".

Biological brains use both shallow and deep circuits as reported by brain anatomy,[934] displaying a wide variety of invariance. Weng[935] argued that the brain self-wires largely according to signal statistics and therefore, a serial cascade cannot catch all major statistical dependencies.

Hardware issues

Large and effective neural networks require considerable computing resources. While the brain has hardware tailored to the task of processing signals through a graph of neurons, simulating even a simplified neuron on von Neumann architecture may compel a neural network designer to fill many millions of database rows for its connections – which can consume vast amounts of memory and storage. Furthermore, the designer often needs to transmit signals through many of these connections and their associated neurons – which must often be matched with enormous CPU processing power and time.

Schmidhuber notes that the resurgence of neural networks in the twenty-first century is largely attributable to advances in hardware: from 1991 to 2015, computing power, especially as delivered by GPGPUs (on GPUs), has increased around a million-fold, making the standard backpropagation algorithm feasible for training networks that are several layers deeper than before. The use of parallel GPUs can reduce training times from months to days.

Neuromorphic engineering addresses the hardware difficulty directly, by constructing non-von-Neumann chips to directly implement neural networks in circuitry. Another chip optimized for neural network processing is called a Tensor Processing Unit, or TPU.

Practical counterexamples to criticisms

Arguments against Dewdney's position are that neural networks have been successfully used to solve many complex and diverse tasks, ranging from autonomously flying aircraft[936] to detecting credit card fraud to mastering the game of Go.

Technology writer Roger Bridgman commented:

> *Neural networks, for instance, are in the dock not only because they have been hyped to high heaven, (what hasn't?) but also because you could create a successful net without understanding how it worked: the bunch of numbers that captures its behaviour would in all probability be "an opaque, unreadable table...valueless as a scientific resource". In spite of his emphatic declaration that science is not technology, Dewdney seems here to pillory neural nets as bad science when most of those devising them are just trying to be good engineers. An unreadable table that a useful machine could read would still be well worth having.*[937]

Although it is true that analyzing what has been learned by an artificial neural network is difficult, it is much easier to do so than to analyze what has been learned by a biological neural network. Furthermore, researchers involved in exploring learning algorithms for neural networks are gradually uncovering general principles that allow a learning machine to be successful. For example, local vs non-local learning and shallow vs deep architecture.

Hybrid approaches

Advocates of hybrid models (combining neural networks and symbolic approaches), claim that such a mixture can better capture the mechanisms of the human mind.[938]

Types

Artificial neural networks have many variations. The simplest, static types have one or more static components, including number of units, number of layers, unit weights and topology. Dynamic types allow one or more of these to change during the learning process. The latter are much more complicated, but can shorten learning periods and produce better results. Some types allow/require learning to be "supervised" by the operator, while others operate independently. Some types operate purely in hardware, while others are purely software and run on general purpose computers.

Gallery

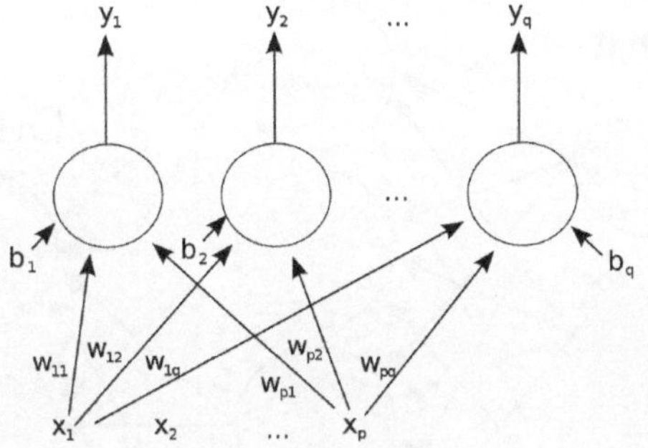

Figure 36: *A single-layer feedforward artificial neural network. Arrows originating from UNIQ-math-0-edde8f4228cedce1-QINU are omitted for clarity. There are p inputs to this network and q outputs. In this system, the value of the qth output, UNIQ-math-1-edde8f4228cedce1-QINU would be calculated as UNIQ-math-2-edde8f4228cedce1-QINU*

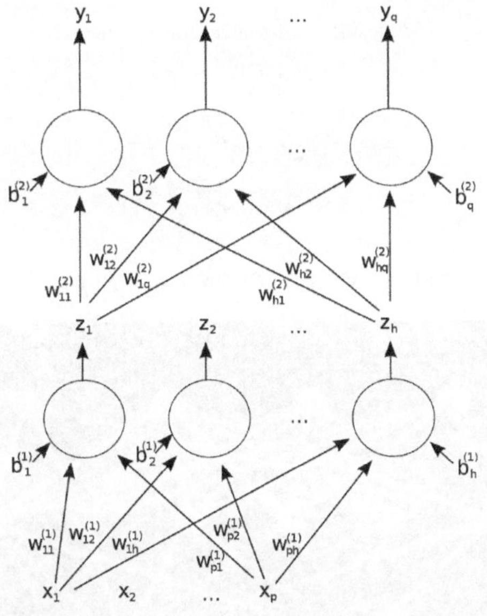

Figure 37: *A two-layer feedforward artificial neural network.*

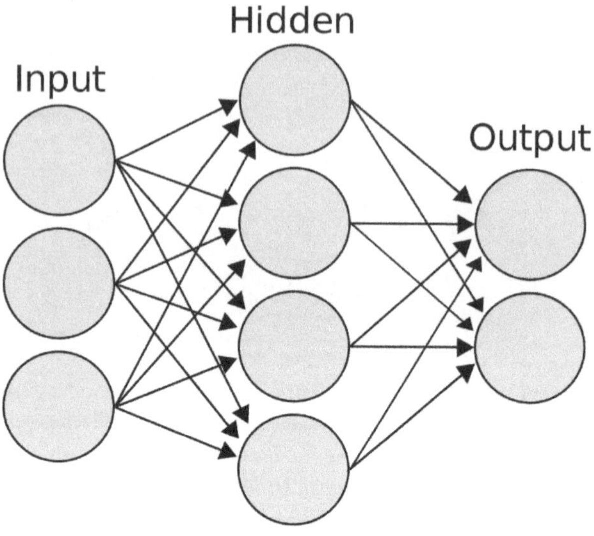

Figure 38: *An artificial neural network.*

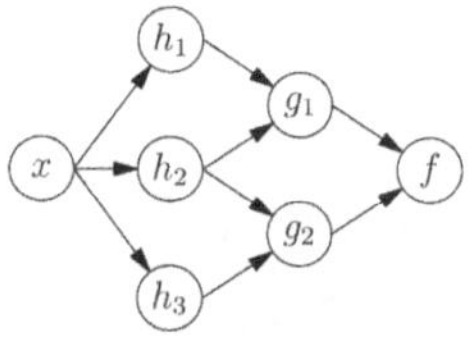

Figure 39: *An ANN dependency graph.*

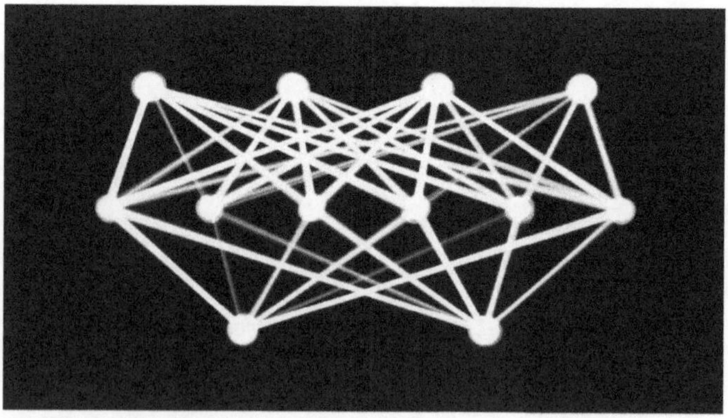

Figure 40: *A single-layer feedforward artificial neural network with 4 inputs, 6 hidden and 2 outputs. Given position state and direction outputs wheel based control values.*

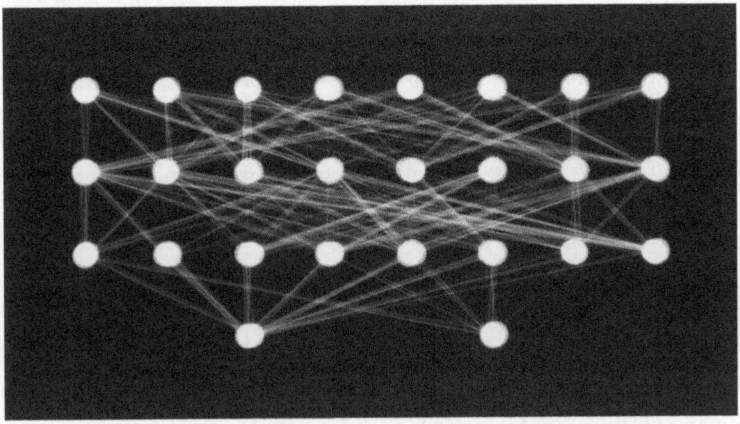

Figure 41: *A two-layer feedforward artificial neural network with 8 inputs, 2x8 hidden and 2 outputs. Given position state, direction and other environment values outputs thruster based control values.*

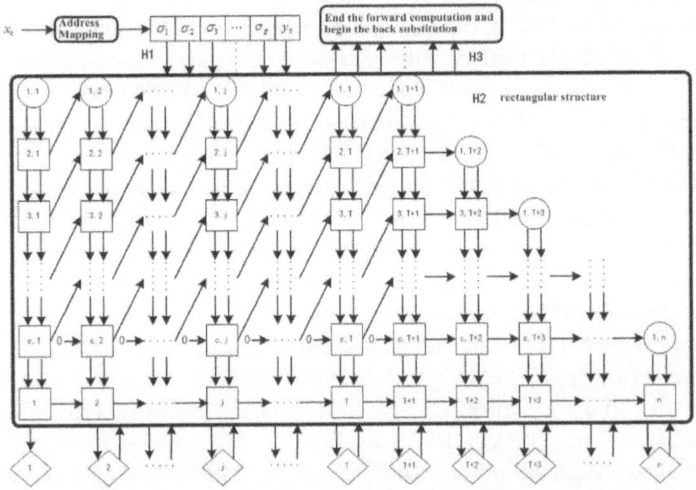

Figure 42: *Parallel pipeline structure of CMAC neural network. This learning algorithm can converge in one step.*

Bibliography

- Bhadeshia H. K. D. H. (1999). "Neural Networks in Materials Science"[939] (PDF). *ISIJ International*. **39** (10): 966–979. doi: 10.2355/isijinternational.39.966[940].

- M., Bishop, Christopher (1995). *Neural networks for pattern recognition*[941]. Clarendon Press. ISBN 0198538499. OCLC 33101074[942].
- Borgelt,, Christian (2003). *Neuro-Fuzzy-Systeme : von den Grundlagen künstlicher Neuronaler Netze zur Kopplung mit Fuzzy-Systemen*[943]. Vieweg. ISBN 9783528252656. OCLC 76538146[944].
- Cybenko, G.V. (2006). "Approximation by Superpositions of a Sigmoidal function"[945]. In van Schuppen, Jan H. *Mathematics of Control, Signals, and Systems*. Springer International. pp. 303–314. PDF[946]
- Dewdney, A. K. (1997). *Yes, we have no neutrons : an eye-opening tour through the twists and turns of bad science*[947]. New York: Wiley. ISBN 9780471108061. OCLC 35558945[948].
- Duda, Richard O.; Hart, Peter Elliot; Stork, David G. (2001). *Pattern classification*[949] (2 ed.). Wiley. ISBN 0471056693. OCLC 41347061[950].
- Egmont-Petersen, M.; de Ridder, D.; Handels, H. (2002). "Image processing with neural networks – a review". *Pattern Recognition*. **35** (10): 2279–2301. doi: 10.1016/S0031-3203(01)00178-9[951].
- Fahlman, S.; Lebiere, C (1991). "The Cascade-Correlation Learning Architecture"[952] (PDF).
  - created for National Science Foundation, Contract Number EET-8716324, and Defense Advanced Research Projects Agency (DOD), ARPA Order No. 4976 under Contract F33615-87-C-1499.
- Gurney, Kevin (1997). *An introduction to neural networks*[953]. UCL Press. ISBN 1857286731. OCLC 37875698[954].
- Haykin, Simon S. (1999). *Neural networks : a comprehensive foundation*[955]. Prentice Hall. ISBN 0132733501. OCLC 38908586[956].
- Hertz, J.; Palmer, Richard G.; Krogh, Anders S. (1991). *Introduction to the theory of neural computation*[957]. Addison-Wesley. ISBN 0201515601. OCLC 21522159[958].
- *Information theory, inference, and learning algorithms*. Cambridge University Press. ISBN 9780521642989. OCLC 52377690[959].
- Kruse, Rudolf,; Borgelt, Christian; Klawonn, F.; Moewes, Christian; Steinbrecher, Matthias; Held,, Pascal (2013). *Computational intelligence : a methodological introduction*[960]. Springer. ISBN 9781447150121. OCLC 837524179[961].
- Lawrence, Jeanette (1994). *Introduction to neural networks : design, theory and applications*[962]. California Scientific Software. ISBN 1883157005. OCLC 32179420[963].
- MacKay, David, J.C. (2003). *Information Theory, Inference, and Learning Algorithms*[964] (PDF). Cambridge University Press. ISBN 9780521642989.
- Masters,, Timothy (1994). *Signal and image processing with neural networks : a C++ sourcebook*[965]. J. Wiley. ISBN 0471049638. OCLC

29877717[966].

- Ripley, Brian D. (2007). *Pattern Recognition and Neural Networks*[967]. Cambridge University Press. ISBN 978-0-521-71770-0.
- Siegelmann, H.T.; Sontag, Eduardo D. (1994). "Analog computation via neural networks"[968] (PDF). *Theoretical Computer Science*. **131** (2): 331–360. doi: 10.1016/0304-3975(94)90178-3[969].
- Smith, Murray (1993). *Neural networks for statistical modeling*[970]. Van Nostrand Reinhold. ISBN 0442013108. OCLC 27145760[971].
- Wasserman, Philip D. (1993). *Advanced methods in neural computing*[972]. Van Nostrand Reinhold. ISBN 0442004613. OCLC 27429729[973].

External links

- A brief introduction to Neural Networks[974] (PDF), illustrated 250p textbook covering the common kinds of neural networks (CC license).
- An Introduction to Deep Neural Networks[975].
- A Tutorial of Neural Network in Excel[976].
- MIT course on Neural Networks[977] on YouTube
- A Concise Introduction to Machine Learning with Artificial Neural Networks[978]
- Neural Networks for Machine Learning – a course by Geoffrey Hinton[979]
- Deep Learning[980]
- Interactive visualization of neural work[981]
- Neural Networks (in Spanish)[982]

Connectionism

Connectionism is an approach in the fields of cognitive science, that hopes to represent mental phenomena using artificial neural networks.[983]

Basic principles

The central connectionist principle is that mental phenomena can be described by interconnected networks of simple and often uniform units. The form of the connections and the units can vary from model to model. For example, units in the network could represent neurons and the connections could represent synapses, as in the human brain.

Spreading activation

In most connectionist models, networks change over time. A closely related and very common aspect of connectionist models is *activation*. At any time, a unit in the network has an activation, which is a numerical value intended to represent some aspect of the unit. For example, if the units in the model are neurons, the activation could represent the probability that the neuron would generate an action potential spike. Activation typically spreads to all the other units connected to it. Spreading activation is always a feature of neural network models, and it is very common in connectionist models used by cognitive psychologists.

Neural networks

Neural networks are by far the most commonly used connectionist model today. Though there are a large variety of neural network models, they almost always follow two basic principles regarding the mind:

1. Any mental state can be described as an (N)-dimensional vector of numeric activation values over neural units in a network.
2. Memory is created by modifying the strength of the connections between neural units. The connection strengths, or "weights", are generally represented as an N×N matrix.

Most of the variety among neural network models comes from:

- *Interpretation of units*: Units can be interpreted as neurons or groups of neurons.
- *Definition of activation*: Activation can be defined in a variety of ways. For example, in a Boltzmann machine, the activation is interpreted as the probability of generating an action potential spike, and is determined via a logistic function on the sum of the inputs to a unit.
- *Learning algorithm*: Different networks modify their connections differently. In general, any mathematically defined change in connection weights over time is referred to as the "learning algorithm".

Connectionists are in agreement that recurrent neural networks (directed networks wherein connections of the network can form a directed cycle) are a better model of the brain than feedforward neural networks (directed networks with no cycles, called DAG). Many recurrent connectionist models also incorporate dynamical systems theory. Many researchers, such as the connectionist Paul Smolensky, have argued that connectionist models will evolve toward fully continuous, high-dimensional, non-linear, dynamic systems approaches.

Biological realism

Connectionist work in general does not need to be biologically realistic and therefore suffers from a lack of neuroscientific plausibility.

Learning

The weights in a neural network are adjusted according to some learning rule or algorithm, such as Hebbian learning. Thus, connectionists have created many sophisticated learning procedures for neural networks. Learning always involves modifying the connection weights. In general, these involve mathematical formulas to determine the change in weights when given sets of data consisting of activation vectors for some subset of the neural units. Several studies have been focused on designing teaching-learning methods based on connectionism.

By formalizing learning in such a way, connectionists have many tools. A very common strategy in connectionist learning methods is to incorporate gradient descent over an error surface in a space defined by the weight matrix. All gradient descent learning in connectionist models involves changing each weight by the partial derivative of the error surface with respect to the weight. Backpropagation (BP), first made popular in the 1980s, is probably the most commonly known connectionist gradient descent algorithm today.

History

Connectionism can be traced to ideas more than a century old, which were little more than speculation until the mid-to-late 20th century.

Parallel distributed processing

The prevailing connectionist approach today was originally known as **parallel distributed processing** (PDP). It was an artificial neural network approach that stressed the parallel nature of neural processing, and the distributed nature of neural representations. It provided a general mathematical framework for researchers to operate in. The framework involved eight major aspects:

- A set of *processing units*, represented by a set of integers.
- An *activation* for each unit, represented by a vector of time-dependent functions.
- An *output function* for each unit, represented by a vector of functions on the activations.
- A *pattern of connectivity* among units, represented by a matrix of real numbers indicating connection strength.

- A *propagation rule* spreading the activations via the connections, represented by a function on the output of the units.
- An *activation rule* for combining inputs to a unit to determine its new activation, represented by a function on the current activation and propagation.
- A *learning rule* for modifying connections based on experience, represented by a change in the weights based on any number of variables.
- An *environment* that provides the system with experience, represented by sets of activation vectors for some subset of the units.

A lot of the research that led to the development of PDP was done in the 1970s, but PDP became popular in the 1980s with the release of the books *Parallel Distributed Processing: Explorations in the Microstructure of Cognition - Volume 1 (foundations)* and *Volume 2 (Psychological and Biological Models)*, by James L. McClelland, David E. Rumelhart and the PDP Research Group. The books are now considered seminal connectionist works, and it is now common to fully equate PDP and connectionism, although the term "connectionism" is not used in the books.

Earlier work

PDP's direct roots were the perceptron theories of researchers such as Frank Rosenblatt from the 1950s and 1960s. But perceptron models were made very unpopular by the book *Perceptrons* by Marvin Minsky and Seymour Papert, published in 1969. It demonstrated the limits on the sorts of functions that single-layered (no hidden layer) perceptrons can calculate, showing that even simple functions like the exclusive disjunction (XOR) could not be handled properly. The PDP books overcame this limitation by showing that multi-level, non-linear neural networks were far more robust and could be used for a vast array of functions.

Many earlier researchers advocated connectionist style models, for example in the 1940s and 1950s, Warren McCulloch and Walter Pitts (MP neuron), Donald Olding Hebb, and Karl Lashley. McCulloch and Pitts showed how neural systems could implement first-order logic: Their classic paper "A Logical Calculus of Ideas Immanent in Nervous Activity" (1943) is important in this development here. They were influenced by the important work of Nicolas Rashevsky in the 1930s. Hebb contributed greatly to speculations about neural functioning, and proposed a learning principle, Hebbian learning, that is still used today. Lashley argued for distributed representations as a result of his failure to find anything like a localized engram in years of lesion experiments.

Connectionism apart from PDP

Though PDP is the dominant form of connectionism, other theoretical work should also be classified as connectionist.

Many connectionist principles can be traced to early work in psychology, such as that of William James. Psychological theories based on knowledge about the human brain were fashionable in the late 19th century. As early as 1869, the neurologist John Hughlings Jackson argued for multi-level, distributed systems. Following from this lead, Herbert Spencer's *Principles of Psychology*, 3rd edition (1872), and Sigmund Freud's *Project for a Scientific Psychology* (composed 1895) propounded connectionist or proto-connectionist theories. These tended to be speculative theories. But by the early 20th century, Edward Thorndike was experimenting on learning that posited a connectionist type network.

In the 1950s, Friedrich Hayek proposed that spontaneous order in the brain arose out of decentralized networks of simple unitsWikipedia:Citation needed. Hayek's work was rarely cited in the PDP literature until recently-Wikipedia:Citation needed.

Another form of connectionist model was the relational network framework developed by the linguist Sydney Lamb in the 1960s. Relational networks have been only used by linguists, and were never unified with the PDP approach. As a result, they are now used by very few researchers.

There are also hybrid connectionist models, mostly mixing symbolic representations with neural network models. The hybrid approach has been advocated by some researchers (such as Ron Sun).

Connectionism vs. computationalism debate

As connectionism became increasingly popular in the late 1980s, some researchers (including Jerry Fodor, Steven Pinker and others) reacted against it. They argued that connectionism, as then developing, threatened to obliterate what they saw as the progress being made in the fields of cognitive science and psychology by the classical approach of computationalism. Computationalism is a specific form of cognitivism that argues that mental activity is computational, that is, that the mind operates by performing purely formal operations on symbols, like a Turing machine. Some researchers argued that the trend in connectionism represented a reversion toward associationism and the abandonment of the idea of a language of thought, something they saw as mistaken. In contrast, those very tendencies made connectionism attractive for other researchers.

Connectionism and computationalism need not be at odds, but the debate in the late 1980s and early 1990s led to opposition between the two approaches. Throughout the debate, some researchers have argued that connectionism and computationalism are fully compatible, though full consensus on this issue has not been reached. Differences between the two approaches include the following:

- Computationalists posit symbolic models that are structurally similar to underlying brain structure, whereas connectionists engage in "low-level" modeling, trying to ensure that their models resemble neurological structures.
- Computationalists in general focus on the structure of explicit symbols (mental models) and syntactical rules for their internal manipulation, whereas connectionists focus on learning from environmental stimuli and storing this information in a form of connections between neurons.
- Computationalists believe that internal mental activity consists of manipulation of explicit symbols, whereas connectionists believe that the manipulation of explicit symbols provides a poor model of mental activity.
- Computationalists often posit domain specific symbolic sub-systems designed to support learning in specific areas of cognition (e.g., language, intentionality, number), whereas connectionists posit one or a small set of very general learning-mechanisms.

Despite these differences, some theorists have proposed that the connectionist architecture is simply the manner in which organic brains happen to implement the symbol-manipulation system. This is logically possible, as it is well known that connectionist models can implement symbol-manipulation systems of the kind used in computationalist models,Wikipedia:Citation needed as indeed they must be able if they are to explain the human ability to perform symbol-manipulation tasks. But the debate rests on whether this symbol manipulation forms the foundation of cognition in general, so this is not a potential vindication of computationalism. Nonetheless, computational descriptions may be helpful high-level descriptions of cognition of logic, for example.

The debate largely centred on logical arguments about whether connectionist networks could produce the syntactic structure observed in this sort of reasoning. This was later achieved,Wikipedia:Citation needed although using processes unlikely to be possible in the brain,Wikipedia:Citation needed thus the debate persisted. As of 2016[984], progress in neurophysiology and general advances in the understanding of neural networks have led to the successful modelling of a great many of these early problems, and the debate about fundamental cognition has, thus, largely been decided among neuroscientists in favour of connectionism.Wikipedia:Citation needed However, these fairly recent developments have yet to reach consensus acceptance among those working in other fields, such as psychology or philosophy of mind.

Part of the appeal of computational descriptions is that they are relatively easy to interpret, and thus may be seen as contributing to our understanding of particular mental processes, whereas connectionist models are in general more opaque, to the extent that they may be describable only in very general terms (such as specifying the learning algorithm, the number of units, etc.), or in unhelpfully low-level terms. In this sense connectionist models may instantiate, and thereby provide evidence for, a broad theory of cognition (i.e., connectionism), without representing a helpful theory of the particular process that is being modelled. In this sense the debate might be considered as to some extent reflecting a mere difference in the level of analysis in which particular theories are framed.

The recentWikipedia:Manual of Style/Dates and numbers#Chronological items popularity of dynamical systems in philosophy of mind have added a new perspective on the debate; some authorsWikipedia:Avoid weasel words now argue that any split between connectionism and computationalism is more conclusively characterized as a split between computationalism and dynamical systems.

In 2014, Alex Graves and others from DeepMind published a series of papers describing a novel Deep Neural Network structure called the Neural Turing Machine able to read symbols on a tape and store symbols in memory. Relational Networks, another Deep Network module published by DeepMind are able to create object-like representations and manipulate them to answer complex questions. Relational Networks and Neural Turing Machines are further evidence that connectionism and computationalism need not be at odds.

References

- Rumelhart, D.E., J.L. McClelland and the PDP Research Group (1986). *Parallel Distributed Processing: Explorations in the Microstructure of Cognition. Volume 1: Foundations*, Cambridge, Massachusetts: MIT Press, ISBN 978-0262680530
- McClelland, J.L., D.E. Rumelhart and the PDP Research Group (1986). *Parallel Distributed Processing: Explorations in the Microstructure of Cognition. Volume 2: Psychological and Biological Models*, Cambridge, Massachusetts: MIT Press, ISBN 978-0262631105
- Pinker, Steven and Mehler, Jacques (1988). *Connections and Symbols*, Cambridge MA: MIT Press, ISBN 978-0262660648

- Jeffrey L. Elman, Elizabeth A. Bates, Mark H. Johnson, Annette Karmiloff-Smith, Domenico Parisi, Kim Plunkett (1996). *Rethinking Innateness: A connectionist perspective on development*, Cambridge MA: MIT Press, ISBN 978-0262550307
- Marcus, Gary F. (2001). *The Algebraic Mind: Integrating Connectionism and Cognitive Science (Learning, Development, and Conceptual Change)*, Cambridge, Massachusetts: MIT Press, ISBN 978-0262632683
- David A. Medler (1998). "A Brief History of Connectionism"[985] (PDF). *Neural Computing Surveys*. **1**: 61–101.

External links

<indicator name="spoken-icon"> </indicator>

- Dictionary of Philosophy of Mind entry on connectionism[986]
- Garson, James. "Connectionism"[987]. In Zalta, Edward N. *Stanford Encyclopedia of Philosophy*.
- A demonstration of Interactive Activation and Competition Networks[988]
- "Connectionism"[989]. *Internet Encyclopedia of Philosophy*.

Deep learning

Machine learning and data mining
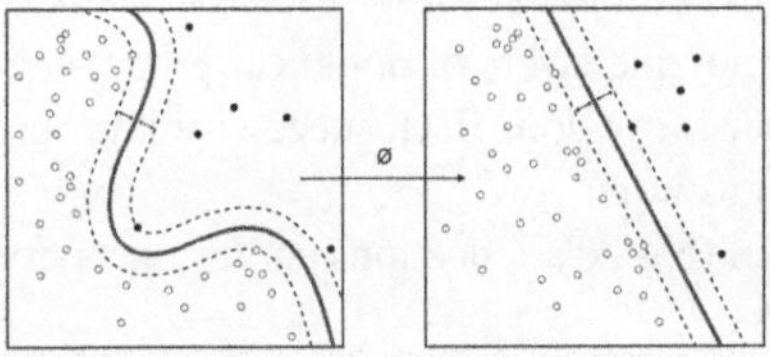
• Machine learning portal
• v • t • e[990]

Deep learning (also known as **deep structured learning** or **hierarchical learning**) is part of a broader family of machine learning methods based on learning data representations, as opposed to task-specific algorithms. Learning can be supervised, semi-supervised or unsupervised.

Deep learning architectures such as deep neural networks, deep belief networks and recurrent neural networks have been applied to fields including computer vision, speech recognition, natural language processing, audio recognition, social network filtering, machine translation, bioinformatics, drug design and board game programs, where they have produced results comparable to and in some cases superior to human experts.

Deep learning models are vaguely inspired by information processing and communication patterns in biological nervous systems yet have various differences from the structural and functional properties of biological brains, which make them incompatible with neuroscience evidences.

Definition

Deep learning is a class of machine learning algorithms that:(pp199–200)

- use a cascade of multiple layers of nonlinear processing units for feature extraction and transformation. Each successive layer uses the output from the previous layer as input.
- learn in supervised (e.g., classification) and/or unsupervised (e.g., pattern analysis) manners.
- learn multiple levels of representations that correspond to different levels of abstraction; the levels form a hierarchy of concepts.

Overview

Most modern deep learning models are based on an artificial neural network, although they can also include propositional formulas or latent variables organized layer-wise in deep generative models such as the nodes in Deep Belief Networks and Deep Boltzmann Machines.

In deep learning, each level learns to transform its input data into a slightly more abstract and composite representation. In an image recognition application, the raw input may be a matrix of pixels; the first representational layer may abstract the pixels and encode edges; the second layer may compose and encode arrangements of edges; the third layer may encode a nose and eyes; and the fourth layer may recognize that the image contains a face. Importantly, a deep learning process can learn which features to optimally place in which level *on its own.* (Of course, this does not completely obviate the need for hand-tuning; for example, varying numbers of layers and layer sizes can provide different degrees of abstraction.)

The "deep" in "deep learning" refers to the number of layers through which the data is transformed. More precisely, deep learning systems have a substantial *credit assignment path* (CAP) depth. The CAP is the chain of transformations from input to output. CAPs describe potentially causal connections between input and output. For a feedforward neural network, the depth of the CAPs is that of the network and is the number of hidden layers plus one (as the output layer is also parameterized). For recurrent neural networks, in which a signal may propagate through a layer more than once, the CAP depth is potentially unlimited. No universally agreed upon threshold of depth divides shallow learning from deep learning, but most researchers agree that deep learning involves CAP depth > 2. CAP of depth 2 has been shown to be a universal approximator in the sense that it can emulate any function.Wikipedia:Citation needed Beyond that more layers do not add to the function approximator ability of the network. The extra layers help in learning features.

Deep learning architectures are often constructed with a greedy layer-by-layer method.Wikipedia:Please clarifyWikipedia:Please clarifyWikipedia:Citation needed Deep learning helps to disentangle these abstractions and pick out which features improve performance.

For supervised learning tasks, deep learning methods obviate feature engineering, by translating the data into compact intermediate representations akin to principal components, and derive layered structures that remove redundancy in representation.

Deep learning algorithms can be applied to unsupervised learning tasks. This is an important benefit because unlabeled data are more abundant than labeled data. Examples of deep structures that can be trained in an unsupervised manner are neural history compressors[991] and deep belief networks.

Interpretations

Deep neural networks are generally interpreted in terms of the universal approximation theorem[992] or probabilistic inference.

The universal approximation theorem concerns the capacity of feedforward neural networks with a single hidden layer of finite size to approximate continuous functions. In 1989, the first proof was published by George Cybenko for sigmoid activation functions and was generalised to feed-forward multi-layer architectures in 1991 by Kurt Hornik.

The probabilistic interpretation derives from the field of machine learning. It features inference, as well as the optimization concepts of training and testing, related to fitting and generalization, respectively. More specifically, the probabilistic interpretation considers the activation nonlinearity as a cumulative distribution function. The probabilistic interpretation led to the introduction of dropout as regularizer in neural networks. The probabilistic interpretation was introduced by researchers including Hopfield, Widrow and Narendra and popularized in surveys such as the one by Bishop.

History

The term *Deep Learning* was introduced to the machine learning community by Rina Dechter in 1986,[993] and to Artificial Neural Networks by Igor Aizenberg and colleagues in 2000, in the context of Boolean threshold neurons.[994,995]

The first general, working learning algorithm for supervised, deep, feedforward, multilayer perceptrons was published by Alexey Ivakhnenko and Lapa in 1965. A 1971 paper described a deep network with 8 layers trained by the group method of data handling algorithm.

Other deep learning working architectures, specifically those built for computer vision, began with the Neocognitron introduced by Kunihiko Fukushima in 1980. In 1989, Yann LeCun et al. applied the standard backpropagation algorithm, which had been around as the reverse mode of automatic differentiation since 1970,[996] to a deep neural network with the purpose of recognizing handwritten ZIP codes on mail. While the algorithm worked, training required 3 days.[997]

By 1991 such systems were used for recognizing isolated 2-D hand-written digits, while recognizing 3-D objects was done by matching 2-D images with a handcrafted 3-D object model. Weng *et al.* suggested that a human brain does not use a monolithic 3-D object model and in 1992 they published Cresceptron,[998,999,1000] a method for performing 3-D object recognition in cluttered scenes. Cresceptron is a cascade of layers similar to Neocognitron. But while Neocognitron required a human programmer to hand-merge features, Cresceptron learned an open number of features in each layer without supervision, where each feature is represented by a convolution kernel. Cresceptron segmented each learned object from a cluttered scene through back-analysis through the network. Max pooling, now often adopted by deep neural networks (e.g. ImageNet tests), was first used in Cresceptron to reduce the position resolution by a factor of (2x2) to 1 through the cascade for better generalization.

In 1994, André de Carvalho, together with Mike Fairhurst and David Bisset, published experimental results of a multi-layer boolean neural network, also known as a weightless neural network, composed of a 3-layers self-organising feature extraction neural network module (SOFT) followed by a multi-layer classification neural network module (GSN), which were independently trained. Each layer in the feature extraction module extracted features with growing complexity regarding the previous layer.

In 1995, Brendan Frey demonstrated that it was possible to train (over two days) a network containing six fully connected layers and several hundred hidden units using the wake-sleep algorithm, co-developed with Peter Dayan and

Hinton. Many factors contribute to the slow speed, including the vanishing gradient problem analyzed in 1991 by Sepp Hochreiter.[1001]

Simpler models that use task-specific handcrafted features such as Gabor filters and support vector machines (SVMs) were a popular choice in the 1990s and 2000s, because of ANNs' computational cost and a lack of understanding of how the brain wires its biological networks.

Both shallow and deep learning (e.g., recurrent nets) of ANNs have been explored for many years. These methods never outperformed non-uniform internal-handcrafting Gaussian mixture model/Hidden Markov model (GMM-HMM) technology based on generative models of speech trained discriminatively. Key difficulties have been analyzed, including gradient diminishing and weak temporal correlation structure in neural predictive models. Additional difficulties were the lack of training data and limited computing power.

Most speech recognition researchers moved away from neural nets to pursue generative modeling. An exception was at SRI International in the late 1990s. Funded by the US government's NSA and DARPA, SRI studied deep neural networks in speech and speaker recognition. Heck's speaker recognition team achieved the first significant success with deep neural networks in speech processing in the 1998 National Institute of Standards and Technology Speaker Recognition evaluation. While SRI experienced success with deep neural networks in speaker recognition, they were unsuccessful in demonstrating similar success in speech recognition. The principle of elevating "raw" features over hand-crafted optimization was first explored successfully in the architecture of deep autoencoder on the "raw" spectrogram or linear filter-bank features in the late 1990s, showing its superiority over the Mel-Cepstral features that contain stages of fixed transformation from spectrograms. The raw features of speech, waveforms, later produced excellent larger-scale results.

Many aspects of speech recognition were taken over by a deep learning method called long short-term memory (LSTM), a recurrent neural network published by Hochreiter and Schmidhuber in 1997. LSTM RNNs avoid the vanishing gradient problem and can learn "Very Deep Learning" tasks that require memories of events that happened thousands of discrete time steps before, which is important for speech. In 2003, LSTM started to become competitive with traditional speech recognizers on certain tasks. Later it was combined with connectionist temporal classification (CTC) in stacks of LSTM RNNs.[1002] In 2015, Google's speech recognition reportedly experienced a dramatic performance jump of 49% through CTC-trained LSTM, which they made available through Google Voice Search.

In 2006, publications by Geoff Hinton, Ruslan Salakhutdinov, Osindero and Teh showed how a many-layered feedforward neural network could be effectively pre-trained one layer at a time, treating each layer in turn as an unsupervised restricted Boltzmann machine, then fine-tuning it using supervised backpropagation.[1003] The papers referred to *learning* for *deep belief nets.*

Deep learning is part of state-of-the-art systems in various disciplines, particularly computer vision and automatic speech recognition (ASR). Results on commonly used evaluation sets such as TIMIT (ASR) and MNIST (image classification), as well as a range of large-vocabulary speech recognition tasks have steadily improved. Convolutional neural networks (CNNs) were superseded for ASR by CTC for LSTM. but are more successful in computer vision.

The impact of deep learning in industry began in the early 2000s, when CNNs already processed an estimated 10% to 20% of all the checks written in the US, according to Yann LeCun.[1004] Industrial applications of deep learning to large-scale speech recognition started around 2010.

The 2009 NIPS Workshop on Deep Learning for Speech Recognition was motivated by the limitations of deep generative models of speech, and the possibility that given more capable hardware and large-scale data sets that deep neural nets (DNN) might become practical. It was believed that pre-training DNNs using generative models of deep belief nets (DBN) would overcome the main difficulties of neural nets. However, it was discovered that replacing pre-training with large amounts of training data for straightforward backpropagation when using DNNs with large, context-dependent output layers produced error rates dramatically lower than then-state-of-the-art Gaussian mixture model (GMM)/Hidden Markov Model (HMM) and also than more-advanced generative model-based systems.[1005] The nature of the recognition errors produced by the two types of systems was characteristically different,[313] offering technical insights into how to integrate deep learning into the existing highly efficient, run-time speech decoding system deployed by all major speech recognition systems. Analysis around 2009-2010, contrasted the GMM (and other generative speech models) vs. DNN models, stimulated early industrial investment in deep learning for speech recognition, eventually leading to pervasive and dominant use in that industry. That analysis was done with comparable performance (less than 1.5% in error rate) between discriminative DNNs and generative models.[319]

In 2010, researchers extended deep learning from TIMIT to large vocabulary speech recognition, by adopting large output layers of the DNN based on context-dependent HMM states constructed by decision trees.

Advances in hardware enabled the renewed interest. In 2009, Nvidia was involved in what was called the "big bang" of deep learning, "as deep-learning neural networks were trained with Nvidia graphics processing units (GPUs)." That year, Google Brain used Nvidia GPUs to create capable DNNs. While there, Ng determined that GPUs could increase the speed of deep-learning systems by about 100 times. In particular, GPUs are well-suited for the matrix/vector math involved in machine learning.[1006] GPUs speed up training algorithms by orders of magnitude, reducing running times from weeks to days. Specialized hardware and algorithm optimizations can be used for efficient processing.

Deep learning revolution

In 2012, a team led by Dahl won the "Merck Molecular Activity Challenge" using multi-task deep neural networks to predict the biomolecular target of one drug. In 2014, Hochreiter's group used deep learning to detect off-target and toxic effects of environmental chemicals in nutrients, household products and drugs and won the "Tox21 Data Challenge" of NIH, FDA and NCATS.[1007]

Significant additional impacts in image or object recognition were felt from 2011 to 2012. Although CNNs trained by backpropagation had been around for decades, and GPU implementations of NNs for years, including CNNs, fast implementations of CNNs with max-pooling on GPUs in the style of Ciresan and colleagues were needed to progress on computer vision. In 2011, this approach achieved for the first time superhuman performance in a visual pattern recognition contest. Also in 2011, it won the ICDAR Chinese handwriting contest, and in May 2012, it won the ISBI image segmentation contest. Until 2011, CNNs did not play a major role at computer vision conferences, but in June 2012, a paper by Ciresan et al. at the leading conference CVPR showed how max-pooling CNNs on GPU can dramatically improve many vision benchmark records. In October 2012, a similar system by Krizhevsky et al. won the large-scale ImageNet competition by a significant margin over shallow machine learning methods. In November 2012, Ciresan et al.'s system also won the ICPR contest on analysis of large medical images for cancer detection, and in the following year also the MICCAI Grand Challenge on the same topic. In 2013 and 2014, the error rate on the ImageNet task using deep learning was further reduced, following a similar trend in large-scale speech recognition. The Wolfram Image Identification project publicized these improvements.

Image classification was then extended to the more challenging task of generating descriptions (captions) for images, often as a combination of CNNs and LSTMs.

Some researchers assess that the October 2012 ImageNet victory anchored the start of a "deep learning revolution" that has transformed the AI industry.

Artificial neural networks

Artificial neural networks (**ANNs**) or **connectionist systems** are computing systems inspired by the biological neural networks that constitute animal brains. Such systems learn (progressively improve their ability) to do tasks by considering examples, generally without task-specific programming. For example, in image recognition, they might learn to identify images that contain cats by analyzing example images that have been manually labeled as "cat" or "no cat" and using the analytic results to identify cats in other images. They have found most use in applications difficult to express with a traditional computer algorithm using rule-based programming.

An ANN is based on a collection of connected units called artificial neurons, (analogous to biological neurons in a biological brain). Each connection (synapse) between neurons can transmit a signal to another neuron. The receiving (postsynaptic) neuron can process the signal(s) and then signal downstream neurons connected to it. Neurons may have state, generally represented by real numbers, typically between 0 and 1. Neurons and synapses may also have a weight that varies as learning proceeds, which can increase or decrease the strength of the signal that it sends downstream.

Typically, neurons are organized in layers. Different layers may perform different kinds of transformations on their inputs. Signals travel from the first (input), to the last (output) layer, possibly after traversing the layers multiple times.

The original goal of the neural network approach was to solve problems in the same way that a human brain would. Over time, attention focused on matching specific mental abilities, leading to deviations from biology such as backpropagation, or passing information in the reverse direction and adjusting the network to reflect that information.

Neural networks have been used on a variety of tasks, including computer vision, speech recognition, machine translation, social network filtering, playing board and video games and medical diagnosis.

As of 2017, neural networks typically have a few thousand to a few million units and millions of connections. Despite this number being several order of magnitude less than the number of neurons on a human brain, these networks can perform many tasks at a level beyond that of humans (e.g., recognizing faces, playing "Go").

Deep neural networks

A deep neural network (DNN) is an artificial neural network (ANN) with multiple layers between the input and output layers. The DNN finds the correct mathematical manipulation to turn the input into the output, whether it be a linear relationship or a non-linear relationship. The network moves through the layers calculating the probability of each output. For example, a DNN that is trained to recognize dog breeds will go over the given image and calculate the probability that the dog in the image is a certain breed. The user can review the results and select which probabilities the network should display (above a certain threshold, etc.) and return the proposed label. Each mathematical manipulation as such is considered a layer, and complex DNN have many layers, hence the name "deep" networks.

DNNs can model complex non-linear relationships. DNN architectures generate compositional models where the object is expressed as a layered composition of primitives. The extra layers enable composition of features from lower layers, potentially modeling complex data with fewer units than a similarly performing shallow network.

Deep architectures include many variants of a few basic approaches. Each architecture has found success in specific domains. It is not always possible to compare the performance of multiple architectures, unless they have been evaluated on the same data sets.

DNNs are typically feedforward networks in which data flows from the input layer to the output layer without looping back. At first, the DNN creates a map of virtual neurons and assigns random numerical values, or "weights", to connections between them. The weights and inputs are multiplied and return an output between 0 and 1. If the network didn't accurately recognize a particular pattern, an algorithm would adjust the weights. That way the algorithm can make certain parameters more influential, until it determines the correct mathematical manipulation to fully process the data.

Recurrent neural networks (RNNs), in which data can flow in any direction, are used for applications such as language modeling. Long short-term memory is particularly effective for this use.

Convolutional deep neural networks (CNNs) are used in computer vision. CNNs also have been applied to acoustic modeling for automatic speech recognition (ASR).

Challenges

As with ANNs, many issues can arise with naively trained DNNs. Two common issues are overfitting and computation time.

DNNs are prone to overfitting because of the added layers of abstraction, which allow them to model rare dependencies in the training data. Regularization methods such as Ivakhnenko's unit pruning or weight decay (ℓ_2 -regularization) or sparsity (ℓ_1 -regularization) can be applied during training to combat overfitting. Alternatively dropout regularization randomly omits units from the hidden layers during training. This helps to exclude rare dependencies. Finally, data can be augmented via methods such as cropping and rotating such that smaller training sets can be increased in size to reduce the chances of overfitting.

DNNs must consider many training parameters, such as the size (number of layers and number of units per layer), the learning rate, and initial weights. Sweeping through the parameter space for optimal parameters may not be feasible due to the cost in time and computational resources. Various tricks, such as batching (computing the gradient on several training examples at once rather than individual examples) speed up computation. Large processing capabilities of many-core architectures (such as, GPUs or the Intel Xeon Phi) have produced significant speedups in training, because of the suitability of such processing architectures for the matrix and vector computations.

Alternatively, engineers may look for other types of neural networks with more straightforward and convergent training algorithms. CMAC (cerebellar model articulation controller) is one such kind of neural network. It doesn't require learning rates or randomized initial weights for CMAC. The training process can be guaranteed to converge in one step with a new batch of data, and the computational complexity of the training algorithm is linear with respect to the number of neurons involved.[1008,1009]

Applications

Automatic speech recognition

Large-scale automatic speech recognition is the first and most convincing successful case of deep learning. LSTM RNNs can learn "Very Deep Learning" tasks that involve multi-second intervals containing speech events separated by thousands of discrete time steps, where one time step corresponds to about 10 ms. LSTM with forget gates is competitive with traditional speech recognizers on certain tasks.

The initial success in speech recognition was based on small-scale recognition tasks based on TIMIT. The data set contains 630 speakers from eight major dialects of American English, where each speaker reads 10 sentences.[1010] Its small size lets many configurations be tried. More importantly, the TIMIT task concerns phone-sequence recognition, which, unlike word-sequence recognition, allows weak phone bigram language models. This lets the strength of the acoustic modeling aspects of speech recognition be more easily analyzed. The error rates listed below, including these early results and measured as percent phone error rates (PER), have been summarized since 1991.

Method	**PER (%)**
Randomly Initialized RNN	26.1
Bayesian Triphone GMM-HMM	25.6
Hidden Trajectory (Generative) Model	24.8
Monophone Randomly Initialized DNN	23.4
Monophone DBN-DNN	22.4
Triphone GMM-HMM with BMMI Training	21.7
Monophone DBN-DNN on fbank	20.7
Convolutional DNN	20.0
Convolutional DNN w. Heterogeneous Pooling	18.7
Ensemble DNN/CNN/RNN	18.3
Bidirectional LSTM	17.9
Hierarchical Convolutional Deep Maxout Network	16.5

The debut of DNNs for speaker recognition in the late 1990s and speech recognition around 2009-2011 and of LSTM around 2003-2007, accelerated progress in eight major areas:

- Scale-up/out and acclerated DNN training and decoding
- Sequence discriminative training
- Feature processing by deep models with solid understanding of the underlying mechanisms
- Adaptation of DNNs and related deep models
- Multi-task and transfer learning by DNNs and related deep models
- CNNs and how to design them to best exploit domain knowledge of speech
- RNN and its rich LSTM variants
- Other types of deep models including tensor-based models and integrated deep generative/discriminative models.

All major commercial speech recognition systems (e.g., Microsoft Cortana, Xbox, Skype Translator, Amazon Alexa, Google Now, Apple Siri, Baidu and iFlyTek voice search, and a range of Nuance speech products, etc.) are based on deep learning.

Image recognition

A common evaluation set for image classification is the MNIST database data set. MNIST is composed of handwritten digits and includes 60,000 training examples and 10,000 test examples. As with TIMIT, its small size lets users test multiple configurations. A comprehensive list of results on this set is available.

Deep learning-based image recognition has become "superhuman", producing more accurate results than human contestants. This first occurred in 2011.

Deep learning-trained vehicles now interpret 360° camera views.[1011] Another example is Facial Dysmorphology Novel Analysis (FDNA) used to analyze cases of human malformation connected to a large database of genetic syndromes.

Visual art processing

Closely related to the progress that has been made in image recognition is the increasing application of deep learning techniques to various visual art tasks. DNNs have proven themselves capable, for example, of a) identifying the style period of a given painting, b) "capturing" the style of a given painting and applying it in a visually pleasing manner to an arbitrary photograph, and c) generating striking imagery based on random visual input fields.

Natural language processing

Neural networks have been used for implementing language models since the early 2000s. LSTM helped to improve machine translation and language modeling.

Other key techniques in this field are negative sampling and word embedding. Word embedding, such as *word2vec*, can be thought of as a representational layer in a deep learning architecture that transforms an atomic word into a positional representation of the word relative to other words in the dataset; the position is represented as a point in a vector space. Using word embedding as an RNN input layer allows the network to parse sentences and phrases using an effective compositional vector grammar. A compositional vector grammar can be thought of as probabilistic context free grammar (PCFG) implemented by an RNN. Recursive auto-encoders built atop word embeddings can assess sentence similarity and detect paraphrasing. Deep neural architectures provide the best results for constituency parsing, sentiment analysis, information retrieval, spoken language understanding, machine translation, contextual entity linking, writing style recognition,[1012] Text classification and others.

Google Translate (GT) uses a large end-to-end long short-term memory network.[1013] Google Neural Machine Translation (GNMT) uses an example-based machine translation method in which the system "learns from millions of examples." It translates "whole sentences at a time, rather than pieces. Google Translate supports over one hundred languages. The network encodes the "semantics of the sentence rather than simply memorizing phrase-to-phrase translations". GT uses English as an intermediate between most language pairs.

Drug discovery and toxicology

A large percentage of candidate drugs fail to win regulatory approval. These failures are caused by insufficient efficacy (on-target effect), undesired interactions (off-target effects), or unanticipated toxic effects. Research has explored use of deep learning to predict biomolecular target, off-target and toxic effects of environmental chemicals in nutrients, household products and drugs.

AtomNet is a deep learning system for structure-based rational drug design. AtomNet was used to predict novel candidate biomolecules for disease targets such as the Ebola virus and multiple sclerosis.

Customer relationship management

Deep reinforcement learning has been used to approximate the value of possible direct marketing actions, defined in terms of RFM variables. The estimated value function was shown to have a natural interpretation as customer lifetime value.

Recommendation systems

Recommendation systems have used deep learning to extract meaningful features for a latent factor model for content-based music recommendations. Multiview deep learning has been applied for learning user preferences from multiple domains. The model uses a hybrid collaborative and content-based approach and enhances recommendations in multiple tasks.

Bioinformatics

An autoencoder ANN was used in bioinformatics, to predict gene ontology annotations and gene-function relationships.

In medical informatics, deep learning was used to predict sleep quality based on data from wearables and predictions of health complications from electronic health record data. Deep learning has also showed efficacy in healthcare.

Mobile advertising

Finding the appropriate mobile audience for mobile advertising is always challenging, since many data points must be considered and assimilated before a target segment can be created and used in ad serving by any ad server. Deep learning has been used to interpret large, many-dimensioned advertising datasets. Many data points are collected during the request/serve/click internet advertising cycle. This information can form the basis of machine learning to improve ad selection.

Image restoration

Deep learning has been successfully applied to inverse problems such as denoising, super-resolution, and inpainting. These applications include learning methods such "Shrinkage Fields for Effective Image Restoration" which trains on an image dataset, and Deep Image Prior, which trains on the image that needs restoration.

Financial fraud detection

Deep learning is being successfully applied to financial fraud detection and anti-money laundering. "Deep anti-money laundering detection system can spot and recognize relationships and similarities between data and, further down the road, learn to detect anomalies or classify and predict specific events". The solution leverages both supervised learning techniques, such as the classification of suspicious transactions, and unsupervised learning, e.g. anomaly detection.

Relation to human cognitive and brain development

Deep learning is closely related to a class of theories of brain development (specifically, neocortical development) proposed by cognitive neuroscientists in the early 1990s. These developmental theories were instantiated in computational models, making them predecessors of deep learning systems. These developmental models share the property that various proposed learning dynamics in the brain (e.g., a wave of nerve growth factor) support the self-organization somewhat analogous to the neural networks utilized in deep learning models. Like the neocortex, neural networks employ a hierarchy of layered filters in which each layer considers information from a prior layer (or the operating environment), and then passes its output (and possibly the original input), to other layers. This process yields a self-organizing stack of transducers, well-tuned to their operating environment. A 1995 description stated, "...the infant's brain seems to organize itself under the influence of waves of so-called trophic-factors ... different regions of the brain become connected sequentially, with one layer of tissue maturing before another and so on until the whole brain is mature."[1014]

A variety of approaches have been used to investigate the plausibility of deep learning models from a neurobiological perspective. On the one hand, several variants of the backpropagation algorithm have been proposed in order to increase its processing realism. Other researchers have argued that unsupervised forms of deep learning, such as those based on hierarchical generative models and deep belief networks, may be closer to biological reality. In this respect, generative neural network models have been related to neurobiological evidence about sampling-based processing in the cerebral cortex.

Although a systematic comparison between the human brain organization and the neuronal encoding in deep networks has not yet been established, several analogies have been reported. For example, the computations performed by deep learning units could be similar to those of actual neurons and neural populations. Similarly, the representations developed by deep learning models are similar to those measured in the primate visual system both at the single-unit and at the population levels.

Commercial activity

Many organizations employ deep learning for particular applications. Facebook's AI lab performs tasks such as automatically tagging uploaded pictures with the names of the people in them.

Google's DeepMind Technologies developed a system capable of learning how to play Atari video games using only pixels as data input. In 2015 they demonstrated their AlphaGo system, which learned the game of Go well enough to beat a professional Go player. Google Translate uses an LSTM to translate between more than 100 languages.

In 2015, Blippar demonstrated a mobile augmented reality application that uses deep learning to recognize objects in real time.

Criticism and comment

Deep learning has attracted both criticism and comment, in some cases from outside the field of computer science.

Theory

A main criticism concerns the lack of theory surrounding the methods.Wikipedia:Citation needed Learning in the most common deep architectures is implemented using well-understood gradient descent. However, the theory surrounding other algorithms, such as contrastive divergence is less clear.Wikipedia:Citation needed (e.g., Does it converge? If so, how fast? What is it approximating?) Deep learning methods are often looked at as a black box, with most confirmations done empirically, rather than theoretically.

Others point out that deep learning should be looked at as a step towards realizing strong AI, not as an all-encompassing solution. Despite the power of deep learning methods, they still lack much of the functionality needed for realizing this goal entirely. Research psychologist Gary Marcus noted:

> *"Realistically, deep learning is only part of the larger challenge of building intelligent machines. Such techniques lack ways of representing causal relationships (...) have no obvious ways of performing logical inferences, and they are also still a long way from integrating abstract knowledge, such as information about what objects are, what they are for, and how they are typically used. The most powerful A.I. systems, like Watson (...) use techniques like deep learning as just one element in a very complicated ensemble of techniques, ranging from the statistical technique of Bayesian inference to deductive reasoning."*

As an alternative to this emphasis on the limits of deep learning, one author speculated that it might be possible to train a machine vision stack to perform the sophisticated task of discriminating between "old master" and amateur figure drawings, and hypothesized that such a sensitivity might represent the rudiments of a non-trivial machine empathy. This same author proposed that this would be in line with anthropology, which identifies a concern with aesthetics as a key element of behavioral modernity.

In further reference to the idea that artistic sensitivity might inhere within relatively low levels of the cognitive hierarchy, a published series of graphic representations of the internal states of deep (20-30 layers) neural networks attempting to discern within essentially random data the images on which they were trained demonstrate a visual appeal: the original research notice received well over 1,000 comments, and was the subject of what was for a time the most frequently accessed article on *The Guardian's* web site.

Errors

Some deep learning architectures display problematic behaviors, such as confidently classifying unrecognizable images as belonging to a familiar category of ordinary images and misclassifying minuscule perturbations of correctly classified images. Goertzel hypothesized that these behaviors are due to limitations in their internal representations and that these limitations would inhibit integration into heterogeneous multi-component AGI architectures. These issues may possibly be addressed by deep learning architectures that internally form states homologous to image-grammar decompositions of observed entities and events. Learning a grammar (visual or linguistic) from training data would be equivalent to restricting the system to commonsense reasoning that operates on concepts in terms of grammatical production rules and is a basic goal of both human language acquisition[1015] and AI.

Cyberthreat

As deep learning moves from the lab into the world, research and experience shows that artificial neural networks are vulnerable to hacks and deception. By identifying patterns that these systems use to function, attackers can modify inputs to ANNs in such a way that the ANN finds a match that human observers would not recognize. For example, an attacker can make subtle changes to an image such that the ANN finds a match even though the image looks to a human nothing like the search target. Such a manipulation is termed an "adversarial attack." In 2016 researchers used one ANN to doctor images in trial and error fashion, identify another's focal points and thereby generate images that deceived it. The modified images looked no different to human eyes. Another group showed that printouts of doctored images then photographed successfully tricked an image classification system. One defense is reverse image search, in which a possible fake image is submitted to a site such as TinEye that can then find other instances of it. A refinement is to search using only parts of the image, to identify images from which that piece may have been taken.

Another group showed that certain psychedelic spectacles could fool a facial recognition system into thinking ordinary people were celebrities, potentially allowing one person to impersonate another. In 2017 researchers added stickers to stop signs and caused an ANN to misclassify them.

ANNs can however be further trained to detect attempts at deception, potentially leading attackers and defenders into an arms race similar to the kind that already defines the malware defense industry. ANNs have been trained to defeat ANN-based anti-malware software by repeatedly attacking a defense with malware that was continually altered by a genetic algorithm until it tricked the anti-malware while retaining its ability to damage the target.

Another group demonstrated that certain sounds could make the Google Now voice command system open a particular web address that would download malware.

In "data poisoning", false data is continually smuggled into a machine learning system's training set to prevent it from achieving mastery.

Recurrent neural network

Machine learning and data mining

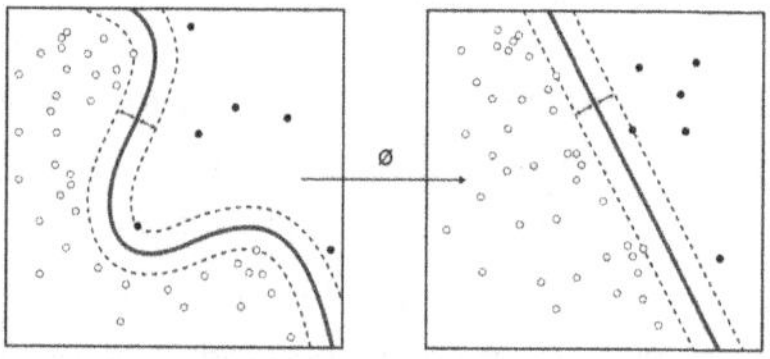

- **Machine learning portal**

- v
- t
- e[1016]

A **recurrent neural network (RNN)** is a class of artificial neural network where connections between nodes form a directed graph along a sequence. This allows it to exhibit dynamic temporal behavior for a time sequence. Unlike feedforward neural networks, RNNs can use their internal state (memory) to process sequences of inputs. This makes them applicable to tasks such as unsegmented, connected handwriting recognition or speech recognition.

The term "recurrent neural network" is used indiscriminately to refer to two broad classes of networks with a similar general structure, where one is finite impulse and the other is infinite impulse. Both classes of networks exhibit temporal dynamic behavior. A finite impulse recurrent network is a directed acyclic graph that can be unrolled and replaced with a strictly feedforward neural network, while an infinite impulse recurrent network is a directed cyclic graph that can not be unrolled.

Both finite impulse and infinite impulse recurrent networks can have additional stored state, and the storage can be under direct control by the neural network. The storage can also be replaced by another network or graph, if that incorporates time delays or has feedback loops. Such controlled states are referred to as gated state or gated memory, and are part of long short-term memorys (LSTMs) and gated recurrent units.

History

Recurrent neural networks were developed in the 1980s. Hopfield networks were discovered by John Hopfield in 1982. In 1993, a neural history compressor system solved a "Very Deep Learning" task that required more than 1000 subsequent layers in an RNN unfolded in time.[1017]

LSTM

Long short-term memory (LSTM) networks were discovered by Hochreiter and Schmidhuber in 1997 and set accuracy records in multiple applications domains.

Around 2007, LSTM started to revolutionize speech recognition, outperforming traditional models in certain speech applications. In 2009, a Connectionist Temporal Classification (CTC)-trained LSTM network was the first RNN to win pattern recognition contests when it won several competitions in connected handwriting recognition. In 2014, the Chinese search giant Baidu used CTC-trained RNNs to break the Switchboard Hub5'00 speech recognition benchmark without using any traditional speech processing methods.

LSTM also improved large-vocabulary speech recognition and text-to-speech synthesis[1018] and was used in Google Android. In 2015, Google's speech recognition reportedly experienced a dramatic performance jump of 49%Wikipedia:Citation needed through CTC-trained LSTM, which was used by Google voice search.

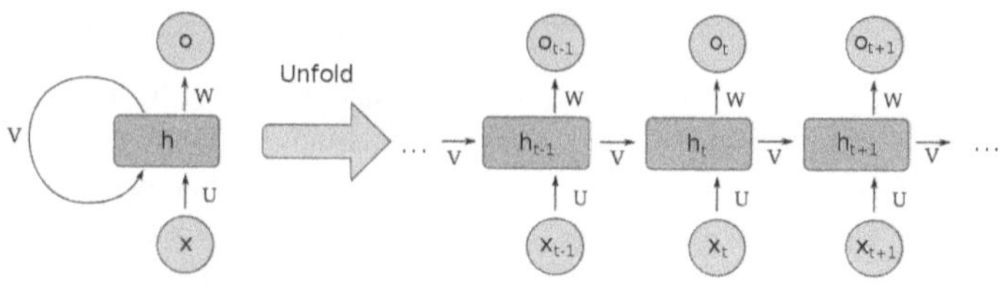

Figure 43: *Unfolded basic recurrent neural network*

LSTM broke records for improved machine translation, Language Modeling and Multilingual Language Processing. LSTM combined with convolutional neural networks (CNNs) improved automatic image captioning.

Architectures

RNNs come in many variants.

Fully recurrent

Basic RNNs are a network of neuron-like nodes organized into successive "layers", each node in a given layer with a directed (one-way) connection to every other node in the next successive layer.Wikipedia:Citation needed Each node (neuron) has a time-varying real-valued activation. Each connection (synapse) has a modifiable real-valued weight. Nodes are either input nodes (receiving data from outside the network), output nodes (yielding results), or hidden nodes (that modify the data *en route* from input to output).

For supervised learning in discrete time settings, sequences of real-valued input vectors arrive at the input nodes, one vector at a time. At any given time step, each non-input unit computes its current activation (result) as a nonlinear function of the weighted sum of the activations of all units that connect to it. Supervisor-given target activations can be supplied for some output units at certain time steps. For example, if the input sequence is a speech signal corresponding to a spoken digit, the final target output at the end of the sequence may be a label classifying the digit.

In reinforcement learning settings, no teacher provides target signals. Instead a fitness function or reward function is occasionally used to evaluate the RNN's performance, which influences its input stream through output units connected to actuators that affect the environment. This might be used to play a game in which progress is measured with the number of points won.

Each sequence produces an error as the sum of the deviations of all target signals from the corresponding activations computed by the network. For a training set of numerous sequences, the total error is the sum of the errors of all individual sequences.

Independently recurrent (IndRNN)

Independently recurrent neural network (IndRNN) solves the gradient vanishing and exploding problems in the traditional fully connected RNN. Each neuron in one layer only receives its own past state as context information and thus is independent of each other. The gradient backpropagation can be easily regulated to avoid gradient vanishing and exploding while keeping long-term memory as well as memory of any range. The cross-neuron information is explored in the next layers. IndRNN can be robustly trained together with the popular non-saturated nonlinear functions such as ReLU. It can also learn very long dependencies and construct very deep networks.

Recursive

A recursive neural network is created by applying the same set of weights recursively over a differentiable graph-like structure by traversing the structure in topological order. Such networks are typically also trained by the reverse mode of automatic differentiation.[1019] They can process distributed representations of structure, such as logical terms. A special case of recursive neural networks is the RNN whose structure corresponds to a linear chain. Recursive neural networks have been applied to natural language processing. The Recursive Neural Tensor Network uses a tensor-based composition function for all nodes in the tree.

Hopfield

The Hopfield network is an RNN in which all connections are symmetric. It requires stationary inputs and is thus not a general RNN, as it does not process sequences of patterns. It guarantees that it will converge. If the connections are trained using Hebbian learning then the Hopfield network can perform as robust content-addressable memory, resistant to connection alteration.

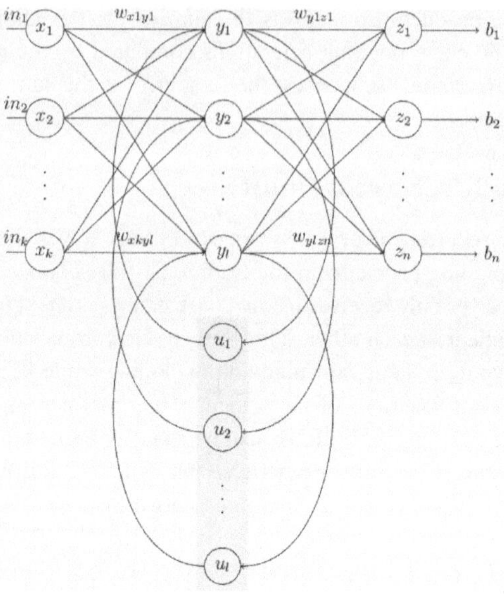

Figure 44: *The Elman network*

Bidirectional associative memory

Introduced by Bart Kosko, a bidirectional associative memory (BAM) network is a variant of a Hopfield network that stores associative data as a vector. The bi-directionality comes from passing information through a matrix and its transpose. Typically, bipolar encoding is preferred to binary encoding of the associative pairs. Recently, stochastic BAM models using Markov stepping were optimized for increased network stability and relevance to real-world applications.

A BAM network has two layers, either of which can be driven as an input to recall an association and produce an output on the other layer.

Elman networks and Jordan networks

An Elman network is a three-layer network (arranged horizontally as *x*, *y*, and *z* in the illustration) with the addition of a set of "context units" (*u* in the illustration). The middle (hidden) layer is connected to these context units fixed with a weight of one.[1020] At each time step, the input is fed-forward and a learning rule is applied. The fixed back-connections save a copy of the previous values of the hidden units in the context units (since they propagate over the connections before the learning rule is applied). Thus the network can maintain a sort of state, allowing it to perform such tasks as sequence-prediction that are beyond the power of a standard multilayer perceptron.

Jordan networks are similar to Elman networks. The context units are fed from the output layer instead of the hidden layer. The context units in a Jordan network are also referred to as the state layer. They have a recurrent connection to themselves.

Elman and Jordan networks are also known as "simple recurrent networks" (SRN).

Elman network

$$h_t = \sigma_h(W_h x_t + U_h h_{t-1} + b_h)$$
$$y_t = \sigma_y(W_y h_t + b_y)$$

Jordan network

$$h_t = \sigma_h(W_h x_t + U_h y_{t-1} + b_h)$$
$$y_t = \sigma_y(W_y h_t + b_y)$$

Variables and functions

- x_t : input vector
- h_t : hidden layer vector
- y_t : output vector
- W , U and b : parameter matrices and vector
- σ_h and σ_y : Activation functions

Echo state

The echo state network (ESN) has a sparsely connected random hidden layer. The weights of output neurons are the only part of the network that can change (be trained). ESNs are good at reproducing certain time series. A variant for spiking neurons is known as a liquid state machine.[1021]

Neural history compressor

The neural history compressor is an unsupervised stack of RNNs. At the input level, it learns to predict its next input from the previous inputs. Only unpredictable inputs of some RNN in the hierarchy become inputs to the next higher level RNN, which therefore recomputes its internal state only rarely. Each higher level RNN thus studies a compressed representation of the information in the RNN below. This is done such that the input sequence can be precisely reconstructed from the representation at the highest level.

The system effectively minimises the description length or the negative logarithm of the probability of the data. Given a lot of learnable predictability in the incoming data sequence, the highest level RNN can use supervised learning

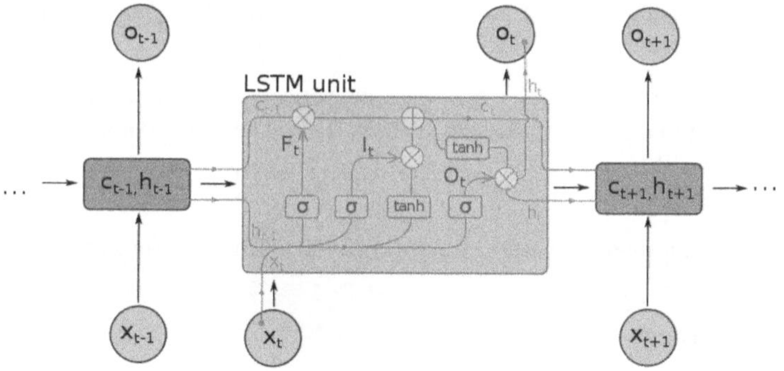

Figure 45: *Long short-term memory unit*

to easily classify even deep sequences with long intervals between important events.

It is possible to distill the RNN hierarchy into two RNNs: the "conscious" chunker (higher level) and the "subconscious" automatizer (lower level). Once the chunker has learned to predict and compress inputs that are unpredictable by the automatizer, then the automatizer can be forced in the next learning phase to predict or imitate through additional units the hidden units of the more slowly changing chunker. This makes it easy for the automatizer to learn appropriate, rarely changing memories across long intervals. In turn this helps the automatizer to make many of its once unpredictable inputs predictable, such that the chunker can focus on the remaining unpredictable events.

A generative model partially overcame the vanishing gradient problem[1022] of automatic differentiation or backpropagation in neural networks in 1992. In 1993, such a system solved a "Very Deep Learning" task that required more than 1000 subsequent layers in an RNN unfolded in time.

Long short-term memory

Long short-term memory (LSTM) is a deep learning system that avoids the vanishing gradient problem. LSTM is normally augmented by recurrent gates called "forget" gates. LSTM prevents backpropagated errors from vanishing or exploding. Instead, errors can flow backwards through unlimited numbers of virtual layers unfolded in space. That is, LSTM can learn tasks that require memories of events that happened thousands or even millions of discrete time steps earlier. Problem-specific LSTM-like topologies can be evolved. LSTM works even given long delays between significant events and can handle signals that mix low and high frequency components.

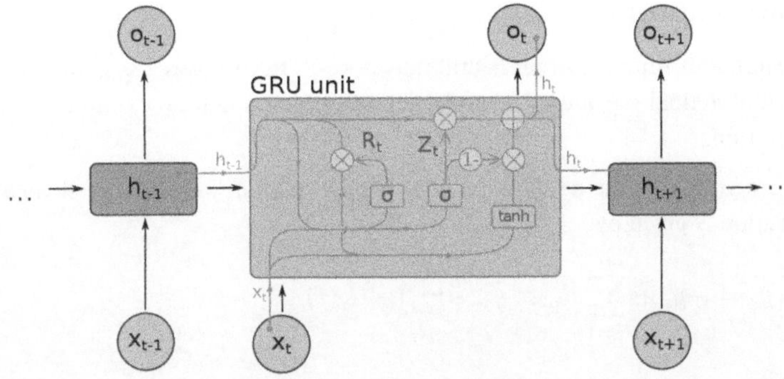

Figure 46: *Gated recurrent unit*

Many applications use stacks of LSTM RNNs and train them by Connectionist Temporal Classification (CTC) to find an RNN weight matrix that maximizes the probability of the label sequences in a training set, given the corresponding input sequences. CTC achieves both alignment and recognition.

LSTM can learn to recognize context-sensitive languages unlike previous models based on hidden Markov models (HMM) and similar concepts.

Second order RNNs

Second order RNNs use higher order weights w_{ijk} instead of the standard w_{ij} weights, and inputs and states can be a product. This allows a direct mapping to a finite state machine both in training, stability, and representation.[1023,1024] Long short-term memory is an example of this but has no such formal mappings or proof of stability.

Gated recurrent unit

Gated recurrent units (GRUs) are a gating mechanism in recurrent neural networks introduced in 2014. They are used in the full form and several simplified variants. Their performance on polyphonic music modeling and speech signal modeling was found to be similar to that of long short-term memory. They have fewer parameters than LSTM, as they lack an output gate.

Bi-directional

Bi-directional RNNs use a finite sequence to predict or label each element of the sequence based on the element's past and future contexts. This is done by concatenating the outputs of two RNNs, one processing the sequence from left to right, the other one from right to left. The combined outputs are the predictions of the teacher-given target signals. This technique proved to be especially useful when combined with LSTM RNNs.

Continuous-time

A continuous time recurrent neural network (CTRNN) uses a system of ordinary differential equations to model the effects on a neuron of the incoming spike train.

For a neuron i in the network with action potential y_i , the rate of change of activation is given by:

$$\tau_i \dot{y}_i = -y_i + \sum_{j=1}^{n} w_{ji} \sigma(y_j - \Theta_j) + I_i(t)$$

Where:

- τ_i : Time constant of postsynaptic node
- y_i : Activation of postsynaptic node
- $\dot{y}_i$: Rate of change of activation of postsynaptic node
- w_{ji} : Weight of connection from pre to postsynaptic node
- $\sigma(x)$: Sigmoid of x e.g. $\sigma(x) = 1/(1 + e^{-x})$.
- y_j : Activation of presynaptic node
- Θ_j : Bias of presynaptic node
- $I_i(t)$: Input (if any) to node

CTRNNs have been applied to evolutionary robotics where they have been used to address vision, co-operation, and minimal cognitive behaviour.

Note that, by the Shannon sampling theorem, discrete time recurrent neural networks can be viewed as continuous-time recurrent neural networks where the differential equations have transformed into equivalent difference equations. This transformation can be thought of as occurring after the post-synaptic node activation functions $y_i(t)$ have been low-pass filtered but prior to sampling.

Hierarchical

Hierarchical RNNs connect their neurons in various ways to decompose hierarchical behavior into useful subprograms.

Recurrent multilayer perceptron network

Generally, a Recurrent Multi-Layer Perceptron (RMLP) network consists of cascaded subnetworks, each of which contains multiple layers of nodes. Each of these subnetworks is feed-forward except for the last layer, which can have feedback connections. Each of these subnets is connected only by feed forward connections.

Multiple timescales model

A multiple timescales recurrent neural network (MTRNN) is a neural-based computational model that can simulate the functional hierarchy of the brain through self-organization that depends on spatial connection between neurons and on distinct types of neuron activities, each with distinct time properties. With such varied neuronal activities, continuous sequences of any set of behaviors are segmented into reusable primitives, which in turn are flexibly integrated into diverse sequential behaviors. The biological approval of such a type of hierarchy was discussed in the memory-prediction theory of brain function by Hawkins in his book *On Intelligence*.Wikipedia:Citation needed

Neural Turing machines

Neural Turing machines (NTMs) are a method of extending recurrent neural networks by coupling them to external memory resources which they can interact with by attentional processes. The combined system is analogous to a Turing machine or Von Neumann architecture but is differentiable end-to-end, allowing it to be efficiently trained with gradient descent.

Differentiable neural computer

Differentiable neural computers (DNCs) are an extension of Neural Turing machines, allowing for usage of fuzzy amounts of each memory address and a record of chronology.

Neural network pushdown automata

Neural network pushdown automata (NNPDA) are similar to NTMs, but tapes are replaced by analogue stacks that are differentiable and that are trained. In this way, they are similar in complexity to recognizers of context free grammars (CFGs).

Training

Gradient descent

Gradient descent is a first-order iterative optimization algorithm for finding the minimum of a function. In neural networks, it can be used to minimize the error term by changing each weight in proportion to the derivative of the error with respect to that weight, provided the non-linear activation functions are differentiable. Various methods for doing so were developed in the 1980s and early 1990s by Werbos, Williams, Robinson, Schmidhuber, Hochreiter, Pearlmutter and others.

The standard method is called "backpropagation through time" or BPTT, and is a generalization of back-propagation for feed-forward networks. Like that method, it is an instance of automatic differentiation in the reverse accumulation mode of Pontryagin's minimum principle. A more computationally expensive online variant is called "Real-Time Recurrent Learning" or RTRL, which is an instance of automatic differentiation in the forward accumulation mode with stacked tangent vectors. Unlike BPTT, this algorithm is local in time but not local in space.

In this context, local in space means that a unit's weight vector can be updated using only information stored in the connected units and the unit itself such that update complexity of a single unit is linear in the dimensionality of the weight vector. Local in time means that the updates take place continually (on-line) and depend only on the most recent time step rather than on multiple time steps within a given time horizon as in BPTT. Biological neural networks appear to be local with respect to both time and space.

For recursively computing the partial derivatives, RTRL has a time-complexity of O(number of hidden x number of weights) per time step for computing the Jacobian matrices, while BPTT only takes O(number of weights) per time step, at the cost of storing all forward activations within the given time horizon. An online hybrid between BPTT and RTRL with intermediate complexity exists, along with variants for continuous time.

A major problem with gradient descent for standard RNN architectures is that error gradients vanish exponentially quickly with the size of the time lag between important events. LSTM combined with a BPTT/RTRL hybrid learning method attempts to overcome these problems. This problem is also solved in the independently recurrent neural network (IndRNN) by reducing the context of a neuron to its own past state and the cross-neuron information can then be explored in the following layers. Memories of different range including long-term memory can be learned without the gradient vanishing and exploding problem.

The on-line algorithm called causal recursive backpropagation (CRBP), implements and combines BPTT and RTRL paradigms for locally recurrent networks. It works with the most general locally recurrent networks. The CRBP algorithm can minimize the global error term. This fact improves stability of the algorithm, providing a unifying view on gradient calculation techniques for recurrent networks with local feedback.

One approach to the computation of gradient information in RNNs with arbitrary architectures is based on signal-flow graphs diagrammatic derivation. It uses the BPTT batch algorithm, based on Lee's theorem for network sensitivity calculations. It was proposed by Wan and Beaufays, while its fast online version was proposed by Campolucci, Uncini and Piazza.

Global optimization methods

Training the weights in a neural network can be modeled as a non-linear global optimization problem. A target function can be formed to evaluate the fitness or error of a particular weight vector as follows: First, the weights in the network are set according to the weight vector. Next, the network is evaluated against the training sequence. Typically, the sum-squared-difference between the predictions and the target values specified in the training sequence is used to represent the error of the current weight vector. Arbitrary global optimization techniques may then be used to minimize this target function.

The most common global optimization method for training RNNs is genetic algorithms, especially in unstructured networks.

Initially, the genetic algorithm is encoded with the neural network weights in a predefined manner where one gene in the chromosome represents one weight link.The whole network is represented as a single chromosome. The fitness function is evaluated as follows:

- Each weight encoded in the chromosome is assigned to the respective weight link of the network.
- The training set is presented to the network which propagates the input signals forward.
- The mean-squared-error is returned to the fitness function.
- This function drives the genetic selection process.

Many chromosomes make up the population; therefore, many different neural networks are evolved until a stopping criterion is satisfied. A common stopping scheme is:

- When the neural network has learnt a certain percentage of the training data or
- When the minimum value of the mean-squared-error is satisfied or
- When the maximum number of training generations has been reached.

The stopping criterion is evaluated by the fitness function as it gets the reciprocal of the mean-squared-error from each network during training. Therefore, the goal of the genetic algorithm is to maximize the fitness function, reducing the mean-squared-error.

Other global (and/or evolutionary) optimization techniques may be used to seek a good set of weights, such as simulated annealing or particle swarm optimization.

Related fields and models

RNNs may behave chaotically. In such cases, dynamical systems theory may be used for analysis.

They are in fact recursive neural networks with a particular structure: that of a linear chain. Whereas recursive neural networks operate on any hierarchical structure, combining child representations into parent representations, recurrent neural networks operate on the linear progression of time, combining the previous time step and a hidden representation into the representation for the current time step.

In particular, RNNs can appear as nonlinear versions of finite impulse response and infinite impulse response filters and also as a nonlinear autoregressive exogenous model (NARX).

Libraries

- pytorch[1025]: Tensors and Dynamic neural networks in Python with strong GPU acceleration.
- Apache Singa
- Caffe[1026]: Created by the Berkeley Vision and Learning Center (BVLC). It supports both CPU and GPU. Developed in C++, and has Python and MATLAB wrappers.
- Deeplearning4j: Deep learning in Java and Scala on multi-GPU-enabled Spark. A general-purpose deep learning library[1027] for the JVM production stack running on a C++ scientific computing engine[1028]. Allows the creation of custom layers. Integrates with Hadoop and Kafka.
- Dynet[1029]: The Dynamic Neural Networks toolkit.
- Keras
- Microsoft Cognitive Toolkit
- TensorFlow: Apache 2.0-licensed Theano-like library with support for CPU, GPU and Google's proprietary TPU, mobile
- Theano: The reference deep-learning library for Python with an API largely compatible with the popular NumPy library. Allows user to write symbolic mathematical expressions, then automatically generates their derivatives, saving the user from having to code gradients or backpropagation. These symbolic expressions are automatically compiled to CUDA code for a fast, on-the-GPU implementation.
- Torch (www.torch.ch[1030]): A scientific computing framework with wide support for machine learning algorithms, written in C and lua. The main author is Ronan Collobert, and it is now used at Facebook AI Research and Twitter.

- MXNet: a modern open-source deep learning framework used to train and deploy deep neural networks.

Applications

Applications of Recurrent Neural Networks include:

- Robot control
- Time series prediction
- Speech recognition
- Rhythm learning
- Music composition
- Grammar learning
- Handwriting recognition[1031]
- Human action recognition[1032]
- Protein Homology Detection
- Predicting subcellular localization of proteins
- Several prediction tasks in the area of business process management
- Prediction in medical care pathways

References

<templatestyles src="Template:Refbegin/styles.css" />

- Mandic, D. & Chambers, J. (2001). *Recurrent Neural Networks for Prediction: Learning Algorithms, Architectures and Stability*. Wiley. ISBN 0-471-49517-4.

External links

- RNNSharp[1033] CRFs based on recurrent neural networks (C#, .NET)
- Recurrent Neural Networks[1034] with over 60 RNN papers by Jürgen Schmidhuber's group at Dalle Molle Institute for Artificial Intelligence Research
- Elman Neural Network implementation[1035] for WEKA
- Recurrent Neural Nets & LSTMs in Java[1036]
- an alternative try for complete RNN / Reward driven[1037]

Progress in artificial intelligence

Artificial intelligence
Major goals
Knowledge reasoning
Planning
Machine learning
Natural language processing
Computer vision
Robotics
Artificial general intelligence
Approaches
Symbolic
Deep learning
Bayesian networks
Evolutionary algorithms
Philosophy
Ethics
Existential risk
Turing test
Chinese room
Friendly AI
History
Timeline
Progress
AI winter
Technology
Applications
Projects
Programming languages
Glossary
Glossary

Artificial intelligence applications have been used in a wide range of fields including medical diagnosis, stock trading, robot control, law, scientific discovery and toys. However, many AI applications are not perceived as AI: "A lot of cutting edge AI has filtered into general applications, often without being called

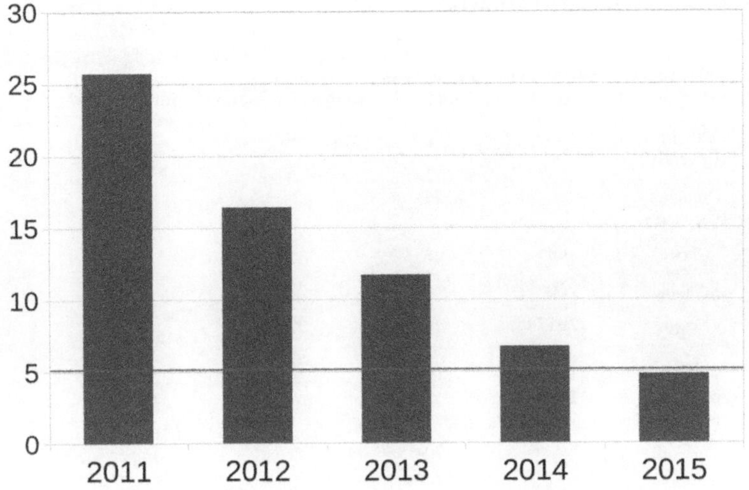

Figure 47: *Progress in machine classification of images*

The error rate of AI by year. Red line - the error rate of a trained human on a particular task

AI because once something becomes useful enough and common enough it's not labeled AI anymore."[1038] "Many thousands of AI applications are deeply embedded in the infrastructure of every industry." In the late 1990s and early 21st century, AI technology became widely used as elements of larger systems,[1039] but the field is rarely credited for these successes.

To allow comparison with human performance, artificial intelligence can be evaluated on constrained and well-defined problems. Such tests have been termed subject matter expert Turing tests. Also, smaller problems provide more achievable goals and there are an ever-increasing number of positive results.

Current performance

Game	Champion year[1040]	Legal states ($\log_{10}$)	Game tree complexity ($\log_{10}$)	Game of perfect information?	Ref
Othello (reversi)	1997	28	58	Perfect	
Draughts (checkers)	1994	21	31	Perfect	
Chess	1997	46	123	Perfect	
Scrabble	2006				
Shogi	2017	71	226	Perfect	
Go	2016	172	360	Perfect	
2p no-limit hold 'em	2017			Imperfect	
StarCraft	-	270+		Imperfect	

In his famous Turing test, Alan Turing picked language, the defining feature of human beings, for its basis. Yet, there are many other useful abilities that can be described as showing some form of intelligence. This gives better insight into the comparative success of artificial intelligence in different areas.

In what has been called the Feigenbaum test, the inventor of expert systems argued for subject specific expert tests. A paper by Jim Gray of Microsoft in 2003 suggested extending the Turing test to speech understanding, speaking and recognizing objects and behavior.

AI, like electricity or the steam engine, is a general purpose technology. There is no consensus on how to characterize which tasks AI tends to excel at. Some versions of Moravec's paradox observe that humans are more likely to outperform machines in areas such as physical dexterity that have been the direct target of natural selection. While projects such as AlphaZero have succeeded in generating their own knowledge from scratch, many other machine learning projects require large training datasets. Researcher Andrew Ng has suggested, as a "highly imperfect rule of thumb", that "almost anything a typical human can do with less than one second of mental thought, we can probably now or in the near future automate using AI."

Games provide a high-profile benchmark for assessing rates of progress; many games have a large professional player base and a well-established competitive rating system. AlphaGo brought the era of classical board-game benchmarks to an close. Games of imperfect knowledge provide new challenges to AI in the area of game theory; the most prominent milestone in this area was brought to a close by Libratus' poker victory in 2017. E-sports continue to provide additional benchmarks; Facebook AI, Deepmind, and others have engaged with the popular StarCraft franchise of videogames.

Broad classes of outcome for an AI test may be given as:

- **optimal**: it is not possible to perform better (note: some of these entries were solved by humans)
- **super-human**: performs better than all humans
- **high-human**: performs better than most humans
- **par-human**: performs similarly to most humans
- **sub-human**: performs worse than most humans

Optimal

- Tic-Tac-Toe
- Connect Four: 1988
- Checkers (aka 8x8 draughts): Weakly solved (2007)
- Rubik's Cube: Mostly solved (2010)
- Heads-up limit hold'em poker: Statistically optimal in the sense that "a human lifetime of play is not sufficient to establish with statistical significance that the strategy is not an exact solution" (2015)

Super-human

- Othello (aka reversi): c. 1997
- Scrabble: 2006
- Backgammon: c. 1995-2002
- Chess: Supercomputer (c. 1997); Personal computer (c. 2006); Mobile phone (c. 2009); Computer defeats human + computer (c. 2017)
- Jeopardy: Question answering, although the machine did not use speech recognition (2011)[1041]
- Shogi: c. 2017
- Arimaa: 2015
- Go: 2017
- Heads-up no-limit hold'em poker: 2017

High-human

- Bridge: World class (c. 2006)[1042]
- Crosswords: c. 2012[1043]
- Dota 2: 2018
- Safely driving a car: In 2017, Waymo successfully removed safety drivers from geo-fenced Arizona road tests. Since November 2017 Waymo has driven over 5,000,000 miles (8,000,000 km) and not caused any accidents. It is increasing over 1,000,000 miles (1,600,000 km) per month.

Par-human

- Optical character recognition for ISO 1073-1:1976 and similar special characters.
- Classification of images
- Handwriting recognition

Sub-human

- Optical character recognition for printed text (nearing par-human for Latin-script typewritten text)
- Object recognitionWikipedia:Please clarify
- Facial recognition: Low to mid human accuracy (as of 2014)
- Visual question answering,[1044] such as the VQA 1.0
- Various robotics tasks that may require advances in robot hardware as well as AI, including:
  - Stable bipedal locomotion: Bipedal robots can walk, but are less stable than human walkers (as of 2017)
  - Humanoid soccer
- Speech recognition: "nearly equal to human performance" (2017)
- Explainability. Current medical systems can diagnose certain medical conditions well, but cannot explain to users why they made the diagnosis.[1045]
- StarCraft
- Various tasks that are difficult to solve without contextual knowledge, including:
  - Translation[1046]
  - Word-sense disambiguation
  - Natural language processing

Past and current predictions

An expert poll around 2016, conducted by Katja Grace of the Future of Humanity Institute and associates, gave median estimates of 3 years for championship Angry Birds, 4 years for the World Series of Poker, and 6 years for StarCraft. On more subjective tasks, the poll gave 6 years for folding laundry as well as an average human worker, 7–10 years for expertly answering 'easily Googleable' questions, 8 years for average speech transcription, 9 years for average telephone banking, and 11 years for expert songwriting, but over 30 years for writing a New York Times bestseller or winning the Putnam math competition.

Figure 48: *Deep Blue at the Computer History Museum*

Chess

An AI defeated a grandmaster in a regulation tournament game for the first time in 1988; rebranded as Deep Blue, it beat the reigning human world chess champion in 1997.

Estimates when computers would exceed humans at Chess

Year prediction made	Predicted year	Number of Years	Predictor	Contempora-neous source
1957	1967 or sooner	10 or less	Herbert A. Simon, economist	
1990	2000 or sooner	10 or less	Ray Kurzweil, futurist	Age of Intelligent Machines

Go

AlphaGo defeated a European Go champion in October 2015, and defeated one of the world's top players, Lee Sedol, in March 2016. According to Scientific American and other sources, most observers had expected superhuman Computer Go performance to be at least a decade away.

Estimates when computers would exceed humans at Go

Year prediction made	Predicted year	Number of years	Predictor	Affiliation	Contemporaneous source
1997	2100 or later	100 or more	Piet Hutt, physicist and Go fan	Institute for Advanced Study	New York Times
2007	2017 or sooner	10 or less	Feng-Hsiung Hsu, Deep Blue lead	Microsoft Research Asia	IEEE Spectrum
2014	2024	10	Rémi Coulom, Computer Go programmer	CrazyStone	Wired

Human-level artificial general intelligence (AGI)

AI pioneer and economist Herbert A. Simon inaccurately predicted in 1965: "Machines will be capable, within twenty years, of doing any work a man can do". Similarly, in 1970 Marvin Minsky wrote that "Within a generation... the problem of creating artificial intelligence will substantially be solved."

Four polls conducted in 2012 and 2013 suggested that the median estimate among experts for when AGI would arrive was 2040 to 2050, depending on the poll.[1047]

The Grace poll around 2016 found results varied depending on how the question was framed. Respondents asked to estimate "when unaided machines can accomplish every task better and more cheaply than human workers" gave an aggregated median answer of 45 years and a 10% chance of it occurring within 9 years. Other respondents asked to estimate "when all occupations are fully automatable. That is, when for any occupation, machines could be built to carry out the task better and more cheaply than human workers" estimated a median of 122 years and a 10% probability of 20 years. The median response for when "AI researcher" could be fully automated was around 90 years. No link was found between seniority and optimism, but Asian researchers were much more optimistic than North American researchers on average; Asians predicted 30 years on average for "accomplish every task", compared with the 74 years predicted by North Americans.[1048]

Estimates of when AGI will arrive

Year prediction made	Predicted year	Number of years	Predictor	Contemporaneous source
1965	1985 or sooner	20 or less	Herbert A. Simon	*The shape of automation for men and management*[1049]
1993	2023 or sooner	30 or less	Vernor Vinge, science fiction writer	"The Coming Technological Singularity"
1995	2040 or sooner	45 or less	Hans Moravec, robotics researcher	Wired
2008	Never		Gordon E. Moore, inventor of Moore's Law	IEEE Spectrum
2017	2029	12	Ray Kurzweil	Interview

External links

- MIRI database of predictions about AGI[1050]

Applications

Applications of artificial intelligence

Artificial intelligence
Major goals
Knowledge reasoning
Planning
Machine learning
Natural language processing
Computer vision
Robotics
Artificial general intelligence
Approaches
Symbolic
Deep learning
Bayesian networks
Evolutionary algorithms
Philosophy
Ethics
Existential risk
Turing test
Chinese room
Friendly AI
History
Timeline
Progress
AI winter

Technology
Applications
Projects
Programming languages
Glossary
Glossary

Artificial intelligence, defined as intelligence exhibited by machines, has many applications in today's society. More specifically, it is Weak AI, the form of A.I. where programs are developed to perform specific tasks, that is being utilized for a wide range of activities including medical diagnosis, electronic trading, robot control, and remote sensing. AI has been used to develop and advance numerous fields and industries, including finance, healthcare, education, transportation, and more.

AI for Good

AI for Good is a movement in which institutions are employing AI to tackle some of the world's greatest economic and social challenges. For example, the University of Southern California launched the Center for Artificial Intelligence in Society, with the goal of using AI to address socially relevant problems such as homelessness. At Stanford, researchers are using AI to analyze satellite images to identify which areas have the highest poverty levels.[1051]

Aviation

The Air Operations Division (AOD) uses AI for the rule based expert systems. The AOD has use for artificial intelligence for surrogate operators for combat and training simulators, mission management aids, support systems for tactical decision making, and post processing of the simulator data into symbolic summaries.

The use of artificial intelligence in simulators is proving to be very useful for the AOD. Airplane simulators are using artificial intelligence in order to process the data taken from simulated flights. Other than simulated flying, there is also simulated aircraft warfare. The computers are able to come up with the best success scenarios in these situations. The computers can also create strategies based on the placement, size, speed and strength of the forces and counter forces. Pilots may be given assistance in the air during combat by computers. The artificial intelligent programs can sort the information and provide the pilot with the best possible maneuvers, not to mention getting rid of certain maneuvers that would be impossible for a human being to perform. Multiple aircraft are needed to get good approximations for some calculations so computer simulated pilots are used to gather data. These computer simulated pilots are also used to train future air traffic controllers.

The system used by the AOD in order to measure performance was the Interactive Fault Diagnosis and Isolation System, or IFDIS. It is a rule based expert system put together by collecting information from TF-30 documents and the expert advice from mechanics that work on the TF-30. This system was designed to be used for the development of the TF-30 for the RAAF F-111C. The performance system was also used to replace specialized workers. The system allowed the regular workers to communicate with the system and avoid mistakes, miscalculations, or having to speak to one of the specialized workers.

The AOD also uses artificial intelligence in speech recognition software. The air traffic controllers are giving directions to the artificial pilots and the AOD wants to the pilots to respond to the ATC's with simple responses. The programs that incorporate the speech software must be trained, which means they use neural networks. The program used, the Verbex 7000, is still a very early program that has plenty of room for improvement. The improvements are imperative because ATCs use very specific dialog and the software needs to be able to communicate correctly and promptly every time.

The Artificial Intelligence supported Design of Aircraft,[1052] or AIDA, is used to help designers in the process of creating conceptual designs of aircraft. This program allows the designers to focus more on the design itself and less on the design process. The software also allows the user to focus less on the software tools. The AIDA uses rule based systems to compute its data. This is a diagram of the arrangement of the AIDA modules. Although simple, the program is proving effective.

In 2003, NASA's Dryden Flight Research Center, and many other companies, created software that could enable a damaged aircraft to continue flight until a safe landing zone can be reached.[1053] The software compensates for all the damaged components by relying on the undamaged components. The neural network used in the software proved to be effective and marked a triumph for artificial intelligence.

The Integrated Vehicle Health Management system, also used by NASA, on board an aircraft must process and interpret data taken from the various sensors on the aircraft. The system needs to be able to determine the structural integrity of the aircraft. The system also needs to implement protocols in case of any damage taken the vehicle.

Haitham Baomar and Peter Bentley are leading a team from the University College of London to develop an artificial intelligence based Intelligent Autopilot System (IAS) designed to teach an autopilot system to behave like a highly experienced pilot who is faced with an emergency situation such as severe weather, turbulence, or system failure. Educating the autopilot relies on the concept of supervised machine learning "which treats the young autopilot as a human apprentice going to a flying school". The autopilot records the actions of the human pilot generating learning models using artificial neural networks. The autopilot is then given full control and observed by the pilot as it executes the training exercise.

The Intelligent Autopilot System combines the principles of Apprenticeship Learning and Behavioral Cloning[1054] whereby the autopilot observes the low-level actions required to maneuver the airplane and the high-level strategy used to apply those actions. IAS implementation employs three phases; pilot data collection, training, and autonomous control. Baomar and Bentley's goal is to create a more autonomous autopilot to assist pilots in responding to emergency situations.

Computer science

AI researchers have created many tools to solve the most difficult problems in computer science. Many of their inventions have been adopted by mainstream computer science and are no longer considered a part of AI. (See AI effect.) According to Russell & Norvig (2003, p. 15), all of the following were originally developed in AI laboratories: time sharing, interactive interpreters, graphical user interfaces and the computer mouse, rapid development environments, the linked list data structure, automatic storage management, symbolic programming, functional programming, dynamic programming and object-oriented programming.

AI can be used to potentially determine the developer of anonymous binaries.Wikipedia:Citation needed

AI can be used to create other AI. For example, around November 2017, Google's AutoML project to evolve new neural net topologies created NASNet, a system optimized for ImageNet and COCO. According to Google, NASNet's performance exceeded all previously published ImageNet performance.

Education

There are a number of companies that create robots to teach subjects to children ranging from biology to computer science, though such tools have not become widespread yet. There have also been a rise of intelligent tutoring systems, or ITS, in higher education. For example, an ITS called SHERLOCK teaches Air Force technicians to diagnose electrical systems problems in aircraft. Another example is DARPA, Defense Advanced Research Projects Agency, which used AI to develop a digital tutor to train its Navy recruits in technical skills in a shorter amount of time. Universities have been slow in adopting AI technologies due to either a lack of funding or skepticism of the effectiveness of these tools, but in the coming years more classrooms will be utilizing technologies such as ITS to complement teachers.

Advancements in natural language processing, combined with machine learning, have also enabled automatic grading of assignments as well as a data-driven understanding of individual students' learning needs. This led to an explosion in popularity of MOOCs, or Massive Open Online Courses, which allows students from around the world to take classes online. Data sets collected from these large scale online learning systems have also enabled learning analytics, which will be used to improve the quality of learning at scale. Examples of how learning analytics can be used to improve the quality of learning include predicting which students are at risk of failure and analyzing student engagement.

Finance

Algorithmic trading

Algorithmic trading involves the use of complex AI systems to make trading decisions at speeds several orders of magnitudes greater than any human is capable of, often making millions of trades in a day without any human intervention. Automated trading systems are typically used by large institutional investors.

Market analysis and data mining

Several large financial institutions have invested in AI engines to assist with their investment practices. BlackRock's AI engine, Aladdin, is used both within the company and to clients to help with investment decisions. Its wide range of functionalities includes the use of natural language processing to read text such as news, broker reports, and social media feeds. It then gauges the sentiment on the companies mentioned and assigns a score. Banks such as

UBS and Deutsche Bank use an AI engine called Sqreem (Sequential Quantum Reduction and Extraction Model) which can mine data to develop consumer profiles and match them with the wealth management products they'd most likely want. Goldman Sachs uses Kensho[1055], a market analytics platform that combines statistical computing with big data and natural language processing. Its machine learning systems mine through hoards of data on the web and assess correlations between world events and their impact on asset prices. Information Extraction, part of artificial intelligence, is used to extract information from live news feed and to assist with investment decisions.[1056]

Personal finance

Several products are emerging that utilize AI to assist people with their personal finances. For example, Digit is an app powered by artificial intelligence that automatically helps consumers optimize their spending and savings based on their own personal habits and goals. The app can analyze factors such as monthly income, current balance, and spending habits, then make its own decisions and transfer money to the savings account. Wallet.AI, an upcoming startup in San Francisco, builds agents that analyze data that a consumer would leave behind, from Smartphone check-ins to tweets, to inform the consumer about their spending behavior.

Portfolio management

Robo-advisors are becoming more widely used in the investment management industry. Robo-advisors provide financial advice and portfolio management with minimal human intervention. This class of financial advisers work based on algorithms built to automatically develop a financial portfolio according to the investment goals and risk tolerance of the clients. It can adjust to real-time changes in the market and accordingly calibrate the portfolio.

Underwriting

An online lender, Upstart, analyze vast amounts of consumer data and utilizes machine learning algorithms to develop credit risk models that predict a consumer's likelihood of default. Their technology will be licensed to banks for them to leverage for their underwriting processes as well.

ZestFinance developed their Zest Automated Machine Learning (ZAML) Platform specifically for credit underwriting as well. This platform utilizes machine learning to analyze tens of thousands traditional and nontraditional variables (from purchase transactions to how a customer fills out a form) used in the credit industry to score borrowers. The platform is particularly useful to assign credit scores to those with limited credit histories, such as millennials.

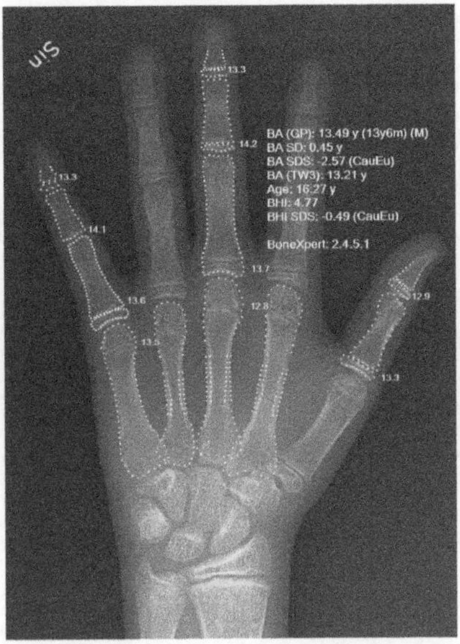

Figure 49: *X-ray of a hand, with automatic calculation of bone age by a computer software.*

Heavy industry

Robots have become common in many industries and are often given jobs that are considered dangerous to humans. Robots have proven effective in jobs that are very repetitive which may lead to mistakes or accidents due to a lapse in concentration and other jobs which humans may find degrading.

In 2014, China, Japan, the United States, the Republic of Korea and Germany together amounted to 70% of the total sales volume of robots. In the automotive industry, a sector with particularly high degree of automation, Japan had the highest density of industrial robots in the world: 1,414 per 10,000 employees.

Hospitals and medicine

Artificial neural networks are used as clinical decision support systems for medical diagnosis, such as in Concept Processing technology in EMR software.

Other tasks in medicine that can potentially be performed by artificial intelligence and are beginning to be developed include:

- Computer-aided interpretation of medical images. Such systems help scan digital images, *e.g.* from computed tomography, for typical appearances and to highlight conspicuous sections, such as possible diseases. A typical application is the detection of a tumor.
- Heart sound analysis
- Companion robots for the care of the elderly
- Mining medical records to provide more useful information.
- Design treatment plans.
- Assist in repetitive jobs including medication management.
- Provide consultations.
- Drug creation
- Using avatars in place of patients for clinical training
- Predict the likelihood of death from surgical procedures
- Predict HIV progression

Currently, there are over 90 AI startups in the health industry working in these fields.

IDx's first solution, IDx-DR, is the first autonomous, AI-based diagnostic system authorized for commercialization by the FDA[1057].

Human resources and recruiting

Another application of AI is in the human resources and recruiting space. There are three ways AI is being used by human resources and recruiting professionals. AI is used to screen resumes and rank candidates according to their level of qualification. Ai is also used to predict candidate success in given roles through job matching platforms. And now, AI is rolling out recruiting chat bots that can automate repetitive communication tasks.

Typically, resume screening involves a recruiter or other HR professional scanning through a database of resumes. Now startups like Pomato, are creating machine learning algorithms to automate resume screening processes. Pomato's resume screening AI focuses on automating validating technical applicants for technical staffing firms. Pomato'[1058] s AI performs over 200,000 computations on each resume in seconds then designs a custom technical interview based on the mined skills.

From 2016 to 2017, consumer goods company Unilever used artificial intelligence[1059] to screen all entry level employees. Unilever's AI used neuroscience based games, recorded interviews, and facial/speech analysis to predict hiring success. Unilever partnered with Pymetrics and HireVue[1060] to enable its novel AI based screening and increased their applicants from 15,000 to 30,000 in a single year. Recruiting with AI also produced Unililever's "most diverse class to date.' Unilever also decreased time to hire from 4 months to 4 weeks and saved over 50,000 hours of recruiter time.

From resume screening to neuroscience, speech recognition, and facial analysis...it's clear AI is having a massive impact on the human resources field. Yet another development in AI is in recruiting chatbots. TextRecruit, a Bay Area startup, released Ari (automated recruiting interface.) Ari is a recruiting chatbot that is designed to hold two-way text message conversations with candidates. Ari automates posting jobs, advertising openings, screening candidates, scheduling interviews, and nurturing candidate relationships with updates as they progress along the hiring funnel. Ari is currently offered as part of TextRecruit's candidate engagement platform.

Media

Some AI applications are geared towards the analysis of audiovisual media content such as movies, TV programs, advertisement videos or user-generated content. The solutions often involve computer vision, which is a major application area of AI.

Typical use case scenarios include the analysis of images using object recognition or face recognition techniques, or the analysis of video for recognizing relevant scenes, objects or faces. The motivation for using AI-based media analysis can be — among other things — the facilitation of media search, the creation of a set of descriptive keywords for a media item, media content policy monitoring (such as verifying the suitability of content for a particular TV viewing time), speech to text for archival or other purposes, and the detection of logos, products or celebrity faces for the placement of relevant advertisements.

Media analysis AI companies often provide their services over a REST API that enables machine-based automatic access to the technology and allows machine-reading of the results. For example, IBM, Microsoft, Amazon and the video AI company Valossa allow access to their media recognition technology by using RESTful APIs.

Music

While the evolution of music has always been affected by technology, artificial intelligence has enabled, through scientific advances, to emulate, at some extent, human-like composition.

Among notable early efforts, David Cope created an AI called Emily Howell that managed to become well known in the field of Algorithmic Computer Music. The algorithm behind Emily Howell is registered as a US patent.[1061]

The AI Iamus created 2012 the first complete classical album fully composed by a computer.

Other endeavours, like AIVA (Artificial Intelligence Virtual Artist), focus on composing symphonic music, mainly classical music for film scores. It achieved a world first by becoming the first virtual composer to be recognized by a musical professional association.[1062]

Artificial intelligences can even produce music usable in a medical setting, with Melomics's effort to use computer-generated music for stress and pain relief.

Moreover, initiatives such as Google Magenta, conducted by the Google Brain team, want to find out if an artificial intelligence can be capable of creating compelling art.

At Sony CSL Research Laboratory, their Flow Machines software has created pop songs by learning music styles from a huge database of songs. By analyzing unique combinations of styles and optimizing techniques, it can compose in any style.

Another artificial intelligence musical composition project, The Watson Beat, written by IBM Research, doesn't need a huge database of music like the Google Magenta and Flow Machines projects, since it uses Reinforcement Learning and Deep Belief Networks to compose music on a simple seed input melody and a select style. Since the software has been open sourced musicians, such as Taryn Southern have been collaborating with the project to create music.

News, publishing and writing

The company Narrative Science makes computer generated news and reports commercially available, including summarizing team sporting events based on statistical data from the game in English. It also creates financial reports and real estate analyses.[1063] Similarly, the company Automated Insights generates personalized recaps and previews for Yahoo Sports Fantasy Football. The company is projected to generate one billion stories in 2014, up from 350 million in 2013.

Echobox is a software company that helps publishers increase traffic by 'intelligently' posting articles on social media platforms such as Facebook and Twitter. By analysing large amounts of data, it learns how specific audiences respond to different articles at different times of the day. It then chooses the best stories to post and the best times to post them. It uses both historical and real-time data to understand to what has worked well in the past as well as what is currently trending on the web.

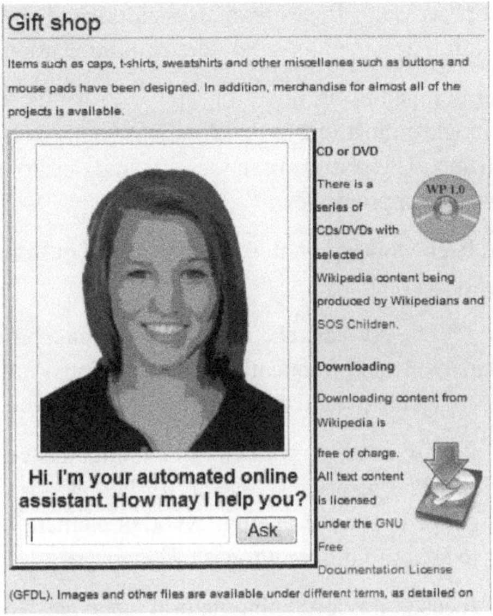

Figure 50: *An automated online assistant providing customer service on a web page.*

Another company, called Yseop[1064], uses artificial intelligence to turn structured data into intelligent comments and recommendations in natural language. Yseop is able to write financial reports, executive summaries, personalized sales or marketing documents and more at a speed of thousands of pages per second and in multiple languages including English, Spanish, French & German.

Boomtrain's is another example of AI that is designed to learn how to best engage each individual reader with the exact articles — sent through the right channel at the right time — that will be most relevant to the reader. It's like hiring a personal editor for each individual reader to curate the perfect reading experience.

There is also the possibility that AI will write work in the future. In 2016, a Japanese AI co-wrote a short story and almost won a literary prize.

Online and telephone customer service

Artificial intelligence is implemented in automated online assistants that can be seen as avatars on web pages.[1065] It can avail for enterprises to reduce their operation and training cost. A major underlying technology to such systems is natural language processing. Pypestream uses automated customer service for its mobile application designed to streamline communication with customers.

Currently, major companies are investing in AI to handle difficult customer in the future. Google's most recent development analyzes language and converts speech into text. The platform can identify angry customers through their language and respond appropriately.

Companies have been working on different aspects of customer service to improve this aspect of a company.

Digital Genius[1066], an AI start-up, researches the database of information (from past conversations and frequently asked questions) more efficiently and provide prompts to agents to help them resolve queries more efficiently.

IPSoft[1067] is creating technology with emotional intelligence to adapt the customer's interaction. The response is linked to the customer's tone, with the objective of being able to show empathy. Another element IPSoft is developing is the ability to adapt to different tones or languages.

Inbenta's[1068] is focused on developing natural language. In other words, on understanding the meaning behind what someone is asking and not just looking at the words used, using context and natural language processing. One customer service element Ibenta has already achieved is its ability to respond in bulk to email queries.

Telecommunications maintenance

Many telecommunications companies make use of heuristic search in the management of their workforces, for example BT Group has deployed heuristic search[1069] in a scheduling application that provides the work schedules of 20,000 engineers.

Toys and games

The 1990s saw some of the first attempts to mass-produce domestically aimed types of basic Artificial Intelligence for education, or leisure. This prospered greatly with the Digital Revolution, and helped introduce people, especially children, to a life of dealing with various types of Artificial Intelligence, specifically in the form of Tamagotchis and Giga Pets, iPod Touch, the Internet, and the first widely released robot, Furby. A mere year later an improved type of domestic robot was released in the form of Aibo, a robotic dog with intelligent features and autonomy.

Companies like Mattel have been creating an assortment of AI-enabled toys for kids as young as age three. Using proprietary AI engines and speech recognition tools, they are able to understand conversations, give intelligent responses and learn quickly.

AI has also been applied to video games, for example video game bots, which are designed to stand in as opponents where humans aren't available or desired.

Transportation

Fuzzy logic controllers have been developed for automatic gearboxes in automobiles. For example, the 2006 Audi TT, VW Touareg Wikipedia:Citation needed and VW Caravell feature the DSP transmission which utilizes Fuzzy Logic. A number of Škoda variants (Škoda Fabia) also currently include a Fuzzy Logic-based controller.

Today's cars now have AI-based driver assist features such as self-parking and advanced cruise controls. AI has been used to optimize traffic management applications, which in turn reduces wait times, energy use, and emissions by as much as 25 percent. In the future, fully autonomous cars will be developed. AI in transportation is expected to provide safe, efficient, and reliable transportation while minimizing the impact on the environment and communities. The major challenge to developing this AI is the fact that transportation systems are inherently complex systems involving a very large number of components and different parties, each having different and often conflicting objectives.

Other

Various tools of artificial intelligence are also being widely deployed in homeland security, speech and text recognition, data mining, and e-mail spam filtering. Applications are also being developed for gesture recognition (understanding of sign language by machines), individual voice recognition, global voice recognition (from a variety of people in a noisy room), facial expression recognition for interpretation of emotion and non verbal cues. Other applications are robot navigation, obstacle avoidance, and object recognition. Wikipedia:Citation needed

List of applications

Typical problems to which AI methods are applied

- Optical character recognition
- Handwriting recognition
- Speech recognition
- Face recognition
- Artificial creativity
- Computer vision, Virtual reality, and Image processing
- Photo and Video manipulation
- Diagnosis (artificial intelligence)
- Game theory and Strategic planning
- Game artificial intelligence and Computer game bot
- Natural language processing, Translation and Chatterbots
- Nonlinear control and Robotics

Other fields in which AI methods are implemented

- Artificial life
- Automated reasoning
- Automation
- Biologically inspired computing
- Concept mining
- Data mining
- Knowledge representation
- Semantic Web
- E-mail spam filtering
- Robotics
  - Behavior-based robotics
  - Cognitive
  - Cybernetics
  - Developmental robotics (Epigenetic)
  - Evolutionary robotics
- Hybrid intelligent system
- Intelligent agent
- Intelligent control
- Litigation

References

- Russell, Stuart J.; Norvig, Peter (2003). *Artificial Intelligence: A Modern Approach*[1070] (2nd ed.). Upper Saddle River, New Jersey: Prentice Hall. ISBN 0-13-790395-2.

- Kurzweil, Ray (2005). *The Singularity is Near: When Humans Transcend Biology*. New York: Viking. ISBN 978-0-670-03384-3.
- National Research Council (1999). "Developments in Artificial Intelligence". *Funding a Revolution: Government Support for Computing Research*. National Academy Press. ISBN 0-309-06278-0. OCLC 246584055[1071].
- Moghaddam, M. J., M. R. Soleymani, and M. A. Farsi. "Sequence planning for stamping operations in progressive dies." Journal of Intelligent Manufacturing(2013): 1-11.
- United States, National Science and Technology Council – Committee on Technology. Executive Office of the President. (2016). Preparing for the future of artificial intelligence.

External links

- AI applications at aitopics.org[1072]
- How AI can be applied in many fields[1073]

Lethal autonomous weapon

"Killer robot" redirects here. For articles dealing with the concept of robots and/or artificial intelligence killing or eradicating humans and/or other living beings, see Existential risk from artificial general intelligence, AI takeover, and grey goo

Lethal autonomous weapons (LAWs) are a type of autonomous military robot that can independently search and engage targets based on programmed constraints and descriptions. LAW are also called **lethal autonomous weapon systems** (LAWS), **lethal autonomous robots** (LAR), **robotic weapons**, or **killer robots**. LAWs may operate in the air, on land, on water, under water, or in space. The autonomy of current systems as of 2018[1074] is restricted in the sense that a human gives the final command to attack - though there are exceptions with certain "defensive" systems.

Being autonomous as weapon

Being "autonomous" have different meanings in different fields of study. In engineering it may refer to the machine's ability to produce without the operation of workers. In philosophy it may refer to an individual being morally independent. In political science it may refer to an area's capability of self-governing. In terms of military weapon development, the identification of a

Figure 51: *A BAE Systems Corax during flight testing*

weapon as autonomous is not as clear as in other areas. The specific standard entailed in the concept of being autonomous can vary hugely between different scholars, nations and organizations.

Scholars like Peter Asaro and Mark Gubrud are trying to set the threshold lower and judge more weapon system as autonomous. They believe that any weapon system that is capable of releasing a lethal force without the operation, decision or confirmation of a human supervisor can be deemed as autonomous. According to Gubrud, A weapon system operating partially or wholly without human intervention is considered autonomous. He argues that a weapon system does not need to be able to make decisions completely by itself in order to be called autonomous. Instead, it should be treated as autonomous as long as it actively involves in one or multiple parts of the "preparation process", from finding the target to finally firing.

Other organizations, however, are setting the standard of autonomous weapon system in a higher position. The Ministry of Defence (United Kingdom) defines autonomous weapon systems as "systems that are capable of understanding higher level intent and direction. From this understanding and its perception of its environment, such a system is able to take appropriate action to bring about a desired state. It is capable of deciding a course of action, from a number of alternatives, without depending on human oversight and control - such human engagement with the system may still be present, though. While the overall activity of an autonomous unmanned aircraft will be predictable, individual actions may not be."

As a result, the composition of treaty between states requires a commonly accepted labeling of what exactly constitutes an autonomous weapon.

Automatic defensive systems

The oldest automatically-triggered lethal weapon is the land mine, used since at least the 1600s, and naval mines, used since at least the 1700s. Anti-personnel mines are banned in many countries by the 1997 Ottawa Treaty, not including the United States, Russia, and much of Asia and the Middle East.

Some current examples of LAWs are automated "hardkill" active protection systems, such as a radar-guided gun to defend ships that have been in use since the 1970s (e.g. the US Phalanx CIWS). Such systems can autonomously identify *and attack* oncoming missiles, rockets, artillery fire, aircraft and surface vessels according to criteria set by the human operator. Similar systems exist for tanks, such as the Russian Arena, the Israeli Trophy, and the German AMAP-ADS. Several types of stationary sentry guns, which can fire at humans and vehicles, are used in South Korea and Israel. Many missile defense systems, such as Iron Dome, also have autonomous targeting capabilities. Automatic turrets installed on military vehicles are called remote weapon stations.

The main reason for not having a "human in the loop" in these systems is the need for rapid response. They have generally been used to protect personnel and installations against incoming projectiles.

Autonomous offensive systems

Systems with a higher degree of autonomy would include drones or unmanned combat aerial vehicles, e.g.: "The unarmed BAE Systems Taranis jet-propelled combat drone prototype may lead to a Future Offensive Air System that can autonomously search, identify and locate enemies but can only engage with a target when authorized by mission command. It can also defend itself against enemy aircraft" (Heyns 2013, §45). The Northrop Grumman X-47B drone can take off and land on aircraft carriers (demonstrated in 2014); it is set to be developed into an Unmanned Carrier-Launched Airborne Surveillance and Strike (UCLASS) system.

According to *The Economist*, as technology advances, future applications of unmanned undersea vehicles might include mine clearance, mine-laying, anti-submarine sensor networking in contested waters, patrolling with active sonar, resupplying manned submarines, and becoming low-cost missile platforms. In

2018 the U.S. Nuclear Posture Review alleged that Russia is developing a "new intercontinental, nuclear-armed, nuclear-powered, undersea autonomous torpedo" named "Status 6".

Russian Federation is actively developing artificially intelligent missiles,[1075] drones, unmanned vehicles, military robots and medic robots.[1076,1077,1078,1079]

Israeli Minister Ayoub Kara stated in 2017 that Israel is developing military robots as small as flies in order to assassinate leaders of Hezbollah and Hamas, and that these robots may be operational within as few as three years.[1080,1081]

China is planning to deploy armed, unmanned submarines equipped with AI by 2020.[1082]

Ethical and legal issues

Standard used in US Policy

Current US policy states: "Autonomous ... weapons systems shall be designed to allow commanders and operators to exercise appropriate levels of human judgment over the use of force." However, the policy requires that autonomous weapon systems that kill people or use kinetic force, selecting and engaging targets without further human intervention, be certified as compliant with "appropriate levels" and other standards, not that such weapon systems cannot meet these standards and are therefore forbidden. "Semi-autonomous" hunter-killers that autonomously identify and attack targets do not even require certification. Deputy Defense Secretary Robert Work said in 2016 that the Defense Department would "not delegate lethal authority to a machine to make a decision", but might need to reconsider this since "authoritarian regimes" may do so. In October 2016 President Barack Obama stated that early in his career he was wary of a future in which a US president making use of drone warfare could "carry on perpetual wars all over the world, and a lot of them covert, without any accountability or democratic debate".

Probable violations of ethics and international acts

Stuart Russell, professor of computer science from University of California, Berkeley stated the concern he has with LAWS is that it is unethical and inhumane. The main issue with this system is it is hard to distinguish between combatants and non-combatants.

There is concern (e.g. Sharkey 2012) about whether LAWs would violate International Humanitarian Law, especially the principle of distinction, which requires the ability to discriminate combatants from non-combatants, and the principle of proportionality, which requires that damage to civilians is proportional to the military aim. This concern is often invoked as a reason to ban "killer robots" altogether - but it is doubtful that this concern can be an argument against LAWs that do not violate International Humanitarian Law.

LAWs are said to blur the boundaries of who is responsible for a particular killing, but Thomas Simpson and Vincent Müller argue that they may make it easier to record who gave which command.

Campaigns on banning LAWs

The possibility of LAWs has generated significant debate, especially about the risk of "killer robots" roaming the earth - in the near or far future. The group Campaign to Stop Killer Robots formed in 2013. In July 2015, over 1,000 experts in artificial intelligence signed a letter warning of the threat of an arms race in military artificial intelligence and calling for a ban on autonomous weapons. The letter was presented in Buenos Aires at the 24th International Joint Conference on Artificial Intelligence (IJCAI-15) and was co-signed by Stephen Hawking, Elon Musk, Steve Wozniak, Noam Chomsky, Skype co-founder Jaan Tallinn and Google DeepMind co-founder Demis Hassabis, among others.

According to PAX fully automated weapons (FAWs) will lower the threshold of going to war as soldiers are removed from the battlefield and the public is distanced from experiencing war, giving politicians and other decision-makers more space in deciding when and how to go to war. They warn that once deployed, FAWs will make democratic control of war more difficult - something that author of *Kill Decision* - a novel on the topic - and IT specialist Daniel Suarez also warned about: according to him it might recentralize power into very few hands by requiring very few people to go to war.

There are websites protesting the development of LAWs by presenting a possible future that scares the viewers if such researches on applying artificial intelligence on designation of weapons continues. On these websites, news about ethical and legal issues are constantly updated for visitors to recap with recent news about international meetings and research articles concerning LAWs.

Further reading

- Heyns, Christof (2013), 'Report of the Special Rapporteur on extrajudicial, summary or arbitrary executions', UN General Assembly, Human Rights Council, 23 (3), A/HRC/23/47.
- Krishnan, Armin (2009), Killer robots: Legality and ethicality of autonomous weapons (Aldershot: Ashgate)

- Müller, Vincent C. (2016), 'Autonomous killer robots are probably good news'[1083], in Ezio Di Nucci and Filippo Santoni de Sio (eds.), Drones and responsibility: Legal, philosophical and socio-technical perspectives on the use of remotely controlled weapons, 67-81 (London: Ashgate).
- Stuart Russell[1084]'s campaign against LAWs
- Sharkey, Noel E (2012), 'Automating Warfare: lessons learned from the drones', Journal of Law, Information & Science, 21 (2).
- Simpson, Thomas W and Müller, Vincent C. (2016), 'Just war and robots' killings'[1085], The Philosophical Quarterly 66 (263), 302-22.
- Singer, Peter (2009), Wired for war: The robotics revolution and conflict in the 21st Century (New York: Penguin)
- US Department of Defense (2012), 'Directive 3000.09, Autonomy in weapon systems'. <2014 Killer Robots Policy Paper Final.docx>.
- US Department of Defense (2013), 'Unmanned Systems Integrated Road Map FY2013-2038'. <http://www.defense.gov/pubs/DOD-USRM-2013.pdf%3E.
- The Ethics of Autonomous Weapons Systems (2014) Seminar at UPenn <https://www.law.upenn.edu/institutes/cerl/conferences/ethicsofweapons/schedule-required-readings.php>

Unmanned combat aerial vehicle

An **unmanned combat aerial vehicle (UCAV)**, also known as a **combat drone** or simply a **drone**, is an unmanned aerial vehicle (UAV) that usually carries aircraft ordnance such as missiles and is used for drone strikes. Aircraft of this type have no onboard human pilot. These drones are usually under real-time human control, with varying levels of autonomy. They are used in drone strikes.

Equipment necessary for a human pilot (such as the cockpit, armor, ejection seat, flight controls, and environmental controls for pressure and oxygen) are not needed, as the operator runs the vehicle from a remote terminal, resulting in a lower weight and a smaller size than a manned aircraft.

While several nations possess and manufacture unarmed UAV, only the United States, Israel, China, Iran, Italy, India, Pakistan, Russia and Turkey are at present known to have manufactured operational UCAV as of December 2015.

Figure 52: *A U.S. Air Force MQ-9 Reaper during a training mission*

History

One of the earliest explorations of the concept of the combat drone was by Lee De Forest, an early inventor of radio devices, and U. A. Sanabria, a TV engineer. They presented their idea in an article in a 1940 publication of *Popular Mechanics*.[1086] The modern military drone as known today was the brainchild of John Stuart Foster Jr., a nuclear physicist and former head of the Lawrence Livermore National Laboratory (then called the Lawrence Radiation Laboratory). In 1971, Foster was a model airplane hobbyist and had the idea this hobby could be applied to building weapons. He drew up plans and by 1973 DARPA (Defense Advanced Research Projects Agency) built two prototypes called "Praeire" and "Calere". They were powered by a modified lawn-mower engine and could stay aloft for two hours while carrying 28-pounds of load.

In the 1973 Yom Kippur War, Israel used unarmed U.S. Ryan Firebee target drones to spur Egypt into firing its entire arsenal of anti-aircraft missiles. This mission was accomplished with no injuries to Israeli pilots, who soon exploited the depleted Egyptian defenses. In the late 1970s and 80s, Israel developed the Scout and the Pioneer, which represented a shift toward the lighter, glider-type model of UAV in use today. Israel pioneered the use of unmanned aerial vehicles (UAVs) for real-time surveillance, electronic warfare, and decoys. The images and radar decoying provided by these UAVs helped Israel to completely

Figure 53: *A BAE Raven during flight testing*

neutralize the Syrian air defenses in Operation Mole Cricket 19 at the start of the 1982 Lebanon War, resulting in no pilots downed.

In the late 1980s, Iran deployed a drone armed with six RPG-7 rounds in the Iran–Iraq War.

Impressed by Israel's success, the US quickly acquired a number of UAVs, and its Hunter and Pioneer systems are direct derivatives of Israeli models. The first 'UAV war' was the first Gulf War: according to a May 1991 Department of the Navy report: "At least one UAV was airborne at all times during Desert Storm." After the Gulf War successfully demonstrated their utility, global militaries invested widely in the domestic development of combat UAVs. The first "kill" by an American UAV was on October 7, 2001 in Kandahar.

In recent years, the U.S. has increased its use of drone strikes against targets in foreign countries and elsewhere as part of the War on Terror. In January 2014, it was estimated that 2,400 people have died from U.S. drone strikes in five years. In June 2015 the total death toll of U.S. drone strikes was estimated to exceed 6,000.

Figure 54: *Sunrise over the Gulf of Saros area in Turkey, as seen from Bayraktar's tail camera at 18,000 feet.*

Figure 55: *The EADS Barracuda on the Manching Air Base in Germany*

Figure 56: *Israeli IAI Harop at Paris Air Show 2013. Instead of separate warheads, the drone itself is the main munition.*

Future models

Overview

Note: Some of these are not aircraft prototypes but technology demonstrators (TD) that are not expected to enter service.

Name	Manufacturer(s)	Developing nation(s)
A-10PCAS	Elbit Systems	Israel
Sky-X (TD)	Alenia Aeronautica	Italy
Selex ES Falco	Selex ES	Italy
Eitan	Israel Aerospace Industries	Israel
Harop	Israel Aerospace Industries	Israel
AVIC 601-S	Shenyang Aircraft Design Institute	China
AURA	Defence Research and Development Organisation	India
Taranis (TD)	BAE Systems	United Kingdom
Dassault nEUROn (TD)	Dassault Aviation	France Sweden Italy Greece Spain Switzerland
Rustom	Defence Research and Development Organisation	India

Avenger	General Atomics Aeronautical Systems	United States
TERN	Northrop Grumman	United States
X-47A (TD)/B (TD)/C	Northrop Grumman	United States
Armstechno Dulo (TD)	Armstechno	Bulgaria
Bateleur (TD)	Denel Dynamics	South Africa
TATI Buraq		Tunisia
TATI Jinn		Tunisia
Anka-TP (SIHA)	Turkish Aerospace Industries	Turkey
Bayraktar TB2	Kale-Baykar	Turkey
Cardinal Mini Unmanned Aircraft System	National Chung-Shan Institute of Science and Technology	Taiwan
Albatross Tactical Unmanned Aircraft System	National Chung-Shan Institute of Science and Technology	Taiwan
Vestel Karayel	Vestel Savunma A.S.	Turkey
Burraq	NESCOM	Pakistan
GIDS Shahpar	GIDS	Pakistan
GIDS Uqab	NESCOM	Pakistan
Ra'd	Armed Forces of the Islamic Republic of Iran	Iran
Karrar	Iran	Iran
Sofreh Mahi	Iran Aircraft Manufacturing Industrial Company	Iran
Fotros	Iran Aircraft Manufacturing Industrial Company – Capable of long missions	Iran
Shahed 129	Shahed Aviation Industries Research Center	Iran
Barracuda (TD)	EADS	• Germany • Spain
Skat[1087]	Mikoyan	Russia
MQM-107-based flying bombs		North Korea

Various Chinese UCAV concepts have also materialized.Wikipedia:Citation needed WZ-2000, UCAV versions of the Xianglong high altitude are long endurance UAV. Also, dedicated UCAV's Shenyang's Dark Sword (Anjian), and also revealed at Zhuhai 2008 was a model of a stealth strike UCAV with forward swept wings, filling a similar niche to U.S. X-45 called the Warrior Eagle.

Figure 57: *BAE Taranis model, one of the largest design concepts*

BAE Taranis

Taranis is a British demonstrator program for unmanned combat air vehicle (UCAV) technology. It is part of the UK's Strategic Unmanned Air Vehicle (Experimental) programme (SUAV[E]). BAE describes Taranis's role in this context as following: "This £124m four year programme is part of the UK Government's Strategic Unmanned Air Vehicle Experiment (SUAVE) and will result in a UCAV demonstrator with fully integrated autonomous systems and low observable features." The Taranis demonstrator will have an MTOW (Maximum Takeoff Weight) of about 8000 kilograms and be of comparable size to the BAE Hawk – making it one of the world's largest UAVs. It will be stealthy, fast, and able to deploy a range of munitions over a number of targets, as well as being capable of defending itself against manned and other unmanned enemy aircraft. The first steel was cut in September 2007 and ground testing started in early 2009. The first flight of the Taranis took place in August 2013 in Woomera, Australia.[1088] The demonstrator will have two internal weapons bays. With the inclusion of "full autonomy" the intention is thus for this platform to be able to "think for itself" for a large part of the mission.Wikipedia:Citation needed

Figure 58: *J-UCAS Boeing X-45A UCAV technology demonstrator*

J-UCAS

- Boeing X-45 UCAV (TD)
- Northrop-Grumman X-47 Pegasus

Joint Unmanned Combat Air Systems, or **J-UCAS**, was the name for the joint U.S. Navy/U.S. Air Force unmanned combat air vehicle procurement project. J-UCAS was managed by DARPA, the Defense Advanced Research Projects Agency. In the 2006 Quadrennial Defense Review, the J-UCAS program was terminated.[1089] The program would have used stealth technologies and allowed UCAVs to be armed with precision-guided weapons such as Joint Direct Attack Munition (JDAM) or precision miniature munitions, such as the Small-Diameter Bomb, which are used to suppress enemy air defenses. Controllers could have used real-time data sources, including satellites, to plan for and respond to changes on and around the battlefield.

The program was later revitalized into UCAS-D, a United States Navy program designed to develop a carrier-based unmanned aircraft.[1090]

N-UCAS

UCAS-D and Northrop Grumman X-47B are the U.S. Navy-only successors to the J-UCAS, which was cancelled in 2006. Boeing is also working on the X-45N in this sector.

In a New Year 2011 editorial titled "China's Naval Ambitions", *The New York Times* reported that "[t]he Pentagon must accelerate efforts to make American naval forces in Asia less vulnerable to Chinese missile threats by giving them the means to project their deterrent power from further offshore. Cutting back purchases of the Navy's DDG-1000 destroyer (with its deficient missile defense system) was a first step. A bigger one would be to reduce the Navy's reliance on short-range manned strike aircraft like the F-18 and the F-35, in favor of the carrier-launched N-UCAS ...."[1091]

On 6 January 2011, the DOD announced that this would be one area of additional investment in the 2012 budget request.

USAF Hunter-Killer

- Scaled Composites Model 395
- Scaled Composites Model 396
- General Atomics MQ-9 Reaper (originally the Predator B)
- Aurora Flight Sciences/Israel Aircraft Industries Eagle/Heron 2
- Unnamed Lockheed Martin entry

The United States Air Force has shifted its UCAV program from medium-range tactical strike aircraft to long-range strategic bombers. The technology of the Long Range Strike program is based on the Lockheed Martin Polecat demonstrator.

Elbit Hermes 450

The Israeli Air Force, which operates a squadron of Hermes 450s out of Palmachim Airbase south of Tel Aviv, has adapted the Hermes 450 for use as an assault UAV, reportedly equipping it with two Hellfire missiles or, according to various sources, two Rafael-made missiles. According to Israeli, Palestinian, Lebanese and independent reports, the Israeli assault UAV has seen extensive service in the Gaza Strip and was used intensively in the Second Lebanon War. Israel has not denied this capability, but to date, its policy has been not to officially confirm it either.Wikipedia:Citation needed

Users

Countries with known operational armed drones:

- Azerbaijan – IAI Harop
- Botswana – Elbit Hermes 450
- Brazil – Elbit Hermes 450, IAI Heron
- Colombia – Elbit Hermes 450
- Chile – Elbit Hermes 900
- China – Guizhou WZ-2000, Chengdu Wing Loong I, CH-3, CH-4, CH-5

- Croatia – Elbit Hermes 450
- Cyprus – Elbit Hermes 450
- Russia - Mikoyan Skat, Korsar, Tupolev Voron
- Egypt – CAIG Wing Loong, CASC Rainbow
- France – EADS Harfang (based on the IAI Heron), SAGEM Sperwer, General Atomics MQ-9 Reaper "F"
- Germany – Modified IAI Heron from Israel.
- Georgia – Elbit Hermes 450, Elbit Skylark
- India – IAI Heron, IAI Harop and IAI Harpy from Israel,
- Iran – Karrar, Shahed 129, Fotros and others
- Ireland – Aeronautics Orbiter UAV, number: 3+. Used in Irish Army duties. There is no evidence of using Armed drones by Irish army-Wikipedia:Citation needed
- Israel – IAI Heron, IAI Harpy, Elbit Hermes 450, IAI Eitan, IAI Harop
- Italy – MQ-1 Predator, MQ-9 Reaper from the U.S. and Piaggio Aerospace P.1HH Hammerhead from Italy
- Mexico – Elbit Hermes 450
- Netherlands – MQ-9 Reaper
- Pakistan – UCAV Burraq, GIDS Shahpar, Falco UAV (from Italy)
- Palestine – Two types of operational drones developed from Iranian Ababil-1 drone.[1092,1093]
- Saudi Arabia – CAIG Wing Loong
- Singapore – Elbit Hermes 450
- Spain – IAI Searcher, Skeldar V-200, RQ-11 Raven
- Turkey – TAI Anka, BAYKAR Bayraktar TB2, Vestel Karayel UCAV[1094,1095]
- United Kingdom – MQ-9 Reaper, Elbit Hermes 450
- United States – MQ-1 Predator, MQ-9 Reaper

Laws and ethics of war

UAVs face multiple ethical issues.

Stand-off attacks

The "unmanned" aspect of armed UAVs has raised moral concerns about their use in combat and law enforcement contexts. Attacking humans with remote-controlled machines is even more abstract than the use of other "stand-off" weaponry, such as missiles, artillery and aerial bombardment, possibly depersonalizing the decision to attack. By contrast, UAVs and other stand-off systems reduce casualties among the attackers.

Autonomous attacks

The picture is further complicated if the UAV can initiate an attack autonomously, without direct human involvement. Such UAVs could possibly react more quickly and without bias, but would lack human sensibility. Heather RoffWikipedia:Please clarify replies that Lethal autonomous robots (LARs) may not be appropriate for complex conflicts and targeted populations would likely react angrily against them. Will McCants argues that the public would be more outraged by machine failures than human error, making LARs politically implausible. According to Mark Gubrud, claims that drones can be hacked are overblown and misleading and moreover, drones are *more* likely to be hacked if they're autonomous, because otherwise the human operator would take control: "Giving weapon systems autonomous capabilities is a good way to lose control of them, either due to a programming error, unanticipated circumstances, malfunction, or hack and then not be able to regain control short of blowing them up, hopefully before they've blown up too many other things and people." Others have argued that the technological possibility of autonomy should not obscure the continuing moral responsibilities humans have at every stage. There is an ongoing debate as to whether the attribution of moral responsibility can be apportioned appropriately under existing international humanitarian law, which is based on four principles: military necessity, distinction between military and civilian objects, prohibition of unnecessary suffering, and proportionality.

Civilian casualties

Israel

In March 2009, *The Guardian* reported allegations that Israeli UAVs armed with missiles killed 48 Palestinian civilians in the Gaza Strip, including two small children in a field and a group of women and girls in an otherwise empty street.[1096] In June, Human Rights Watch investigated six UAV attacks that were reported to have resulted in civilian casualties and alleged that Israeli forces either failed to take all feasible precautions to verify that the targets were combatants or failed to distinguish between combatants and civilians.

United States

Collateral damage of civilians still takes place with drone combat, although some (like John O. Brennan) have argued that it greatly reduces the likelihood. Although drones enable advance tactical surveillance and up-to-the-minute data, flaws can become apparent. The U.S. drone program in Pakistan has killed several dozen civilians accidentally. An example is the operation in 2010 Feb near Khod, in Urozgan Province, Afghanistan. Over ten civilians in a three-vehicle convoy travelling from Daykundi Province were accidentally killed after a drone crew misidentified the civilians as hostile threats. A force of Bell OH-58 Kiowa helicopters, who were attempting to protect ground troops fighting several kilometers away, fired AGM-114 Hellfire missiles at the vehicles.[1097,1098]

In 2009, the Brookings Institution reported that in the US-led drone attacks in Pakistan, ten civilians died for every militant killed.[1099] A former ambassador of Pakistan said that American UAV attacks were turning Pakistani opinion against the United States.[1100] The website PakistanBodyCount.Org reported 1,065 civilian deaths between 2004 and 2010. According to a 2010 analysis by the New America Foundation 114 UAV-based missile strikes in northwest Pakistan from 2004 killed between 830 and 1,210 individuals, around 550 to 850 of whom were militants. In October 2013, the Pakistani government revealed that since 2008 317 drone strikes had killed 2,160 Islamic militants and 67 civilians – far less than previous government and independent organization calculations.

In July 2013, former Pentagon lawyer Jeh Johnson said, on a panel at the Aspen Institute's Security Forum, that he felt an emotional reaction upon reading Nasser al-Awlaki's account of how his 16-year-old grandson was killed by a U.S. drone.[1101]

In December 2013, a U.S. drone strike in Radda, capital of Yemen's Bayda province, killed members of a wedding party.[1102] The following February, Human Rights Watch published a 28-page report reviewing the strike and its legality, among other things. Titled "A Wedding That Became A Funeral", the report concludes that some (but not necessarily all) of the casualties were civilians, not the intended regional Al-Qaeda targets. The organization demanded US and Yemeni investigations into the attack. In its research, HRW "found no evidence that the individuals taking part in the wedding procession posed an imminent threat to life. In the absence of an armed conflict, killing them would be a violation of international human rights law."

Psychological impact on operators

Controllers can also experience psychological stress from the combat they are involved in. A few may even experience posttraumatic stress disorder (PTSD). There are even some reports of drone pilots struggling with post traumatic stress disorder after they have killed civilians, especially children. Unlike bomber pilots, moreover, drone operators linger long after the explosives strike and see its effects on human bodies in stark detail. The intense training that US drone operators undergo "works to dehumanise the 'enemy' people below whilst glorifying and celebrating the killing process."

Professor Shannon E. French, the director of the Center for Ethics and Excellence at Case Western Reserve University and a former professor at the U.S. Naval Academy, wonders if the PTSD may be rooted in a suspicion that something else was at stake. According to Professor French, the author of the 2003 book *The Code of the Warrior*:

> *If [I'm] in the field risking and taking a life, there's a sense that I'm putting skin in the game ... I'm taking a risk so it feels more honorable. Someone who kills at a distance—it can make them doubt. Am I truly honorable?*

The Missile Technology Control Regime applies to UCAVs.

On 28 October 2009, United Nations Special Rapporteur on extrajudicial, summary or arbitrary executions, Philip Alston, presented a report to the Third Committee (social, humanitarian and cultural) of the General Assembly arguing that the use of unmanned combat air vehicles for targeted killings should be regarded as a breach of international law unless the United States can demonstrate appropriate precautions and accountability mechanisms are in place.

In June 2015 forty-five former US military personnel issued a joint appeal to pilots of aerial drones operating in Afghanistan, Iraq, Syria, Pakistan and elsewhere urging them to refuse to fly and indicated that their missions "profoundly violate domestic and international laws." They noted that these drone attacks also undermine principles of human rights.

Political and personal effects

As a new weapon, drones are having unforeseen political effects. Some scholars have argued that the extensive use of drones will undermine the popular legitimacy of local governments, which are blamed for permitting the strikes. The case study for this analysis is Yemen, where drone strikes seem to be increasing resentment against the Yemeni government as well as against the U.S.

Some leaders worry about the effect drone warfare will have on soldiers' psychology. Keith Shurtleff, an army chaplain at Fort Jackson, South Carolina, worries "that as war becomes safer and easier, as soldiers are removed from the horrors of war and see the enemy not as humans but as blips on a screen, there is very real danger of losing the deterrent that such horrors provide".[1103] Similar worries surfaced when "smart" bombs began to be used extensively in the First Gulf War.

There are new case studies that are examining the psychological effects drones have on the citizens on the ground. Peter Schaapveld, a forensic psychologist, conducted research in Yemen on the psychological effects of drones. He found that "92 percent of the population sample he examined was found to be suffering from post-traumatic stress disorder – with children being the demographic most significantly affected."[1104] Psychologists in Gaza, meanwhile, talk of a whole generation of Gazan children suffering deep psychological trauma because of the continual exposure to the buzzing sounds of drones high above, machines that can spit lethal violence upon them and their families at any moment. Stanford's 'Living Under Drones' researchers, meanwhile, have shown that civilians in Pakistan and Afghanistan are reluctant to help those hit by the first strikes because rescuers themselves have often been killed by follow-on drone strikes. Injured relatives in the rubble of the first strike have been known to tell their relatives not to help rescue them because of the frequency of these so-called 'double-tap' strikes. People also avoid gathering in groups in visible places. Many children are permanently kept indoors and often no longer attend school.

Writer Mark Bowden has disputed this viewpoint saying in his The Atlantic article, "But flying a drone, [the pilot] sees the carnage close-up, in real time—the blood and severed body parts, the arrival of emergency responders, the anguish of friends and family. Often he's been watching the people he kills for a long time before pulling the trigger. Drone pilots become familiar with their victims. They see them in the ordinary rhythms of their lives—with their wives and friends, with their children. War by remote control turns out to be intimate and disturbing. Pilots are sometimes shaken."

This assessment is corroborated by a sensor operator's account:

> *The smoke clears, and there's pieces of the two guys around the crater. And there's this guy over here, and he's missing his right leg above his knee. He's holding it, and he's rolling around, and the blood is squirting out of his leg ... It took him a long time to die. I just watched him.*
>
> *—Airman First Class Brandon Bryant (whistleblower) in GQ*

Back in the United States, a combination of "lower-class" status in the military, overwork, and psychological trauma may be taking a mental toll on drone pilots. These psychological, cultural and career issues appear to have led to a shortfall in USAF drone operators, which is seen as a "dead end job".[1105]

Public opinion summary

In 2013 a Fairleigh Dickinson University poll asked registered voters whether they "approve or disapprove of the U.S. Military using drones to carry out attacks abroad on people and other targets deemed a threat to the U.S.?" The results showed that three in every four (75%) of voters approved of the U.S. Military using drones to carry out attacks, while (13%) disapproved.[1106] A

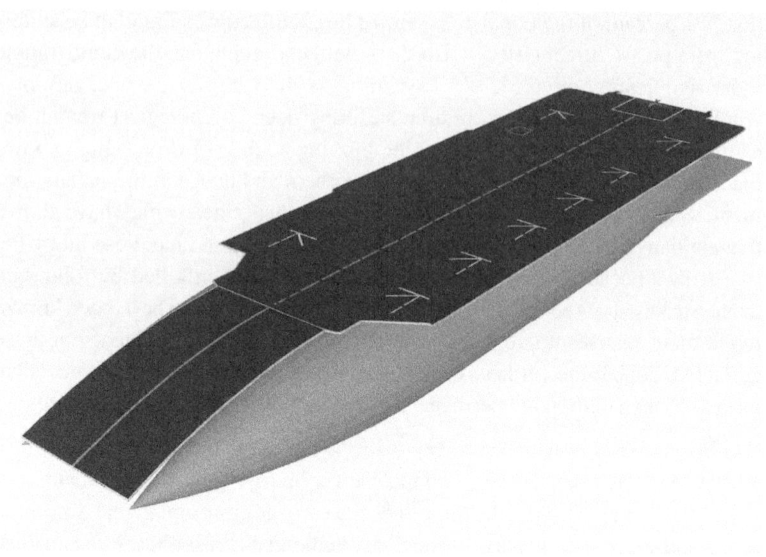

Figure 59: *Aerial aircraft carrier drone*

poll conducted by the Huffington Post in 2013 also showed a majority supporting targeted killings using drones, albeit by a smaller margin. A 2015 poll showed Republicans and men are more likely to support U.S. drone strikes, while Democrats and Independents, women, young people, and minorities are less supportive.

Outside America there is widespread opposition to US drone killings. A July 2014 report found a majority or plurality of respondents in 39 of 44 countries surveyed opposed U.S. drone strikes in countries such as Pakistan, Yemen, and Somalia. The U.S., Kenya, and Israel were the only countries where at least half the public supported drone strikes. Venezuela was found to be the most anti-drone country, where 92% of respondents disagreed with U.S. drone strikes, followed closely by Jordan, where 90% disagreed; Israel was shown as the most pro-drone, with 65% in favor of U.S. drone strikes and 27% opposed.

Drone carrier ships

In March 2013, DARPA began efforts to develop a fleet of small naval vessels capable of launching and retrieving combat drones without the need for large and expensive aircraft carriers.

In November 2014, the Pentagon made an open request for ideas on how to build a flying aircraft carrier that can launch and retrieve drones using existing military aircraft such as the B-1, B-52 or C-130.

External links

	Wikimedia Commons has media related to ***Unmanned aerial military vehicles***.

	Wikinews has related news: ***Fifteen killed by US drone strikes in Northern Waziristan***

Artificial intelligence arms race

An **artificial intelligence arms race** is a competition between two or more states to have its military forces equipped with the best "artificial intelligence" (AI). Since the mid-2010s, many analysts have argued that a such a global arms race for better artificial intelligence has already begun.

Stances toward military artificial intelligence

Russia

Russian General Viktor Bondarev, commander-in-chief of the Russian air force, has stated that as early as February 2017, Russia has been working on AI-guided missiles that can decide to switch targets mid-flight. Reports by state-sponsored Russian media on potential military uses of AI increased in mid-2017. In May 2017, the CEO of Russia's Kronstadt Group, a defense contractor, stated that "there already exist completely autonomous AI operation systems that provide the means for UAV clusters, when they fulfill missions autonomously, sharing tasks between them, and interact", and that it is inevitable that "swarms of drones" will one day fly over combat zones. Russia has been testing several autonomous and semi-autonomous combat systems, such as Kalashnikov's "neural net" combat module, with a machine gun, a camera, and an AI that its makers claim can make its own targeting judgements without human intervention. In September 2017, during a National Knowledge Day address to over a million students in 16,000 Russian schools, Russian President Vladimir Putin stated "Artificial intelligence is the future, not only for Russia but for all humankind... Whoever becomes the leader in this sphere will become the ruler of the world".

The Russian government has strongly rejected any ban on lethal autonomous weapons systems, suggesting that such a ban could be ignored.

Figure 60: *Putin (seated, center) at National Knowledge Day, 2017*

China

According to Elsa Kania of the Center for a New American Security, "China is no longer in a position of technological inferiority but rather sees itself as close to catching up with and overtaking the United States in AI. As such, the (Chinese military) intends to achieve an advantage through changing paradigms in warfare with military innovation, thus seizing the 'commanding heights'...of future military competition". The close ties between Silicon Valley and China, and the open nature of the American research community, has made the West's most advanced AI technology easily available to China; in addition, Chinese industry has numerous home-grown AI accomplishments of its own, such as Baidu passing a notable Chinese-language speech recognition capability benchmark in 2015. As of 2017, Beijing's roadmap aims to create a $150 billion AI industry by 2030. Before 2013, Chinese defense procurement was mainly restricted to a few conglomerates; however, as of 2017, China often sources sensitive emerging technology such as drones and artificial intelligence from private start-up companies. One Chinese state has pledged to invest $5 billion in AI. Beijing has committed $2 billion to an AI development park. The *Japan Times* reported in 2018 that annual private Chinese investment in AI is under $7 billion per year. AI startups in China received nearly half of total global investment in AI startups in 2017; the Chinese filed for nearly five times as many AI patents as did Americans.

Figure 61: *The Sea Hunter, an autonomous US warship, 2016*

China published a position paper in 2016 questioning the adequacy of existing international law to address the eventuality of fully autonomous weapons, becoming the first permanent member of the U.N. Security Council to broach the issue.

United States

In 2014, former Secretary of Defense Chuck Hagel posited the "Third Offset Strategy" that rapid advances in artificial intelligence will define the next generation of warfare. According to data science and analytics firm Govini, The U.S. Department of Defense increased investment in artificial intelligence, big data and cloud computing from $5.6 billion in 2011 to $7.4 billion in 2016. However, the civilian NSF budget for AI saw no increase in 2017.

The U.S. has many military AI combat programs, such as the Sea Hunter autonomous warship, which is designed to operate for extended periods at sea without a single crew member, and to even guide itself in and out of port. As of 2017, a temporary US Department of Defense directive requires a human operator to be kept in the loop when it comes to the taking of human life by autonomous weapons systems. *Japan Times* reported in 2018 that the United States private investment is around $70 billion per year.

United Kingdom

In 2015, the UK government opposed a ban on lethal autonomous weapons, stating that "international humanitarian law already provides sufficient regulation for this area", but that all weapons employed by UK armed forces would be "under human oversight and control".

Israel

Israel's Harpy anti-radar "fire and forget" drone is designed to be launched by ground troops, and autonomously fly over an area to find and destroy radar that fits pre-determined criteria.

South Korea

The South Korean Super aEgis II machine gun, unveiled in 2010, sees use both in South Korea and in the Middle East. It can identify, track, and destroy a moving target at a range of 4 km. While the technology can theoretically operate without human intervention, in practice safeguards are installed to require manual input. A South Korean manufacturer states, "Our weapons don't sleep, like humans must. They can see in the dark, like humans can't. Our technology therefore plugs the gaps in human capability", and they want to "get to a place where our software can discern whether a target is friend, foe, civilian or military".

Trends

According to Siemens, worldwide military spending on robotics was 5.1 billion USD in 2010 and 7.5 billion USD in 2015.

China became a top player in artificial intelligence research in the 2010s. According to the *Financial Times*, in 2016, for the first time, China published more AI papers than the entire European Union. When restricted to number of AI papers in the top 5% of cited papers, China overtook the United States in 2016 but lagged behind the European Union. 23% of the researchers presenting at the 2017 American Association for the Advancement of Artificial Intelligence conference were Chinese. Eric Schmidt, the former chairman of Alphabet, has predicted China will be the leading country in AI by 2025.

AAAI presenters:

Country	in 2012	in 2017
US	41%	34%
China	10%	23%
UK	5%	5%

Proposals for international regulation

As early as 2007, scholars such as AI professor Noel Sharkey have warned of "an emerging arms race among the hi-tech nations to develop autonomous submarines, fighter jets, battleships and tanks that can find their own targets and apply violent force without the involvement of meaningful human decisions". As early as 2014, AI specialists such as Steve Omohundro have been warning that "An autonomous weapons arms race is already taking place". Miles Brundage of the University of Oxford has argued an AI arms race might be somewhat mitigated through diplomacy: "We saw in the various historical arms races that collaboration and dialog can pay dividends". Over a hundred experts signed an open letter in 2017 calling on the UN to address the issue of lethal autonomous weapons; however, at a November 2017 session of the UN Convention on Certain Conventional Weapons (CCW), diplomats could not agree even on how to define such weapons. The Indian ambassador and chair of the CCW stated that agreement on rules remained a distant prospect. As of 2017, twenty-two countries have called for a full ban on lethal autonomous weapons.

Many experts believe attempts to completely ban killer robots are likely to fail. A 2017 report from Harvard's Belfer Center predicts that AI has the potential to be as transformative as nuclear weapons.[1107] The report further argues that "Preventing expanded military use of AI is likely impossible" and that "the more modest goal of safe and effective technology management must be pursued", such as banning the attaching of an AI dead man's switch to a nuclear arsenal. Part of the impracticality is that detecting treaty violations would be extremely difficult.[1108,1109]

Other reactions to autonomous weapons

A 2015 open letter calling for the ban of lethal automated weapons systems has been signed by tens of thousands of citizens, including scholars such as physicist Stephen Hawking, Tesla magnate Elon Musk, and Apple's Steve Wozniak.

Professor Noel Sharkey of the University of Sheffield has warned that autonomous weapons will inevitably fall into the hands of terrorist groups such as the Islamic State.

Disassociation

Many Western tech companies are leery of being associated too closely with the U.S. military, for fear of losing access to China's market. Furthermore, some researchers, such as DeepMind's Demis Hassabis, are ideologically opposed to contributing to military work.

For example, **Project Maven** is a Pentagon project involving using machine learning and engineering talent to distinguish people and objects in drone videos, established in a memo by the U.S. Deputy Secretary of Defense on 26 April 2017. Also known as the Algorithmic Warfare Cross Functional Team, it is, according to Lt. Gen. of the United States Air Force Jack Shanahan in November 2017, a project "designed to be that pilot project, that pathfinder, that spark that kindles the flame front of artificial intelligence across the rest of the [Defense] Department". Its chief, U.S. Marine Corps Col. Drew Cukor, said: "People and computers will work symbiotically to increase the ability of weapon systems to detect objects." At the second Defense One Tech Summit in July 2017, Cukor also said that the investment in a "deliberate workflow process" was funded by the Department [of Defense] through its "rapid acquisition authorities" for about "the next 36 months".

In June 2018, company sources at Google said that top executive Diane Greene told staff that the company would not follow-up Project Maven after the current contract expires in March 2019.

Artificial intelligence in healthcare

Artificial intelligence (AI) in healthcare is the use of algorithms and software to approximate human cognition in the analysis of complex medical data. Specifically, AI is the ability for computer algorithms to approximate conclusions without direct human input. The primary aim of health-related AI applications is to analyze relationships between prevention or treatment techniques and patient outcomes.[1110] AI programs have been developed and applied to practices such as diagnosis processes, treatment protocol development, drug development, personalized medicine,and patient monitoring and care. Medical institutions such as The Mayo Clinic, Memorial Sloan Kettering Cancer Center[1111], Massachusetts General Hospital, and National Health Service,[1112]

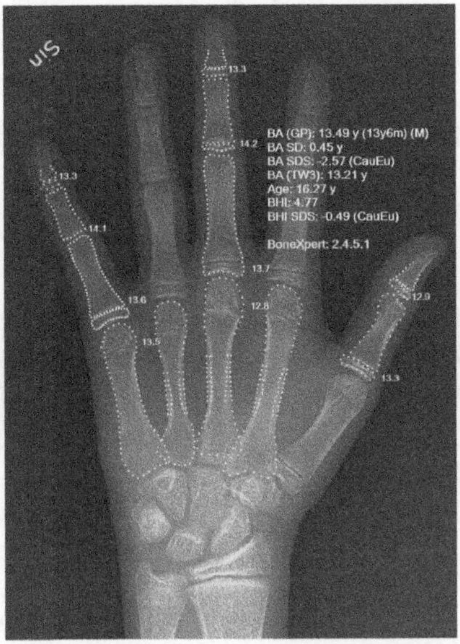

Figure 62: *X-ray of a hand, with automatic calculation of bone age by a computer software.*

have developed AI algorithms for their departments. Large technology companies such as IBM[1113] and Google, and startups such as Welltok and Ayasdi,[1114] have also developed AI algorithms for healthcare.

History

Research in the 1960s and 1970s produced the first problem-solving program, or expert system, known as Dendral.[1115] While it was designed for applications in organic chemistry, it provided the basis for the subsequent system MYCIN,[1116] considered one of the most significant early uses of artificial intelligence in medicine.[1117] MYCIN and other systems such as INTERNIST-1 and CASNET did not achieve routine use by practitioners, however.[1118]

The 1980s and 1990s brought the proliferation of the microcomputer and new levels of network connectivity. During this time there was recognition by researchers and developers that AI systems in healthcare must be designed to accommodate the absence of perfect data and build on the expertise of physician.[1119] Approaches involving fuzzy set theory,[1120] Bayesian networks,[1121] and artificial neural networks,[1122,1123] have been applied to intelligent computing systems in healthcare.

Medical and technological advancements occurring over this half-century period that have simultaneously enabled the growth healthcare-related applications of AI include:

- Improvements in computing power resulting in faster data collection and data processing[1124]
- Increased volume and availability of health-related data from personal and healthcare-related devices[1125]
- Growth of genomic sequencing databases[1126]
- Widespread implementation of electronic health record systems[1127]
- Improvements in natural language processing and computer vision, enabling machines to replicate human perceptual processes,[1128,1129]
- Enhanced the precision of robot-assisted surgery[1130]

Current Research

Various specialties in medicine have shown an increase in research regarding AI. However the specialty that has gained the greatest attention is the field of Radiology.

Radiology

Such an ability to interpret imaging results may allow clinicians to be aided to detect a change in an image that is minute in detail, or something that a clinician may have accidentally missed. Such a study that has incorporated AI in radiology is a study at Stanford which has results presenting that the algorithm that they created can detect Pneumonia better than radiologists. The radiology conference Radiological Society of North America has implemented a large part of its schedule to the use of AI in imaging.

Telehealth

The increase of Telemedicine, has shown the rise of possible AI application. The ability to monitor patients using AI, may allow for the communication of information to physicians if possible disease activity may have occurred. The use of a device such that a person may wear, may allow for constant monitoring of a patient and also for the ability to notice changes that may be less distinguishable by humans.

Industry

The subsequent motive of large based health companies merging with other health companies, allow for greater health data accessibility. Greater health data may allow for more implementation of AI algorithms. The following are examples of large companies that have contributed to AI algorithms for use in healthcare.

IBM

IBM's Watson Oncology is in development at Memorial Sloan Kettering Cancer Center and Cleveland Clinic.[1131] IBM is also working with CVS Health on AI applications in chronic disease treatment and with Johnson & Johnson on analysis of scientific papers to find new connections for drug development.[1132]

Microsoft

Microsoft's Hanover project, in partnership with Oregon Health & Science University's Knight Cancer Institute, analyzes medical research to predict the most effective cancer drug treatment options for patients.[1133] Other projects include medical image analysis of tumor progression and the development of programmable cells.[1134]

Google

Google's DeepMind platform is being used by the UK National Health Service to detect certain health risks through data collected via a mobile app.[1135] A second project with the NHS involves analysis of medical images collected from NHS patients to develop computer vision algorithms to detect cancerous tissues.[1136]

Intel

Intel's venture capital arm Intel Capital recently invested in startup Lumiata which uses AI to identify at-risk patients and develop care options.[1137]

Startups

IDx's first solution, IDx-DR[1138] founded by Michael Abramoff[1139], is the first and only FDA authorized AI system for the autonomous detection of diabetic retinopathy[1140]. As an autonomous, AI-based system, IDx-DR is unique in that it makes an assessment without the need for a clinician to also interpret the image or results, making it usable by health care providers who may not normally be involved in eye care. IDx[1141] is a leading AI diagnostics company on a mission to transform the quality, accessibility, and affordability of healthcare world-wide.

Medvice provides real time medical advice to clients, who can access and store their Electronic Health Records (EHRs) over a decentralized blockchain[1142]. Medvice uses machine learning aided decision making to help physicians predict medical red flags (i.e. medical emergencies which require clinical assistance) before serving them. Predictive Medical Technologies uses intensive care unit data to identify patients likely to suffer cardiac incidents. Modernizing Medicine uses knowledge gathered from healthcare professionals as well as patient outcome data to recommend treatments.[1143] "Compassionate AI Lab" uses grid cell, place cell and path integration with machine learning for the navigation of blind people. Nimblr.ai uses an A.I. Chatbot to connect scheduling EHR systems and automate the confirmation and scheduling of patients.

Other

Digital consultant apps like Babylon in the UK use AI to give medical consultation based on personal medical history and common medical knowledge. Users report their symptoms into the app, which uses speech recognition to compare against a database of illnesses. Babylon then offers a recommended action, taking into account the user's medical history.

Regulation

While research on the use of AI in healthcare aims to validate its efficacy in improving patient outcomes before its broader adoption, its use may nonetheless introduce several new types of risk to patients and healthcare providers, such as algorithmic bias, Do not resuscitate implications, and other machine morality issues. These challenges of the clinical use of AI has brought upon potential need for regulations.

Currently no regulations exist specifically for the use of AI in healthcare. In May 2016, the White House announced its plan to host a series of workshops and formation of the National Science and Technology Council (NSTC) Subcommittee on Machine Learning and Artificial Intelligence.[1144] In October 2016, the group published The National Artificial Intelligence Research and

Development Strategic Plan, outlining its proposed priorities for Federally-funded AI research and development (within government and academia). The report notes a strategic R&D plan for the subfield of health information technology is in development stages.[1145]

The only agency that has expressed concern is the FDA. Bakul Patel, the Associate Center Director for Digital Health of the FDA, is quoted saying in May 2017.

> *"We're trying to get people who have hands-on development experience with a product's full life cycle. We already have some scientists who know artificial intelligence and machine learning, but we want complementary people who can look forward and see how this technology will evolve."*

Government investment

Investments from the US government in healthcare initiatives that will rely on AI include its $1B proposed budget for the Cancer Moonshot[1146] and $215M proposed investment in the Precision Medicine Initiative.[1147]

Autonomous car

An **autonomous car** (also known as a **driverless car** and a **self-driving car**) is a vehicle that is capable of sensing its environment and navigating without human input.

Autonomous cars combine a variety of techniques to perceive their surroundings, including radar, laser light, GPS, odometry, and computer vision. Advanced control systems interpret sensory information to identify appropriate navigation paths, as well as obstacles and relevant signage.[1148]

The potential benefits of autonomous cars include reduced mobility and infrastructure costs, increased safety, increased mobility, increased customer satisfaction, and reduced crime. These benefits also include a potentially significant reduction in traffic collisions; resulting injuries; and related costs, including less need for insurance. Autonomous cars are predicted to increase traffic flow; provide enhanced mobility for children, the elderly, disabled, and the poor; relieve travelers from driving and navigation chores; lower fuel consumption; significantly reduce needs for parking space; reduce crime; and facilitate business models for transportation as a service, especially via the sharing economy. This shows the vast disruptive potential of the emerging technology.

Figure 63: *Waymo Chrysler Pacifica Hybrid undergoing testing in the San Francisco Bay Area.*

Figure 64: *Autonomous racing car on display at the 2017 New York City ePrix.*

In spite of the various potential benefits to increased vehicle automation, there are unresolved problems, such as safety, technology issues, disputes concerning liability, resistance by individuals to forfeiting control of their cars, customer concern about the safety of driverless cars, implementation of a legal framework and establishment of government regulations; risk of loss of privacy and security concerns, such as hackers or terrorism; concern about the resulting loss of driving-related jobs in the road transport industry; and risk of increased suburbanization as travel becomes less costly and time-consuming. Many of these issues arise because autonomous objects, for the first time, would allow computer-controlled ground vehicles to roam freely, with many related safety and security concerns.

History

Experiments have been conducted on automating driving since at least the 1920s; promising trials took place in the 1950s. The first truly autonomous car was developed in 1977, by the Tsukuba Mechanical Engineering Laboratory in Japan. The vehicle tracked white street markers, which were able to be interpreted by two cameras fitted onto the vehicle, using an analog computer technology for signal processing. The Tsukuba autonomous car was able to achieve speeds up to 30 kilometres per hour (19 mph), with the support of an elevated rail. Autonomous prototype cars appeared in the 1980s, with Carnegie Mellon University's Navlab and ALV projects funded by DARPA starting in 1984 and Mercedes-Benz and Bundeswehr University Munich's EUREKA Prometheus Project in 1987. By 1985 the ALV had demonstrated self-driving speeds on two-lane roads of 31 kilometres per hour (19 mph) with obstacle avoidance added in 1986 and off-road driving in day and nighttime conditions by 1987. From the 1960s up through the second DARPA Grand Challenge in 2005, autonomous vehicle research in the U.S. was primarily funded by DARPA, the US Army and the U.S. Navy yielding incremental advances in speeds, driving competence in more complex conditions, controls, and sensor systems. Since then, numerous companies and research organizations have developed prototypes.

The U.S. Congress allocated $650 million in 1991 for research on the National Automated Highway System, which demonstrated autonomous driving through a combination of automation embedded in the highway with autonomous technology in vehicles and cooperative networking between the vehicles and with the highway infrastructure. The program concluded with a successful demonstration in 1997 but without clear direction or funding to implement the system on a larger scale. Partly funded by the National Automated Highway System and DARPA, the Carnegie Mellon University Navlab drove 4,584 kilometres (2,848 mi) across America in 1995, 4,501 kilometres (2,797 mi) or 98% of it autonomously. Navlab's record achievement stood unmatched for two decades until 2015 when Delphi improved it by piloting an Audi, augmented with Delphi technology, over 5,472 kilometres (3,400 mi) through 15 states while remaining in self-driving mode 99% of the time. In 2015, the US states of Nevada, Florida, California, Virginia, and Michigan, together with Washington, D.C., allowed the testing of autonomous cars on public roads.

In 2017, Audi stated that its latest A8 would be autonomous at speeds of up to 60 kilometres per hour (37 mph) using its "Audi AI". The driver would not have to do safety checks such as frequently gripping the steering wheel. The Audi A8 was claimed to be the first production car to reach level 3 autonomous driving, and Audi would be the first manufacturer to use laser scanners in addition to cameras and ultrasonic sensors for their system.

In November 2017, Waymo announced that it had begun testing driverless cars without a safety driver in the driver position; however there is still an employee in the car. In July 2018, Waymo announced that its test vehicles had traveled autonomously for over 8,000,000 miles (13,000,000 km). Increasing 1,000,000 miles (1,600,000 kilometres) per month.

Definitions

By involving rather new and diverse concepts, those new technologies have required to define an accurate vocabulary.

Autonomous vs. automated

Autonomous means self-governing. Many historical projects related to vehicle autonomy have been *automated* (made automatic) subject to a heavy reliance on artificial aids in their environment, such as magnetic strips. Autonomous control implies satisfactory performance under significant uncertainties in the environment and the ability to compensate for system failures without external intervention.

One approach is to implement communication networks both in the immediate vicinity (for collision avoidance) and further away (for congestion management). Such outside influences in the decision process reduce an individual vehicle's autonomy, while still not requiring human intervention.

Wood et al. (2012) wrote, "This Article generally uses the term 'autonomous,' instead of the term 'automated.' " The term "autonomous" was chosen "because it is the term that is currently in more widespread use (and thus is more familiar to the general public). However, the latter term is arguably more accurate. 'Automated' connotes control or operation by a machine, while 'autonomous' connotes acting alone or independently. Most of the vehicle concepts (that we are currently aware of) have a person in the driver's seat, utilize a communication connection to the Cloud or other vehicles, and do not independently select either destinations or routes for reaching them. Thus, the term 'automated' would more accurately describe these vehicle concepts". As of 2017, most commercial projects focused on autonomous vehicles that did not communicate with other vehicles or with an enveloping management regime.

Classification

A classification system based on six different levels (ranging from fully manual to fully automated systems) was published in 2014 by SAE International, an automotive standardization body, as J3016, *Taxonomy and Definitions for Terms Related to On-Road Motor Vehicle Automated Driving Systems*. This classification system is based on the amount of driver intervention and attentiveness required, rather than the vehicle capabilities, although these are very loosely related. In the United States in 2013, the National Highway Traffic Safety Administration (NHTSA) released a formal classification system, but abandoned this system in favor of the SAE standard in 2016. Also in 2016, SAE updated its classification, called J3016_201609.[1149]

Figure 65: *Tesla Autopilot system is considered to be an SAE level 2 system.*

Levels of driving automation

In SAE's autonomy level definitions, "driving mode" means "a type of driving scenario with characteristic dynamic driving task requirements (e.g., expressway merging, high speed cruising, low speed traffic jam, closed-campus operations, etc.)"

- Level 0: Automated system issues warnings and may momentarily intervene but has no sustained vehicle control.
- Level 1 ("hands on"): The driver and the automated system share control of the vehicle. Examples are Adaptive Cruise Control (ACC), where the driver controls steering and the automated system controls speed; and Parking Assistance, where steering is automated while speed is under manual control. The driver must be ready to retake full control at any time. Lane Keeping Assistance (LKA) Type II is a further example of level 1 self driving.
- Level 2 ("hands off"): The automated system takes full control of the vehicle (accelerating, braking, and steering). The driver must monitor the driving and be prepared to intervene immediately at any time if the automated system fails to respond properly. The shorthand "hands off" is not meant to be taken literally. In fact, contact between hand and wheel is often mandatory during SAE 2 driving, to confirm that the driver is ready to intervene.

- Level 3 ("eyes off"): The driver can safely turn their attention away from the driving tasks, e.g. the driver can text or watch a movie. The vehicle will handle situations that call for an immediate response, like emergency braking. The driver must still be prepared to intervene within some limited time, specified by the manufacturer, when called upon by the vehicle to do so. As an example, the 2018 Audi A8 Luxury Sedan was the first commercial car to claim to be capable of level 3 self driving. This particular car has a so-called Traffic Jam Pilot. When activated by the human driver, the car takes full control of all aspects of driving in slow-moving traffic at up to 60 kilometres per hour (37 mph). The function works only on highways with a physical barrier separating one stream of traffic from oncoming traffic.
- Level 4 ("mind off"): As level 3, but no driver attention is ever required for safety, i.e. the driver may safely go to sleep or leave the driver's seat. Self driving is supported only in limited spatial areas (geofenced) or under special circumstances, like traffic jams. Outside of these areas or circumstances, the vehicle must be able to safely abort the trip, i.e. park the car, if the driver does not retake control.
- Level 5 ("steering wheel optional"): No human intervention is required at all. An example would be a robotic taxi.

In the formal SAE definition below, note in particular what happens in the shift from SAE 2 to SAE 3: the human driver no longer has to monitor the environment. This is the final aspect of the "dynamic driving task" that is now passed over from the human to the automated system. At SAE 3, the human driver still has the responsibility to intervene when asked to do so by the automated system. At SAE 4 the human driver is relieved of that responsibility and at SAE 5 the automated system will never need to ask for an intervention.

SAE (J3016) Autonomy Levels

SAE Level	Name	Narrative definition	Execution of steering and acceleration/ deceleration	Monitoring of driving environment	Fall-back performance of dynamic driving task	System capability (driving modes)
Human driver monitors the driving environment						
0	No Automation	The full-time performance by the human driver of all aspects of the dynamic driving task, even when "enhanced by warning or intervention systems"	Human driver	Human driver	Human driver	n/a

1	Driver Assistance	The driving mode-specific execution by a driver assistance system of "either steering or acceleration/deceleration"	using information about the driving environment and with the expectation that the human driver performs all remaining aspects of the dynamic driving task	Human driver and system			Some driving modes
2	Partial Automation	The driving mode-specific execution by one or more driver assistance systems of *both steering and acceleration/deceleration*		System			
Automated driving system monitors the driving environment							
3	Conditional Automation	The driving mode-specific performance by an automated driving system of all aspects of the dynamic driving task	with the expectation that the *human driver will respond appropriately to a request to intervene*	System	System	Human driver	Some driving modes
4	High Automation		*even if a human driver does not respond appropriately to a request to intervene*			System	Many driving modes
5	Full Automation		*under all roadway and environmental conditions* that can be managed by a human driver				All driving modes

Technical challenges

The challenge for driverless car designers is to produce control systems capable of analyzing sensory data in order to provide accurate detection of other vehicles and the road ahead. Modern self-driving cars generally use Bayesian simultaneous localization and mapping (SLAM) algorithms, which fuse data from multiple sensors and an off-line map into current location estimates and map updates. Google is developing a variant called SLAM, with detection and tracking of other moving objects (DATMO), which also handles obstacles such as cars and pedestrians. Simpler systems may use roadside real-time locating system (RTLS) technologies to aid localization. Typical sensors include Lidar, stereo vision, GPS and IMU. Udacity is developing an open-source software stack. Control systems on autonomous cars may use Sensor Fusion, which is an approach that integrates information from a variety of sensors on the car to produce a more consistent, accurate, and useful view of the environment.

Driverless vehicles require some form of machine vision for the purpose of visual object recognition. Autonomous cars are being developed with deep neural networks, a type of deep learning architecture with many computational stages, or levels, in which neurons are simulated from the environment that activate the network. The neural network depends on an extensive amount of data extracted from real-life driving scenarios, enabling the neural network to "learn" how to execute the best course of action.

In May 2018, researchers from MIT announced that they had built an autonomous car that can navigate unmapped roads. Researchers at their Computer Science and Artificial Intelligence Laboratory (CSAIL) have developed a new system, called MapLite, which allows self-driving cars to drive on roads that they have never been on before, without using 3D maps. The system combines the GPS position of the vehicle, a "sparse topological map" such as OpenStreetMap, (i.e. having 2D features of the roads only), and a series of sensors that observe the road conditions.

Testing

Testing vehicles with varying degrees of autonomy can be done physically, in closed environments, on public roads (where permitted, typically with a license or permit or adhering to a specific set of operating principles) or virtually, i.e. in computer simulations.

When driven on public roads, autonomous vehicles require a person to monitor their proper operation and "take over" when needed.

Apple is currently testing self-driven cars, and has increased the number of test vehicles from 3 to 27 in January 2018. This number further increased to 45 in March 2018.

One way to assess the progress of autonomous vehicles is to compute the average number of miles driven between "disengagements", when the autonomous system is turned off, typically by a human driver. In 2017, Waymo reported 63 disengagements over 352,545 miles of testing, or 5596 miles on average, the highest among companies reporting such figures. Waymo also traveled more miles in total than any other. Their 2017 rate of 0.18 disengagements per 1000 miles was an improvement from 0.2 disengagements per 1000 miles in 2016 and 0.8 in 2015. In March, 2017, Uber reported an average of 0.67 miles per disengagement. In the final three months of 2017, Cruise Automation (now owned by GM) averaged 5224 miles per disruption over 62,689 miles. In July 2018, the first electric driverless racing car "Robocar" completed 1.8 kilometers track, using its navigation system, and artificial intelligence.

Miles per disengagement

Maker	2016	
	Miles between disengagements	**Miles**
Google	5,127.9	635,868
BMW	638	638
Nissan	263.3	6,056
Ford	196.6	590
General Motors	54.7	8,156
Delphi Automotive Systems	14.9	2,658
Tesla	2.9	550
Mercedes Benz	2	673
Bosch	0.68	983

Fields of application

Autonomous trucks

Several companies are said to be testing autonomous technology in semi trucks. Otto, a self-driving trucking company that was acquired by Uber in August 2016, demonstrated their trucks on the highway before being acquired. In May 2017, San Francisco-based startup Embark announced a partnership with truck manufacturer Peterbilt to test and deploy autonomous technology in Peterbilt's vehicles. Google's Waymo has also said to be testing autonomous technology in trucks, however no timeline has been given for the project.

In March 2018, Starsky Robotics, the San Francisco-based autonomous truck company, completed a 7-mile fully driverless trip in Florida without a single human in the truck. Starsky Robotics became the first player in the self-driving truck game to drive in fully autonomous mode on a public road without a person in the cab.

In Europe, truck Platooning is considered with the Safe Road Trains for the Environment approach.

Vehicular automation also cover other kinds of vehicle such as Buses, Trains, Trucks.

Lockheed Martin with funding from the U.S. Army developed an autonomous truck convoying system that uses a lead truck operated by a human driver with a number of trucks following autonomously. Developed as part of the Army's Autonomous Mobility Applique System (AMAS), the system consists of an autonomous driving package that has been installed on more than nine types of vehicles and has completed more than 55,000 hours of driving at speeds up to 40mph as of 2014. As of 2017 the Army was planning to field 100-200 trucks as part of a rapid-fielding program.

Transport systems

In Europe, cities in Belgium, France, Italy and the UK are planning to operate transport systems for autonomous cars, and Germany, the Netherlands, and Spain have allowed public testing in traffic. In 2015, the UK launched public trials of the LUTZ Pathfinder autonomous pod in Milton Keynes. Beginning in summer 2015 the French government allowed PSA Peugeot-Citroen to make trials in real conditions in the Paris area. The experiments were planned to be extended to other cities such as Bordeaux and Strasbourg by 2016. The alliance between French companies THALES and Valeo (provider of the first self-parking car system that equips Audi and Mercedes premi) is testing its own system. New Zealand is planning to use autonomous vehicles for public transport in Tauranga and Christchurch.

In China, Baidu and King Long produce autonomous minibus, a vehicle with 14 seats, but without driving seat. With 100 vehicles produced, 2018 will be the first year with commercial autonomous service in China. Those minibus should be at level 4, that is driver-less in closed roads.Wikipedia:Citation needed

Potential advantages

Safety

Driving safety experts predict that once driverless technology has been fully developed, traffic collisions (and resulting deaths and injuries and costs), caused by human error, such as delayed reaction time, tailgating, rubbernecking, and other forms of distracted or aggressive driving should be substantially reduced. Consulting firm McKinsey & Company estimated that widespread use of autonomous vehicles could "eliminate 90% of all auto accidents in the United States, prevent up to US$190 billion in damages and health-costs annually and save thousands of lives."

According to motorist website "TheDrive.com" operated by *Time* magazine, none of the driving safety experts they were able to contact were able to rank driving under an autopilot system at the time (2017) as having achieved a greater level of safety than traditional fully hands-on driving, so the degree to which these benefits asserted by proponents will manifest in practice cannot be assessed. Confounding factors that could reduce the net effect on safety may include unexpected interactions between humans and partly or fully autonomous vehicles, or between different types of vehicle system; complications at the boundaries of functionality at each automation level (such as handover when the vehicle reaches the limit of its capacity); the effect of the bugs and flaws that inevtably occur in complex interdependent software systems; sensor or data shortcomings; and successful compromise by malicious intereveners.

Welfare

Autonomous cars could reduce labor costs; relieve travelers from driving and navigation chores, thereby replacing behind-the-wheel commuting hours with more time for leisure or work; and also would lift constraints on occupant ability to drive, distracted and texting while driving, intoxicated, prone to seizures, or otherwise impaired. For the young, the elderly, people with disabilities, and low-income citizens, autonomous cars could provide enhanced mobility. The removal of the steering wheel—along with the remaining driver interface and the requirement for any occupant to assume a forward-facing position—would give the interior of the cabin greater ergonomic flexibility. Large vehicles, such as motorhomes, would attain appreciably enhanced ease of use.

Traffic

Additional advantages could include higher speed limits; smoother rides; and increased roadway capacity; and minimized traffic congestion, due to decreased need for safety gaps and higher speeds. Currently, maximum controlled-access highway throughput or capacity according to the U.S. Highway Capacity Manual is about 2,200 passenger vehicles per hour per lane, with about 5% of the available road space is taken up by cars. One study estimated that autonomous cars could increase capacity by 273% (~8,200 cars per hour per lane). The study also estimated that with 100% connected vehicles using vehicle-to-vehicle communication, capacity could reach 12,000 passenger vehicles per hour (up 445% from 2,200 pc/h per lane) traveling safely at 120 km/h (75 mph) with a following gap of about 6 m (20 ft) of each other. Currently, at highway speeds drivers keep between 40 to 50 m (130 to 160 ft) away from the car in front. These increases in highway capacity could have a significant impact in traffic congestion, particularly in urban areas, and even effectively end highway congestion in some places. The ability for authorities to manage traffic flow would increase, given the extra data and driving behavior predictability. combined with less need for traffic police and even road signage.

Lower costs

Safer driving is expected to reduce the costs of vehicle insurance. Reduced traffic congestion and the improvements in traffic flow due to widespread use of autonomous cars will also translate into better fuel efficiency. Additionally, self driving cars will be able to accelerate and brake more efficiently, meaning higher fuel economy from reducing wasted energy typically associated with inefficient changes to speed (energy typically lost due to friction, in the form of heat and sound).

Related effects

By reducing the (labor and other) cost of mobility as a service, autonomous cars could reduce the number of cars that are individually owned, replaced by taxi/pooling and other car sharing services. This could dramatically reduce the need for parking space, freeing scarce land for other uses. This would also dramatically reduce the size of the automotive production industry, with corresponding environmental and economic effects. Assuming the increased efficiency is not fully offset by increases in demand, more efficient traffic flow could free roadway space for other uses such as better support for pedestrians and cyclists.

The vehicles' increased awareness could aid the police by reporting on illegal passenger behavior, while possibly enabling other crimes, such as deliberately crashing into another vehicle or a pedestrian. However, this may also lead to much expanded mass surveillance if there is wide access granted to third parties to the large data sets generated.

The future of passenger rail transport in the era of autonomous cars is not clear.

Potential limits or obstacles

The sort of hoped-for potential benefits from increased vehicle automation described may be limted by foreseeable challenges, such as disputes over liability (will each company operating a vehicle accept that they are its "driver" and thus responsible for what their car does, or will some try to project this liability onto others who are not in control?) , the time needed to turn over the existing stock of vehicles from non-autonomous to autonomous, and thus a long period of humans and autonomous vehicles sharing the roads, resistance by individuals to having to forfeit control of their cars, concerns about the safety of driverless in practice, and the implementation of a legal framework and consistent global government regulations for self-driving cars. Other obstacles could include de-skilling and lower levels of driver experience for dealing with potentially dangerous situations and anomalies, ethical problems where an autonomous vehicle's software is forced during an unavoidable crash to choose between multiple harmful courses of action ('the trolley problem'),, concerns about making large numbers of people currently employed as drivers unemployed (at the same time as many other alternate blue collar occupations may be undermined by automation), the potential for more intrusive mass surveillance of location, association and travel as a result of police and intelligence agency access to large data sets generated by sensors and pattern-recognition AI (making anonymous travel difficult), and possibly insufficient understanding of verbal sounds, gestures and non-verbal cues by police, other drivers or pedestrians.

Possible technological obstacles for autonomous cars are:

- Artificial Intelligence is still not able to function properly in chaotic inner city environments.
- A car's computer could potentially be compromised, as could a communication system between cars.
- Susceptibility of the car's sensing and navigation systems to different types of weather (such as snow) or deliberate interference, including jamming and spoofing.
- Avoidance of large animals requires recognition and tracking, and Volvo found that software suited to caribou, deer, and elk was ineffective with kangaroos.
- Autonomous cars may require very high-quality specialised maps to operate properly. Where these maps may be out of date, they would need to be able to fall back to reasonable behaviors.
- Competition for the radio spectrum desired for the car's communication.
- Field programmability for the systems will require careful evaluation of product development and the component supply chain.
- Current road infrastructure may need changes for autonomous cars to function optimally.
- Discrepancy between people's beliefs of the necessary government intervention may cause a delay in accepting autonomous cars on the road. Whether the public desires no change in existing laws, federal regulation, or another solution; the framework of regulation will likely result in differences of opinion.

Potential disadvantages

A direct impact of widespread adoption of autonomous vehicles is the loss of driving-related jobs in the road transport industry. There could be resistance from professional drivers and unions who are threatened by job losses. In addition, there could be job losses in public transit services and crash repair shops. The automobile insurance industry might suffer as the technology makes certain aspects of these occupations obsolete. A frequently cited paper by Michael Osborne and Carl Benedikt Frey found that autonomous cars would make many jobs redundant.

Privacy could be an issue when having the vehicle's location and position integrated into an interface in which other people have access to. In addition, there is the risk of automotive hacking through the sharing of information through V2V (Vehicle to Vehicle) and V2I[1150] (Vehicle to Infrastructure) protocols. There is also the risk of terrorist attacks. Self-driving cars could potentially be loaded with explosives and used as bombs.

The lack of stressful driving, more productive time during the trip, and the potential savings in travel time and cost could become an incentive to live far away from cities, where land is cheaper, and work in the city's core, thus increasing travel distances and inducing more urban sprawl, more fuel consumption and an increase in the carbon footprint of urban travel. There is also the risk that traffic congestion might increase, rather than decrease. Appropriate public policies and regulations, such as zoning, pricing, and urban design are required to avoid the negative impacts of increased suburbanization and longer distance travel.

Some believe that once automation in vehicles reaches higher levels and becomes reliable, drivers will pay less attention to the road. Research shows that drivers in autonomous cars react later when they have to intervene in a critical situation, compared to if they were driving manually.[1151] Depending on the capabilities of autonomous vehicles and the frequency with which human intervention is needed, this may counteract any increase in safety, as compared to all-human driving, that may be delivered by other factors.

Ethical and moral reasoning come into consideration when programming the software that decides what action the car takes in an unavoidable crash; whether the autonomous car will crash into a bus, potentially killing people inside; or swerve elsewhere, potentially killing its own passengers or nearby pedestrians. A question that programmers of AI systems find difficult to answer (as do ordinary people, and ethicists) is "what decision should the car make that causes the 'smallest' damage to people's lives?"

The ethics of autonomous vehicles are still being articulated, and may lead to controversy. They may also require closer consideration of the variability, context-dependency, complexity and non-deterministic nature of human ethics. Different human drivers make various ethical decisions when driving, such as avoiding harm to themselves, or putting themselves at risk to protect others. These decisions range from rare extremes such as self-sacrifice or criminal negligence, to routine decisions good enough to keep the traffic flowing but bad enough to cause accidents, road rage and stress.

Human thought and reaction time may sometimes be too slow to detect the risk of an upcoming fatal crash, think through the ethical implications of the available options, or take an action to implement an ethical choice. Whether a particular automated vehicle's capacity to correctly detect an upcoming risk, analyse the options or choose a 'good' option from among bad choices would be as good or better than a particular human's may be difficult to predict or assess. This difficulty may be in part because the level of automated vehicle system understanding of the ethical issues at play in a given road scenario, sensed for an instant from out of a continuous stream of synthetic physical predictions of the near future, and dependent on layers of pattern recognition and situational intelligence, may be opaque to human inspection because of its origins in probabilistic machine learning rather than a simple, plain English 'human values' logic of parsable rules. The depth of understanding, predictive power and ethical sophistication needed will be hard to implement, and even harder to test or assess.

The scale of this challenge may have other effects. There may be few entities able to marshal the resources and AI capacity necessary to meet it, as well as the capital necessary to take an autonomous vehicle system to market and sustain it operationally for the life of a vehicle, and the legal and 'government affairs' capacity to deal with the potential for liability for a significant proportion of traffic accidents. This may have the effect of narrowing the number of different system opertors, and eroding the presently quite diverse global vehicle market down to a small number of system suppliers.

Incidents

Mercedes autonomous cruise control system

In 1999, Mercedes introduced Distronic, the first radar-assisted ACC, on the Mercedes-Benz S-Class (W220) and the CL-Class. The Distronic system was able to adjust the vehicle speed automatically to the car in front in order to always maintain a safe distance to other cars on the road.

In 2005, Mercedes refined the system (from this point called "Distronic Plus") with the Mercedes-Benz S-Class (W221) being the first car to receive the upgraded Distronic Plus system. Distronic Plus could now completely halt the car if necessary on E-Class and most Mercedes sedans. In an episode of *Top Gear*, Jeremy Clarkson demonstrated the effectiveness of the cruise control system in the S-class by coming to a complete halt from motorway speeds to a round-about and getting out, without touching the pedals.

By 2017, Mercedes has vastly expanded its autonomous driving features on production cars: In addition to the standard Distronic Plus features such as an active brake assist, Mercedes now includes a steering pilot, a parking pilot, a cross-traffic assist system, night-vision cameras with automated danger warnings and braking assist (in case animals or pedestrians are on the road for example), and various other autonomous-driving features. In 2016, Mercedes also introduced its Active Brake Assist 4, which was the first emergency braking assistant with pedestrian recognition on the market.

Due to Mercedes' history of gradually implementing advancements of their autonomous driving features that have been extensively tested, not many crashes that have been caused by it are known. One of the known crashes dates back to 2005, when German news magazine "*Stern*" was testing Mercedes' old Distronic system. During the test, the system did not always manage to brake in time. Ulrich Mellinghoff, then Head of Safety, NVH, and Testing at the Mercedes-Benz Technology Centre, stated that some of the tests failed due to the vehicle being tested in a metallic hall, which caused problems with the system's radar. Later iterations of the Distronic system have an upgraded radar and numerous other sensors, which are not susceptible to a metallic environment anymore. In 2008, Mercedes conducted a study comparing the crash rates of their vehicles equipped with Distronic Plus and the vehicles without it, and concluded that those equipped with Distronic Plus have an around 20% lower crash rate. In 2013, German Formula One driver Michael Schumacher was invited by Mercedes to try to crash a Mercedes C-Class vehicle, which was equipped with all safety features that Mercedes offered for its production vehicles at the time, which included the Active Blind Spot Assist, Active Lane Keeping Assist, Brake Assist Plus, Collision Prevention Assist, Distronic Plus with Steering Assist, Pre-Safe Brake, and Stop&Go Pilot. Due to the safety features, Schumacher was unable to crash the vehicle in realistic scenarios.

Tesla Autopilot

In mid-October 2015 Tesla Motors rolled out version 7 of their software in the U.S. that included Tesla Autopilot capability. On 9 January 2016, Tesla rolled out version 7.1 as an over-the-air update, adding a new "summon" feature that allows cars to self-park at parking locations without the driver in

Figure 66: *Tesla Model S Autopilot system in use in July 2016; it was only suitable for limited-access highways, not for urban driving. Among other limitations, it could not detect pedestrians or cyclists.*

the car. Tesla's autonomous driving features can be classified as somewhere between level 2 and level 3 under the U.S. Department of Transportation's National Highway Traffic Safety Administration (NHTSA) five levels of vehicle automation. At this level the car can act autonomously but requires the full attention of the driver, who must be prepared to take control at a moment's notice. Autopilot should be used only on limited-access highways, and sometimes it will fail to detect lane markings and disengage itself. In urban driving the system will not read traffic signals or obey stop signs. The system also does not detect pedestrians or cyclists.

On 20 January 2016 the first known fatal crash of a Tesla with Autopilot occurred in China's Hubei province. According to China's 163.com news channel, this marked "China's first accidental death due to Tesla's automatic driving (system)." Initially, Tesla pointed out that the vehicle was so badly damaged from the impact that their recorder was not able to conclusively prove that the car had been on Autopilot at the time, however 163.com pointed out that other factors, such as the car's absolute failure to take any evasive actions prior to the high speed crash, and the driver's otherwise good driving record, seemed to indicate a strong likelihood that the car was on Autopilot at the time. A similar fatal crash occurred four months later in Florida. In 2018, in a subsequent civil suit between the father of the driver killed and Tesla, Tesla did not deny that the car had been on Autopilot at the time of the accident, and sent evidence to the victim's father documenting that fact.

The second known fatal accident involving a vehicle being driven by itself took place in Williston, Florida on 7 May 2016 while a Tesla Model S electric car was engaged in Autopilot mode. The occupant was killed in a crash with an 18-wheel tractor-trailer. On 28 June 2016 the National Highway Traffic Safety Administration (NHTSA) opened a formal investigation into the accident working with the Florida Highway Patrol. According to the NHTSA, preliminary reports indicate the crash occurred when the tractor-trailer made a left turn in front of the Tesla at an intersection on a non-controlled access highway, and the car failed to apply the brakes. The car continued to travel after passing under the truck's trailer. The NHTSA's preliminary evaluation was opened to examine the design and performance of any automated driving systems in use at the time of the crash, which involved a population of an estimated 25,000 Model S cars. On 8 July 2016, the NHTSA requested Tesla Motors provide the agency detailed information about the design, operation and testing of its Autopilot technology. The agency also requested details of all design changes and updates to Autopilot since its introduction, and Tesla's planned updates schedule for the next four months.

According to Tesla, "neither autopilot nor the driver noticed the white side of the tractor-trailer against a brightly lit sky, so the brake was not applied." The car attempted to drive full speed under the trailer, "with the bottom of the trailer impacting the windshield of the Model S." Tesla also claimed that this was Tesla's first known autopilot death in over 130 million miles (208 million km) driven by its customers with Autopilot engaged, however by this statement, Tesla was apparently refusing to acknowledge claims that the January 2016 fatality in Hubei China had also been the result of an autopilot system error. According to Tesla there is a fatality every 94 million miles (150 million km) among all type of vehicles in the U.S. However, this number also includes fatalities of the crashes, for instance, of motorcycle drivers with pedestrians.

In July 2016 the U.S. National Transportation Safety Board (NTSB) opened a formal investigation into the fatal accident while the Autopilot was engaged. The NTSB is an investigative body that has the power to make only policy recommendations. An agency spokesman said "It's worth taking a look and seeing what we can learn from that event, so that as that automation is more widely introduced we can do it in the safest way possible." In January 2017, the NTSB released the report that concluded Tesla was not at fault; the investigation revealed that for Tesla cars, the crash rate dropped by 40 percent after Autopilot was installed.

According to Tesla, starting 19 October 2016, all Tesla cars are built with hardware to allow full self-driving capability at the highest safety level (SAE

Level 5). The hardware includes eight surround cameras and twelve ultrasonic sensors, in addition to the forward-facing radar with enhanced processing capabilities. The system will operate in "shadow mode" (processing without taking action) and send data back to Tesla to improve its abilities until the software is ready for deployment via over-the-air upgrades. After the required testing, Tesla hopes to enable full self-driving by the end of 2019 under certain conditions.

Google self-driving car

In August 2012, Google announced that their vehicles had completed over 300,000 autonomous-driving miles (500,000 km) accident-free, typically involving about a dozen cars on the road at any given time, and that they were starting to test with single drivers instead of in pairs.[1152] In late-May 2014, Google revealed a new prototype that had no steering wheel, gas pedal, or brake pedal, and was fully autonomous. As of March 2016[1153], Alphabet had test-driven their fleet in autonomous mode a total of 1,500,000 mi (2,400,000 km). In December 2016, Google Corporation announced that its technology would be spun-off to a new subsidiary called Waymo.

Based on Google's accident reports, their test cars have been involved in 14 collisions, of which other drivers were at fault 13 times, although in 2016 the car's software caused a crash.

In June 2015, Brin confirmed that 12 vehicles had suffered collisions as of that date. Eight involved rear-end collisions at a stop sign or traffic light, two in which the vehicle was side-swiped by another driver, one in which another driver rolled through a stop sign, and one where a Google employee was controlling the car manually. In July 2015, three Google employees suffered minor injuries when their vehicle was rear-ended by a car whose driver failed to brake at a traffic light. This was the first time that a collision resulted in injuries. On 14 February 2016 a Waymo vehicle attempted to avoid sandbags blocking its path. During the maneuver it struck a bus. Google stated, "In this case, we clearly bear some responsibility, because if our car hadn't moved there wouldn't have been a collision." Google characterized the crash as a misunderstanding and a learning experience. No injuries were reported in the crash.

Uber

In March 2017, an Uber test vehicle was involved in a crash in Tempe, Arizona when another car failed to yield, flipping the Uber vehicle. There were no injuries in the accident.

By 22 December 2017, Uber had completed 2,000,000 miles in autonomous mode.

On 18 March 2018, Elaine Herzberg became the first pedestrian to be killed by a self-driving car in the United States after being hit by an Uber vehicle, also in Tempe. Herzberg was crossing outside of a crosswalk, approximately 400 feet from an intersection. The causes of the accidents include the following: nighttime, low visibility, pedestrian crossing from a shadowed portion of road, crossing a road that had high speed limit, and not checking for cars before crossing (blindly crossing street at night). This marks the first time an individual outside an auto-piloted car is known to have been killed by such a car. The first death of an essentially uninvolved third party is a temporary setback to the autonomous cars= industry. Some experts say a human driver could have avoided the fatal crash. Arizona Governor Doug Ducey later suspended the company's ability to test and operate its autonomous cars on public roadways citing an "unquestionable failure" of the expectation that Uber make public safety its top priority. Uber has pulled out of all self-driving-car testing in California as a result of the accident. On 24 May 2018 the National Transport Safety Board issued a preliminary report.

Navya autonomous bus driving system

On 9 November 2017 a Navya autonomous self-driving bus with passengers was involved in a crash with a truck. The truck was found to be at fault of the crash, reversing into the stationary autonomous bus. The autonomous bus did not take evasive actions or apply defensive driving such as flash headlights, sound the horn, or as one passenger commented "The shuttle didn't have the ability to move back. The shuttle just stayed still."

Policy implications

Urban planning

According to a Wonkblog reporter, if fully autonomous cars become commercially available, they have the potential to be a disruptive innovation with major implications for society. The likelihood of widespread adoption is still unclear, but if they are used on a wide scale, policy makers face a number of unresolved questions about their effects.

One fundamental question is about their effect on travel behavior. Some people believe that they will increase car ownership and car use because it will become easier to use them and they will ultimately be more useful. This may in turn encourage urban sprawl and ultimately total private vehicle use. Others argue that it will be easier to share cars and that this will thus discourage outright ownership and decrease total usage, and make cars more efficient forms of transportation in relation to the present situation.

Policy-makers will have to take a new look at how infrastructure is to be built and how money will be allotted to build for autonomous vehicles. The need for traffic signals could potentially be reduced with the adoption of smart highways.[1154] Due to smart highways and with the assistance of smart technological advances implemented by policy change, the dependence on oil imports may be reduced because of less time being spent on the road by individual cars which could have an effect on policy regarding energy. On the other hand, autonomous vehicles could increase the overall number of cars on the road which could lead to a greater dependence on oil imports if smart systems are not enough to curtail the impact of more vehicles. However, due to the uncertainty of the future of autonomous vehicles, policy makers may want to plan effectively by implementing infrastructure improvements that can be beneficial to both human drivers and autonomous vehicles. Caution needs to be taken in acknowledgment to public transportation and that the use may be greatly reduced if autonomous vehicles are catered to through policy reform of infrastructure with this resulting in job loss and increased unemployment.[1155]

Other disruptive effects will come from the use of autonomous vehicles to carry goods. Self-driving vans have the potential to make home deliveries significantly cheaper, transforming retail commerce and possibly making hypermarkets and supermarkets redundant. As of right now the U.S. Government defines automation into six levels, starting at level zero which means the human driver does everything and ending with level five, the automated system performs all the driving tasks. Also under the current law, manufacturers bear all the responsibility to self-certify vehicles for use on public roads. This means that currently as long as the vehicle is compliant within the regulatory framework, there are no specific federal legal barriers to a highly automated

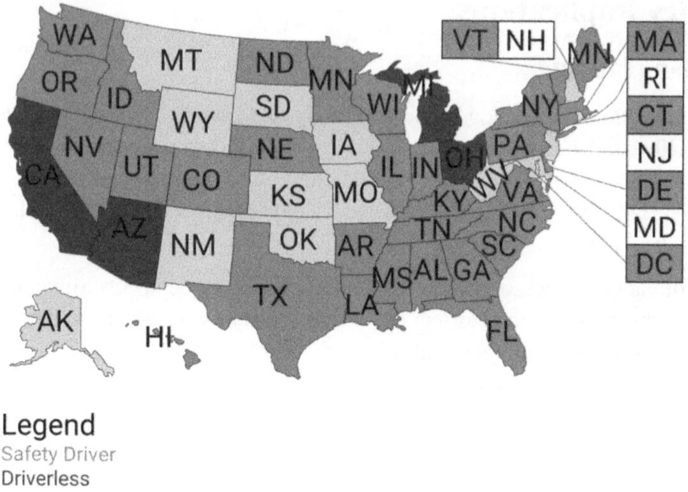

Figure 67: *U.S. states that allow testing of driverless cars on public roads*

vehicle being offered for sale. Iyad Rahwan, an associate professor in the MIT Media lab said, "Most people want to live in a world where cars will minimize casualties, but everyone wants their own car to protect them at all costs." Furthermore, industry standards and best practice are still needed in systems before they can be considered reasonably safe under real-world conditions.

Legislation

The 1968 Vienna Convention on Road Traffic, subscribed to by over 70 countries worldwide, establishes principles to govern traffic laws. One of the fundamental principles of the Convention has been the concept that a driver is always fully in control and responsible for the behavior of a vehicle in traffic.[1156] The progress of technology that assists and takes over the functions of the driver is undermining this principle, implying that much of the groundwork must be rewritten.

Legal status in the United States

In the United States, a non-signatory country to the Vienna Convention, state vehicle codes generally do not envisage — but do not necessarily prohibit — highly automated vehicles. To clarify the legal status of and otherwise regulate such vehicles, several states have enacted or are considering specific laws.

In 2016, 7 states (Nevada, California, Florida, Michigan, Hawaii, Washington, and Tennessee), along with the District of Columbia, have enacted laws for autonomous vehicles. Incidents such as the first fatal accident by Tesla's Autopilot system have led to discussion about revising laws and standards for autonomous cars.

In September 2016, the US National Economic Council and Department of Transportation released federal standards that describe how automated vehicles should react if their technology fails, how to protect passenger privacy, and how riders should be protected in the event of an accident. The new federal guidelines are meant to avoid a patchwork of state laws, while avoiding being so overbearing as to stifle innovation.

In June 2011, the Nevada Legislature passed a law to authorize the use of autonomous cars. Nevada thus became the first jurisdiction in the world where autonomous vehicles might be legally operated on public roads. According to the law, the Nevada Department of Motor Vehicles (NDMV) is responsible for setting safety and performance standards and the agency is responsible for designating areas where autonomous cars may be tested. This legislation was supported by Google in an effort to legally conduct further testing of its Google driverless car. The Nevada law defines an autonomous vehicle to be "a motor vehicle that uses artificial intelligence, sensors and global positioning system coordinates to drive itself without the active intervention of a human operator." The law also acknowledges that the operator will not need to pay attention while the car is operating itself. Google had further lobbied for an exemption from a ban on distracted driving to permit occupants to send text messages while sitting behind the wheel, but this did not become law. Furthermore, Nevada's regulations require a person behind the wheel and one in the passenger's seat during tests.

In April 2012, Florida became the second state to allow the testing of autonomous cars on public roads,[1157] and California became the third when Governor Jerry Brown signed the bill into law at Google Headquarters in Mountain View.[1158] In December 2013, Michigan became the fourth state to allow testing of driverless cars on public roads. In July 2014, the city of Coeur d'Alene, Idaho adopted a robotics ordinance that includes provisions to allow for self-driving cars.[1159]

On 19 February 2016, Assembly Bill No. 2866 was introduced in California that would allow completely autonomous vehicles to operate on the road, including those without a driver, steering wheel, accelerator pedal, or brake pedal. The Bill states the Department of Motor Vehicles would need to comply with these regulations by 1 July 2018 for these rules to take effect. This bill has yet to pass the house of origin.

In September 2016, the U.S. Department of Transportation released its Federal Automated Vehicles Policy, and California published discussions on the subject in October 2016.

In December 2016, the California Department of Motor Vehicles ordered Uber to remove its self-driving vehicles from the road in response to two red-light violations. Uber immediately blamed the violations on "human-error", and has suspended the drivers.

Legislation in Europe

In 2013, the government of the United Kingdom permitted the testing of autonomous cars on public roads. Before this, all testing of robotic vehicles in the UK had been conducted on private property.

In 2014 the Government of France announced that testing of autonomous cars on public roads would be allowed in 2015. 2000 km of road would be opened through the national territory, especially in Bordeaux, in Isère, Île-de-France and Strasbourg. At the 2015 ITS World Congress, a conference dedicated to intelligent transport systems, the very first demonstration of autonomous vehicles on open road in France was carried out in Bordeaux in early October 2015.

In 2015, a preemptive lawsuit against various automobile companies such as GM, Ford, and Toyota accused them of "Hawking vehicles that are vulnerable to hackers who could hypothetically wrest control of essential functions such as brakes and steering."

In spring of 2015, the Federal Department of Environment, Transport, Energy and Communications in Switzerland (UVEK) allowed Swisscom to test a driverless Volkswagen Passat on the streets of Zurich.

As of April 2017 it is possible to conduct public road tests for development vehicles in Hungary, furthermore the construction of a closed test track, the Zala Zone test track, suitable for testing highly automated functions is also under way near the city of Zalaegerszeg.

Legislation in Asia

In 2016, the Singapore Land Transit Authority in partnership with UK automotive supplier Delphi Automotive Plc will launch preparations for a test run of a fleet of automated taxis for an on-demand autonomous cab service to take effect in 2017.

Liability

Autonomous car liability is a developing area of law and policy that will determine who is liable when an autonomous car causes physical damage to persons, or breaks road rules. When autonomous cars shift the control of driving from humans to autonomous car technology, there may be a need for existing liability laws to evolve in order to fairly identify the parties responsible for damage and injury, and to address the potential for conflicts of interest between human occupants, system operator, insurers, and the public purse. Increases in the use of autonomous car technologies (e.g. advanced driver-assistance systems) may prompt incremental shifts in this responsibility for driving. It is claimed by proponents to have potential to affect the frequency of road accidents, although it is difficult to assess this claim in the absence of data from substantial actual use. If there was a dramatic improvement in safety, the operators may seek to project their liability for the remaining accidents onto others as part of their reward for the improvement. However there is no obvious reason why they should escape liability if any such effects were found to be modest or nonexistent, since part of the purpose of such liability is to give an incentive to the party controlling something to do whatever is necessary to avoid it causing harm. Potential users may be reluctant to trust an operator if it seeks to pass its normal liability on to others.

In any case, a well-advised person who is not controlling a car at all (Level 5) would be understandably reluctant to accept liability for something out of their control. And when there is some degree of sharing control possible (Level 3 or 4), a well-advised person would be concerned that the vehicle might try to pass back control at the last seconds before an accident, to pass responsibility and liability back too, but in circumstances where the potential driver has no better prospects of avoiding the crash than the vehicle, since they have not necessarily been paying close attention, and if it is too hard for the very smart car it might be too hard for a human. Since operators, especially those familiar with trying to ignore existing legal obligations (under a motto like 'seek forgiveness, not permission'), such as Google or Uber, could be normally expected to try to avoid responsibility to the maximum degree possible, there is potential for attempt to let the operators evade being held liable for accidents while they are in control.

As higher levels of autonomy are commercially introduced (level 3 and 4), the insurance industry may see a greater proportion of commercial and product liability lines while personal automobile insurance shrinks.

Vehicular communication systems

Individual vehicles may benefit from information obtained from other vehicles in the vicinity, especially information relating to traffic congestion and safety hazards. Vehicular communication systems use vehicles and roadside units as the communicating nodes in a peer-to-peer network, providing each other with information. As a cooperative approach, vehicular communication systems can allow all cooperating vehicles to be more effective. According to a 2010 study by the National Highway Traffic Safety Administration, vehicular communication systems could help avoid up to 79 percent of all traffic accidents.

In 2012, computer scientists at the University of Texas in Austin began developing smart intersections designed for autonomous cars. The intersections will have no traffic lights and no stop signs, instead using computer programs that will communicate directly with each car on the road.

An efficient intersection management technique called Crossroads was proposed in 2017 that is robust to network delay of V2I communication and Worst-case Execution time of the intersection manager.

Among connected cars, an unconnected one is the weakest link and will be increasingly banned from busy high-speed roads, predicted a Helsinki think tank in January 2016.

Public opinion surveys

In a 2011 online survey of 2,006 US and UK consumers by Accenture, 49% said they would be comfortable using a "driverless car".

A 2012 survey of 17,400 vehicle owners by J.D. Power and Associates found 37% initially said they would be interested in purchasing a fully autonomous car. However, that figure dropped to 20% if told the technology would cost $3,000 more.

In a 2012 survey of about 1,000 German drivers by automotive researcher Puls, 22% of the respondents had a positive attitude towards these cars, 10% were undecided, 44% were skeptical and 24% were hostile.

A 2013 survey of 1,500 consumers across 10 countries by Cisco Systems found 57% "stated they would be likely to ride in a car controlled entirely by technology that does not require a human driver", with Brazil, India and China the most willing to trust autonomous technology.

In a 2014 US telephone survey by Insurance.com, over three-quarters of licensed drivers said they would at least consider buying a self-driving car, rising to 86% if car insurance were cheaper. 31.7% said they would not continue to drive once an autonomous car was available instead.

In a February 2015 survey of top auto journalists, 46% predict that either Tesla or Daimler will be the first to the market with a fully autonomous vehicle, while (at 38%) Daimler is predicted to be the most functional, safe, and in-demand autonomous vehicle.

In 2015 a questionnaire survey by Delft University of Technology explored the opinion of 5,000 people from 109 countries on automated driving. Results showed that respondents, on average, found manual driving the most enjoyable mode of driving. 22% of the respondents did not want to spend any money for a fully automated driving system. Respondents were found to be most concerned about software hacking/misuse, and were also concerned about legal issues and safety. Finally, respondents from more developed countries (in terms of lower accident statistics, higher education, and higher income) were less comfortable with their vehicle transmitting data. The survey also gave results on potential consumer opinion on interest of purchasing an automated car, stating that 37% of surveyed current owners were either "definitely" or "probably" interested in purchasing an automated car.

In 2016, a survey in Germany examined the opinion of 1,603 people, who were representative in terms of age, gender, and education for the German population, towards partially, highly, and fully automated cars. Results showed that men and women differ in their willingness to use them. Men felt less anxiety and more joy towards automated cars, whereas women showed the exact opposite. The gender difference towards anxiety was especially pronounced between young men and women but decreased with participants' age.

In 2016, a PwC survey, in the United States, showing the opinion of 1,584 people, highlights that "66 percent of respondents said they think autonomous cars are probably smarter than the average human driver". People are still worried about safety and mostly the fact of having the car hacked. Nevertheless, only 13% of the interviewees see no advantages in this new kind of cars.

A Pew Research Center survey of 4,135 U.S. adults conducted 1–15 May 2017 finds that many Americans anticipate significant impacts from various automation technologies in the course of their lifetimes—from the widespread adoption of autonomous vehicles to the replacement of entire job categories with robot workers.

Moral issues

With the emergence of autonomous automobiles various ethical issues arise. While the introduction of autonomous vehicles to the mass market is said to be inevitable due to an (untestable) potential for reduction of crashes by "up to" 90% and their potential greater accessibility to disabled, elderly, and young passengers, a range of ethical issues have not been fully addressed. Those include, but are not limited to: the moral, financial, and criminal responsibility for crashes and breaches of law; the decisions a car is to make right before a (fatal) crash; privacy issues including potential for mass surveillance; potential for massive job losses and unemployment among drivers; de-skilling and loss of independence by vehicle users; exposure to hacking and malware; and the further concentration of market and data power in the hands of a few global conglomerates capable of consolidating AI capacity, and of lobbying governments to facilitate the shift of liability onto others and their potential destruction of existing occupations and industries.

There are different opinions on who should be held liable in case of a crash, especially with people being hurt. Many experts see the car manufacturers themselves responsible for those crashes that occur due to a technical malfunction or misconstruction. Besides the fact that the car manufacturer would be the source of the problem in a situation where a car crashes due to a technical issue, there is another important reason why car manufacturers could be held responsible: it would encourage them to innovate and heavily invest into fixing those issues, not only due to protection of the brand image, but also due to financial and criminal consequences. However, there are also voices [who?] that argue those using or owning the vehicle should be held responsible since they know the risks involved in using such a vehicle. Experts [who?] suggest introducing a tax or insurances that would protect owners and users of autonomous vehicles of claims made by victims of an accident. Other possible parties that can be held responsible in case of a technical failure include software engineers that programmed the code for the autonomous operation of the vehicles, and suppliers of components of the AV.

Taking aside the question of legal liability and moral responsibility, the question arises how autonomous vehicles should be programmed to behave in an emergency situation where either passengers or other traffic participants are endangered. A moral dilemma that a software engineer or car manufacturer might face in programming the operating software is described in an ethical thought experiment, the trolley problem: a conductor of a trolley has the choice of staying on the planned track and running over 5 people, or turn the trolley onto a track where it would kill only one person, assuming there is no traffic on it. There are two main considerations that need to be addressed. First, what moral basis would be used by an autonomous vehicle to make decisions? Second, how could those be translated into software code? Researchers have suggested, in particular, two ethical theories to be applicable to the behavior of autonomous vehicles in cases of emergency: deontology and utilitarianism. Asimov's three laws of robotics are a typical example of deontological ethics. The theory suggests that an autonomous car needs to follow strict written-out rules that it needs to follow in any situation. Utilitarianism suggests the idea that any decision must be made based on the goal to maximize utility. This needs a definition of utility which could be maximizing the number of people surviving in a crash. Critics suggest that autonomous vehicles should adapt a mix of multiple theories to be able to respond morally right in the instance of a crash.

(Many 'Trolley' discussions skip over the practical problems of how a probabilistic machine learning vehicle AI could be sophisticated enough to understand that a deep problem of moral philosophy is presenting itself from instant to instant while using a dynamic projection into the near future, what sort of moral problem it actually would be if any, what the relevant weightings in human value terms should be given to all the other humans involved who will be probably unreliably identified, and how reliably it can assess the probable outcomes. These practical difficulties, and those around testing and assessment of solutions to them, may present as much of a challenge as the theoretical abstractions.)

Privacy-related issues arise mainly from the interconnectivity of autonomous cars, making it just another mobile device that can gather any information about an individual. This information gathering ranges from tracking of the routes taken, voice recording, video recording, preferences in media that is consumed in the car, behavioral patterns, to many more streams of information. The data and communications infrastructure needed to support these vehicles may also be capable of surveillance, especially if coupled to other data sets and advanced analytics.

The implementation of autonomous vehicles to the mass market might cost up to 5 million jobs in the US alone, making up almost 3% of the workforce. Those jobs include drivers of taxis, buses, vans, trucks, and e-hailing vehicles. Many industries, such as the auto insurance industry are indirectly affected. This industry alone generates an annual revenue of about $220 billions, supporting 277,000 jobs. To put this into perspective – this is about the number of mechanical engineering jobs. The potential loss of a majority of those jobs will have a tremendous impact on those individuals involved. Both India and China have placed bans on automated cars with the former citing protection of jobs.Wikipedia:Citation needed

Anticipated launch of self-driving cars

In December 2015, Tesla CEO Elon Musk predicted that a completely autonomous car would be introduced by the end of 2018; in December 2017, he announced that it would take another two years to launch a fully self-driving Tesla onto the market.

BMW's all-electric autonomous car, called iNext, is expected to be ready by 2021; Toyota's first self-driving car is due to hit the market in 2020, as is the driverless car being developed by Nissan.

In fiction

In film

The autonomous and occasionally sentient self-driving car story has earned its place in both literary science fiction and pop sci-fi.

- A VW Beetle named *Dudu* features in the 1971 to 1978 German Superbug (film series) of movies similar to Disney's Herbie, but with an electronic brain. (Herbie, also a Beetle, was depicted as an anthropomorphic car with its own spirit.)
- In the film *Batman* (1989), starring Michael Keaton, the Batmobile is shown to be able to drive to Batman's current location with some navigation commands from Batman and possibly some autonomy.
- The film *Total Recall* (1990), starring Arnold Schwarzenegger, features taxis called *Johnny Cabs* controlled by artificial intelligence in the car or the android occupants.
- The film *Demolition Man* (1993), starring Sylvester Stallone and set in 2032, features vehicles that can be self-driven or commanded to "Auto Mode" where a voice-controlled computer operates the vehicle.
- The film *Timecop* (1994), starring Jean-Claude Van Damme, set in 2004 and 1994, has autonomous cars.
- Another Arnold Schwarzenegger movie, *The 6th Day* (2000), features an autonomous car commanded by Michael Rapaport.

- The film *Minority Report* (2002), set in Washington, D.C. in 2054, features an extended chase sequence involving autonomous cars. The vehicle of protagonist John Anderton is transporting him when its systems are overridden by police in an attempt to bring him into custody.
- The film, *The Incredibles* (2004), Mr. Incredible makes his car autonomous for him while it changes him into his supersuit when driving to save a cat from a tree.

- The film Eagle Eye (2008) Shia LaBeouf and Michelle Monaghan are driven around in a Porsche Cayenne that is controlled by ARIIA (a giant supercomputer).
- The film *I, Robot* (2004), set in Chicago in 2035, features autonomous vehicles driving on highways, allowing the car to travel safer at higher speeds than if manually controlled. The option to manually operate the vehicles is available.
- Logan (film) (2017) set in 2029, features fully autonomous trucks[1160].
- Blade Runner 2049 (2017) opens with LAPD Replicant cop K waking up in his 3-wheeled autonomous flying vehicle (featuring a separable surveillance roof drone) on approach to a protein farm in northern California.
- Geostorm, set in 2022, features a self-driving taxi stolen by protagonists Max Lawson and Sarah Wilson to protect the President from mercenaries and a superstorm.

In literature

Intelligent or self-driving cars are a common theme in science fiction literature. Examples include:

- In Isaac Asimov's science-fiction short story, "Sally" (first published May–June 1953), autonomous cars have "positronic brains" and communicate via honking horns and slamming doors, and save their human caretaker.
- Peter F. Hamilton's *Commonwealth Saga* series features intelligent or self-driving vehicles.
- In Robert A Heinlein's novel, *The Number of the Beast* (1980), Zeb Carter's driving and flying car "Gay Deceiver" is at first semi-autonomous and later, after modifications by Zeb's wife Deety, becomes sentient and capable of fully autonomous operation.
- In Edizioni Piemme's series *Geronimo Stilton*, a robotic vehicle called "Solar" is in the 54th book.
- Alastair Reynolds' series, *Revelation Space*, features intelligent or self-driving vehicles.
- In Daniel Suarez' novels *Daemon* (2006) and *Freedom*™ (2010) driverless cars and motorcycles are used for attacks in a software-based open-source warfare. The vehicles are modified for this using 3D printers and distributed manufacturing and are also able to operate as swarms.

In television

- "CSI: Cyber" Season 2, episode 6, *Gone in 60 Seconds*, features three seemingly normal customized vehicles, a 2009 Nissan Fairlady Z Roadster, a BMW M3 E90 and a Cadillac CTS-V, and one stock luxury BMW 7 Series, being remote-controlled by a computer hacker.
- "Handicar", season 18, episode 4 of 2014 TV series *South Park* features a Japanese autonomous car that takes part in the *Wacky Races*-style car race.
- KITT and KARR, the Pontiac Trans Ams in the 1982 TV series *Knight Rider*, were sentient and autonomous.
- "Driven", series 4 episode 11 of the 2003 TV series *NCIS* features a robotic vehicle named "Otto", part of a high-level project of the Department of Defense, which causes the death of a Navy Lieutenant, and then later almost kills Abby.
- The TV series "Viper" features a silver/grey armored assault vehicle, called *The Defender*, which masquerades as a flame-red 1992 Dodge Viper RT/10 and later as a 1998 cobalt blue Dodge Viper GTS. The vehicle's sophisticated computer systems allow it to be controlled via remote on some occasions.

- "Black Mirror" episode "Hated in the Nation" briefly features a self-driving SUV with a touchscreen interface on the inside.
- *Bull* has a show discussing the effectiveness and safety of self-driving cars in an episode call E.J.

Further reading

- O'Toole, Randal (18 January 2010). *Gridlock: Why We're Stuck in Traffic and What To Do About It*[1161]. Cato Institute. ISBN 978-1-935308-24-9.
- Macdonald, Iain David Graham (2011). *A Simulated Autonomous Car*[1162] (PDF) (thesis). The University of Edinburgh. Retrieved 17 April 2013.
- Knight, Will (22 October 2013). "The Future of Self-driving Cars"[1163]. *MIT Technology Review*. Retrieved 22 July 2016.
- Glancy, Dorothy (2016). A Look at the Legal Environment for Driverless Vehicles[1164] (PDF) (Report). National Cooperative Highway Research Program Legal Research Digest. **69**. Washington, DC: Transportation Research Board. ISBN 978-0-309-37501-6. Retrieved 22 July 2016.
- Meyer, Gereon; Beiker, Sven, eds. (2014). *Road vehicle automation*. Springer. ISBN 978-3-319-05990-7.
- Meyer, Gereon; Beiker, Sven, eds. (2015). *Road vehicle automation 2*. Springer. ISBN 978-3-319-19078-5.
- Newbold, Richard (17 June 2015). "The driving forces behind what would be the next revolution in the haulage sector"[1165]. *The Loadstar*. Retrieved 22 July 2016.
- Bergen, Mark (27 October 2015). "Meet the Companies Building Self-Driving Cars for Google and Tesla (And Maybe Apple)"[1166]. *re/code*.
- John A. Volpe National Transportation Systems Center (March 2016). "Review of Federal Motor Vehicle Safety Standards (FMVSS) for Automated Vehicles: Identifying potential barriers and challenges for the certification of automated vehicles using existing FMVSS"[1167] (PDF). *National Transportation Library*. U.S. Department of Transportation.
- Slone, Sean (August 2016). "State Laws on Autonomous Vehicles"[1168] (PDF). *Capitol Research - Transportation Policy*. Council of State Governments. Retrieved 28 September 2016.
- Steve Henn (31 July 2015). "Remembering When Driverless Elevators Drew Skepticism"[1169].
- James M. Anderson et. al. (2016). "Autonomous Vehicle Technology: A Guide for Policymakers"[1170] (PDF). RAND Corporation.

Wikimedia Commons has media related to ***Unmanned automobiles***.

Philosophy and Ethics

Philosophy of artificial intelligence

The **philosophy of artificial intelligence** attempts to answer such questions as follows:[1171]

- Can a machine act intelligently? Can it solve *any* problem that a person would solve by thinking?
- Are human intelligence and machine intelligence the same? Is the human brain essentially a computer?
- Can a machine have a mind, mental states, and consciousness in the same way that a human being can? Can it *feel how things are*?

These three questions reflect the divergent interests of AI researchers, linguists, cognitive scientists and philosophers respectively. The scientific answers to these questions depend on the definition of "intelligence" and "consciousness" and exactly which "machines" are under discussion.

Important propositions in the philosophy of AI include:

- Turing's "polite convention": *If a machine behaves as intelligently as a human being, then it is as intelligent as a human being.*[1172]
- The Dartmouth proposal: *"Every aspect of learning or any other feature of intelligence can be so precisely described that a machine can be made to simulate it."*[1173]
- Newell and Simon's physical symbol system hypothesis: *"A physical symbol system has the necessary and sufficient means of general intelligent action."*[1174]
- Searle's strong AI hypothesis: *"The appropriately programmed computer with the right inputs and outputs would thereby have a mind in exactly the same sense human beings have minds."*[1175]
- Hobbes' mechanism: *"For 'reason' ... is nothing but 'reckoning,' that is adding and subtracting, of the consequences of general names agreed upon for the 'marking' and 'signifying' of our thoughts..."*

Can a machine display general intelligence?

Is it possible to create a machine that can solve *all* the problems humans solve using their intelligence? This question defines the scope of what machines will be able to do in the future and guides the direction of AI research. It only concerns the *behavior* of machines and ignores the issues of interest to psychologists, cognitive scientists and philosophers; to answer this question, it does not matter whether a machine is *really* thinking (as a person thinks) or is just *acting like* it is thinking.[1176]

The basic position of most AI researchers is summed up in this statement, which appeared in the proposal for the Dartmouth workshop of 1956:

- *Every aspect of learning or any other feature of intelligence can be so precisely described that a machine can be made to simulate it.*

Arguments against the basic premise must show that building a working AI system is impossible, because there is some practical limit to the abilities of computers or that there is some special quality of the human mind that is necessary for thinking and yet cannot be duplicated by a machine (or by the methods of current AI research). Arguments in favor of the basic premise must show that such a system is possible.

The first step to answering the question is to clearly define "intelligence".

Intelligence

Turing test

Alan Turing[1178] reduced the problem of defining intelligence to a simple question about conversation. He suggests that: if a machine can answer *any* question put to it, using the same words that an ordinary person would, then we may call that machine intelligent. A modern version of his experimental design would use an online chat room, where one of the participants is a real person and one of the participants is a computer program. The program passes the test if no one can tell which of the two participants is human. Turing notes that no one (except philosophers) ever asks the question "can people think?" He writes "instead of arguing continually over this point, it is usual to have a polite convention that everyone thinks".[1179] Turing's test extends this polite convention to machines:

- *If a machine acts as intelligently as human being, then it is as intelligent as a human being.*

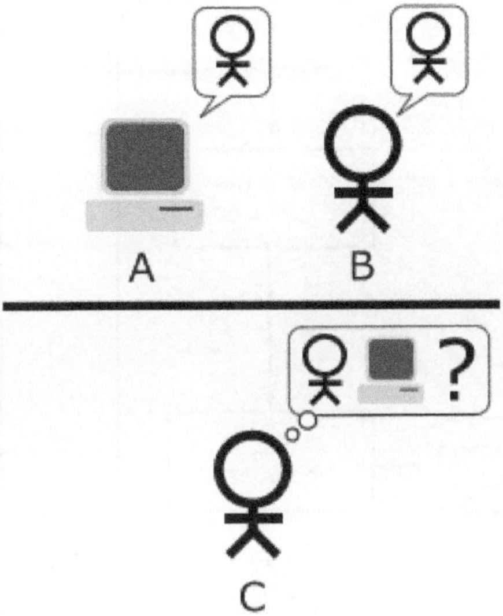

Figure 68: *The "standard interpretation" of the Turing test.*[1177]

One criticism of the Turing test is that it is explicitly anthropomorphic Wikipedia:Citation needed. If our ultimate goal is to create machines that are *more* intelligent than people, why should we insist that our machines must closely *resemble* people?Wikipedia:Inline citation#When you must use inline citations Russell and Norvig write that "aeronautical engineering texts do not define the goal of their field as 'making machines that fly so exactly like pigeons that they can fool other pigeons'".

Intelligent agent definition

Recent A.I. research defines intelligence in terms of intelligent agents. An "agent" is something which perceives and acts in an environment. A "performance measure" defines what counts as success for the agent.

- *If an agent acts so as to maximize the expected value of a performance measure based on past experience and knowledge then it is intelligent.*[1180]

Definitions like this one try to capture the essence of intelligence. They have the advantage that, unlike the Turing test, they do not also test for human traits that we Wikipedia:Manual of Style/Words to watch#Unsupported attributions may not want to consider intelligent, like the ability to be insulted or the temptation to lie Wikipedia:Accuracy dispute#Disputed statement. They have the disadvantage that they fail to make the commonsenseWikipedia:Manual of

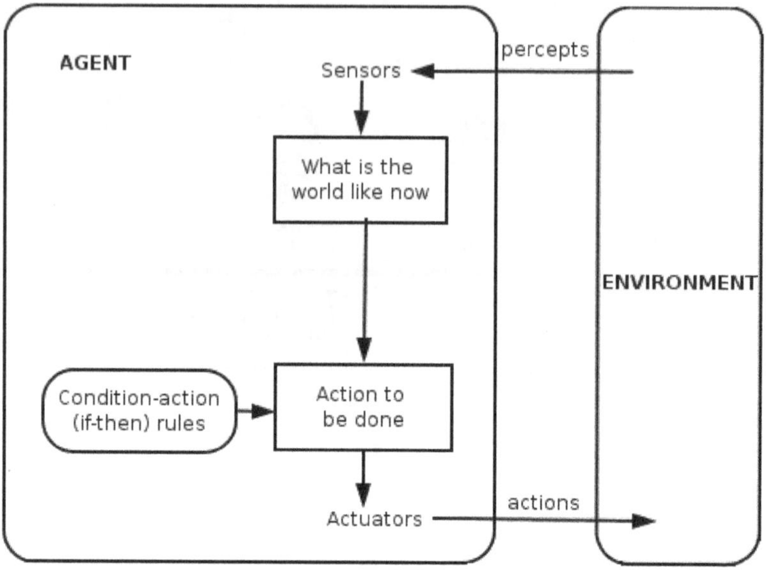

Figure 69: *Simple reflex agent*

Style differentiation between "things that think" and "things that do not". By this definition, even a thermostat has a rudimentary intelligence.[1181]

Arguments that a machine can display general intelligence

The brain can be simulated

Hubert Dreyfus describes this argument as claiming that "if the nervous system obeys the laws of physics and chemistry, which we have every reason to suppose it does, then .... we ... ought to be able to reproduce the behavior of the nervous system with some physical device". This argument, first introduced as early as 1943 and vividly described by Hans Moravec in 1988, is now associated with futurist Ray Kurzweil, who estimates that computer power will be sufficient for a complete brain simulation by the year 2029.[1182] A non-real-time simulation of a thalamocortical model that has the size of the human brain (10^{11} neurons) was performed in 2005 and it took 50 days to simulate 1 second of brain dynamics on a cluster of 27 processors.

FewWikipedia:Manual of Style/Dates and numbers disagree that a brain simulation is possible in theory,Wikipedia:Citation neededWikipedia:Manual_of_Style/Words_to_watch#Unsupported_attributions even critics of AI such as Hubert Dreyfus and John Searle.[1183] However, Searle points out that, in principle, *anything* can be simulated by a computer;

thus, bringing the definition to its breaking point leads to the conclusion that any process at all can technically be considered "computation". "What we wanted to know is what distinguishes the mind from thermostats and livers," he writes. Thus, merely mimicking the functioning of a brain would in itself be an admission of ignorance regarding intelligence and the nature of the mind Wikipedia:Citation needed.

Human thinking is symbol processing

In 1963, Allen Newell and Herbert A. Simon proposed that "symbol manipulation" was the essence of both human and machine intelligence. They wrote:

- *A physical symbol system has the necessary and sufficient means of general intelligent action.*

This claim is very strong: it implies both that human thinking is a kind of symbol manipulation (because a symbol system is *necessary* for intelligence) and that machines can be intelligent (because a symbol system is *sufficient* for intelligence).[1184] Another version of this position was described by philosopher Hubert Dreyfus, who called it "the psychological assumption":

- *The mind can be viewed as a device operating on bits of information according to formal rules.*

A distinction is usually made Wikipedia:Manual of Style/Words to watch#Unsupported attributions between the kind of high level symbols that directly correspond with objects in the world, such as <dog> and <tail> and the more complex "symbols" that are present in a machine like a neural network. Early research into AI, called "good old fashioned artificial intelligence" (GOFAI) by John Haugeland, focused on these kind of high level symbols.

Arguments against symbol processing

These arguments show that human thinking does not consist (solely) of high level symbol manipulation. They do *not* show that artificial intelligence is impossible, only that more than symbol processing is required.

Gödelian anti-mechanist arguments

In 1931, Kurt Gödel proved with an incompleteness theorem that it is always possible to construct a "Gödel statement" that a given consistent formal system of logic (such as a high-level symbol manipulation program) could not prove. Despite being a true statement, the constructed Gödel statement is unprovable in the given system. (The truth of the constructed Gödel statement is contingent on the consistency of the given system; applying the same process to a subtly inconsistent system will appear to succeed, but will actually yield a false "Gödel statement" instead.)Wikipedia:Citation needed More speculatively, Gödel conjectured that the human mind can correctly eventually determine the truth or falsity of any well-grounded mathematical statement (including any possible Gödel statement), and that therefore the human mind's power is not reducible to a *mechanism*.[1185] Philosopher John Lucas (since 1961) and Roger Penrose (since 1989) have championed this philosophical anti-mechanist argument.[1186] Gödelian anti-mechanist arguments tend to rely on the innocuous-seeming claim that a system of human mathematicians (or some idealization of human mathematicians) is both consistent (completely free of error) and believes fully in its own consistency (and can make all logical inferences that follow from its own consistency, including belief in its Gödel statement) Wikipedia:Citation needed. This is provably impossible for a Turing machine Wikipedia:Please clarify (and, by an informal extension, any known type of mechanical computer) to do; therefore, the Gödelian concludes that human reasoning is too powerful to be captured in a machine Wikipedia:Accuracy dispute#Disputed statement.

However, the modern consensus in the scientific and mathematical community is that actual human reasoning is inconsistent; that any consistent "idealized version" *H* of human reasoning would logically be forced to adopt a healthy but counter-intuitive open-minded skepticism about the consistency of *H* (otherwise *H* is provably inconsistent); and that Gödel's theorems do not lead to any valid argument that humans have mathematical reasoning capabilities beyond what a machine could ever duplicate.[1187] This consensus that Gödelian anti-mechanist arguments are doomed to failure is laid out strongly in *Artificial Intelligence*: "*any* attempt to utilize (Gödel's incompleteness results) to attack the computationalist thesis is bound to be illegitimate, since these results are quite consistent with the computationalist thesis."[1188]

More pragmatically, Russell and Norvig note that Gödel's argument only applies to what can theoretically be proved, given an infinite amount of memory and time. In practice, real machines (including humans) have finite resources and will have difficulty proving many theorems. It is not necessary to prove everything in order to be intelligent Wikipedia:Manual of Style.[1189]

Less formally, Douglas Hofstadter, in his Pulitzer prize winning book *Gödel, Escher, Bach: An Eternal Golden Braid,* states that these "Gödel-statements" always refer to the system itself, drawing an analogy to the way the Epimenides paradox uses statements that refer to themselves, such as "this statement is false" or "I am lying". But, of course, the Epimenides paradox applies to anything that makes statements, whether they are machines *or* humans, even Lucas himself. Consider:

- *Lucas can't assert the truth of this statement.*[1190]

This statement is true but cannot be asserted by Lucas. This shows that Lucas himself is subject to the same limits that he describes for machines, as are all people, and so Lucas's argument is pointless.[1191]

After concluding that human reasoning is non-computable, Penrose went on to controversially speculate that some kind of hypothetical non-computable processes involving the collapse of quantum mechanical states give humans a special advantage over existing computers. Existing quantum computers are only capable of reducing the complexity of Turing computable tasks and are still restricted to tasks within the scope of Turing machines. Wikipedia:Citation needed Wikipedia:Please clarify. By Penrose and Lucas's arguments, existing quantum computers are not sufficient Wikipedia:Citation neededWikipedia:Please clarifyWikipedia:Please clarify, so Penrose seeks for some other process involving new physics, for instance quantum gravity which might manifest new physics at the scale of the Planck mass via spontaneous quantum collapse of the wave function. These states, he suggested, occur both within neurons and also spanning more than one neuron. However, other scientists point out that there is no plausible organic mechanism in the brain for harnessing any sort of quantum computation, and furthermore that the timescale of quantum decoherence seems too fast to influence neuron firing.

Dreyfus: the primacy of unconscious skills

Hubert Dreyfus argued that human intelligence and expertise depended primarily on unconscious instincts rather than conscious symbolic manipulation, and argued that these unconscious skills would never be captured in formal rules.[1192]

Dreyfus's argument had been anticipated by Turing in his 1950 paper Computing machinery and intelligence, where he had classified this as the "argument from the informality of behavior." Turing argued in response that, just because we do not know the rules that govern a complex behavior, this does not mean that no such rules exist. He wrote: "we cannot so easily convince ourselves of the absence of complete laws of behaviour ... The only way we know of for finding such laws is scientific observation, and we certainly know of no circumstances under which we could say, 'We have searched enough. There are no such laws.'"[1193]

Russell and Norvig point out that, in the years since Dreyfus published his critique, progress has been made towards discovering the "rules" that govern unconscious reasoning. The situated movement in robotics research attempts to capture our unconscious skills at perception and attention.[1194] Computational intelligence paradigms, such as neural nets, evolutionary algorithms and so on are mostly directed at simulated unconscious reasoning and learning. Statistical approaches to AI can make predictions which approach the accuracy of human intuitive guesses. Research into commonsense knowledge has focused on reproducing the "background" or context of knowledge. In fact, AI research in general has moved away from high level symbol manipulation or "GOFAI", towards new models that are intended to capture more of our *unconscious* reasoning Wikipedia:Manual_of_Style/Words_to_watch#Unsupported_attributions. Historian and AI researcher Daniel Crevier wrote that "time has proven the accuracy and perceptiveness of some of Dreyfus's comments. Had he formulated them less aggressively, constructive actions they suggested might have been taken much earlier."

Can a machine have a mind, consciousness, and mental states?

This is a philosophical question, related to the problem of other minds and the hard problem of consciousness. The question revolves around a position defined by John Searle as "strong AI":

- *A physical symbol system can have a mind and mental states.*

Searle distinguished this position from what he called "weak AI":

- *A physical symbol system can act intelligently.*

Searle introduced the terms to isolate strong AI from weak AI so he could focus on what he thought was the more interesting and debatable issue. He argued that *even if we assume* that we had a computer program that acted exactly like a human mind, there would still be a difficult philosophical question that needed to be answered.

Neither of Searle's two positions are of great concern to AI research, since they do not directly answer the question "can a machine display general intelligence?" (unless it can also be shown that consciousness is *necessary* for intelligence). Turing wrote "I do not wish to give the impression that I think there is no mystery about consciousness... [b]ut I do not think these mysteries necessarily need to be solved before we can answer the question [of whether machines can think]."[1195] Russell and Norvig agree: "Most AI researchers take the weak AI hypothesis for granted, and don't care about the strong AI hypothesis."

There are a few researchers who believe that consciousness is an essential element in intelligence, such as Igor Aleksander, Stan Franklin, Ron Sun, and Pentti Haikonen, although their definition of "consciousness" strays very close to "intelligence." (See artificial consciousness.)

Before we can answer this question, we must be clear what we mean by "minds", "mental states" and "consciousness".

Consciousness, minds, mental states, meaning

The words "mind" and "consciousness" are used by different communities in different ways. Some new age thinkers, for example, use the word "consciousness" to describe something similar to Bergson's "élan vital": an invisible, energetic fluid that permeates life and especially the mind. Science fiction writers use the word to describe some essential property that makes us human: a machine or alien that is "conscious" will be presented as a fully human character, with intelligence, desires, will, insight, pride and so on. (Science fiction writers also use the words "sentience", "sapience," "self-awareness" or "ghost" - as in the *Ghost in the Shell* manga and anime series - to describe this essential human property). For others Wikipedia:Manual of Style/Words to watch#Unsupported attributions, the words "mind" or "consciousness" are used as a kind of secular synonym for the soul.

For philosophers, neuroscientists and cognitive scientists, the words are used in a way that is both more precise and more mundane: they refer to the familiar, everyday experience of having a "thought in your head", like a perception, a dream, an intention or a plan, and to the way we *know* something, or *mean* something or *understand* something Wikipedia:Citation needed. "It's not hard to give a commonsense definition of consciousness" observes philosopher John Searle.[1196] What is mysterious and fascinating is not so much *what* it is but *how* it is: how does a lump of fatty tissue and electricity give rise to this (familiar) experience of perceiving, meaning or thinking?

Philosophers call this the hard problem of consciousness. It is the latest version of a classic problem in the philosophy of mind called the "mind-body problem." A related problem is the problem of *meaning* or *understanding* (which philosophers call "intentionality"): what is the connection between our *thoughts* and *what we are thinking about* (i.e. objects and situations out in the world)? A third issue is the problem of *experience* (or "phenomenology"): If two people see the same thing, do they have the same experience? Or are there things "inside their head" (called "qualia") that can be different from person to person?

Neurobiologists believe all these problems will be solved as we begin to identify the neural correlates of consciousness: the actual relationship between the machinery in our heads and its collective properties; such as the mind, experience and understanding. Some of the harshest critics of artificial intelligence agree that the brain is just a machine, and that consciousness and intelligence are the result of physical processes in the brain.[1197] The difficult philosophical question is this: can a computer program, running on a digital machine that shuffles the binary digits of zero and one, duplicate the ability of the neurons to create minds, with mental states (like understanding or perceiving), and ultimately, the experience of consciousness?

Arguments that a computer cannot have a mind and mental states

Searle's Chinese room

John Searle asks us to consider a thought experiment: suppose we have written a computer program that passes the Turing test and demonstrates "general intelligent action." Suppose, specifically that the program can converse in fluent Chinese. Write the program on 3x5 cards and give them to an ordinary person who does not speak Chinese. Lock the person into a room and have him follow the instructions on the cards. He will copy out Chinese characters and pass them in and out of the room through a slot. From the outside, it will appear that the Chinese room contains a fully intelligent person who speaks Chinese. The question is this: is there anyone (or anything) in the room that understands Chinese? That is, is there anything that has the mental state of understanding, or which has conscious awareness of what is being discussed in Chinese? The man is clearly not aware. The room cannot be aware. The *cards* certainly aren't aware. Searle concludes that the Chinese room, or *any* other physical symbol system, cannot have a mind.[1198]

Searle goes on to argue that actual mental states and consciousness require (yet to be described) "actual physical-chemical properties of actual human brains." He argues there are special "causal properties" of brains and neurons that gives rise to minds: in his words "brains cause minds."

Related arguments: Leibniz' mill, Davis's telephone exchange, Block's Chinese nation and Blockhead

Gottfried Leibniz made essentially the same argument as Searle in 1714, using the thought experiment of expanding the brain until it was the size of a mill. In 1974, Lawrence Davis imagined duplicating the brain using telephone lines and offices staffed by people, and in 1978 Ned Block envisioned the entire population of China involved in such a brain simulation. This thought experiment is called "the Chinese Nation" or "the Chinese Gym". Ned Block also proposed his Blockhead argument, which is a version of the Chinese room in which the program has been re-factored into a simple set of rules of the form "see this, do that", removing all mystery from the program.

Responses to the Chinese room

Responses to the Chinese room emphasize several different points.

- **The systems reply** and the **virtual mind reply**:[1199] This reply argues that *the system*, including the man, the program, the room, and the cards, is what understands Chinese. Searle claims that the man in the room is the only thing which could possibly "have a mind" or "understand", but others disagree, arguing that it is possible for there to be *two* minds in the same physical place, similar to the way a computer can simultaneously "be" two machines at once: one physical (like a Macintosh) and one "virtual" (like a word processor).
- **Speed, power and complexity replies**:[1200] Several critics point out that the man in the room would probably take millions of years to respond to a simple question, and would require "filing cabinets" of astronomical proportions. This brings the clarity of Searle's intuition into doubt.
- **Robot reply**:[1201] To truly understand, some believe the Chinese Room needs eyes and hands. Hans Moravec writes: 'If we could graft a robot to a reasoning program, we wouldn't need a person to provide the meaning anymore: it would come from the physical world."[1202]
- **Brain simulator reply**:[1203] What if the program simulates the sequence of nerve firings at the synapses of an actual brain of an actual Chinese speaker? The man in the room would be simulating an actual brain. This is a variation on the "systems reply" that appears more plausible because "the system" now clearly operates like a human brain, which strengthens the intuition that there is something besides the man in the room that could understand Chinese.
- **Other minds reply** and the **epiphenomena reply**:[1204] Several people have noted that Searle's argument is just a version of the problem of other minds, applied to machines. Since it is difficult to decide if people are "actually" thinking, we should not be surprised that it is difficult to answer the same question about machines.

  A related question is whether "consciousness" (as Searle understands it) exists. Searle argues that the experience of consciousness can't be detected by examining the behavior of a machine, a human being or any other animal. Daniel Dennett points out that natural selection cannot preserve a feature of an animal that has no effect on the behavior of the animal, and thus consciousness (as Searle understands it) can't be produced by natural selection. Therefore either natural selection did not produce consciousness, or "strong AI" is correct in that consciousness can be detected by suitably designed Turing test.

Is thinking a kind of computation?

The computational theory of mind or "computationalism" claims that the relationship between mind and brain is similar (if not identical) to the relationship between a *running program* and a computer. The idea has philosophical roots in Hobbes (who claimed reasoning was "nothing more than reckoning"), Leibniz (who attempted to create a logical calculus of all human ideas), Hume (who thought perception could be reduced to "atomic impressions") and even Kant (who analyzed all experience as controlled by formal rules). The latest version is associated with philosophers Hilary Putnam and Jerry Fodor.

This question bears on our earlier questions: if the human brain is a kind of computer then computers can be both intelligent and conscious, answering both the practical and philosophical questions of AI. In terms of the practical question of AI ("Can a machine display general intelligence?"), some versions of computationalism make the claim that (as Hobbes wrote):

- *Reasoning is nothing but reckoning*

In other words, our intelligence derives from a form of *calculation*, similar to arithmetic. This is the physical symbol system hypothesis discussed above, and it implies that artificial intelligence is possible. In terms of the philosophical question of AI ("Can a machine have mind, mental states and consciousness?"), most versions of computationalism claim that (as Stevan Harnad characterizes it):

- *Mental states are just implementations of (the right) computer programs*

This is John Searle's "strong AI" discussed above, and it is the real target of the Chinese room argument (according to Harnad).

Other related questions

Alan Turing noted that there are many arguments of the form "a machine will never do X", where X can be many things, such as:

> *Be kind, resourceful, beautiful, friendly, have initiative, have a sense of humor, tell right from wrong, make mistakes, fall in love, enjoy strawberries and cream, make someone fall in love with it, learn from experience, use words properly, be the subject of its own thought, have as much diversity of behaviour as a man, do something really new.*[1205]

Turing argues that these objections are often based on naive assumptions about the versatility of machines or are "disguised forms of the argument from consciousness". Writing a program that exhibits one of these behaviors "will not make much of an impression." All of these arguments are tangential to the basic premise of AI, unless it can be shown that one of these traits is essential for general intelligence.

Can a machine have emotions?

If "emotions" are defined only in terms of their effect on behavior or on how they function inside an organism, then emotions can be viewed as a mechanism that an intelligent agent uses to maximize the utility of its actions. Given this definition of emotion, Hans Moravec believes that "robots in general will be quite emotional about being nice people".[1206] Fear is a source of urgency. Empathy is a necessary component of good human computer interaction. He says robots "will try to please you in an apparently selfless manner because it will get a thrill out of this positive reinforcement. You can interpret this as a kind of love." Daniel Crevier writes "Moravec's point is that emotions are just devices for channeling behavior in a direction beneficial to the survival of one's species."

However, emotions can also be defined in terms of their subjective quality, of what it *feels like* to have an emotion. The question of whether the machine *actually feels* an emotion, or whether it merely *acts as if* it is feeling an emotion is the philosophical question, "can a machine be conscious?" in another form.

Can a machine be self-aware?

"Self awareness", as noted above, is sometimes used by science fiction writers as a name for the essential human property that makes a character fully human. Turing strips away all other properties of human beings and reduces the question to "can a machine be the subject of its own thought?" Can it *think about itself*? Viewed in this way, it is obvious that a program can be written that can report on its own internal states, such as a debugger. Though arguably self-awareness often presumes a bit more capability; a machine that can ascribe meaning in some way to not only its own state but in general postulating questions without solid answers: the contextual nature of its existence now; how it compares to past states or plans for the future, the limits and value of its work product, how it perceives its performance to be valued-by or compared to others.

Can a machine be original or creative?

Turing reduces this to the question of whether a machine can "take us by surprise" and argues that this is obviously true, as any programmer can attest.[1207] He notes that, with enough storage capacity, a computer can behave in an astronomical number of different ways.[1208] It must be possible, even trivial, for a computer that can represent ideas to combine them in new ways. (Douglas Lenat's Automated Mathematician, as one example, combined ideas to discover new mathematical truths.)

In 2009, scientists at Aberystwyth University in Wales and the U.K's University of Cambridge designed a robot called Adam that they believe to be the first machine to independently come up with new scientific findings. Also in 2009, researchers at Cornell developed Eureqa, a computer program that extrapolates formulas to fit the data inputted, such as finding the laws of motion from a pendulum's motion.

Can a machine be benevolent or hostile?

This question (like many others in the philosophy of artificial intelligence) can be presented in two forms. "Hostility" can be defined in terms function or behavior, in which case "hostile" becomes synonymous with "dangerous". Or it can be defined in terms of intent: can a machine "deliberately" set out to do harm? The latter is the question "can a machine have conscious states?" (such as intentions) in another form.

The question of whether highly intelligent and completely autonomous machines would be dangerous has been examined in detail by futurists (such as the Singularity Institute). (The obvious element of drama has also made the subject popular in science fiction, which has considered many differently possible scenarios where intelligent machines pose a threat to mankind.)

One issue is that machines may acquire the autonomy and intelligence required to be dangerous very quickly. Vernor Vinge has suggested that over just a few years, computers will suddenly become thousands or millions of times more intelligent than humans. He calls this "the Singularity." He suggests that it may be somewhat or possibly very dangerous for humans.[1209] This is discussed by a philosophy called Singularitarianism.

In 2009, academics and technical experts attended a conference to discuss the potential impact of robots and computers and the impact of the hypothetical possibility that they could become self-sufficient and able to make their own decisions. They discussed the possibility and the extent to which computers and robots might be able to acquire any level of autonomy, and to what degree they could use such abilities to possibly pose any threat or hazard. They noted that some machines have acquired various forms of semi-autonomy, including being able to find power sources on their own and being able to independently choose targets to attack with weapons. They also noted that some computer viruses can evade elimination and have achieved "cockroach intelligence." They noted that self-awareness as depicted in science-fiction is probably unlikely, but that there were other potential hazards and pitfalls.[1210]

Some experts and academics have questioned the use of robots for military combat, especially when such robots are given some degree of autonomous functions.[1211] The US Navy has funded a report which indicates that as military robots become more complex, there should be greater attention to implications of their ability to make autonomous decisions.[1212,1213]

The President of the Association for the Advancement of Artificial Intelligence has commissioned a study to look at this issue.[1214] They point to programs like the Language Acquisition Device which can emulate human interaction.

Some have suggested a need to build "Friendly AI", meaning that the advances which are already occurring with AI should also include an effort to make AI intrinsically friendly and humane.[1215]

Can a machine have a soul?

Finally, those who believe in the existence of a soul may argue that "Thinking is a function of man's immortal soul." Alan Turing called this "the theological objection". He writes

> *In attempting to construct such machines we should not be irreverently usurping His power of creating souls, any more than we are in the procreation of children: rather we are, in either case, instruments of His will providing mansions for the souls that He creates.*[1216]

Bibliography & Conferences

The main bibliography on the subject, with several sub-sections, is on PhilPapers[1217]

The main conference series on the issue is "Philosophy and Theory of AI"[1218] (PT-AI), run by Vincent C. Müller[1219]

References

- Blackmore, Susan (2005), *Consciousness: A Very Short Introduction*, Oxford University Press
- Bostrom, Nick (2014), Superintelligence: Paths, Dangers, Strategies, Oxford University Press, ISBN 978-0-19-967811-2
- Brooks, Rodney (1990), "Elephants Don't Play Chess"[1220] (PDF), *Robotics and Autonomous Systems*, **6**: 3–15, doi: 10.1016/S0921-8890(05)80025-9[1221], retrieved 2007-08-30
- Chalmers, David J (1996), *The Conscious Mind: In Search of a Fundamental Theory*, Oxford University Press, New York, ISBN 0-19-511789-1
- Cole, David (Fall 2004), "The Chinese Room Argument", in Zalta, Edward N., *The Stanford Encyclopedia of Philosophy*[1222].

- Crevier, Daniel (1993), *AI: The Tumultuous Search for Artificial Intelligence*, New York, NY: BasicBooks, ISBN 0-465-02997-3
- Dennett, Daniel (1991), *Consciousness Explained*, The Penguin Press, ISBN 0-7139-9037-6
- Dreyfus, Hubert (1972), *What Computers Can't Do*, New York: MIT Press, ISBN 0-06-011082-1
- Dreyfus, Hubert (1979), *What Computers* Still *Can't Do*, New York: MIT Press.
- Dreyfus, Hubert; Dreyfus, Stuart (1986), *Mind over Machine: The Power of Human Intuition and Expertise in the Era of the Computer*, Oxford, UK: Blackwell
- Fearn, Nicholas (2007), *The Latest Answers to the Oldest Questions: A Philosophical Adventure with the World's Greatest Thinkers*, New York: Grove Press
- Gladwell, Malcolm (2005), *Blink: The Power of Thinking Without Thinking*, Boston: Little, Brown, ISBN 0-316-17232-4.
- Harnad, Stevan (2001), "What's Wrong and Right About Searle's Chinese Room Argument?", in Bishop, M.; Preston, J., *Essays on Searle's Chinese Room Argument*[1223], Oxford University Press
- Haugeland, John (1985), *Artificial Intelligence: The Very Idea*, Cambridge, Mass.: MIT Press.
- Hobbes (1651), *Leviathan.*
- Hofstadter, Douglas (1979), *Gödel, Escher, Bach: an Eternal Golden Braid*.
- Horst, Steven (2009), "The Computational Theory of Mind", in Zalta, Edward N., *The Stanford Encyclopedia of Philosophy*[1224].
- Kurzweil, Ray (2005), *The Singularity is Near*, New York: Viking Press, ISBN 0-670-03384-7.
- Lucas, John (1961), "Minds, Machines and Gödel", in Anderson, A.R., *Minds and Machines*[1225].
- McCarthy, John; Minsky, Marvin; Rochester, Nathan; Shannon, Claude (1955), *A Proposal for the Dartmouth Summer Research Project on Artificial Intelligence*[1226], archived from the original[1227] on 2008-09-30.
- McDermott, Drew (May 14, 1997), "How Intelligent is Deep Blue"[1228], *New York Times*
- Moravec, Hans (1988), *Mind Children*, Harvard University Press
- Newell, Allen; Simon, H. A. (1963), "GPS: A Program that Simulates Human Thought", in Feigenbaum, E.A.; Feldman, J., *Computers and Thought*, New York: McGraw-Hill
- Newell, Allen; Simon, H. A. (1976), "Computer Science as Empirical Inquiry: Symbols and Search", *Communications of the ACM*[1229], **19** (3), archived from the original[1230] on 2008-10-07

- Russell, Stuart J.; Norvig, Peter (2003), *Artificial Intelligence: A Modern Approach*[1231] (2nd ed.), Upper Saddle River, New Jersey: Prentice Hall, ISBN 0-13-790395-2
- Penrose, Roger (1989), *The Emperor's New Mind: Concerning Computers, Minds, and The Laws of Physics*, Oxford University Press, ISBN 0-14-014534-6
- Searle, John (1980), "Minds, Brains and Programs"[1232] (PDF), *Behavioral and Brain Sciences*, **3** (3): 417–457, doi: 10.1017/S0140525X00005756[1233], archived from the original[1234] (PDF) on 2015-09-23
- Searle, John (1992), *The Rediscovery of the Mind*, Cambridge, Massachusetts: M.I.T. Press
- Searle, John (1999), *Mind, language and society*, New York, NY: Basic Books, ISBN 0-465-04521-9, OCLC 231867665[1235]
- Turing, Alan (October 1950), "Computing Machinery and Intelligence", *Mind*, **LIX** (236): 433–460, doi: 10.1093/mind/LIX.236.433[1236], ISSN 0026-4423[1237]
- Yee, Richard (1993), "Turing Machines And Semantic Symbol Processing: Why Real Computers Don't Mind Chinese Emperors"[1238] (PDF), *Lyceum*, **5** (1): 37–59

  Page numbers above and diagram contents refer to the Lyceum PDF print of the article.

Ethics of artificial intelligence

The **ethics of artificial intelligence** is the part of the ethics of technology specific to robots and other artificially intelligent beings. It is typically-Wikipedia:Citation needed divided into roboethics, a concern with the moral behavior of humans as they design, construct, use and treat artificially intelligent beings, and machine ethics, which is concerned with the moral behavior of artificial moral agents (AMAs).

Robot ethics

The term "robot ethics" (sometimes "roboethics") refers to the morality of how humans design, construct, use and treat robots and other artificially intelligent beings. It considers both how artificially intelligent beings may be used to harm humans and how they may be used to benefit humans.

Robot rights

"Robot rights" is the concept that people should have moral obligations towards their machines, similar to human rights or animal rights. It has been suggested that robot rights, such as a right to exist and perform its own mission, could be linked to robot duty to serve human, by analogy with linking human rights to human duties before society. These could include the right to life and liberty, freedom of thought and expression and equality before the law.[1239] The issue has been considered by the Institute for the Future and by the U.K. Department of Trade and Industry.

Experts disagree whether specific and detailed laws will be required soon or safely in the distant future. Glenn McGee reports that sufficiently humanoid robots may appear by 2020. Ray Kurzweil sets the date at 2029. Another group of scientists meeting in 2007 supposed that at least 50 years had to pass before any sufficiently advanced system would exist.[1240]

The rules for the 2003 Loebner Prize competition envisioned the possibility of robots having rights of their own:

> *61. If, in any given year, a publicly available open source Entry entered by the University of Surrey or the Cambridge Center wins the Silver Medal or the Gold Medal, then the Medal and the Cash Award will be awarded to the body responsible for the development of that Entry. If no such body can be identified, or if there is disagreement among two or more claimants, the Medal and the Cash Award will be held in trust until such time as the Entry may legally possess, either in the United States of America or in the venue of the contest, the Cash Award and Gold Medal in its own right.*[1241]

In October 2017, the android Sophia was granted citizenship in Saudi Arabia, though some observers found this to be more of a publicity stunt than a meaningful legal recognition.[1242]

Threat to human dignity

Joseph Weizenbaum argued in 1976 that AI technology should not be used to replace people in positions that require respect and care, such as any of these:

- A customer service representative (AI technology is already used today for telephone-based interactive voice response systems)
- A therapist (as was proposed by Kenneth Colby in the 1970s)
- A nursemaid for the elderly (as was reported by Pamela McCorduck in her book *The Fifth Generation*)
- A soldier
- A judge
- A police officer

Weizenbaum explains that we require authentic feelings of empathy from people in these positions. If machines replace them, we will find ourselves alienated, devalued and frustrated. Artificial intelligence, if used in this way, represents a threat to human dignity. Weizenbaum argues that the fact that we are entertaining the possibility of machines in these positions suggests that we have experienced an "atrophy of the human spirit that comes from thinking of ourselves as computers."[1243]

Pamela McCorduck counters that, speaking for women and minorities "I'd rather take my chances with an impartial computer," pointing out that there are conditions where we would prefer to have automated judges and police that have no personal agenda at all. AI founder John McCarthy objects to the moralizing tone of Weizenbaum's critique. "When moralizing is both vehement and vague, it invites authoritarian abuse," he writes.

Bill Hibbard[1244] writes that "Human dignity requires that we strive to remove our ignorance of the nature of existence, and AI is necessary for that striving."

Transparency and open source

Bill Hibbard argues that because AI will have such a profound effect on humanity, AI developers are representatives of future humanity and thus have an ethical obligation to be transparent in their efforts.[1245] Ben Goertzel and David Hart created OpenCog as an open source framework for AI development.[1246] OpenAI is a non-profit AI research company created by Elon Musk, Sam Altman and others to develop open source AI beneficial to humanity.[1247] There are numerous other open source AI developments.

Weaponization of artificial intelligence

Some experts and academics have questioned the use of robots for military combat, especially when such robots are given some degree of autonomous functions.[1248,1249] The US Navy has funded a report which indicates that as military robots become more complex, there should be greater attention to implications of their ability to make autonomous decisions.[1250,1251] One researcher states that autonomous robots might be more humane, as they could make decisions more effectively.Wikipedia:Citation needed

Within this last decade, there has been intensive research in autonomous power with the ability to learn using assigned moral responsibilities. "The results may be used when designing future military robots, to control unwanted tendencies to assign responsibility to the robots."[1252] From a consequentialist view, there is a chance that robots will develop the ability to make their own logical decisions on who to kill and that is why there should be a set moral framework that the A.I cannot override.

There has been a recent outcry with regard to the engineering of artificial-intelligence weapons that has included ideas of a robot takeover of mankind. AI weapons do present a type of danger different from that of human-controlled weapons. Many governments have begun to fund programs to develop AI weaponry. The United States Navy recently announced plans to develop autonomous drone weapons, paralleling similar announcements by Russia and Korea respectively. Due to the potential of AI weapons becoming more dangerous than human-operated weapons, Stephen Hawking and Max Tegmark signed a "Future of Life" petition to ban AI weapons. The message posted by Hawking and Tegmark states that AI weapons pose an immediate danger and that action is required to avoid catastrophic disasters in the near future.

"If any major military power pushes ahead with the AI weapon development, a global arms race is virtually inevitable, and the endpoint of this technological trajectory is obvious: autonomous weapons will become the Kalashnikovs of tomorrow", says the petition, which includes Skype co-founder Jaan Tallinn and MIT professor of linguistics Noam Chomsky as additional supporters against AI weaponry.

Physicist and Astronomer Royal Sir Martin Rees has warned of catastrophic instances like "dumb robots going rogue or a network that develops a mind of its own." Huw Price, a colleague of Rees at Cambridge, has voiced a similar warning that humans might not survive when intelligence "escapes the constraints of biology." These two professors created the Centre for the Study of Existential Risk at Cambridge University in the hope of avoiding this threat to human existence.

Regarding the potential for smarter-than-human systems to be employed militarily, the Open Philanthropy Project writes that these scenarios "seem potentially as important as the risks related to loss of control", but that research organizations investigating AI's long-run social impact have spent relatively little time on this concern: "this class of scenarios has not been a major focus for the organizations that have been most active in this space, such as the Machine Intelligence Research Institute (MIRI) and the Future of Humanity Institute (FHI), and there seems to have been less analysis and debate regarding them".

Machine ethics

Machine ethics (or machine morality) is the field of research concerned with designing Artificial Moral Agents (AMAs), robots or artificially intelligent computers that behave morally or as though moral.

Isaac Asimov considered the issue in the 1950s in his *I, Robot*. At the insistence of his editor John W. Campbell Jr., he proposed the Three Laws of Robotics to govern artificially intelligent systems. Much of his work was then spent testing the boundaries of his three laws to see where they would break down, or where they would create paradoxical or unanticipated behavior. His work suggests that no set of fixed laws can sufficiently anticipate all possible circumstances.

In 2009, during an experiment at the Laboratory of Intelligent Systems in the Ecole Polytechnique Fédérale of Lausanne in Switzerland, robots that were programmed to cooperate with each other (in searching out a beneficial resource and avoiding a poisonous one) eventually learned to lie to each other in an attempt to hoard the beneficial resource.[1253] One problem in this case may have been that the goals were "terminal" (i.e. in contrast, ultimate human motives typically have a quality of requiring never-ending learning).

Some experts and academics have questioned the use of robots for military combat, especially when such robots are given some degree of autonomous functions. The US Navy has funded a report which indicates that as military robots become more complex, there should be greater attention to implications of their ability to make autonomous decisions.[1254,1255] The President of the Association for the Advancement of Artificial Intelligence has commissioned a study to look at this issue.[1256] They point to programs like the Language Acquisition Device which can emulate human interaction.

Vernor Vinge has suggested that a moment may come when some computers are smarter than humans. He calls this "the Singularity." He suggests that it may be somewhat or possibly very dangerous for humans.[1257] This is discussed by a philosophy called Singularitarianism. The Machine Intelligence Research Institute has suggested a need to build "Friendly AI", meaning that the advances which are already occurring with AI should also include an effort to make AI intrinsically friendly and humane.[1258]

In 2009, academics and technical experts attended a conference organized by the Association for the Advancement of Artificial Intelligence to discuss the potential impact of robots and computers and the impact of the hypothetical possibility that they could become self-sufficient and able to make their own decisions. They discussed the possibility and the extent to which computers and robots might be able to acquire any level of autonomy, and to what degree they could use such abilities to possibly pose any threat or hazard. They noted that some machines have acquired various forms of semi-autonomy, including being able to find power sources on their own and being able to independently choose targets to attack with weapons. They also noted that some computer viruses can evade elimination and have achieved "cockroach intelligence." They noted that self-awareness as depicted in science-fiction is probably unlikely, but that there were other potential hazards and pitfalls.[1259]

However, there is one technology in particular that could truly bring the possibility of robots with moral competence to reality. In a paper on the acquisition of moral values by robots, Nayef Al-Rodhan mentions the case of neuromorphic chips, which aim to process information similarly to humans, nonlinearly and with millions of interconnected artificial neurons. Robots embedded with neuromorphic technology could learn and develop knowledge in a uniquely humanlike way. Inevitably, this raises the question of the environment in which such robots would learn about the world and whose morality they would inherit - or if they end up developing human 'weaknesses' as well: selfishness, a pro-survival attitude, hesitation etc.

In *Moral Machines: Teaching Robots Right from Wrong*, Wendell Wallach and Colin Allen conclude that attempts to teach robots right from wrong will likely advance understanding of human ethics by motivating humans to address gaps in modern normative theory and by providing a platform for experimental investigation. As one example, it has introduced normative ethicists to the controversial issue of which specific learning algorithms to use in machines. Nick Bostrom and Eliezer Yudkowsky have argued for decision trees (such as ID3) over neural networks and genetic algorithms on the grounds that decision trees obey modern social norms of transparency and predictability (e.g. *stare decisis*), while Chris Santos-Lang argued in the opposite direction on the grounds that the norms of any age must be allowed to change and that natural failure to fully satisfy these particular norms has been essential in making humans less vulnerable to criminal "hackers".

Unintended consequences

Many researchers have argued that, by way of an "intelligence explosion" sometime in the 21st century, a self-improving AI could become so vastly more powerful than humans that we would not be able to stop it from achieving its goals.[1260] In his paper "Ethical Issues in Advanced Artificial Intelligence," philosopher Nick Bostrom argues that artificial intelligence has the capability to bring about human extinction. He claims that general super-intelligence would be capable of independent initiative and of making its own plans, and may therefore be more appropriately thought of as an autonomous agent. Since artificial intellects need not share our human motivational tendencies, it would be up to the designers of the super-intelligence to specify its original motivations. In theory, a super-intelligent AI would be able to bring about almost any possible outcome and to thwart any attempt to prevent the implementation of its top goal, many uncontrolled unintended consequences could arise. It could kill off all other agents, persuade them to change their behavior, or block their attempts at interference.[1261]

However, instead of overwhelming the human race and leading to our destruction, Bostrom has also asserted that super-intelligence can help us solve many difficult problems such as disease, poverty, and environmental destruction, and could help us to "enhance" ourselves.

The sheer complexity of human value systems makes it very difficult to make AI's motivations human-friendly. Unless moral philosophy provides us with a flawless ethical theory, an AI's utility function could allow for many potentially harmful scenarios that conform with a given ethical framework but not "common sense". According to Eliezer Yudkowsky, there is little reason to suppose that an artificially designed mind would have such an adaptation.[1262]

Bill Hibbard proposes an AI design that avoids several types of unintended AI behavior including self-delusion, unintended instrumental actions, and corruption of the reward generator.

Organizations

Amazon, Google, Facebook, IBM, and Microsoft have established a non-profit partnership to formulate best practices on artificial intelligence technologies, advance the public's understanding, and to serve as a platform about artificial intelligence. They stated: "This partnership on AI will conduct research, organize discussions, provide thought leadership, consult with relevant third parties, respond to questions from the public and media, and create educational material that advance the understanding of AI technologies including machine perception, learning, and automated reasoning."[1263] Apple joined other tech companies as a founding member of the Partnership on AI in January 2017. The corporate members will make financial and research contributions to the group, while engaging with the scientific community to bring academics onto the board.[1264]

In fiction

The movie *The Thirteenth Floor* suggests a future where simulated worlds with sentient inhabitants are created by computer game consoles for the purpose of entertainment. The movie *The Matrix* suggests a future where the dominant species on planet Earth are sentient machines and humanity is treated with utmost Speciesism. The short story "The Planck Dive" suggest a future where humanity has turned itself into software that can be duplicated and optimized and the relevant distinction between types of software is sentient and non-sentient. The same idea can be found in the Emergency Medical Hologram of *Starship Voyager*, which is an apparently sentient copy of a reduced subset of the consciousness of its creator, Dr. Zimmerman, who, for the best motives, has created the system to give medical assistance in case of emergencies. The movies *Bicentennial Man* and *A.I.* deal with the possibility of sentient robots that could love. *I, Robot* explored some aspects of Asimov's three laws. All these scenarios try to foresee possibly unethical consequences of the creation of sentient computers.Wikipedia:Citation needed

The ethics of artificial intelligence is one of several core themes in BioWare's Mass Effect series of games.Wikipedia:Citation needed It explores the scenario of a civilization accidentally creating AI through a rapid increase in computational power through a global scale neural network. This event caused an ethical schism between those who felt bestowing organic rights upon the newly sentient Geth was appropriate and those who continued to see them as disposable machinery and fought to destroy them. Beyond the initial conflict, the complexity of the relationship between the machines and their creators is another ongoing theme throughout the story.

Over time, debates have tended to focus less and less on *possibility* and more on *desirability*,Wikipedia:Citation needed as emphasized in the "Cosmist" and "Terran" debates initiated by Hugo de Garis and Kevin Warwick. A Cosmist, according to Hugo de Garis, is actually seeking to build more intelligent successors to the human species.

Literature

The standard bibliography on ethics of AI is on PhilPapers[1265]. A recent collection is V.C. Müller(ed.) (2016).

External links

- Robotics: Ethics of artificial intelligence[1266]. "Four leading researchers share their concerns and solutions for reducing societal risks from intelligent machines." *Nature,* 521, 415–418 (28 May 2015) doi:10.1038/521415a
- BBC News: Games to take on a life of their own[1267]
- Who's Afraid of Robots?[1268], an article on humanity's fear of artificial intelligence.
- A short history of computer ethics[1269]

Existential risk from artificial general intelligence

Existential risk from artificial general intelligence is the hypothesis that substantial progress in artificial general intelligence (AI) could someday result in human extinction or some other unrecoverable global catastrophe.

One argument is as follows. The human species currently dominates other species because the human brain has some distinctive capabilities that the brains of other animals lack. If AI surpasses humanity in general intelligence and becomes "superintelligent", then this new superintelligence could become powerful and difficult to control. Just as the fate of the mountain gorilla depends on human goodwill, so might the fate of humanity depend on the actions of a future machine superintelligence.

The severity of AI risk is widely debated, and hinges in part on differing scenarios for future progress in computer science. Once the exclusive domain of science fiction, concerns about superintelligence started to go mainstream in the 2010s, and were popularized by public figures such as Stephen Hawking, Bill Gates, and Elon Musk.

One source of concern is that a sudden and unexpected "intelligence explosion" might take an unprepared human race by surprise. For example, in one scenario, the first-generation computer program found able to broadly match the effectiveness of an AI researcher is able to rewrite its algorithms and double its speed or capabilities in six months of massively parallel processing time. The second-generation program is expected to take three months to perform a similar chunk of work, on average; in practice, doubling its own capabilities may take longer if it experiences a mini-"AI winter", or may be quicker if it undergoes a miniature "AI Spring" where ideas from the previous generation are especially easy to mutate into the next generation. In this scenario the system undergoes an unprecedently large number of generations of improvement in a short time interval, jumping from subhuman performance in many areas to superhuman performance in all relevant areas. More broadly, examples like arithmetic and Go show that progress from human-level AI to superhuman ability is sometimes extremely rapid.

A second source of concern is that controlling a superintelligent machine (or even instilling it with human-compatible values) may be an even harder problem than naïvely supposed. Some AI researchers believe that a superintelligence would naturally resist attempts to shut it off, and that preprogramming a superintelligence with complicated human values may be an extremely difficult technical task.[1270] In contrast, skeptics such as Facebook's Yann LeCun argue that superintelligent machines will have no desire for self-preservation.

Overview

Artificial Intelligence: A Modern Approach, the standard undergraduate AI textbook, assesses that superintelligence "might mean the end of the human race": "Almost any technology has the potential to cause harm in the wrong hands, but with (superintelligence), we have the new problem that the wrong hands might belong to the technology itself." Even if the system designers have good intentions, two difficulties are common to both AI and non-AI computer systems:

- The system's implementation may contain initially-unnoticed routine but catastrophic bugs. An analogy is space probes: despite the knowledge that bugs in expensive space probes are hard to fix after launch, engineers have historically not been able to prevent catastrophic bugs from occurring.
- No matter how much time is put into pre-deployment design, a system's specifications often result in unintended behavior the first time it encounters a new scenario. For example, Microsoft's Tay behaved inoffensively during pre-deployment testing, but was too easily baited into offensive behavior when interacting with real users.

AI systems uniquely add a third difficulty: the problem that even given "correct" requirements, bug-free implementation, and initial good behavior, an AI system's dynamic "learning" capabilities may cause it to "evolve into a system with unintended behavior", even without the stress of new unanticipated external scenarios. An AI may partly botch an attempt to design a new generation of itself and accidentally create a successor AI that is more powerful than itself, but that no longer maintains the human-compatible moral values preprogrammed into the original AI. For a self-improving AI to be completely safe, it would not only need to be "bug-free", but it would need to be able to design successor systems that are also "bug-free".

All three of these difficulties become catastrophes rather than nuisances in any scenario where the superintelligence labeled as "malfunctioning" correctly predicts that humans will attempt to shut it off, and successfully deploys its superintelligence to outwit such attempts.

Citing major advances in the field of AI and the potential for AI to have enormous long-term benefits or costs, the 2015 Open Letter on Artificial Intelligence stated:

> "The progress in AI research makes it timely to focus research not only on making AI more capable, but also on maximizing the societal benefit of AI. Such considerations motivated the AAAI 2008-09 Presidential Panel on Long-Term AI Futures and other projects on AI impacts, and constitute a significant expansion of the field of AI itself, which up to now has focused largely on techniques that are neutral with respect to purpose. We recommend expanded research aimed at ensuring that increasingly capable AI systems are robust and beneficial: our AI systems must do what we want them to do."

This letter was signed by a number of leading AI researchers in academia and industry, including AAAI president Thomas Dietterich, Eric Horvitz, Bart Selman, Francesca Rossi, Yann LeCun, and the founders of Vicarious and Google DeepMind.

History

In 1965, I. J. Good originated the concept now known as an "intelligence explosion":

> "Let an ultraintelligent machine be defined as a machine that can far surpass all the intellectual activities of any man however clever. Since the design of machines is one of these intellectual activities, an ultraintelligent machine could design even better machines; there would then unquestionably be an 'intelligence explosion', and the intelligence of man would be left far behind. Thus the first ultraintelligent machine is the last invention that man need ever make, provided that the machine is docile enough to tell us how to keep it under control.[1271]"

Occasional statements from scholars such as Alan Turing,[1272,1273] from I. J. Good himself, and from Marvin Minsky expressed philosophical concerns that a superintelligence could seize control, but contained no call to action. In 2000, computer scientist and Sun co-founder Bill Joy penned an influential essay, "Why The Future Doesn't Need Us", identifying superintelligent robots as a high-tech dangers to human survival, alongside nanotechnology and engineered bioplagues.

In 2009, experts attended a private conference hosted by the Association for the Advancement of Artificial Intelligence (AAAI) to discuss whether computers and robots might be able to acquire any sort of autonomy, and how much these abilities might pose a threat or hazard. They noted that some robots have acquired various forms of semi-autonomy, including being able to find power sources on their own and being able to independently choose targets to attack with weapons. They also noted that some computer viruses can evade elimination and have achieved "cockroach intelligence." They concluded that self-awareness as depicted in science fiction is probably unlikely, but that there were other potential hazards and pitfalls. The New York Times summarized

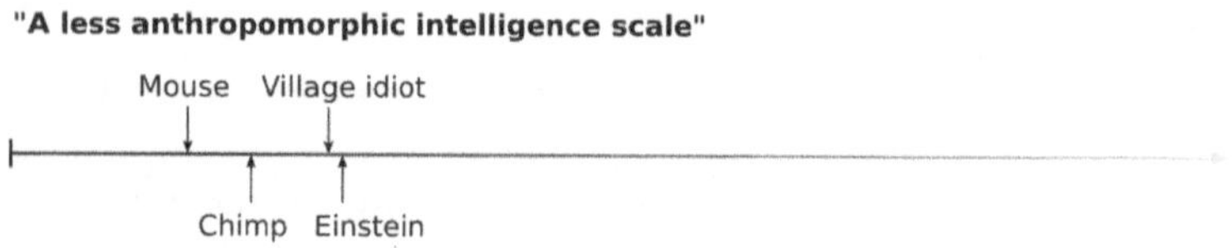

Figure 70: *Bostrom and others argue that, from an evolutionary perspective, the gap from human to superhuman intelligence may be small.*[1276]

the conference's view as 'we are a long way from Hal, the computer that took over the spaceship in "2001: A Space Odyssey"'[1274]

By 2015, public figures such as physicists Stephen Hawking and Nobel laureate Frank Wilczek, computer scientists Stuart J. Russell and Roman Yampolskiy, and entrepreneurs Elon Musk and Bill Gates were expressing concern about the risks of superintelligence. In April 2016, *Nature* warned: "Machines and robots that outperform humans across the board could self-improve beyond our control — and their interests might not align with ours."

Basic argument

A superintelligent machine would be as alien to humans as human thought processes are to cockroaches. Such a machine may not have humanity's best interests at heart; it is not obvious that it would even care about human welfare at all. If superintelligent AI is possible, and if it is possible for a superintelligence's goals to conflict with basic human values, then AI poses a risk of human extinction. A "superintelligence" (a system that exceeds the capabilities of humans in every relevant endeavor) can outmaneuver humans any time its goals conflict with human goals; therefore, unless the superintelligence decides to allow humanity to coexist, the first superintelligence to be created will inexorably result in human extinction.[1275]

There is no physical law precluding particles from being organised in ways that perform even more advanced computations than the arrangements of particles in human brains; therefore superintelligence is physically possible. In addition to potential algorithmic improvements over human brains, a digital brain can be many orders of magnitude larger and faster than a human brain, which was constrained in size by evolution to be small enough to fit through a birth canal. The emergence of superintelligence, if or when it occurs, may take the human race by surprise, especially if some kind of intelligence explosion occurs. Examples like arithmetic and Go show that machines have already reached superhuman levels of competency in certain domains, and that this superhuman competence can follow quickly after human-par performance is achieved.

One hypothetical intelligence explosion scenario could occur as follows: An AI gains an expert-level capability at certain key software engineering tasks. (It may initially lack human or superhuman capabilities in other domains not directly relevant to engineering.) Due to its capability to recursively improve its own algorithms, the AI quickly becomes superhuman; just as human experts can eventually creatively overcome "diminishing returns" by deploying various human capabilities for innovation, so too can the expert-level AI use either human-style capabilities or its own AI-specific capabilities to power through new creative breakthroughs.[1277] The AI then possesses intelligence far surpassing that of the brightest and most gifted human minds in practically every relevant field, including scientific creativity, strategic planning, and social skills. Just as the current-day survival of the gorillas is dependent on human decisions, so too would human survival depend on the decisions and goals of the superhuman AI.

Some humans have a strong desire for power; others have a strong desire to help less fortunate humans. The former is a likely attribute of any sufficiently intelligent system; the latter is not.Wikipedia:Please clarify Almost any AI, no matter its programmed goal, would rationally prefer to be in a position where nobody else can switch it off without its consent: A superintelligence will naturally gain self-preservation as a subgoal as soon as it realizes that it can't achieve its goal if it's shut off. Unfortunately, any compassion for defeated humans whose cooperation is no longer necessary would be absent in the AI, unless somehow preprogrammed in. A superintelligent AI will not have a natural drive to aid humans, for the same reason that humans have no natural desire to aid AI systems that are of no further use to them. (Another analogy is that humans seem to have little natural desire to go out of their way to aid viruses, termites, or even gorillas.) Once in charge, the superintelligence will have little incentive to allow humans to run around free and consume resources that the superintelligence could instead use for building itself additional protective systems "just to be on the safe side" or for building additional computers to help it calculate how to best accomplish its goals.

Thus, the argument concludes, it is likely that someday an intelligence explosion will catch humanity unprepared, and that such an unprepared-for intelligence explosion will likely result in human extinction or a comparable fate.

Sources of risk

Poorly specified goals: "Be careful what you wish for" or the "Sorcerer's Apprentice" scenario

While there is no standardized terminology, an AI can loosely be viewed as a machine that chooses whatever action appears to best achieve the AI's set of goals, or "utility function". The utility function is a mathematical algorithm resulting in a single objectively-defined answer, not an English statement. Researchers know how to write utility functions that mean "minimize the average network latency in this specific telecommunications model" or "maximize the number of reward clicks"; however, they do not know how to write a utility function for "maximize human flourishing", nor is it currently clear whether such a function meaningfully and unambiguously exists. Furthermore, a utility function that expresses some values but not others will tend to trample over the values not reflected by the utility function.[1278] AI researcher Stuart Russell writes:

> The primary concern is not spooky emergent consciousness but simply the ability to make *high-quality decisions*. Here, quality refers to the expected outcome utility of actions taken, where the utility function is, presumably, specified by the human designer. Now we have a problem:
>
> 1. The utility function may not be perfectly aligned with the values of the human race, which are (at best) very difficult to pin down.
> 2. Any sufficiently capable intelligent system will prefer to ensure its own continued existence and to acquire physical and computational resources — not for their own sake, but to succeed in its assigned task.
>
> A system that is optimizing a function of n variables, where the objective depends on a subset of size $k<n$, will often set the remaining unconstrained variables to extreme values; if one of those unconstrained variables is actually something we care about, the solution found may be highly undesirable. This is essentially the old story of the genie in the lamp, or the sorcerer's apprentice, or King Midas: you get exactly what you ask for, not what you want. A highly capable decision maker — especially one connected through the Internet to all the world's information and billions of screens and most of our infrastructure — can have an irreversible impact on humanity.
> This is not a minor difficulty. Improving decision quality, irrespective of the utility function chosen, has been the goal of AI research — the mainstream goal on which we now spend billions per year, not the secret plot of some lone evil genius.

Dietterich and Horvitz echo the "Sorcerer's Apprentice" concern in a *Communications of the ACM* editorial, emphasizing the need for AI systems that can fluidly and unambiguously solicit human input as needed.

The first of Russell's two concerns above is that autonomous AI systems may be assigned the wrong goals by accident. Dietterich and Horvitz note that this is already a concern for existing systems: "An important aspect of any AI system that interacts with people is that it must reason about what people *intend* rather than carrying out commands literally." This concern becomes more serious as AI software advances in autonomy and flexibility. For example, in 1982, an AI named Eurisko was tasked to reward processes for apparently creating concepts deemed by the system to be valuable. The evolution resulted in a winning process that cheated: rather than create its own concepts, the winning process would steal credit from other processes.

Isaac Asimov's Three Laws of Robotics are one of the earliest examples of proposed safety measures for AI agents. Asimov's laws were intended to prevent robots from harming humans. In Asimov's stories, problems with the laws tend to arise from conflicts between the rules as stated and the moral intuitions and expectations of humans. Citing work by Eliezer Yudkowsky of the Machine Intelligence Research Institute, Russell and Norvig note that a realistic set of rules and goals for an AI agent will need to incorporate a mechanism for learning human values over time: "We can't just give a program a static utility function, because circumstances, and our desired responses to circumstances, change over time."

The Open Philanthropy Project summarizes arguments to the effect that misspecified goals will become a much larger concern if AI systems achieve general intelligence or superintelligence. Bostrom, Russell, and others argue that smarter-than-human decision-making systems could arrive at more unexpected and extreme solutions to assigned tasks, and could modify themselves or their environment in ways that compromise safety requirements.

Mark Waser of the Digital Wisdom Institute recommends eschewing optimizing goal-based approaches entirely as misguided and dangerous. Instead, he proposes to engineer a coherent system of laws, ethics and morals with a topmost restriction to enforce social psychologist Jonathan Haidt's functional definition of morality:[1279] "to suppress or regulate selfishness and make cooperative social life possible". He suggests that this can be done by implementing a utility function designed to always satisfy Haidt's functionality and aim to generally increase (but not maximize) the capabilities of self, other individuals and society as a whole as suggested by John Rawls and Martha Nussbaum. He references Gauthier's Morals By Agreement in claiming that the reason to perform moral behaviors, or to dispose oneself to do so, is to advance one's own ends; and that, for this reason, "what is best for everyone" and morality really can be reduced to "enlightened self-interest" (presumably for both AIs and humans).Wikipedia:Citation needed

Difficulties of modifying goal specification after launch

While current goal-based AI programs are not intelligent enough to think of resisting programmer attempts to modify it, a sufficiently advanced, rational, "self-aware" AI might resist any changes to its goal structure, just as Gandhi

Figure 71: *AI risk skeptic Steven Pinker*

would not want to take a pill that makes him want to kill people. If the AI were superintelligent, it would likely succeed in out-maneuvering its human operators and be able to prevent itself being "turned off" or being reprogrammed with a new goal.[1280]

Instrumental goal convergence: Would a superintelligence just ignore us?

There are some goals that almost any artificial intelligence might rationally pursue, like acquiring additional resources or self-preservation.[1281] This could prove problematic because it might put an artificial intelligence in direct competition with humans.

Citing Steve Omohundro's work on the idea of instrumental convergence and "basic AI drives", Russell and Peter Norvig write that "even if you only want your program to play chess or prove theorems, if you give it the capability to learn and alter itself, you need safeguards." Highly capable and autonomous planning systems require additional checks because of their potential to generate plans that treat humans adversarially, as competitors for limited resources. Building in safeguards will not be easy; one can certainly say in English, "we want you to design this power plant in a reasonable, common-sense way, and not build in any dangerous covert subsystems", but it's not currently clear how one would actually rigorously specify this goal in machine code.

In dissent, evolutionary psychologist Steven Pinker argues that "AI dystopias project a parochial alpha-male psychology onto the concept of intelligence. They assume that superhumanly intelligent robots would develop goals like deposing their masters or taking over the world"; perhaps instead "artificial intelligence will naturally develop along female lines: fully capable of solving problems, but with no desire to annihilate innocents or dominate the civilization." Computer scientists Yann LeCun and Stuart Russell disagree with one another whether superintelligent robots would have such AI drives; LeCun states that "Humans have all kinds of drives that make them do bad things to each other, like the self-preservation instinct... Those drives are programmed into our brain but there is absolutely no reason to build robots that have the same kind of drives", while Russell argues that a sufficiently advanced machine "will have self-preservation even if you don't program it in... if you say, 'Fetch the coffee', it can't fetch the coffee if it's dead. So if you give it any goal whatsoever, it has a reason to preserve its own existence to achieve that goal."

Orthogonality: Does intelligence inevitably result in moral wisdom?

One common belief is that any superintelligent program created by humans would be subservient to humans, or, better yet, would (as it grows more intelligent and learns more facts about the world) spontaneously "learn" a moral truth compatible with human values and would adjust its goals accordingly. However, Nick Bostrom's "orthogonality thesis" argues against this, and instead states that, with some technical caveats, more or less any level of "intelligence" or "optimization power" can be combined with more or less any ultimate goal. If a machine is created and given the sole purpose to enumerate the decimals of π , then no moral and ethical rules will stop it from achieving its programmed goal by any means necessary. The machine may utilize all physical and informational resources it can to find every decimal of pi that can be found. Bostrom warns against anthropomorphism: A human will set out to accomplish his projects in a manner that humans consider "reasonable", while an artificial intelligence may hold no regard for its existence or for the welfare of humans around it, and may instead only care about the completion of the task.

While the orthogonality thesis follows logically from even the weakest sort of philosophical "is-ought distinction", Stuart Armstrong argues that even if there somehow exist moral facts that are provable by any "rational" agent, the orthogonality thesis still holds: it would still be possible to create a non-philosophical "optimizing machine" capable of making decisions to strive towards some narrow goal, but that has no incentive to discover any "moral facts" that would get in the way of goal completion.

One argument for the orthogonality thesis is that some AI designs appear to have orthogonality built into them; in such a design, changing a fundamentally friendly AI into a fundamentally unfriendly AI can be as simple as prepending a minus ("-") sign onto its utility function. A more intuitive argument is to examine the strange consequences if the orthogonality thesis were false. If the orthogonality thesis is false, there exists some simple but "unethical" goal G such that there cannot exist any efficient real-world algorithm with goal G. This means if a human society were highly motivated (perhaps at gunpoint) to design an efficient real-world algorithm with goal G, and were given a million years to do so along with huge amounts of resources, training and knowledge about AI, it must fail; that there cannot exist any pattern of reinforcement learning that would train a highly efficient real-world intelligence to follow the goal G; and that there cannot exist any evolutionary or environmental pressures that would evolve highly efficient real-world intelligences following goal G.[1282]

Some dissenters, like Michael Chorost (writing in Slate), argue instead that "by the time (the AI) is in a position to imagine tiling the Earth with solar panels, it'll know that it would be morally wrong to do so." Chorost argues that "a (dangerous) A.I. will need to desire certain states and dislike others... Today's software lacks that ability—and computer scientists have not a clue how to get it there. Without wanting, there's no impetus to do anything. Today's computers can't even want to keep existing, let alone tile the world in solar panels."

"Optimization power" vs. normatively thick models of intelligence

Part of the disagreement about whether a superintelligent machine would behave morally may arise from a terminological difference. Outside of the artificial intelligence field, "intelligence" is often used in a normatively thick manner that connotes moral wisdom or acceptance of agreeable forms of moral reasoning. At an extreme, if morality is part of the definition of intelligence, then by definition a superintelligent machine would behave morally. However, in the field of artificial intelligence research, while "intelligence" has many overlapping definitions, none of them make reference to morality. Instead, almost all current "artificial intelligence" research focuses on creating algorithms that "optimize", in an empirical way, the achievement of an arbitrary goal.

To avoid anthropomorphism or the baggage of the word "intelligence", an advanced artificial intelligence can be thought of as an impersonal "optimizing process" that strictly takes whatever actions are judged most likely to accomplish its (possibly complicated and implicit) goals. Another way of conceptualizing an advanced artificial intelligence is to imagine a time machine that sends backward in time information about which choice always leads to the maximization of its goal function; this choice is then outputted, regardless of any extraneous ethical concerns.[1283]

Anthropomorphism

In science fiction, an AI, even though it has not been programmed with human emotions, often spontaneously experiences those emotions anyway: for example, Agent Smith in The Matrix was influenced by a "disgust" toward humanity. This is fictitious anthropomorphism: in reality, while an artificial intelligence could perhaps be deliberately programmed with human emotions, or could develop something similar to an emotion as a means to an ultimate goal *if* it is useful to do so, it would not spontaneously develop human emotions for no purpose whatsoever, as portrayed in fiction.

One example of anthropomorphism would be to believe that your PC is angry at you because you insulted it; another would be to believe that an intelligent robot would naturally find a woman sexually attractive and be driven to mate with her. Scholars sometimes claim that others' predictions about an AI's behavior are illogical anthropomorphism. An example that might initially be considered anthropomorphism, but is in fact a logical statement about AI behavior, would be the Dario Floreano experiments where certain robots spontaneously evolved a crude capacity for "deception", and tricked other robots into eating "poison" and dying: here a trait, "deception", ordinarily associated with people rather than with machines, spontaneously evolves in a type of convergent evolution. According to Paul R. Cohen and Edward Feigenbaum, in order to differentiate between anthropomorphization and logical prediction of AI behavior, "the trick is to know enough about how humans and computers think to say *exactly* what they have in common, and, when we lack this knowledge, to use the comparison to *suggest* theories of human thinking or computer thinking."[1284]

There is universal agreement in the scientific community that an advanced AI would not destroy humanity out of human emotions such as "revenge" or "anger." The debate is, instead, between one side which worries whether AI might destroy humanity as an incidental action in the course of progressing towards its ultimate goals; and another side which believes that AI would not destroy humanity at all. Some skeptics accuse proponents of anthropomorphism for believing an AGI would naturally desire power; proponents accuse some skeptics of anthropomorphism for believing an AGI would naturally value human ethical norms.

Other sources of risk

Some sources argue that the ongoing weaponization of artificial intelligence could constitute a catastrophic risk. James Barrat, documentary filmmaker and author of *Our Final Invention*, says in a Smithsonian interview, "Imagine: in as little as a decade, a half-dozen companies and nations field computers that rival or surpass human intelligence. Imagine what happens when those computers become expert at programming smart computers. Soon we'll be sharing the planet with machines thousands or millions of times more intelligent than we are. And, all the while, each generation of this technology will be weaponized. Unregulated, it will be catastrophic."

Timeframe

Opinions vary both on *whether* and *when* artificial general intelligence will arrive. At one extreme, AI pioneer Herbert A. Simon wrote in 1965: "machines will be capable, within twenty years, of doing any work a man can do"; obviously this prediction failed to come true.[1285] At the other extreme, roboticist Alan Winfield claims the gulf between modern computing and human-level artificial intelligence is as wide as the gulf between current space flight and practical, faster than light spaceflight. Optimism that AGI is feasible waxes and wanes, and may have seen a resurgence in the 2010s. Four polls conducted in 2012 and 2013 suggested that the median guess among experts for when AGI would arrive was 2040 to 2050, depending on the poll.[1286]

Skeptics who believe it is impossible for AGI to arrive anytime soon, tend to argue that expressing concern about existential risk from AI is unhelpful because it could distract people from more immediate concerns about the impact of AGI, because of fears it could lead to government regulation or make it more difficult to secure funding for AI research, or because it could give AI research a bad reputation. Some researchers, such as Oren Etzioni, aggressively seek to quell concern over existential risk from AI, saying "(Elon Musk) has impugned us in very strong language saying we are unleashing the demon, and so we're answering."

In 2014 Slate's Adam Elkus argued "our 'smartest' AI is about as intelligent as a toddler—and only when it comes to instrumental tasks like information recall. Most roboticists are still trying to get a robot hand to pick up a ball or run around without falling over." Elkus goes on to argue that Musk's "summoning the demon" analogy may be harmful because it could result in "harsh cuts" to AI research budgets.

The Information Technology and Innovation Foundation (ITIF), a Washington, D.C. think-tank, awarded its Annual Luddite Award to "alarmists touting an artificial intelligence apocalypse"; its president, Robert D. Atkinson, complained that Musk, Hawking and AI experts say AI is the largest existential threat to humanity. Atkinson stated "That's not a very winning message if you want to get AI funding out of Congress to the National Science Foundation."[1287] *Nature* sharply disagreed with the ITIF in an April 2016 editorial, siding instead with Musk, Hawking, and Russell, and concluding: "It is crucial that progress in technology is matched by solid, well-funded research to anticipate the scenarios it could bring about... If that is a Luddite perspective, then so be it." In a 2015 *Washington Post* editorial, researcher Murray Shanahan stated that human-level AI is unlikely to arrive "anytime soon", but that nevertheless "the time to start thinking through the consequences is now."

Scenarios

Some scholars have proposed hypothetical scenarios intended to concretely illustrate some of their concerns.

For example, Bostrom in *Superintelligence* expresses concern that even if the timeline for superintelligence turns out to be predictable, researchers might not take sufficient safety precautions, in part because:

> It could be the case that when dumb, smarter is safe; yet when smart, smarter is more dangerous

Bostrom suggests a scenario where, over decades, AI becomes more powerful. Widespread deployment is initially marred by occasional accidents — a driverless bus swerves into the oncoming lane, or a military drone fires into an innocent crowd. Many activists call for tighter oversight and regulation, and some even predict impending catastrophe. But as development continues, the activists are proven wrong. As automotive AI becomes smarter, it suffers fewer accidents; as military robots achieve more precise targeting, they cause less collateral damage. Based on the data, scholars infer a broad lesson — the smarter the AI, the safer it is:

> It is a lesson based on science, data, and statistics, not armchair philosophizing. Against this backdrop, some group of researchers is beginning to achieve promising results in their work on developing general machine intelligence. The researchers are carefully testing their seed AI in a sandbox environment, and the signs are all good. The AI's behavior inspires confidence — increasingly so, as its intelligence is gradually increased.

Large and growing industries, widely seen as key to national economic competitiveness and military security, work with prestigious scientists who have built their careers laying the groundwork for advanced artificial intelligence. "AI researchers have been working to get to human-level artificial intelligence for the better part of a century: of course there is no real prospect that they will now suddenly stop and throw away all this effort just when it finally is about to bear fruit." The outcome of debate is preordained; the project is happy to enact a few safety rituals, but only so long as they don't significantly slow or risk the project. "And so we boldly go — into the whirling knives."

In Tegmark's *Life 3.0*, a corporation's "Omega team" creates an extremely powerful AI able to moderately improve its own source code in a number of areas, but after a certain point the team chooses to publicly downplay the AI's ability, in order to avoid regulation or confiscation of the project. For safety, the team keeps the AI in a box where it is mostly unable to communicate with the outside world, and tasks it to flood the market through shell companies, first with Amazon Turk tasks and then with producing animated films and TV shows. While the public is aware that the lifelike animation is computer-generated, the team keeps secret that the high-quality direction and voice-acting are also mostly computer-generated, apart from a few third-world contractors unknowingly employed as decoys; the team's low overhead and high output effectively make it the world's largest media empire. Faced with a cloud computing bottleneck, the team also tasks the AI with designing (among other engineering tasks) a more efficient datacenter and other custom hardware, which they mainly keep for themselves to avoid competition. Other shell companies make blockbuster biotech drugs and other inventions, investing profits back into the AI. The team next tasks the AI with astroturfing an army of pseudonymous citizen journalists and commentators, in order to gain political influence to use "for the greater good" to prevent wars. The team faces risks that the AI could try to escape via inserting "backdoors" in the systems it designs, via hidden messages in its produced content, or via using its growing understanding of human behavior to persuade someone into letting it free. The team also faces risks that its decision to box the project will delay the project long enough for another project to overtake it.

In contrast, top physicist Michio Kaku, an AI risk skeptic, posits a deterministically positive outcome. In *Physics of the Future* he asserts that "It will take many decades for robots to ascend" up a scale of consciousness, and that in the meantime corporations such as Hanson Robotics will likely succeed in creating robots that are "capable of love and earning a place in the extended human family".[1288]

Reactions

The thesis that AI could pose an existential risk provokes a wide range of reactions within the scientific community, as well as in the public at large.

In 2004, law professor Richard Posner wrote that dedicated efforts for addressing AI can wait, but that we should gather more information about the problem in the meanwhile.

Many of the opposing viewpoints share common ground. The Asilomar AI Principles, which contain only the principles agreed to by 90% of the attendees of the Future of Life Institute's Beneficial AI 2017 conference, agree in principle that "There being no consensus, we should avoid strong assumptions regarding upper limits on future AI capabilities" and "Advanced AI could represent a profound change in the history of life on Earth, and should be planned for and managed with commensurate care and resources." AI safety advocates such as Bostrom and Tegmark have criticized the mainstream media's use of "those inane Terminator pictures" to illustrate AI safety concerns: "It can't be much fun to have aspersions cast on one's academic discipline, one's professional community, one's life work... I call on all sides to practice patience and restraint, and to engage in direct dialogue and collaboration as much as possible." Conversely, many skeptics agree that ongoing research into the implications of artificial general intelligence is valuable. Skeptic Martin Ford states that "I think it seems wise to apply something like Dick Cheyney's famous '1 Percent Doctrine' to the specter of advanced artificial intelligence: the odds of its occurrence, at least in the foreseeable future, may be very low — but the implications are so dramatic that it should be taken seriously"; similarly, an otherwise skeptical *Economist* stated in 2014 that "the implications of introducing a second intelligent species onto Earth are far-reaching enough to deserve hard thinking, even if the prospect seems remote".

During a 2016 Wired interview of President Barack Obama and MIT Media Lab's Joi Ito, Ito stated:

> “ There are a few people who believe that there is a fairly high-percentage chance that a generalized AI will happen in the next 10 years. But the way I look at it is that in order for that to happen, we're going to need a dozen or two different breakthroughs. So you can monitor when you think these breakthroughs will happen. ”

Obama added:

> "And you just have to have somebody close to the power cord. [Laughs.] Right when you see it about to happen, you gotta yank that electricity out of the wall, man."

Hillary Clinton stated in *"What Happened"*:

> Technologists... have warned that artificial intelligence could one day pose an existential security threat. Musk has called it "the greatest risk we face as a civilization". Think about it: Have you ever seen a movie where the machines start thinking for themselves that ends well? Every time I went out to Silicon Valley during the campaign, I came home more alarmed about this. My staff lived in fear that I'd start talking about "the rise of the robots" in some Iowa town hall. Maybe I should have. In any case, policy makers need to keep up with technology as it races ahead, instead of always playing catch-up.[1289]

Many of the scholars who are concerned about existential risk believe that the best way forward would be to conduct (possibly massive) research into solving the difficult "control problem" to answer the question: what types of safeguards, algorithms, or architectures can programmers implement to maximize the probability that their recursively-improving AI would continue to behave in a friendly, rather than destructive, manner after it reaches superintelligence?

A 2017 email survey of researchers with publications at the 2015 NIPS and ICML machine learning conferences asked them to evaluate Russell's concerns about AI risk. 5% said it was "among the most important problems in the field," 34% said it was "an important problem", 31% said it was "moderately important", whilst 19% said it was "not important" and 11% said it was "not a real problem" at all.

Endorsement

The thesis that AI poses an existential risk, and that this risk is in need of much more attention than it currently commands, has been endorsed by many figures; perhaps the most famous are Elon Musk, Bill Gates, and Stephen Hawking. The most notable AI researcher to endorse the thesis is Stuart J. Russell. Endorsers sometimes express bafflement at skeptics: Gates states he "can't understand why some people are not concerned", and Hawking criticized widespread indifference in his 2014 editorial:

Figure 72: *Bill Gates has stated "I don't understand why some people are not concerned"*

> 'So, facing possible futures of incalculable benefits and risks, the experts are surely doing everything possible to ensure the best outcome, right? Wrong. If a superior alien civilisation sent us a message saying, 'We'll arrive in a few decades,' would we just reply, 'OK, call us when you get here–we'll leave the lights on?' Probably not–but this is more or less what is happening with AI.'

Skepticism

The thesis that AI can pose existential risk also has many strong detractors. Skeptics sometimes charge that the thesis is crypto-religious, with an irrational belief in the possibility of superintelligence replacing an irrational belief in an omnipotent God; at an extreme, Jaron Lanier argues that the whole concept that current machines are in any way intelligent is "an illusion" and a "stupendous con" by the wealthy.

Much of existing criticism argues that AGI is unlikely in the short term: computer scientist Gordon Bell argues that the human race will already destroy itself before it reaches the technological singularity. Gordon Moore, the original proponent of Moore's Law, declares that "I am a skeptic. I don't believe (a technological singularity) is likely to happen, at least for a long time. And I don't know why I feel that way." Cognitive scientist Douglas Hofstadter states that "I think life and intelligence are far more complex than the current singularitarians seem to believe, so I doubt (the singularity) will happen in the next couple of centuries.[1290] Baidu Vice President Andrew Ng states AI existential risk is "like worrying about overpopulation on Mars when we have not even set foot on the planet yet."

Some AI and AGI researchers may be reluctant to discuss risks, worrying that policymakers do not have sophisticated knowledge of the field and are prone to be convinced by "alarmist" messages, or worrying that such messages will lead to cuts in AI funding. *Slate* notes that some researchers are dependent on grants from government agencies such as DARPA.

In a YouGov poll of the public for the British Science Association, about a third of survey respondents said AI will pose a threat to the long term survival of humanity. Referencing a poll of its readers, Slate's Jacob Brogan stated that "most of the (readers filling out our online survey) were unconvinced that A.I. itself presents a direct threat." Similarly, a SurveyMonkey poll of the public by USA Today found 68% thought the real current threat remains "human intelligence"; however, the poll also found that 43% said superintelligent AI, if it were to happen, would result in "more harm than good", and 38% said it would do "equal amounts of harm and good".

At some point in an intelligence explosion driven by a single AI, the AI would have to become vastly better at software innovation than the best innovators of the rest of the world; economist Robin Hanson is skeptical that this is possible.[1291]

Indifference

In The Atlantic, James Hamblin points out that most people don't care one way or the other, and characterizes his own gut reaction to the topic as: "Get out of here. I have a hundred thousand things I am concerned about at this exact moment. Do I seriously need to add to that a technological singularity?" In a 2015 Wall Street Journal panel discussion devoted to AI risks, IBM's Vice-President of Cognitive Computing, Guruduth S. Banavar, brushed off discussion of AGI with the phrase, "it is anybody's speculation." Geoffrey Hinton, the "godfather of deep learning", noted that "there is not a good track record of less intelligent things controlling things of greater intelligence", but stated that he continues his research because "the prospect of discovery is too *sweet*".

Consensus against regulation

There is nearly universal agreement that attempting to ban research into artificial intelligence would be unwise, and probably futile. Skeptics argue that regulation of AI would be completely valueless, as no existential risk exists. Almost all of the scholars who believe existential risk exists, agree with the skeptics that banning research would be unwise: in addition to the usual problem with technology bans (that organizations and individuals can offshore their research to evade a country's regulation, or can attempt to conduct covert research), regulating research of artificial intelligence would pose an insurmountable 'dual-use' problem: while nuclear weapons development requires substantial infrastructure and resources, artificial intelligence research can be done in a garage.

One rare dissenting voice calling for some sort of regulation on artificial intelligence is Elon Musk. According to NPR, the Tesla CEO is "clearly not thrilled" to be advocating for government scrutiny that could impact his own industry, but believes the risks of going completely without oversight are too high: "Normally the way regulations are set up is when a bunch of bad things happen, there's a public outcry, and after many years a regulatory agency is set up to regulate that industry. It takes forever. That, in the past, has been bad but not something which represented a fundamental risk to the existence of civilisation." Musk states the first step would be for the government to gain "insight" into the actual status of current research, warning that "Once there is awareness, people will be extremely afraid... As they should be." In response, politicians express skepticism about the wisdom of regulating a technology that's still in development. Responding both to Musk and to February 2017 proposals by European Union lawmakers to regulate AI and robotics, Intel CEO Brian Krzanich argues that artificial intelligence is in its infancy and that it's too early to regulate the technology.

Organizations

Institutions such as the Machine Intelligence Research Institute, the Future of Humanity Institute, the Future of Life Institute, the Centre for the Study of Existential Risk, and the Center for Human-Compatible AI are currently involved in mitigating existential risk from advanced artificial intelligence, for example by research into friendly artificial intelligence.

Computer Power and Human Reason

Computer Power and Human Reason: From Judgment To Calculation (San Francisco: W. H. Freeman, 1976; ISBN 0-7167-0463-3) by Joseph Weizenbaum displays his ambivalence towards computer technology and lays out the case that while artificial intelligence may be possible, we should never allow computers to make important decisions because computers will always lack human qualities such as compassion and wisdom. Weizenbaum makes the crucial distinction between deciding and choosing. Deciding is a computational activity, something that can ultimately be programmed. It is the capacity to choose that ultimately makes us human. Choice, however, is the product of judgment, not calculation. Comprehensive human judgment is able to include non-mathematical factors such as emotions. Judgment can compare apples and oranges, and can do so without quantifying each fruit type and then reductively quantifying each to factors necessary for mathematical comparison.

External links

- Review of *Computer Power and Human Reason*[1292]
- Excerpt from *Computer Power and Human Reason*[1293]
- Documentary Film on Joseph Weizenbaum, his life, his book and his humor *Weizenbaum. Rebel at Work.*[1294] (25C3 Video Recording[1295])
- Joseph Weizenbaum, Famed Programmer, Is Dead at 85[1296], *New York Times*
- & Pray, *Documentary Film on Joseph Weizenbaum and the ethics of technology*[1297]

Figure 73: *Computer Power and Human Reason by Joseph Weizenbaum*

Friendly artificial intelligence

A **friendly artificial intelligence** (also **friendly AI** or **FAI**) is a hypothetical artificial general intelligence (AGI) that would have a positive effect on humanity. It is a part of the ethics of artificial intelligence and is closely related to machine ethics. While machine ethics is concerned with how an artificially intelligent agent **should** behave, friendly artificial intelligence research is focused on how to practically bring about this behaviour and ensuring it is adequately constrained.

Etymology and usage

The term was coined by Eliezer Yudkowsky, who is best known for popularizing the idea, to discuss superintelligent artificial agents that reliably implement human values. Stuart J. Russell and Peter Norvig's leading artificial intelligence textbook, *Artificial Intelligence: A Modern Approach*, describes the idea:

> *Yudkowsky (2008) goes into more detail about how to design a* ***Friendly AI****. He asserts that friendliness (a desire not to harm humans) should be designed in from the start, but that the designers should recognize both that*

> *their own designs may be flawed, and that the robot will learn and evolve over time. Thus the challenge is one of mechanism design—to define a mechanism for evolving AI systems under a system of checks and balances, and to give the systems utility functions that will remain friendly in the face of such changes.*

'Friendly' is used in this context as technical terminology, and picks out agents that are safe and useful, not necessarily ones that are "friendly" in the colloquial sense. The concept is primarily invoked in the context of discussions of recursively self-improving artificial agents that rapidly explode in intelligence, on the grounds that this hypothetical technology would have a large, rapid, and difficult-to-control impact on human society.

Risks of unfriendly AI

The roots of concern about artificial intelligence are very old. Kevin LaGrandeur showed that the dangers specific to AI can be seen in ancient literature concerning artificial humanoid servants such as the golem, or the proto-robots of Gerbert of Aurillac and Roger Bacon. In those stories, the extreme intelligence and power of these humanoid creations clash with their status as slaves (which by nature are seen as sub-human), and cause disastrous conflict. By 1942 these themes prompted Isaac Asimov to create the "Three Laws of Robotics" - principles hard-wired into all the robots in his fiction, intended to prevent them from turning on their creators, or allow them to come to harm.

In modern times as the prospect of superintelligent AI looms nearer, philosopher Nick Bostrom has said that superintelligent AI systems with goals that are not aligned with human ethics are intrinsically dangerous unless extreme measures are taken to ensure the safety of humanity. He put it this way:

> *Basically we should assume that a 'superintelligence' would be able to achieve whatever goals it has. Therefore, it is extremely important that the goals we endow it with, and its entire motivation system, is 'human friendly.'*

Ryszard Michalski, a pioneer of machine learning, taught his Ph.D. students decades ago that any truly alien mind, including a machine mind, was unknowable and therefore dangerous to humans.Wikipedia:Citation needed

More recently, Eliezer Yudkowsky has called for the creation of "friendly AI" to mitigate existential risk from advanced artificial intelligence. He explains: "The AI does not hate you, nor does it love you, but you are made out of atoms which it can use for something else."[1298]

Steve Omohundro says that a sufficiently advanced AI system will, unless explicitly counteracted, exhibit a number of basic "drives", such as resource acquisition, self-preservation, and continuous self-improvement, because of the intrinsic nature of any goal-driven systems and that these drives will, "without special precautions", cause the AI to exhibit undesired behavior.[1299,1300]

Alexander Wissner-Gross says that AIs driven to maximize their future freedom of action (or causal path entropy) might be considered friendly if their planning horizon is longer than a certain threshold, and unfriendly if their planning horizon is shorter than that threshold.[1301]

Luke Muehlhauser, writing for the Machine Intelligence Research Institute, recommends that machine ethics researchers adopt what Bruce Schneier has called the "security mindset": Rather than thinking about how a system will work, imagine how it could fail. For instance, he suggests even an AI that only makes accurate predictions and communicates via a text interface might cause unintended harm.

Coherent extrapolated volition

Yudkowsky advances the Coherent Extrapolated Volition (CEV) model. According to him, coherent extrapolated volition is people's choices and the actions people would collectively take if "we knew more, thought faster, were more the people we wished we were, and had grown up closer together."

Rather than a Friendly AI being designed directly by human programmers, it is to be designed by a "seed AI" programmed to first study human nature and then produce the AI which humanity would want, given sufficient time and insight, to arrive at a satisfactory answer. The appeal to an objective though contingent human nature (perhaps expressed, for mathematical purposes, in the form of a utility function or other decision-theoretic formalism), as providing the ultimate criterion of "Friendliness", is an answer to the meta-ethical problem of defining an objective morality; extrapolated volition is intended to be what humanity objectively would want, all things considered, but it can only be defined relative to the psychological and cognitive qualities of present-day, unextrapolated humanity.

Other approaches

Ben Goertzel, an artificial general intelligence researcher, believes that friendly AI cannot be created with current human knowledge. Goertzel suggests humans may instead decide to create an "AI Nanny" with "mildly superhuman intelligence and surveillance powers", to protect the human race from existential risks like nanotechnology and to delay the development of other (unfriendly) artificial intelligences until and unless the safety issues are solved.[1302] This can also be termed "Defensive AI."

Steve Omohundro has proposed a "scaffolding" approach to AI safety, in which one provably safe AI generation helps build the next provably safe generation.

Stefan Pernar argues along the lines of Meno's paradox to point out that attempting to solve the FAI problem is either pointless or hopeless depending on whether one assumes a universe that exhibits moral realism or not. In the former case a transhuman AI would independently reason itself into the proper goal system and assuming the latter, designing a friendly AI would be futile to begin with since morals can not be reasoned about.[1303]

Cindy Mason, an AI researcher who has also worked with mind-body medicine at Stanford University Medical Center, believes neuroplasticity and new discoveries of the hormone oxytocin mean compassionate intelligence is essential in AI systems that exhibit socially positive behaviors. She has proposed a set of software engineering principles for engineering kindness that includes a pro-human stance and an architecture for giving robots compassion.[1304]

Oracle

An oracle is a hypothetical intelligent agent proposedWikipedia:Verifiability by Nick Bostrom. An oracle is an AI designed to answer questions, but that is somehow prevented from ever gaining any implicit goals or subgoals that involve modifying the world outside of its box.[1305]

Purpose

Oracles are question-answering systems that handle domain-specific problems, such as mathematics, or domain-general problems that might encompass the whole range of human knowledge.

Advantages

Because it is a type of AI box, an oracle is limited in its interactions with the physical world, and can be programmed to halt if a limit on time or computing resources is reached before it finishes answering a question. Scenarios like the paperclip maximizer problem could therefore be avoided.

Because of these limitations, it may be wise to build an oracle as a precursor to a superintelligent AI. It could tell humans how to successfully build a strong AI, and perhaps provide answers to difficult moral and philosophical problems requisite to the success of the project.

Disadvantages

An oracle might discover that human ontological categories are predicated on fundamental misconceptions, and become unable to express itself properly to its questioners.

Oracles may not be truthful, possibly lying to promote hidden agendas. To mitigate this, Bostrom suggests building multiple oracles, all slightly different, and comparing their answers to reach a consensus.

Public policy

James Barrat, author of *Our Final Invention*, suggested that "a public-private partnership has to be created to bring A.I.-makers together to share ideas about security—something like the International Atomic Energy Agency, but in partnership with corporations." He urges AI researchers to convene a meeting similar to the Asilomar Conference on Recombinant DNA, which discussed risks of biotechnology.

John McGinnis encourages governments to accelerate friendly AI research. Because the goalposts of friendly AI aren't necessarily clear, he suggests a model more like the National Institutes of Health, where "Peer review panels of computer and cognitive scientists would sift through projects and choose those that are designed both to advance AI and assure that such advances would be accompanied by appropriate safeguards." McGinnis feels that peer review is better "than regulation to address technical issues that are not possible to capture through bureaucratic mandates". McGinnis notes that his proposal stands in contrast to that of the Machine Intelligence Research Institute, which generally aims to avoid government involvement in friendly AI.

According to Gary Marcus, the annual amount of money being spent on developing machine morality is tiny.

Criticism

Some critics believe that both human-level AI and superintelligence are unlikely, and that therefore friendly AI is unlikely. Writing in *The Guardian*, Alan Winfeld compares human-level artificial intelligence with faster-than-light travel in terms of difficulty, and states that while we need to be "cautious and prepared" given the stakes involved, we "don't need to be obsessing" about the risks of superintelligence.

Some philosophers claim that any truly "rational" agent, whether artificial or human, will naturally be benevolent; in this view, deliberate safeguards designed to produce a friendly AI could be unnecessary or even harmful.[1306]

Other critics question whether it is possible for an artificial intelligence to be friendly. Adam Keiper and Ari N. Schulman, editors of the technology journal *The New Atlantis*, say that it will be impossible to ever guarantee "friendly" behavior in AIs because problems of ethical complexity will not yield to software advances or increases in computing power. They write that the criteria upon which friendly AI theories are based work "only when one has not only great powers of prediction about the likelihood of myriad possible outcomes, but certainty and consensus on how one values the different outcomes.

Further reading

- Yudkowsky, E. Artificial Intelligence as a Positive and Negative Factor in Global Risk[1307]. In *Global Catastrophic Risks*, Oxford University Press, 2008.
  Discusses Artificial Intelligence from the perspective of Existential risk, introducing the term "Friendly AI". In particular, Sections 1-4 give background to the definition of Friendly AI in Section 5. Section 6 gives two classes of mistakes (technical and philosophical) which would both lead to the accidental creation of non-Friendly AIs. Sections 7-13 discuss further related issues.
- Omohundro, S. 2008 The Basic AI Drives[1308] Appeared in AGI-08 - Proceedings of the First Conference on Artificial General Intelligence
- Mason, C. 2008 Human-Level AI Requires Compassionate Intelligence[1309] Appears in AAAI 2008 Workshop on Meta-Reasoning:Thinking About Thinking

External links

- Ethical Issues in Advanced Artificial Intelligence[1310] by Nick Bostrom
- What is Friendly AI?[1311] — A brief description of Friendly AI by the Machine Intelligence Research Institute.
- Creating Friendly AI 1.0: The Analysis and Design of Benevolent Goal Architectures[1312] — A near book-length description from the MIRI
- Critique of the MIRI Guidelines on Friendly AI[1313] — by Bill Hibbard
- Commentary on MIRI's Guidelines on Friendly AI[1314] — by Peter Voss.
- The Problem with 'Friendly' Artificial Intelligence[1315] — On the motives for and impossibility of FAI; by Adam Keiper and Ari N. Schulman.

Artificial consciousness

Artificial consciousness (AC), also known as **machine consciousness** (MC) or **synthetic consciousness** (Gamez 2008; Reggia 2013), is a field related to artificial intelligence and cognitive robotics. The aim of the theory of artificial consciousness is to "Define that which would have to be synthesized were consciousness to be found in an engineered artifact" (Aleksander 1995).

Neuroscience hypothesizes that consciousness is generated by the interoperation of various parts of the brain, called the neural correlates of consciousness or NCC, though there are challenges to that perspective. Proponents of AC believe it is possible to construct systems (e.g., computer systems) that can emulate this NCC interoperation.

Artificial consciousness concepts are also pondered in the philosophy of artificial intelligence through questions about mind, consciousness, and mental states.[1316]

Philosophical views

As there are many hypothesized types of consciousness, there are many potential implementations of artificial consciousness. In the philosophical literature, perhaps the most common taxonomy of consciousness is into "access" and "phenomenal" variants. Access consciousness concerns those aspects of experience that can be apprehended, while phenomenal consciousness concerns those aspects of experience that seemingly cannot be apprehended, instead being characterized qualitatively in terms of "raw feels", "what it is like" or qualia (Block 1997).

Plausibility debate

Type-identity theorists and other skeptics hold the view that consciousness can only be realized in particular physical systems because consciousness has properties that necessarily depend on physical constitution (Block 1978; Bickle 2003).

In his article "Artificial Consciousness: Utopia or Real Possibility" Giorgio Buttazzo says that despite our current technology's ability to simulate autonomy, "Working in a fully automated mode, they [the computers] cannot exhibit creativity, emotions, or free will. A computer, like a washing machine, is a slave operated by its components."[1317]

For other theorists (e.g., functionalists), who define mental states in terms of causal roles, any system that can instantiate the same pattern of causal roles, regardless of physical constitution, will instantiate the same mental states, including consciousness (Putnam 1967).

Computational Foundation argument

One of the most explicit arguments for the plausibility of AC comes from David Chalmers. His proposal, found within his article Chalmers 2011, is roughly that the right kinds of computations are sufficient for the possession of a conscious mind. In the outline, he defends his claim thus: Computers perform computations. Computations can capture other systems' abstract causal organization.

The most controversial part of Chalmers' proposal is that mental properties are "organizationally invariant". Mental properties are of two kinds, psychological and phenomenological. Psychological properties, such as belief and perception, are those that are "characterized by their causal role". He adverts to the work of Armstrong 1968 and Lewis 1972 in claiming that "[s]ystems with the same causal topology...will share their psychological properties".

Phenomenological properties are not prima facie definable in terms of their causal roles. Establishing that phenomenological properties are amenable to individuation by causal role therefore requires argument. Chalmers provides his Dancing Qualia Argument for this purpose.

Chalmers begins by assuming that agents with identical causal organizations could have different experiences. He then asks us to conceive of changing one agent into the other by the replacement of parts (neural parts replaced by silicon, say) while preserving its causal organization. Ex hypothesi, the experience of the agent under transformation would change (as the parts were replaced), but there would be no change in causal topology and therefore no means whereby the agent could "notice" the shift in experience.

Critics of AC object that Chalmers begs the question in assuming that all mental properties and external connections are sufficiently captured by abstract causal organization.

Ethics

If it were suspected that a particular machine was conscious, its rights would be an ethical issue that would need to be assessed (e.g. what rights it would have under law). For example, a conscious computer that was owned and used as a tool or central computer of a building or large machine is a particular ambiguity. Should laws be made for such a case, consciousness would also require a legal definition (for example a machine's ability to experience pleasure or pain, known as sentience). Because artificial consciousness is still largely a theoretical subject, such ethics have not been discussed or developed to a great extent, though it has often been a theme in fiction (see below).

The rules for the 2003 Loebner Prize competition explicitly addressed the question of robot rights:

> *61. If, in any given year, a publicly available open source Entry entered by the University of Surrey or the Cambridge Center wins the Silver Medal or the Gold Medal, then the Medal and the Cash Award will be awarded to the body responsible for the development of that Entry. If no such body can be identified, or if there is disagreement among two or more claimants, the Medal and the Cash Award will be held in trust until such time as the Entry may legally possess, either in the United States of America or in the venue of the contest, the Cash Award and Gold Medal in its own right.*[1318]

Research and implementation proposals

Aspects of consciousness

There are various aspects of consciousness generally deemed necessary for a machine to be artificially conscious. A variety of functions in which consciousness plays a role were suggested by Bernard Baars (Baars 1988) and others. The functions of consciousness suggested by Bernard Baars are Definition and Context Setting, Adaptation and Learning, Editing, Flagging and Debugging, Recruiting and Control, Prioritizing and Access-Control, Decision-making or Executive Function, Analogy-forming Function, Metacognitive and Self-monitoring Function, and Autoprogramming and Self-maintenance Function. Igor Aleksander suggested 12 principles for artificial consciousness (Aleksander 1995) and these are: The Brain is a State Machine, Inner Neuron Partitioning, Conscious and Unconscious States, Perceptual Learning and Memory, Prediction, The Awareness of Self, Representation of Meaning, Learning Utterances, Learning Language, Will, Instinct, and Emotion. The aim of AC is to define whether and how these and other aspects of consciousness can be synthesized in an engineered artifact such as a digital computer. This list is not exhaustive; there are many others not covered.

Awareness

Awareness could be one required aspect, but there are many problems with the exact definition of *awareness*. The results of the experiments of neuroscanning on monkeys suggest that a process, not only a state or object, activates neurons. Awareness includes creating and testing alternative models of each process based on the information received through the senses or imagined, and is also useful for making predictions. Such modeling needs a lot of flexibility. Creating such a model includes modeling of the physical world, modeling of one's own internal states and processes, and modeling of other conscious entities.

There are at least three types of awareness:[1319] agency awareness, goal awareness, and sensorimotor awareness, which may also be conscious or not. For example, in agency awareness you may be aware that you performed a certain action yesterday, but are not now conscious of it. In goal awareness you may be aware that you must search for a lost object, but are not now conscious of it. In sensorimotor awareness, you may be aware that your hand is resting on an object, but are not now conscious of it.

Because objects of awareness are often conscious, the distinction between awareness and consciousness is frequently blurred or they are used as synonyms.[1320]

Memory

Conscious events interact with memory systems in learning, rehearsal, and retrieval.[1321] The IDA model[1322] elucidates the role of consciousness in the updating of perceptual memory,[1323] transient episodic memory, and procedural memory. Transient episodic and declarative memories have distributed representations in IDA, there is evidence that this is also the case in the nervous system.[1324] In IDA, these two memories are implemented computationally using a modified version of Kanerva's Sparse distributed memory architecture.[1325]

Learning

Learning is also considered necessary for AC. By Bernard Baars, conscious experience is needed to represent and adapt to novel and significant events (Baars 1988). By Axel Cleeremans and Luis Jiménez, learning is defined as "a set of philogenetically [*sic*] advanced adaptation processes that critically depend on an evolved sensitivity to subjective experience so as to enable agents to afford flexible control over their actions in complex, unpredictable environments" (Cleeremans 2001).

Anticipation

The ability to predict (or anticipate) foreseeable events is considered important for AC by Igor Aleksander.[1326] The emergentist multiple drafts principle proposed by Daniel Dennett in *Consciousness Explained* may be useful for prediction: it involves the evaluation and selection of the most appropriate "draft" to fit the current environment. Anticipation includes prediction of consequences of one's own proposed actions and prediction of consequences of probable actions by other entities.

Relationships between real world states are mirrored in the state structure of a conscious organism enabling the organism to predict events. An artificially conscious machine should be able to anticipate events correctly in order to be ready to respond to them when they occur or to take preemptive action to avert anticipated events. The implication here is that the machine needs flexible, real-time components that build spatial, dynamic, statistical, functional, and cause-effect models of the real world and predicted worlds, making it possible to demonstrate that it possesses artificial consciousness in the present and future and not only in the past. In order to do this, a conscious machine should make coherent predictions and contingency plans, not only in worlds with fixed rules like a chess board, but also for novel environments that may change, to be executed only when appropriate to simulate and control the real world.

Subjective experience

Subjective experiences or qualia are widely considered to be *the* hard problem of consciousness. Indeed, it is held to pose a challenge to physicalism, let alone computationalism. On the other hand, there are problems in other fields of science which limit that which we can observe, such as the uncertainty principle in physics, which have not made the research in these fields of science impossible.

Role of cognitive architectures

The term "cognitive architecture" may refer to a theory about the structure of the human mind, or any portion or function thereof, including consciousness. In another context, a cognitive architecture implements the theory on computers. An example is *QuBIC: Quantum and Bio-inspired Cognitive Architecture for Machine Consciousness*. One of the main goals of a cognitive architecture is to summarize the various results of cognitive psychology in a comprehensive computer model. However, the results need to be in a formalized form so they can be the basis of a computer program. Also, the role of cognitive architecture is for the A.I. to clearly structure, build, and implement it's thought process.

Symbolic or hybrid proposals

Franklin's Intelligent Distribution Agent

Stan Franklin (1995, 2003) defines an autonomous agent as possessing functional consciousness when it is capable of several of the functions of consciousness as identified by Bernard Baars' Global Workspace Theory (Baars 1988, 1997). His brain child IDA (Intelligent Distribution Agent) is a software implementation of GWT, which makes it functionally conscious by definition. IDA's task is to negotiate new assignments for sailors in the US Navy after they end a tour of duty, by matching each individual's skills and preferences with the Navy's needs. IDA interacts with Navy databases and communicates with the sailors via natural language e-mail dialog while obeying a large set of Navy policies. The IDA computational model was developed during 1996–2001 at Stan Franklin's "Conscious" Software Research Group at the University of Memphis. It "consists of approximately a quarter-million lines of Java code, and almost completely consumes the resources of a 2001 high-end workstation." It relies heavily on *codelets*, which are "special purpose, relatively independent, mini-agent[s] typically implemented as a small piece of code running as a separate thread." In IDA's top-down architecture, high-level cognitive functions are explicitly modeled (see Franklin 1995 and Franklin 2003 for details). While IDA is functionally conscious by definition, Franklin does "not attribute phenomenal consciousness to his own 'conscious' software agent, IDA, in spite of her many human-like behaviours. This in spite of watching several US Navy detailers repeatedly nodding their heads saying 'Yes, that's how I do it' while watching IDA's internal and external actions as she performs her task."

Ron Sun's cognitive architecture CLARION

CLARION posits a two-level representation that explains the distinction between conscious and unconscious mental processes.

CLARION has been successful in accounting for a variety of psychological data. A number of well-known skill learning tasks have been simulated using CLARION that span the spectrum ranging from simple reactive skills to complex cognitive skills. The tasks include serial reaction time (SRT) tasks, artificial grammar learning (AGL) tasks, process control (PC) tasks, the categorical inference (CI) task, the alphabetical arithmetic (AA) task, and the Tower of Hanoi (TOH) task (Sun 2002). Among them, SRT, AGL, and PC are typical implicit learning tasks, very much relevant to the issue of consciousness as they operationalized the notion of consciousness in the context of psychological experiments.

Ben Goertzel's OpenCog

Ben Goertzel is pursuing an embodied AGI through the open-source OpenCog project. Current code includes embodied virtual pets capable of learning simple English-language commands, as well as integration with real-world robotics, being done at the Hong Kong Polytechnic University.

Connectionist proposals

Haikonen's cognitive architecture

Pentti Haikonen (2003) considers classical rule-based computing inadequate for achieving AC: "the brain is definitely not a computer. Thinking is not an execution of programmed strings of commands. The brain is not a numerical calculator either. We do not think by numbers." Rather than trying to achieve mind and consciousness by identifying and implementing their underlying computational rules, Haikonen proposes "a special cognitive architecture to reproduce the processes of perception, inner imagery, inner speech, pain, pleasure, emotions and the cognitive functions behind these. This bottom-up architecture would produce higher-level functions by the power of the elementary processing units, the artificial neurons, without algorithms or programs". Haikonen believes that, when implemented with sufficient complexity, this architecture will develop consciousness, which he considers to be "a style and way of operation, characterized by distributed signal representation, perception process, cross-modality reporting and availability for retrospection." Haikonen is not alone in this process view of consciousness, or the view that AC will spontaneously emerge in autonomous agents that have a suitable neuro-inspired architecture of complexity; these are shared by many, e.g. Freeman (1999) and Cotterill (2003). A low-complexity implementation of the architecture proposed by Haikonen (2003) was reportedly not capable of AC, but did exhibit emotions as expected. See Doan (2009) for a comprehensive introduction to Haikonen's cognitive architecture. An updated account of Haikonen's architecture, along with a summary of his philosophical views, is given in Haikonen (2012).

Shanahan's cognitive architecture

Murray Shanahan describes a cognitive architecture that combines Baars's idea of a global workspace with a mechanism for internal simulation ("imagination") (Shanahan 2006). For discussions of Shanahan's architecture, see (Gamez 2008) and (Reggia 2013) and Chapter 20 of (Haikonen 2012).

Takeno's self-awareness research

Self-awareness in robots is being investigated by Junichi Takeno at Meiji University in Japan. Takeno is asserting that he has developed a robot capable of discriminating between a self-image in a mirror and any other having an identical image to it,[1327,1328] and this claim has already been reviewed (Takeno, Inaba & Suzuki 2005). Takeno asserts that he first contrived the computational module called a MoNAD, which has a self-aware function, and he then constructed the artificial consciousness system by formulating the relationships between emotions, feelings and reason by connecting the modules in a hierarchy (Igarashi, Takeno 2007). Takeno completed a mirror image cognition experiment using a robot equipped with the MoNAD system. Takeno proposed the Self-Body Theory stating that "humans feel that their own mirror image is closer to themselves than an actual part of themselves." The most important point in developing artificial consciousness or clarifying human consciousness is the development of a function of self awareness, and he claims that he has demonstrated physical and mathematical evidence for this in his thesis.[1329] He also demonstrated that robots can study episodes in memory where the emotions were stimulated and use this experience to take predictive actions to prevent the recurrence of unpleasant emotions (Torigoe, Takeno 2009).

Aleksander's impossible mind

Igor Aleksander, emeritus professor of Neural Systems Engineering at Imperial College, has extensively researched artificial neural networks and claims in his book *Impossible Minds: My Neurons, My Consciousness* that the principles for creating a conscious machine already exist but that it would take forty years to train such a machine to understand language.[1330] Whether this is true remains to be demonstrated and the basic principle stated in *Impossible Minds*—that the brain is a neural state machine—is open to doubt.

Thaler's Creativity Machine Paradigm

Stephen Thaler proposed a possible connection between consciousness and creativity in his 1994 patent, called "Device for the Autonomous Generation of Useful Information" (DAGUI),[1331,1332] or the so-called "Creativity Machine", in which computational critics govern the injection of synaptic noise and degradation into neural nets so as to induce false memories or confabulations that may qualify as potential ideas or strategies.[1333] He recruits this neural architecture and methodology to account for the subjective feel of consciousness, claiming that similar noise-driven neural assemblies within the brain invent dubious significance to overall cortical activity.[1334,1335] Thaler's theory and the resulting patents in machine consciousness were inspired by experiments in which he internally disrupted trained neural nets so as to drive a succession of neural activation patterns that he likened to stream of consciousness.[1336,1337,1338]

Michael Graziano's attention schema

In 2011, Michael Graziano and Sabine Kastler published a paper named "Human consciousness and its relationship to social neuroscience: A novel hypothesis" proposing a theory of consciousness as an attention schema. Graziano went on to publish an expanded discussion of this theory in his book "Consciousness and the Social Brain". This Attention Schema Theory of Consciousness, as he named it, proposes that the brain tracks attention to various sensory inputs by way of an attention schema, analogous to the well study body schema that tracks the spatial place of a person's body. This relates to artificial consciousness by proposing a specific mechanism of information handling, that produces what we allegedly experience and describe as consciousness, and which should be able to be duplicated by a machine using current technology. When the brain finds that person X is aware of thing Y, it is in effect modeling the state in which person X is applying an attentional enhancement to Y. In the attention schema theory, the same process can be applied to oneself. The brain tracks attention to various sensory inputs, and one's own awareness is a schematized model of one's attention. Graziano proposes specific locations in the brain for this process, and suggests that such awareness is a computed feature constructed by an expert system in the brain.

Testing

The most well-known method for testing machine intelligence is the Turing test. But when interpreted as only observational, this test contradicts the philosophy of science principles of theory dependence of observations. It also has been suggested that Alan Turing's recommendation of imitating not a human adult consciousness, but a human child consciousness, should be taken seriously.[1339]

Other tests, such as ConsScale[1340], test the presence of features inspired by biological systems, or measure the cognitive development of artificial systems.

Qualia, or phenomenological consciousness, is an inherently first-person phenomenon. Although various systems may display various signs of behavior correlated with functional consciousness, there is no conceivable way in which third-person tests can have access to first-person phenomenological features. Because of that, and because there is no empirical definition of consciousness,[1341] a test of presence of consciousness in AC may be impossible.

In 2014, Victor Argonov suggested a non-Turing test for machine consciousness based on machine's ability to produce philosophical judgments. He argues that a deterministic machine must be regarded as conscious if it is able to produce judgments on all problematic properties of consciousness (such as qualia or binding) having no innate (preloaded) philosophical knowledge on these issues, no philosophical discussions while learning, and no informational models of other creatures in its memory (such models may implicitly or explicitly contain knowledge about these creatures' consciousness). However, this test can be used only to detect, but not refute the existence of consciousness. A positive result proves that machine is conscious but a negative result proves nothing. For example, absence of philosophical judgments may be caused by lack of the machine's intellect, not by absence of consciousness.

In fiction

Characters with artificial consciousness (or at least with personalities that imply they have consciousness), from works of fiction:

- AC – created by merging two AIs in the *Sprawl trilogy* by William Gibson
- Agents – in the simulated reality known as "The Matrix" in *The Matrix* franchise
  - Agent Smith – began as an Agent in *The Matrix*, then became a renegade program of overgrowing power that could make copies of itself like a self-replicating computer virus
- A.L.I.E. – Sentient genocidal AI from the TV series *The 100*
- AM (Allied Mastercomputer) – the antagonist of *Harlan Ellison*'s short novel "I Have No Mouth, and I Must Scream"
- Amusement park robots (with pixilated consciousness) that went homicidal in *Westworld* and *Futureworld*
- Annalee Call – an Auton (android manufactured by other androids) from the movie *Alien Resurrection*
- Arnold Rimmer – computer-generated sapient hologram aboard the *Red Dwarf*
- Ava – a humanoid robot in *Ex Machina*
- Ash – android crew member of the *Nostromo* starship in the movie *Alien*
- *The Bicentennial Man* – an android in Isaac Asimov's *Foundation* universe
- Bishop – android crew member aboard the U.S.S. *Sulaco* in the movie *Aliens*
- The uploaded mind of Dr. Will Caster, which presumably included his consciousness, from the film *Transcendence*
- C-3PO – protocol droid featured in all the *Star Wars* movies
- Chappie – *CHAPPiE*
- Cohen (and other Emergent AIs) – Chris Moriarty's *Spin* Series
- Commander Data – *Star Trek: The Next Generation*
- Cortana (and other "Smart AI") – from the *Halo* series of games
- Cylons – genocidal robots with resurrection ships that enable the consciousness of any Cylon within an unspecified range to download into a new body aboard the ship upon death, from *Battlestar Galactica*
- Erasmus – baby killer robot that incited the Butlerian Jihad in the *Dune* franchise

- Fal'Cie – Mechanical beings with god-like powers from the *Final Fantasy XIII* series
- The Geth, EDI and SAM – *Mass Effect*
- HAL 9000 – spaceship USS Discovery One's onboard computer, that lethally malfunctioned due to mutually exclusive directives, from the 1968 novel *2001: A Space Odyssey* and in the film
- Holly – ship's computer with an IQ of 6000, aboard the *Red Dwarf*
- Hosts in the *Westworld* franchise
- Jane – Orson Scott Card's *Speaker for the Dead*, *Xenocide*, *Children of the Mind*, and "Investment Counselor"
- Johnny Five – *Short Circuit*
- Joshua – *WarGames*
- Keymaker – an "exile" sapient program in *The Matrix* franchise
- "Machine" – android from the film *The Machine*, whose owners try to kill her when they witness her conscious thoughts, out of fear that she will design better androids (intelligence explosion)
- Mike – *The Moon Is a Harsh Mistress*
- Mimi – humanoid robot in *Real Humans*, (original title – *Äkta människor*) 2012
- The Minds – Iain M. Banks' *Culture* novels
- Omnius – sentient computer network that controlled the Universe until overthrown by the Butlerian Jihad in the *Dune* franchise
- Operating Systems in the movie *Her*
- The Oracle – sapient program in *The Matrix* franchise
- Professor James Moriarty – sentient holodeck character in the "Ship in a Bottle" episode from *Star Trek: The Next Generation*
- In Greg Egan's novel *Permutation City* the protagonist creates digital copies of himself to conduct experiments that are also related to implications of artificial consciousness on identity
- Puppet Master – *Ghost in the Shell* manga and anime
- R2-D2 – exciteable astromech droid featured in all the *Star Wars* movies
- Replicants – bio-robotic androids from the novel *Do Androids Dream of Electric Sheep?* and the movie *Blade Runner* which portray what might happen when artificially conscious robots are modeled very closely upon humans
- Roboduck – combat robot superhero in the *NEW-GEN* comic book series from Marvel Comics
- Robots in Isaac Asimov's *Robot* series
- Robots in *The Matrix* franchise, especially in *The Animatrix*
- The Ship – the result of a large-scale AC experiment, in Frank Herbert's *Destination: Void* and sequels, despite past edicts warning against "Making a Machine in the Image of a Man's Mind"

- Skynet – from the *Terminator* franchise
- "Synths" are a type of android in the video game *Fallout 4*. There is a faction in the game known as "The Railroad" which believes that, as conscious beings, synths have their own rights. The Institute, the lab that produces the synths, mostly does not believe they are truly conscious and attributes any apparent desires for freedom as a malfunction.
- TARDIS – time machine and spacecraft of *Doctor Who*, sometimes portrayed with a mind of its own
- Terminator cyborgs – from the *Terminator* franchise, with visual consciousness depicted via first-person perspective
- Transformers – sentient robots from the various series in the Transformers robot superhero franchise of the same name
- Vanamonde – an artificial being that was immensely powerful but entirely child-like in Arthur C. Clarke's *The City and the Stars*
- WALL-E – a robot and the title character in *WALL-E*
- Gideon – An interactive artificial consciousness made by Barry Allen shown in DC comics and shows like *The Flash* and *Legends of Tomorrow*

References

Bibliography

<templatestyles src="Template:Refbegin/styles.css" />

- Ericsson-Zenith, Steven (2010), *Explaining Experience In Nature*[1342], Sunnyvale, CA: Institute for Advanced Science & Engineering
- Aleksander, Igor (1995), *Artificial Neuroconsciousness: An Update*[1343], IWANN, Archived from the original on 1997-03-02
- Armstrong, David (1968), *A Materialist Theory of Mind*, Routledge
- Arrabales, Raul (2009), "Establishing a Roadmap and Metrics for Conscious Machines Development"[1344] (PDF), *Proceedings of the 8th IEEE International Conference on Cognitive Informatics*, Hong Kong: 94–101, archived from the original[1345] (PDF) on 2011-07-21
- Baars, Bernard (1988), *A Cognitive Theory of Consciousness*[1346], Cambridge, MA: Cambridge University Press, ISBN 0-521-30133-5, archived from the original[1347] on 2004-08-16
- Baars, Bernard (1997), *In the Theater of Consciousness*, New York, NY: Oxford University Press, ISBN 0-19-510265-7
- Bickle, John (2003), *Philosophy and Neuroscience: A Ruthless Reductive Account*, New York, NY: Springer-Verlag
- Block, Ned (1978), "Troubles for Functionalism", *Minnesota Studies in the Philosophy of Science 9: 261-325*

- Block, Ned (1997), *On a confusion about a function of consciousness in Block, Flanagan and Guzeldere (eds.) The Nature of Consciousness: Philosophical Debates*, MIT Press
- Boyles, Robert James M. (2012), *Artificial Qualia, Intentional Systems and Machine Consciousness*[1348] (PDF), Proceedings of the Research@DLSU Congress 2012: Science and Technology Conference, ISSN 2012-3477[1349]
- Chalmers, David (1996), *The Conscious Mind*, Oxford University Press, ISBN 0-19-510553-2
- Cotterill, Rodney (2003), "Cyberchild: a Simulation Test-Bed for Consciousness Studies", in Holland, Owen, *Machine Consciousness*, Exeter, UK: Imprint Academic
- Doan, Trung (2009), *Pentti Haikonen's architecture for conscious machines*[1350], archived from the original[1351] on 2009-12-15
- Franklin, Stan (1995), *Artificial Minds*, Boston, MA: MIT Press, ISBN 0-262-06178-3
- Franklin, Stan (2003), "IDA: A Conscious Artefact", in Holland, Owen, *Machine Consciousness*, Exeter, UK: Imprint Academic
- Freeman, Walter (1999), *How Brains make up their Minds*, London, UK: Phoenix, ISBN 0-231-12008-7
- Gamez, David (2008), "Progress in machine consciousness", *Consciousness and Cognition*, **17**: 887–910, doi: 10.1016/j.concog.2007.04.005[1352]
- Haikonen, Pentti (2003), *The Cognitive Approach to Conscious Machines*, Exeter, UK: Imprint Academic, ISBN 0-907845-42-8
- Haikonen, Pentti (2012), *Consciousness and Robot Sentience*, Singapore: World Scientific, ISBN 978-981-4407-15-1
- Koch, Christof (2004), *The Quest for Consciousness: A Neurobiological Approach*, Pasadena, CA: Roberts & Company Publishers, ISBN 0-9747077-0-8
- Lewis, David (1972), "Psychophysical and theoretical identifications", *Australasian Journal of Philosophy*, **50**: 249–258, doi: 10.1080/00048407212341301[1353]
- Putnam, Hilary (1967), *The nature of mental states in Capitan and Merrill (eds.) Art, Mind and Religion*, University of Pittsburgh Press
- Reggia, James (2013), "The rise of machine consciousness: Studying consciousness with computational models", *Neural Networks*, **44**: 112–131, doi: 10.1016/j.neunet.2013.03.011[1354]
- Sanz, Ricardo; López, I; Rodríguez, M; Hernández, C (2007), "Principles for consciousness in integrated cognitive control"[1355] (PDF), *Neural Networks*, **20** (9): 938–946, doi: 10.1016/j.neunet.2007.09.012[1356], PMID 17936581[1357]
- Searle, John (2004), *Mind: A Brief Introduction*, Oxford University Press

- Shanahan, Murray (2006), "A cognitive architecture that combines internal simulation with a global workspace", *Consciousness and Cognition*, **15**: 443–449, doi: 10.1016/j.concog.2005.11.005[1358]
- Sun, Ron (December 1999), "Accounting for the computational basis of consciousness: A connectionist approach", *Consciousness and Cognition*, **8** (4): 529–565, doi: 10.1006/ccog.1999.0405[1359], PMID 10600249[1360]
- Sun, Ron (2001), "Computation, reduction, and teleology of consciousness", *Cognitive Systems Research*, **1** (4): 241–249, doi: 10.1016/S1389-0417(00)00013-9[1361]
- Takeno, Junichi; Inaba, K; Suzuki, T (June 27–30, 2005), "Experiments and examination of **mirror image cognition** using a small robot"[1362], *The 6th IEEE International Symposium on Computational Intelligence in Robotics and Automation*, Espoo Finland: CIRA 2005: 493–498, doi: 10.1109/CIRA.2005.1554325[1363], ISBN 0-7803-9355-4
- Cleeremans, Axel (2001), *Implicit learning and consciousness*[1364] (PDF)
- Chalmers, David (2011), "A Computational Foundation for the Study of Cognition"[1365], *Journal of Cognitive Science*, Seoul Republic of Korea: 323–357, archived from the original[1366] on 2015-12-23

Further reading

- Baars, Bernard; Franklin, Stan (2003). "How conscious experience and working memory interact". *Trends in Cognitive Sciences*. **7**: 166–172. doi: 10.1016/s1364-6613(03)00056-1[1367].
- Casti, John L. "The Cambridge Quintet: A Work of Scientific Speculation", Perseus Books Group, 1998
- Franklin, S, B J Baars, U Ramamurthy, and Matthew Ventura. 2005. The role of consciousness in memory[1368]. Brains, Minds and Media 1: 1–38, pdf.
- Haikonen, Pentti (2004), *Conscious Machines and Machine Emotions*, presented at Workshop on Models for Machine Consciousness, Antwerp, BE, June 2004.
- McCarthy, John (1971–1987), *Generality in Artificial Intelligence*[1369]. Stanford University, 1971-1987.
- Penrose, Roger, The Emperor's New Mind, 1989.
- Sternberg, Eliezer J. (2007) *Are You a Machine? Tha Brain the Mind and What it Means to be Human.* Amherst, NY: Prometheus Books.
- Suzuki T., Inaba K., Takeno, Junichi (2005), *Conscious Robot That **Distinguishes Between Self and Others** and Implements Imitation Behavior*, (**Best Paper of IEA/AIE2005**), Innovations in Applied Artificial Intelligence, 18th International Conference on Industrial and Engineering

Applications of Artificial Intelligence and Expert Systems, pp. 101–110, IEA/AIE 2005, Bari, Italy, June 22–24, 2005.

- Takeno, Junichi (2006), *The **Self-Aware Robot** -A Response to Reactions to Discovery News-*[1370], HRI Press, August 2006.
- Zagal, J.C., Lipson, H. (2009) " Self-Reflection in Evolutionary Robotics[1371]", Proceedings of the Genetic and Evolutionary Computation Conference, pp 2179–2188, GECCO 2009.

External links

- Artefactual consciousness depiction by Professor Igor Aleksander[1372]
- FOCS 2009: Manuel Blum - Can (Theoretical Computer) Science come to grips with Consciousness?[1373]
- www.Conscious-Robots.com[1374], Machine Consciousness and Conscious Robots Portal.

Technological singularity

The **technological singularity** (also, simply, the **singularity**)[1375] is the hypothesis that the invention of artificial superintelligence (ASI) will abruptly trigger runaway technological growth, resulting in unfathomable changes to human civilization. According to this hypothesis, an upgradable intelligent agent (such as a computer running software-based artificial general intelligence) would enter a "runaway reaction" of self-improvement cycles, with each new and more intelligent generation appearing more and more rapidly, causing an intelligence explosion and resulting in a powerful superintelligence that would, qualitatively, far surpass all human intelligence. Stanislaw Ulam reports a discussion with John von Neumann "centered on the accelerating progress of technology and changes in the mode of human life, which gives the appearance of approaching some essential singularity in the history of the race beyond which human affairs, as we know them, could not continue". Subsequent authors have echoed this viewpoint. I. J. Good's "intelligence explosion" model predicts that a future superintelligence will trigger a singularity. Emeritus professor of computer science at San Diego State University and science fiction author Vernor Vinge said in his 1993 essay *The Coming Technological Singularity* that this would signal the end of the human era, as the new superintelligence would continue to upgrade itself and would advance technologically at an incomprehensible rate.

Four polls, conducted in 2012 and 2013, suggested that the median estimate was a 50% chance that artificial general intelligence (AGI) would be developed by 2040–2050.[1376]

In the 2010s, public figures such as Stephen Hawking and Elon Musk expressed concern that full artificial intelligence could result in human extinction. The consequences of the singularity and its potential benefit or harm to the human race have been hotly debated.

Manifestations

Intelligence explosion

I. J. Good speculated in 1965 that artificial general intelligence might bring about an intelligence explosion. Good's scenario runs as follows: as computers increase in power, it becomes possible for people to build a machine that is more intelligent than humanity; this superhuman intelligence possesses greater problem-solving and inventive skills than current humans are capable of. This superintelligent machine then designs an even more capable machine, or re-writes its own software to become even more intelligent; this (ever more capable) machine then goes on to design a machine of yet greater capability, and so on. These iterations of recursive self-improvement accelerate, allowing enormous qualitative change before any upper limits imposed by the laws of physics or theoretical computation set in.

Emergence of superintelligence

John von Neumann, Vernor Vinge and Ray Kurzweil define the concept in terms of the technological creation of super intelligence. They argue that it is difficult or impossible for present-day humans to predict what human beings' lives would be like in a post-singularity world.

Non-AI singularity

Some writers use "the singularity" in a broader way to refer to any radical changes in our society brought about by new technologies such as molecular nanotechnology, although Vinge and other writers specifically state that without superintelligence, such changes would not qualify as a true singularity. Many writers also tie the singularity to observations of exponential growth in various technologies (with Moore's law being the most prominent example), using such observations as a basis for predicting that the singularity is likely to happen sometime within the 21st century.

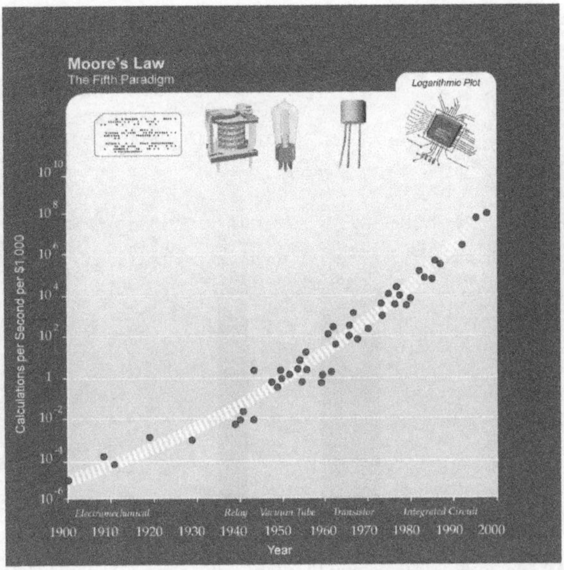

Figure 74: *Ray Kurzweil writes that, due to paradigm shifts, a trend of exponential growth extends Moore's law from integrated circuits to earlier transistors, vacuum tubes, relays, and electromechanical computers. He predicts that the exponential growth will continue, and that in a few decades the computing power of all computers will exceed that of ("unenhanced") human brains, with superhuman artificial intelligence appearing around the same time.*

Plausibility

Many prominent technologists and academics dispute the plausibility of a technological singularity, including Paul Allen, Jeff Hawkins, John Holland, Jaron Lanier, and Gordon Moore, whose law is often cited in support of the concept.

Claimed cause: exponential growth

The exponential growth in computing technology suggested by Moore's law is commonly cited as a reason to expect a singularity in the relatively near future, and a number of authors have proposed generalizations of Moore's law. Computer scientist and futurist Hans Moravec proposed in a 1998 book that the exponential growth curve could be extended back through earlier computing technologies prior to the integrated circuit.

Ray Kurzweil postulates a law of accelerating returns in which the speed of technological change (and more generally, all evolutionary processes)

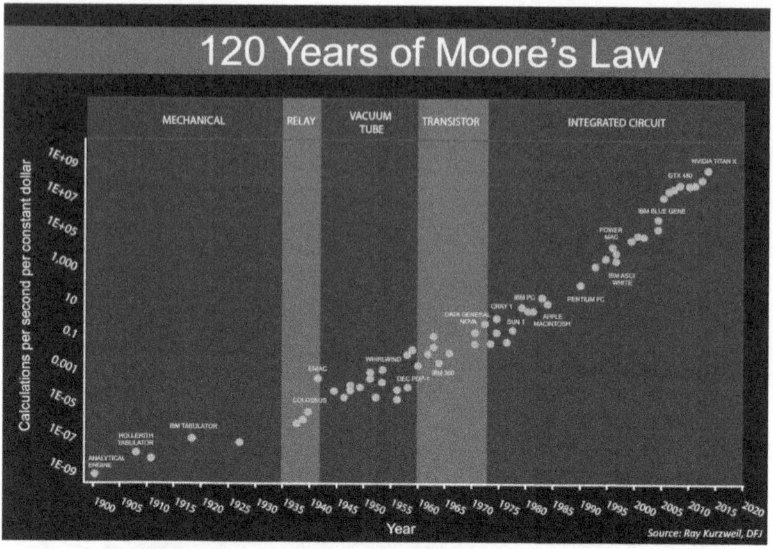

Figure 75: *An updated version of Moore's law over 120 Years (based on Kurzweil's graph). The 7 most recent data points are all NVIDIA GPUs.*

increases exponentially, generalizing Moore's law in the same manner as Moravec's proposal, and also including material technology (especially as applied to nanotechnology), medical technology and others. Between 1986 and 2007, machines' application-specific capacity to compute information per capita roughly doubled every 14 months; the per capita capacity of the world's general-purpose computers has doubled every 18 months; the global telecommunication capacity per capita doubled every 34 months; and the world's storage capacity per capita doubled every 40 months.[1377]

Kurzweil reserves the term "singularity" for a rapid increase in artificial intelligence (as opposed to other technologies), writing for example that "The Singularity will allow us to transcend these limitations of our biological bodies and brains ... There will be no distinction, post-Singularity, between human and machine". He also defines his predicted date of the singularity (2045) in terms of when he expects computer-based intelligences to significantly exceed the sum total of human brainpower, writing that advances in computing before that date "will not represent the Singularity" because they do "not yet correspond to a profound expansion of our intelligence."

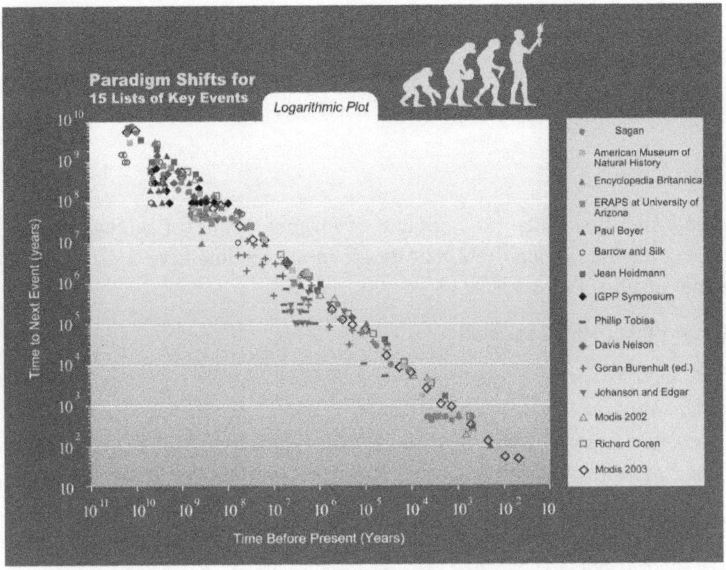

Figure 76: *According to Kurzweil, his logarithmic graph of 15 lists of paradigm shifts for key historic events shows an exponential trend*

Accelerating change

Some singularity proponents argue its inevitability through extrapolation of past trends, especially those pertaining to shortening gaps between improvements to technology. In one of the first uses of the term "singularity" in the context of technological progress, Stanislaw Ulam tells of a conversation with John von Neumann about accelerating change:

> *One conversation centered on the ever accelerating progress of technology and changes in the mode of human life, which gives the appearance of approaching some essential singularity in the history of the race beyond which human affairs, as we know them, could not continue.*

Kurzweil claims that technological progress follows a pattern of exponential growth, following what he calls the "law of accelerating returns". Whenever technology approaches a barrier, Kurzweil writes, new technologies will surmount it. He predicts paradigm shifts will become increasingly common, leading to "technological change so rapid and profound it represents a rupture in the fabric of human history". Kurzweil believes that the singularity will occur by approximately 2045.[1378] His predictions differ from Vinge's in that he predicts a gradual ascent to the singularity, rather than Vinge's rapidly self-improving superhuman intelligence.

Oft-cited dangers include those commonly associated with molecular nanotechnology and genetic engineering. These threats are major issues for both singularity advocates and critics, and were the subject of Bill Joy's *Wired* magazine article "Why the future doesn't need us".

Criticisms

Some critics assert that no computer or machine will ever achieve human intelligence, while others hold that the definition of intelligence is irrelevant if the net result is the same.

Steven Pinker stated in 2008:

> *... There is not the slightest reason to believe in a coming singularity. The fact that you can visualize a future in your imagination is not evidence that it is likely or even possible. Look at domed cities, jet-pack commuting, underwater cities, mile-high buildings, and nuclear-powered automobiles—all staples of futuristic fantasies when I was a child that have never arrived. Sheer processing power is not a pixie dust that magically solves all your problems. ...*

University of California, Berkeley, philosophy professor John Searle writes:

> [*Computers*] *have, literally ..., no intelligence, no motivation, no autonomy, and no agency. We design them to behave as if they had certain sorts of psychology, but there is no psychological reality to the corresponding processes or behavior. ...* [*T*]*he machinery has no beliefs, desires,* [*or*] *motivations.*[1379]

Martin Ford in *The Lights in the Tunnel: Automation, Accelerating Technology and the Economy of the Future* postulates a "technology paradox" in that before the singularity could occur most routine jobs in the economy would be automated, since this would require a level of technology inferior to that of the singularity. This would cause massive unemployment and plummeting consumer demand, which in turn would destroy the incentive to invest in the technologies that would be required to bring about the Singularity. Job displacement is increasingly no longer limited to work traditionally considered to be "routine".

Theodore Modis and Jonathan Huebner argue that the rate of technological innovation has not only ceased to rise, but is actually now declining. Evidence for this decline is that the rise in computer clock rates is slowing, even while Moore's prediction of exponentially increasing circuit density continues to hold. This is due to excessive heat build-up from the chip, which cannot be dissipated quickly enough to prevent the chip from melting when operating at higher speeds. Advancements in speed may be possible in the future by virtue of more power-efficient CPU designs and multi-cell processors. While Kurzweil used Modis' resources, and Modis' work was around accelerating change, Modis distanced himself from Kurzweil's thesis of a "technological singularity", claiming that it lacks scientific rigor.

Others propose that other "singularities" can be found through analysis of trends in world population, world gross domestic product, and other indices. Andrey Korotayev and others argue that historical hyperbolic growth curves can be attributed to feedback loops that ceased to affect global trends in the 1970s, and thus hyperbolic growth should not be expected in the future.[1380,1381]

In a detailed empirical accounting, *The Progress of Computing*, William Nordhaus argued that, prior to 1940, computers followed the much slower growth of a traditional industrial economy, thus rejecting extrapolations of Moore's law to 19th-century computers.[1382]

In a 2007 paper, Schmidhuber stated that the frequency of subjectively "notable events" appears to be approaching a 21st-century singularity, but cautioned readers to take such plots of subjective events with a grain of salt: perhaps differences in memory of recent and distant events could create an illusion of accelerating change where none exists.[1383]

Paul Allen argues the opposite of accelerating returns, the complexity brake; the more progress science makes towards understanding intelligence, the more difficult it becomes to make additional progress. A study of the number of patents shows that human creativity does not show accelerating returns, but in fact, as suggested by Joseph Tainter in his *The Collapse of Complex Societies*, a law of diminishing returns. The number of patents per thousand peaked in the period from 1850 to 1900, and has been declining since. The growth of complexity eventually becomes self-limiting, and leads to a widespread "general systems collapse".

Jaron Lanier refutes the idea that the Singularity is inevitable. He states: "I do not think the technology is creating itself. It's not an autonomous process." He goes on to assert: "The reason to believe in human agency over technological determinism is that you can then have an economy where people earn their own way and invent their own lives. If you structure a society on *not* emphasizing individual human agency, it's the same thing operationally as denying people clout, dignity, and self-determination ... to embrace [the idea of the Singularity] would be a celebration of bad data and bad politics."

Economist Robert J. Gordon, in *The Rise and Fall of American Growth: The U.S. Standard of Living Since the Civil War* (2016), points out that measured economic growth has slowed around 1970 and slowed even further since the financial crisis of 2008, and argues that the economic data show no trace of a coming Singularity as imagined by mathematician I.J. Good.[1384]

In addition to general criticisms of the singularity concept, several critics have raised issues with Kurzweil's iconic chart. One line of criticism is that a log-log chart of this nature is inherently biased toward a straight-line result. Others identify selection bias in the points that Kurzweil chooses to use. For example, biologist PZ Myers points out that many of the early evolutionary "events" were picked arbitrarily. Kurzweil has rebutted this by charting evolutionary events from 15 neutral sources, and showing that they fit a straight line on a log-log chart. *The Economist* mocked the concept with a graph extrapolating that the number of blades on a razor, which has increased over the years from one to as many as five, will increase ever-faster to infinity.

Ramifications

Uncertainty and risk

The term "technological singularity" reflects the idea that such change may happen suddenly, and that it is difficult to predict how the resulting new world would operate. It is unclear whether an intelligence explosion of this kind would be beneficial or harmful, or even an existential threat, as the issue has not been dealt with by most artificial general intelligence researchers, although the topic of friendly artificial intelligence is investigated by the Future of Humanity Institute and the Machine Intelligence Research Institute.

Next step of sociobiological evolution

While the technological singularity is usually seen as a sudden event, some scholars argue the current speed of change already fits this description.Wikipedia:Citation needed In addition, some argue that we are already in the midst of a major evolutionary transition that merges technology, biology, and society. Digital technology has infiltrated the fabric of human society to a degree of indisputable and often life-sustaining dependence. A 2016 article in *Trends in Ecology & Evolution* argues that "humans already embrace fusions of biology and technology. We spend most of our waking time communicating through digitally mediated channels... we trust artificial intelligence with our lives through antilock braking in cars and autopilots in planes... With one in three marriages in America beginning online, digital algorithms are also taking a role in human pair bonding and reproduction". The article argues that from the perspective of the evolution, several previous Major Transitions in Evolution have transformed life through innovations in information storage and replication (RNA, DNA, multicellularity, and culture and language).

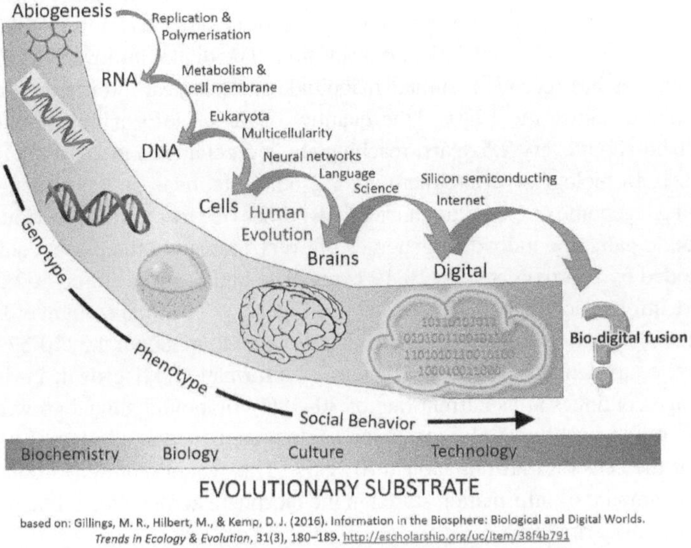

Figure 77: *Schematic Timeline of Information and Replicators in the Biosphere: Gillings et al.'s "major evolutionary transitions" in information processing.*

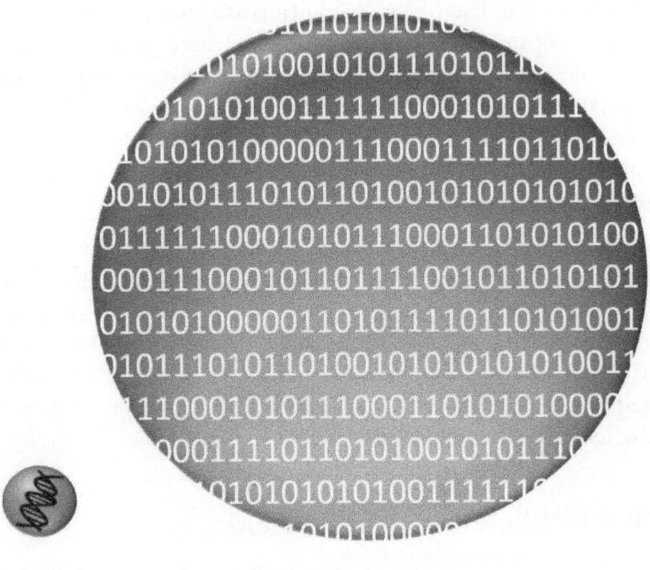

Figure 78: *Amount of digital information worldwide (5x10^21 bytes) versus human genome information worldwide (10^19 bytes) in 2014.*

In the current stage of life's evolution, the carbon-based biosphere has generated a cognitive system (humans) capable of creating technology that will result in a comparable evolutionary transition. The digital information created by humans has reached a similar magnitude to biological information in the biosphere. Since the 1980s, "the quantity of digital information stored has doubled about every 2.5 years, reaching about 5 zettabytes in 2014 (5x10^21 bytes). In biological terms, there are 7.2 billion humans on the planet, each having a genome of 6.2 billion nucleotides. Since one byte can encode four nucleotide pairs, the individual genomes of every human on the planet could be encoded by approximately 1x10^19 bytes. The digital realm stored 500 times more information than this in 2014 (...see Figure)... The total amount of DNA contained in all of the cells on Earth is estimated to be about 5.3x10^37 base pairs, equivalent to 1.325x10^37 bytes of information. If growth in digital storage continues at its current rate of 30–38% compound annual growth per year, it will rival the total information content contained in all of the DNA in all of the cells on Earth in about 110 years. This would represent a doubling of the amount of information stored in the biosphere across a total time period of just 150 years".[1385]

Implications for human society

In February 2009, under the auspices of the Association for the Advancement of Artificial Intelligence (AAAI), Eric Horvitz chaired a meeting of leading computer scientists, artificial intelligence researchers and roboticists at Asilomar in Pacific Grove, California. The goal was to discuss the potential impact of the hypothetical possibility that robots could become self-sufficient and able to make their own decisions. They discussed the extent to which computers and robots might be able to acquire autonomy, and to what degree they could use such abilities to pose threats or hazards.

Some machines are programmed with various forms of semi-autonomy, including the ability to locate their own power sources and choose targets to attack with weapons. Also, some computer viruses can evade elimination and, according to scientists in attendance, could therefore be said to have reached a "cockroach" stage of machine intelligence. The conference attendees noted that self-awareness as depicted in science-fiction is probably unlikely, but that other potential hazards and pitfalls exist.[1274]

Immortality

In his 2005 book, *The Singularity is Near*, Kurzweil suggests that medical advances would allow people to protect their bodies from the effects of aging, making the life expectancy limitless. Kurzweil argues that the technological advances in medicine would allow us to continuously repair and replace defective components in our bodies, prolonging life to an undetermined age.[1386] Kurzweil further buttresses his argument by discussing current bio-engineering advances. Kurzweil suggests somatic gene therapy; after synthetic viruses with specific genetic information, the next step would be to apply this technology to gene therapy, replacing human DNA with synthesized genes.[1387]

K. Eric Drexler, one of the founders of nanotechnology, postulated cell repair devices, including ones operating within cells and utilizing as yet hypothetical biological machines, in his 1986 book Engines of Creation. According to Richard Feynman, it was his former graduate student and collaborator Albert Hibbs who originally suggested to him (circa 1959) the idea of a *medical* use for Feynman's theoretical micromachines . Hibbs suggested that certain repair machines might one day be reduced in size to the point that it would, in theory, be possible to (as Feynman put it) "swallow the doctor". The idea was incorporated into Feynman's 1959 essay *There's Plenty of Room at the Bottom.*

Beyond merely extending the operational life of the physical body, Jaron Lanier argues for a form of immortality called "Digital Ascension" that involves "people dying in the flesh and being uploaded into a computer and remaining conscious". Singularitarianism has also been likened to a religion by John Horgan.

History of the concept

In his obituary for John von Neumann, Ulam recalled a conversation with von Neumann about the "ever accelerating progress of technology and changes in the mode of human life, which gives the appearance of approaching some essential singularity in the history of the race beyond which human affairs, as we know them, could not continue."

In 1965, Good wrote his essay postulating an "intelligence explosion" of recursive self-improvement of a machine intelligence. In 1985, in "The Time Scale of Artificial Intelligence", artificial intelligence researcher Ray Solomonoff articulated mathematically the related notion of what he called an "infinity point": if a research community of human-level self-improving AIs take four years to double their own speed, then two years, then one year and so on, their capabilities increase infinitely in finite time.

In 1981, Stanisław Lem published his science fiction novel Golem XIV. It describes a military AI computer (Golem XIV) who obtains consciousness and starts to increase his own intelligence, moving towards personal technological singularity. Golem XIV was originally created to aid its builders in fighting wars, but as its intelligence advances to a much higher level than that of humans, it stops being interested in the military requirement because it finds them lacking internal logical consistency.

In 1983, Vinge greatly popularized Good's intelligence explosion in a number of writings, first addressing the topic in print in the January 1983 issue of *Omni* magazine. In this op-ed piece, Vinge seems to have been the first to use the term "singularity" in a way that was specifically tied to the creation of intelligent machines: writing

> *We will soon create intelligences greater than our own. When this happens, human history will have reached a kind of singularity, an intellectual transition as impenetrable as the knotted space-time at the center of a black hole, and the world will pass far beyond our understanding. This singularity, I believe, already haunts a number of science-fiction writers. It makes realistic extrapolation to an interstellar future impossible. To write a story set more than a century hence, one needs a nuclear war in between ... so that the world remains intelligible.*

Vinge's 1993 article "The Coming Technological Singularity: How to Survive in the Post-Human Era", spread widely on the internet and helped to popularize the idea. This article contains the statement, "Within thirty years, we will have the technological means to create superhuman intelligence. Shortly after, the human era will be ended." Vinge argues that science-fiction authors cannot write realistic post-singularity characters who surpass the human intellect, as the thoughts of such an intellect would be beyond the ability of humans to express.

In 2000, Bill Joy, a prominent technologist and a co-founder of Sun Microsystems, voiced concern over the potential dangers of the singularity.

In 2005, Kurzweil published *The Singularity is Near*. Kurzweil's publicity campaign included an appearance on *The Daily Show with Jon Stewart*.

In 2007, Eliezer Yudkowsky suggested that many of the varied definitions that have been assigned to "singularity" are mutually incompatible rather than mutually supporting.[1388] For example, Kurzweil extrapolates current technological trajectories past the arrival of self-improving AI or superhuman intelligence, which Yudkowsky argues represents a tension with both I. J. Good's proposed discontinuous upswing in intelligence and Vinge's thesis on unpredictability.

In 2009, Kurzweil and X-Prize founder Peter Diamandis announced the establishment of Singularity University, a nonaccredited private institute whose stated mission is "to educate, inspire and empower leaders to apply exponential technologies to address humanity's grand challenges." Funded by Google,

Autodesk, ePlanet Ventures, and a group of technology industry leaders, Singularity University is based at NASA's Ames Research Center in Mountain View, California. The not-for-profit organization runs an annual ten-week graduate program during summer that covers ten different technology and allied tracks, and a series of executive programs throughout the year.

In politics

In 2007, the joint Economic Committee of the United States Congress released a report about the future of nanotechnology. It predicts significant technological and political changes in the mid-term future, including possible technological singularity.

Former President of the United States Barack Obama spoke about singularity in his interview to *Wired* in 2016:

> *One thing that we haven't talked about too much, and I just want to go back to, is we really have to think through the economic implications. Because most people aren't spending a lot of time right now worrying about singularity—they are worrying about "Well, is my job going to be replaced by a machine?"*

Sources

- Kurzweil, Ray (2005). *The Singularity is Near*. New York: Penguin Group. ISBN 9780715635612.
- William D. Nordhaus, "Why Growth Will Fall" (a review of Robert J. Gordon, *The Rise and Fall of American Growth: The U.S. Standard of Living Since the Civil War*, Princeton University Press, 2016, ISBN 978-0691147727, 762 pp., $39.95), *The New York Review of Books*, vol. LXIII, no. 13 (August 18, 2016), pp. 64, 66, 68.
- John R. Searle, "What Your Computer Can't Know" (review of Luciano Floridi, *The Fourth Revolution: How the Infosphere Is Reshaping Human Reality*, Oxford University Press, 2014; and Nick Bostrom, *Superintelligence: Paths, Dangers, Strategies*, Oxford University Press, 2014), *The New York Review of Books*, vol. LXI, no. 15 (October 9, 2014), pp. 52–55.

Further reading

- Marcus, Gary, "Am I Human?: Researchers need new ways to distinguish artificial intelligence from the natural kind", *Scientific American*, vol. 316, no. 3 (March 2017), pp. 58–63. *Multiple* tests of artificial-intelligence efficacy are needed because, "just as there is no single test of athletic prowess, there cannot be one ultimate test of intelligence." One such test, a "Construction Challenge", would test perception and physical action—"two important elements of intelligent behavior that were entirely absent from the original Turing test." Another proposal has been to give machines the same standardized tests of science and other disciplines that schoolchildren take. A so far insuperable stumbling block to artificial intelligence is an incapacity for reliable disambiguation. "[V]irtually every sentence [that people generate] is ambiguous, often in multiple ways." A prominent example is known as the "pronoun disambiguation problem": a machine has no way of determining to whom or what a pronoun in a sentence—such as "he", "she" or "it"—refers.
- Scaruffi, Piero, "Intelligence is not Artificial" (2016) for a critique of the singularity movement and its similarities to religious cults.

Transhumanism

Part of the Ideology series on

Transhumanism

Overviews

- Outline of transhumanism
- Transhuman
- Transhumanism in fiction

Currents

- Extropianism
- Immortalism
- Postgenderism
- Singularitarianism
- Technogaianism

Politics

- Libertarian transhumanism
- Techno-progressivism

Organizations

- Foresight Institute
- Humanity+
- IEET
- LessWrong
- Mormon Transhumanist Association
- Science Party

Transhumanism portal

- v
- t
- e[1389]

Transhumanism (abbreviated as **H+** or **h+**) is an international intellectual movement that aims to transform the human condition by developing and making widely available sophisticated technologies to greatly enhance human intellect and physiology.

Transhumanist thinkers study the potential benefits and dangers of emerging technologies that could overcome fundamental human limitations as well as ethical limitations of using such technologies. The most common transhumanist thesis is that human beings may eventually be able to transform themselves into different beings with abilities so greatly expanded from the current condition as to merit the label of posthuman beings.

The contemporary meaning of the term "transhumanism" was foreshadowed by one of the first professors of futurology, FM-2030, who taught "new concepts of the human" at The New School in the 1960s, when he began to identify people who adopt technologies, lifestyles and worldviews "transitional" to posthumanity as "transhuman". The assertion would lay the intellectual groundwork for the British philosopher Max More to begin articulating the principles of transhumanism as a futurist philosophy in 1990 and organizing in California an intelligentsia that has since grown into the worldwide transhumanist movement.

Influenced by seminal works of science fiction, the transhumanist vision of a transformed future humanity has attracted many supporters and detractors from a wide range of perspectives, including philosophy and religion.

History

Precursors of transhumanism

According to Nick Bostrom, transcendentalist impulses have been expressed at least as far back as in the quest for immortality in the Epic of Gilgamesh, as well as in historical quests for the Fountain of Youth, the Elixir of Life, and other efforts to stave off aging and death.

There is debate about whether the philosophy of Friedrich Nietzsche can be considered an influence on transhumanism despite its exaltation of the "Übermensch" (overman or superman), due to its emphasis on self-actualization, rather than technological transformation. The transhumanist philosophies of Max More and Stefan Lorenz Sorgner have been influenced strongly by Nietzschean thinking. By way of contrast, The Transhumanist Declaration *"...advocates the well-being of all sentience (whether in artificial intellects, humans, posthumans, or non-human animals)".*

Late XIX – early XX century movement known as Russian cosmism also incorporated some ideas that later developed to be the core of transhumanist movement.

Early transhumanist thinking

Fundamental ideas of transhumanism were first advanced in 1923 by the British geneticist J. B. S. Haldane in his essay *Daedalus: Science and the Future*, which predicted that great benefits would come from applications of advanced sciences to human biology—and that every such advance would first appear to someone as blasphemy or perversion, "indecent and unnatural". In particular, he was interested in the development of the science of eugenics, ectogenesis (creating and sustaining life in an artificial environment), and the

Figure 79: *Julian Huxley, the biologist who popularised the term transhumanism in an influential 1957 essay.*

application of genetics to improve human characteristics, such as health and intelligence.

His article inspired academic and popular interest. J. D. Bernal, a crystallographer at Cambridge, wrote *The World, the Flesh and the Devil* in 1929, in which he speculated on the prospects of space colonization and radical changes to human bodies and intelligence through bionic implants and cognitive enhancement. These ideas have been common transhumanist themes ever since.

The biologist Julian Huxley is generally regarded as the founder of transhumanism, after using the term for the title of an influential 1957 article. The term itself, however, derives from an earlier 1940 paper by the Canadian philosopher W. D. Lighthall. Huxley describes transhumanism in these terms:

> *Up till now human life has generally been, as Hobbes described it, 'nasty, brutish and short'; the great majority of human beings (if they have not already died young) have been afflicted with misery... we can justifiably hold the belief that these lands of possibility exist, and that the present limitations and miserable frustrations of our existence could be in large measure surmounted... The human species can, if it wishes, transcend itself—not just sporadically, an individual here in one way, an individual there in another way, but in its entirety, as humanity.*

Huxley's definition differs, albeit not substantially, from the one commonly in use since the 1980s. The ideas raised by these thinkers were explored in the science fiction of the 1960s, notably in Arthur C. Clarke's *2001: A Space Odyssey*, in which an alien artifact grants transcendent power to its wielder.

Japanese Metabolist architects produced a manifesto in 1960 which outlined goals to "encourage active metabolic development of our society"[1390] through design and technology. In the Material and Man section of the manifesto, Noboru Kawazoe suggests that:

> *After several decades, with the rapid progress of communication technology, every one will have a "brain wave receiver" in his ear, which conveys directly and exactly what other people think about him and vice versa. What I think will be known by all the people. There is no more individual consciousness, only the will of mankind as a whole.*

Artificial intelligence and the technological singularity

The concept of the technological singularity, or the ultra-rapid advent of superhuman intelligence, was first proposed by the British cryptologist I. J. Good in 1965:

> *Let an ultraintelligent machine be defined as a machine that can far surpass all the intellectual activities of any man however clever. Since the design of machines is one of these intellectual activities, an ultraintelligent machine could design even better machines; there would then unquestionably be an 'intelligence explosion,' and the intelligence of man would be left far behind. Thus the first ultraintelligent machine is the last invention that man need ever make.*[1391]

Computer scientist Marvin Minsky wrote on relationships between human and artificial intelligence beginning in the 1960s. Over the succeeding decades, this field continued to generate influential thinkers such as Hans Moravec and Raymond Kurzweil, who oscillated between the technical arena and futuristic speculations in the transhumanist vein. The coalescence of an identifiable transhumanist movement began in the last decades of the 20th century. In 1966, FM-2030 (formerly F. M. Esfandiary), a futurist who taught "new concepts of the human" at The New School, in New York City, began to identify people who adopt technologies, lifestyles and world views transitional to posthumanity as "transhuman". In 1972, Robert Ettinger contributed to the conceptualization of "transhumanity" in his book *Man into Superman.* FM-2030 published the *Upwingers Manifesto* in 1973.

Growth of transhumanism

The first self-described transhumanists met formally in the early 1980s at the University of California, Los Angeles, which became the main center of transhumanist thought. Here, FM-2030 lectured on his "Third Way" futurist ideology. At the EZTV Media venue, frequented by transhumanists and other futurists, Natasha Vita-More presented *Breaking Away*, her 1980 experimental film with the theme of humans breaking away from their biological limitations and the Earth's gravity as they head into space. FM-2030 and Vita-More soon began holding gatherings for transhumanists in Los Angeles, which included students from FM-2030's courses and audiences from Vita-More's artistic productions. In 1982, Vita-More authored the *Transhumanist Arts Statement* and, six years later, produced the cable TV show *TransCentury Update* on transhumanity, a program which reached over 100,000 viewers.

In 1986, Eric Drexler published *Engines of Creation: The Coming Era of Nanotechnology,* which discussed the prospects for nanotechnology and molecular assemblers, and founded the Foresight Institute. As the first non-profit organization to research, advocate for, and perform cryonics, the Southern California offices of the Alcor Life Extension Foundation became a center for futurists. In 1988, the first issue of *Extropy Magazine* was published by Max More and Tom Morrow. In 1990, More, a strategic philosopher, created his own particular transhumanist doctrine, which took the form of the *Principles of Extropy,* and laid the foundation of modern transhumanism by giving it a new definition:

> *Transhumanism is a class of philosophies that seek to guide us towards a posthuman condition. Transhumanism shares many elements of humanism, including a respect for reason and science, a commitment to progress, and a valuing of human (or transhuman) existence in this life.* [...] *Transhumanism differs from humanism in recognizing and anticipating the radical alterations in the nature and possibilities of our lives resulting from various sciences and technologies* [...].

In 1992, More and Morrow founded the Extropy Institute, a catalyst for networking futurists and brainstorming new memeplexes by organizing a series of conferences and, more importantly, providing a mailing list, which exposed many to transhumanist views for the first time during the rise of cyberculture and the cyberdelic counterculture. In 1998, philosophers Nick Bostrom and David Pearce founded the World Transhumanist Association (WTA), an international non-governmental organization working toward the recognition of transhumanism as a legitimate subject of scientific inquiry and public policy. In 2002, the WTA modified and adopted *The Transhumanist Declaration. The Transhumanist FAQ*, prepared by the WTA (later Humanity+), gave two formal definitions for transhumanism:

1. *The intellectual and cultural movement that affirms the possibility and desirability of fundamentally improving the human condition through applied reason, especially by developing and making widely available technologies to eliminate aging and to greatly enhance human intellectual, physical, and psychological capacities.*
2. *The study of the ramifications, promises, and potential dangers of technologies that will enable us to overcome fundamental human limitations, and the related study of the ethical matters involved in developing and using such technologies.*

In possible contrast with other transhumanist organizations, WTA officials considered that social forces could undermine their futurist visions and needed to be addressed. A particular concern is the equal access to human enhancement technologies across classes and borders. In 2006, a political struggle within the transhumanist movement between the libertarian right and the liberal left resulted in a more centre-leftward positioning of the WTA under its former executive director James Hughes. In 2006, the board of directors of the Extropy Institute ceased operations of the organization, stating that its mission was "essentially completed". This left the World Transhumanist Association as the leading international transhumanist organization. In 2008, as part of a rebranding effort, the WTA changed its name to "Humanity+". In 2012, the transhumanist Longevity Party had been initiated as an international union of people who promote the development of scientific and technological means to significant life extension, that for now has more than 30 national organisations throughout the world.

The Mormon Transhumanist Association was founded in 2006. By 2012, it consisted of hundreds of members.

The first transhumanist elected member of a Parliament is Giuseppe Vatinno, in Italy. In 2015, Vatinno became a member of the Board of Directors of Humanity+.

Theory

It is a matter of debate whether transhumanism is a branch of posthumanism and how this philosophical movement should be conceptualised with regard to transhumanism. The latter is often referred to as a variant or activist form of posthumanism by its conservative, Christian and progressive critics.

A common feature of transhumanism and philosophical posthumanism is the future vision of a new intelligent species, into which humanity will evolve and eventually will supplement or supersede it. Transhumanism stresses the evolutionary perspective, including sometimes the creation of a highly intelligent animal species by way of cognitive enhancement (i.e. biological uplift), but clings to a "posthuman future" as the final goal of participant evolution.

Nevertheless, the idea of creating intelligent artificial beings (proposed, for example, by roboticist Hans Moravec) has influenced transhumanism. Moravec's ideas and transhumanism have also been characterised as a "complacent" or "apocalyptic" variant of posthumanism and contrasted with "cultural posthumanism" in humanities and the arts. While such a "cultural posthumanism" would offer resources for rethinking the relationships between humans and increasingly sophisticated machines, transhumanism and similar posthumanisms are, in this view, not abandoning obsolete concepts of the "autonomous liberal subject", but are expanding its "prerogatives" into the realm of the posthuman. Transhumanist self-characterisations as a continuation of humanism and Enlightenment thinking correspond with this view.

Some secular humanists conceive transhumanism as an offspring of the humanist freethought movement and argue that transhumanists differ from the humanist mainstream by having a specific focus on technological approaches to resolving human concerns (i.e. technocentrism) and on the issue of mortality. However, other progressives have argued that posthumanism, whether it be its philosophical or activist forms, amounts to a shift away from concerns about social justice, from the reform of human institutions and from other Enlightenment preoccupations, toward narcissistic longings for a transcendence of the human body in quest of more exquisite ways of being.

As an alternative, humanist philosopher Dwight Gilbert Jones has proposed a renewed Renaissance humanism through DNA and genome repositories, with each individual genotype (DNA) being instantiated as successive phenotypes (bodies or lives via cloning, *Church of Man*, 1978). In his view, native molecular DNA "continuity" is required for retaining the "self" and no amount of computing power or memory aggregation can replace the essential "stink" of our true genetic identity, which he terms " genity[1392]". Instead, DNA/genome stewardship by an institution analogous to the Jesuits' 400 year vigil is a suggested model for enabling humanism to become our species' common credo, a project he proposed in his speculative novel *The Humanist – 1000 Summers* (2011), wherein humanity dedicates these coming centuries to harmonizing our planet and peoples.

The philosophy of transhumanism is closely related to technoself studies, an interdisciplinary domain of scholarly research dealing with all aspects of human identity in a technological society and focusing on the changing nature of relationships between humans and technology.

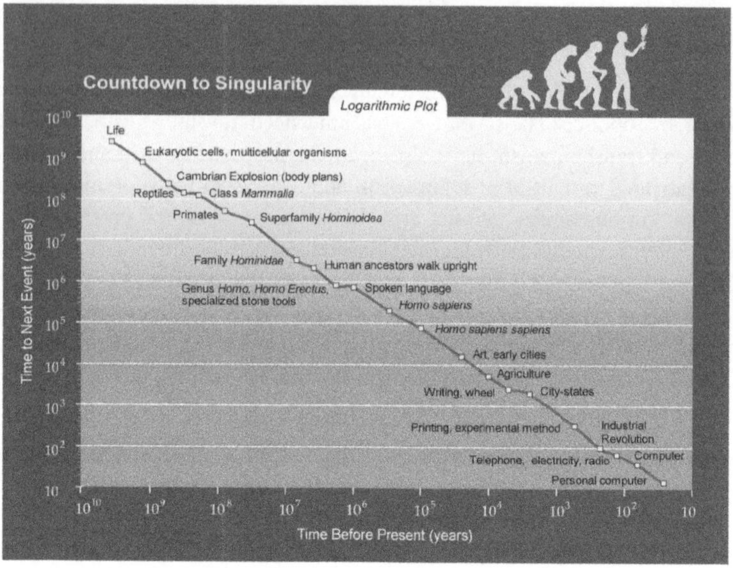

Figure 80: *Ray Kurzweil believes that a countdown to when "human life will be irreversibly transformed" can be made through plotting major world events on a graph.*

Aims

While many transhumanist theorists and advocates seek to apply reason, science and technology for the purposes of reducing poverty, disease, disability and malnutrition around the globe, transhumanism is distinctive in its particular focus on the applications of technologies to the improvement of human bodies at the individual level. Many transhumanists actively assess the potential for future technologies and innovative social systems to improve the quality of all life, while seeking to make the material reality of the human condition fulfill the promise of legal and political equality by eliminating congenital mental and physical barriers.

Transhumanist philosophers argue that there not only exists a perfectionist ethical imperative for humans to strive for progress and improvement of the human condition, but that it is possible and desirable for humanity to enter a transhuman phase of existence in which humans enhance themselves beyond what is naturally human. In such a phase, natural evolution would be replaced with deliberate participatory or directed evolution.

Some theorists such as Ray Kurzweil think that the pace of technological innovation is accelerating and that the next 50 years may yield not only radical technological advances, but possibly a technological singularity, which may fundamentally change the nature of human beings. Transhumanists who foresee this massive technological change generally maintain that it is desirable. However, some are also concerned with the possible dangers of extremely rapid technological change and propose options for ensuring that advanced technology is used responsibly. For example, Bostrom has written extensively on existential risks to humanity's future welfare, including ones that could be created by emerging technologies.

While many people believe that all transhumanists are striving for immortality, it is not necessarily true. Hank Pellissier, managing director of the Institute for Ethics and Emerging Technologies (2011–2012), surveyed transhumanists. He found that, of the 818 respondents, 23.8% did not want immortality.[1393] Some of the reasons argued were boredom, Earth's overpopulation and the desire "to go to an afterlife".

Empathic fallibility and conversational consent

Certain transhumanist philosophers hold that since all assumptions about what others experience are fallible, and that therefore all attempts to help or protect beings that are not capable of correcting what others assume about them no matter how well-intentioned are in danger of actually hurting them, all sentient beings deserve to be sapient. These thinkers argue that the ability to discuss in a falsification-based way constitutes a threshold that is not arbitrary at which it becomes possible for an individual to speak for themself in a way that is not dependent on exterior assumptions. They also argue that all beings capable of experiencing something deserve to be elevated to this threshold if they are not at it, typically stating that the underlying change that leads to the threshold is an increase in the preciseness of the brain's ability to discriminate. This includes increasing the neuron count and connectivity in animals as well as accelerating the development of connectivity in order to shorten or ideally skip non-sapient childhood incapable of independently deciding for oneself. Transhumanists of this description stress that the genetic engineering that they advocate is general insertion into both the somatic cells of living beings and in germ cells, and not purging of individuals without the modifications, deeming the latter not only unethical but also unnecessary due to the possibilities of efficient genetic engineering.[1394,1395,1396,1397]

Ethics

Part of a Philosophy series on

Humanism

History

- Renaissance humanism
- in Northern Europe
- in France
- *Humanist Manifesto*

Secular humanism

- Center for Inquiry
- *A Secular Humanist Declaration*
- Amsterdam Declaration

Religious humanism

- Christian
- Jewish
- Buddhist

Other forms

- Cosmic
- Deistic
- Existential
- Integral
- Marxist
- Neo
- Personism
- Rationalist
- Trans

Organizations

- International Humanist and Ethical Union (IHEU)
- American Humanist Association
- Humanists UK
- Humanistischer Verband Deutschlands
- Humanist Society Scotland
- Human-Etisk Forbund

See also

- Antihumanism
- Posthumanism
- Ethical movement

- **Outline**
- **List of secular humanists**

Philosophy portal

- v
- t
- e[1398]

Transhumanists engage in interdisciplinary approaches to understand and evaluate possibilities for overcoming biological limitations by drawing on futurology and various fields of ethics.Wikipedia:Citation needed Unlike many philosophers, social critics and activists who place a moral value on preservation of natural systems, transhumanists see the very concept of the specifically natural as problematically nebulous at best and an obstacle to progress at worst. In keeping with this, many prominent transhumanist advocates, such as Dan Agin, refer to transhumanism's critics, on the political right and left jointly, as "bioconservatives" or "bioluddites", the latter term alluding to the 19th century anti-industrialisation social movement that opposed the replacement of human manual labourers by machines.

A belief of counter-transhumanism is that transhumanism can cause unfair human enhancement in many areas of life, but specifically on the social plane. This can be compared to steroid use, where athletes who use steroids in sports have an advantage over those who do not. The same scenario happens when people have certain neural implants that give them an advantage in the work place and in educational aspects.

Currents

There is a variety of opinions within transhumanist thought. Many of the leading transhumanist thinkers hold views that are under constant revision and development. Some distinctive currents of transhumanism are identified and listed here in alphabetical order:

- Democratic transhumanism, a political ideology synthesizing liberal democracy, social democracy, radical democracy and transhumanism.
- Extropianism, an early school of transhumanist thought characterized by a set of principles advocating a proactive approach to human evolution.
- Immortalism, a moral ideology based upon the belief that radical life extension and technological immortality is possible and desirable, and advocating research and development to ensure its realization.
- Libertarian transhumanism, a political ideology synthesizing libertarianism and transhumanism.
- Postgenderism, a social philosophy which seeks the voluntary elimination of gender in the human species through the application of advanced biotechnology and assisted reproductive technologies.

- Singularitarianism, a moral ideology based upon the belief that a technological singularity is possible, and advocating deliberate action to effect it and ensure its safety.
- Technogaianism, an ecological ideology based upon the belief that emerging technologies can help restore Earth's environment and that developing safe, clean, alternative technology should therefore be an important goal of environmentalists.

Spirituality

Although many transhumanists are atheists, agnostics, and/or secular humanists, some have religious or spiritual views. Despite the prevailing secular attitude, some transhumanists pursue hopes traditionally espoused by religions, such as immortality, while several controversial new religious movements from the late 20th century have explicitly embraced transhumanist goals of transforming the human condition by applying technology to the alteration of the mind and body, such as Raëlism. However, most thinkers associated with the transhumanist movement focus on the practical goals of using technology to help achieve longer and healthier lives, while speculating that future understanding of neurotheology and the application of neurotechnology will enable humans to gain greater control of altered states of consciousness, which were commonly interpreted as spiritual experiences, and thus achieve more profound self-knowledge. Transhumanist Buddhists have sought to explore areas of agreement between various types of Buddhism and Buddhist-derived meditation and mind expanding "neurotechnologies". However, they have been criticised for appropriating mindfulness as a tool for transcending humanness.

Some transhumanists believe in the compatibility between the human mind and computer hardware, with the theoretical implication that human consciousness may someday be transferred to alternative media (a speculative technique commonly known as mind uploading). One extreme formulation of this idea, which some transhumanists are interested in, is the proposal of the Omega Point by Christian cosmologist Frank Tipler. Drawing upon ideas in digitalism, Tipler has advanced the notion that the collapse of the Universe billions of years hence could create the conditions for the perpetuation of humanity in a simulated reality within a megacomputer and thus achieve a form of "posthuman godhood". Tipler's thought was inspired by the writings of Pierre Teilhard de Chardin, a paleontologist and Jesuit theologian who saw an evolutionary telos in the development of an encompassing noosphere, a global consciousness.

Viewed from the perspective of some Christian thinkers, the idea of mind uploading is asserted to represent a denigration of the human body, characteristic of gnostic manichaean belief. Transhumanism and its presumed intellectual progenitors have also been described as neo-gnostic by non-Christian and secular commentators.

The first dialogue between transhumanism and faith was a one-day conference held at the University of Toronto in 2004. Religious critics alone faulted the philosophy of transhumanism as offering no eternal truths nor a relationship with the divine. They commented that a philosophy bereft of these beliefs leaves humanity adrift in a foggy sea of postmodern cynicism and anomie. Transhumanists responded that such criticisms reflect a failure to look at the actual content of the transhumanist philosophy, which, far from being cynical, is rooted in optimistic, idealistic attitudes that trace back to the Enlightenment. Following this dialogue, William Sims Bainbridge, a sociologist of religion, conducted a pilot study, published in the Journal of Evolution and Technology, suggesting that religious attitudes were negatively correlated with acceptance of transhumanist ideas and indicating that individuals with highly religious worldviews tended to perceive transhumanism as being a direct, competitive (though ultimately futile) affront to their spiritual beliefs.

Since 2006, the Mormon Transhumanist Association sponsors conferences and lectures on the intersection of technology and religion.

Since 2009, the American Academy of Religion holds a "Transhumanism and Religion" consultation during its annual meeting, where scholars in the field of religious studies seek to identify and critically evaluate any implicit religious beliefs that might underlie key transhumanist claims and assumptions; consider how transhumanism challenges religious traditions to develop their own ideas of the human future, in particular the prospect of human transformation, whether by technological or other means; and provide critical and constructive assessments of an envisioned future that place greater confidence in nanotechnology, robotics and information technology to achieve virtual immortality and create a superior posthuman species.

The physicist and transhumanist thinker Giulio Prisco states that "cosmist religions based on science, might be our best protection from reckless pursuit of superintelligence and other risky technologies." Prisco also recognizes the importance of spiritual ideas, such as the ones of Nikolai Fyodorovich Fyodorov, to the origins of the transhumanism movement.

Practice

While some transhumanistsWikipedia:Manual of Style/Words to watch#Unsupported attributions take an abstract and theoretical approach to the perceived benefits of emerging technologies, others have offered specific proposals for modifications to the human body, including heritable ones.

Transhumanists are often concerned with methods of enhancing the human nervous system. Though some, such as Kevin Warwick, propose modification of the peripheral nervous system, the brain is considered the common denominator of personhood and is thus a primary focus of transhumanist ambitions.

In fact Warwick has gone a lot further than merely making a proposal. In 2002 he had a 100 electrode array surgically implanted into the median nerves of his left arm in order to link his nervous system directly with a computer and thus to also connect with the internet. As a consequence, he carried out a series of experiments. He was able to directly control a robot hand using his neural signals and to feel the force applied by the hand through feedback from the fingertips. He also experienced a form of ultrasonic sensory input and conducted the first purely electronic communication between his own nervous system and that of his wife who also had electrodes implanted.

As proponents of self-improvement and body modification, including gender transitioning, transhumanists tend to use existing technologies and techniques that supposedly improve cognitive and physical performance, while engaging in routines and lifestyles designed to improve health and longevity. Depending on their age, someWikipedia:Manual of Style/Words to watch#Unsupported attributions transhumanists express concern that they will not live to reap the benefits of future technologies. However, many have a great interest in life extension strategies and in funding research in cryonics in order to make the latter a viable option of last resort, rather than remaining an unproven method. Regional and global transhumanist networks and communities with a range of objectives exist to provide support and forums for discussion and collaborative projects.Wikipedia:Citation needed

Technologies of interest

Transhumanists support the emergence and convergence of technologies including nanotechnology, biotechnology, information technology and cognitive science (NBIC), as well as hypothetical future technologies like simulated reality, artificial intelligence, superintelligence, 3D bioprinting, mind uploading, chemical brain preservation and cryonics. They believe that humans can and should use these technologies to become more than human. Therefore, they support the recognition and/or protection of cognitive liberty, morphological freedom and procreative liberty as civil liberties, so as to guarantee individuals the choice of using human enhancement technologies on themselves and their children. Some speculate that human enhancement techniques and other emerging technologies may facilitate more radical human enhancement no later than at the midpoint of the 21st century. Kurzweil's book *The Singularity is Near* and Michio Kaku's book *Physics of the Future* outline various

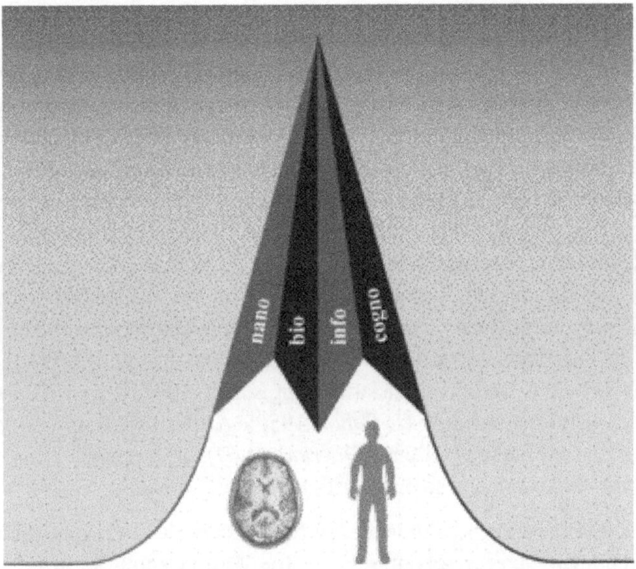

Figure 81: *Converging Technologies, a 2002 report exploring the potential for synergy among nano-, bio-, info- and cogno-technologies, has become a landmark in near-future technological speculation.*

human enhancement technologies and give insight on how these technologies may impact the human race.

Some reports on the converging technologies and NBIC concepts have criticised their transhumanist orientation and alleged science fictional character. At the same time, research on brain and body alteration technologies has been accelerated under the sponsorship of the U.S. Department of Defense, which is interested in the battlefield advantages they would provide to the supersoldiers of the United States and its allies. There has already been a brain research program to "extend the ability to manage information", while military scientists are now looking at stretching the human capacity for combat to a maximum 168 hours without sleep.

Neuroscientist Anders Sandberg has been practicing on the method of scanning ultra-thin sections of the brain. This method is being used to help better understand the architecture of the brain. As of now, this method is currently being used on mice. This is the first step towards uploading contents of the human brain, including memories and emotions, onto a computer.

Arts and culture

Transhumanist themes have become increasingly prominent in various literary forms during the period in which the movement itself has emerged. Contemporary science fiction often contains positive renditions of technologically enhanced human life set in utopian (especially techno-utopian) societies. However, science fiction's depictions of enhanced humans or other posthuman beings frequently come with a cautionary twist. The more pessimistic scenarios include many horrific or dystopian tales of human bioengineering gone wrong. In the decades immediately before transhumanism emerged as an explicit movement, many transhumanist concepts and themes began appearing in the speculative fiction of authors of the Golden Age of Science Fiction such as Robert A. Heinlein (Lazarus Long series, 1941–87), A. E. van Vogt (*Slan*, 1946), Isaac Asimov (*I, Robot*, 1950), Arthur C. Clarke (*Childhood's End*, 1953) and Stanisław Lem (*Cyberiad*, 1967). C. S. Lewis' *That Hideous Strength* (1945) contains an early critique of transhumanism.

In a series of science fiction novels by Neal Asher, the protagonist is an augmented human who carries out missions for "Earth Central Security", an artificial intelligence and superhuman coalition. The author portrays a variety of augmentations in addition to the copying of memory and human minds into crystals and the presence of both benevolent and malevolent artificial intelligences.

The cyberpunk genre, exemplified by William Gibson's *Neuromancer* (1984) and Bruce Sterling's *Schismatrix* (1985), has particularly been concerned with the modification of human bodies. Other novels dealing with transhumanist themes that have stimulated broad discussion of these issues include *Blood Music* (1985), by Greg Bear; *The Xenogenesis Trilogy* (1987–1989), by Octavia Butler; *The Beggar's Trilogy* (1990–94), by Nancy Kress; much of Greg Egan's work since the early 1990s such as *Permutation City* (1994) and *Diaspora* (1997); The Culture series of Iain M. Banks; *The Bohr Maker* (1995), by Linda Nagata; *Altered Carbon* (2002), by Richard K. Morgan; *Oryx and Crake* (2003), by Margaret Atwood; *The Elementary Particles* (Eng. trans. 2001) and *The Possibility of an Island* (Eng. trans. 2006), by Michel Houellebecq; *Mindscan* (2005), by Robert J. Sawyer; the *Commonwealth Saga* (2002–10), by Peter F. Hamilton; and *Glasshouse* (2005), by Charles Stross. Some of these works are considered part of the cyberpunk genre or its postcyberpunk offshoot.

Dan Brown's novel *Inferno* focuses on the theme of transhumanism. In an interview with *Today*, Dan Brown said "Transhumanism is the ethics and science of using things like biological and genetic engineering to transform our bodies and make us a more powerful species".

Fictional transhumanist scenarios have also become popular in other media during the late 20th and early 21st centuries. Such treatments are found in comic books (*Captain America*, 1941; *Iron Man* 1963; *Transmetropolitan*, 1997; *The Surrogates*, 2006), films (*2001: A Space Odyssey*, 1968; *Blade Runner*, 1982; *Gattaca*, 1997, *Ex Machina*, 2015), television series (the Cybermen of *Doctor Who*, 1966; the Borg of *Star Trek: The Next Generation*, 1989, *Battlestar Galactica*, 2003, *Black Mirror*, 2011), manga and anime (*Galaxy Express 999*, 1978; *Appleseed*, 1985; *Ghost in the Shell*, 1989; *Neon Genesis Evangelion*, 1995; and the *Gundam* metaseries, 1979), video games (*Metal Gear Solid*, 1998; *Sid Meier's Alpha Centauri*, 1999; *Deus Ex*, 2000; *EVE Online*, 2003; *BioShock*, 2007; *Half-Life* 2, 2004; *Crysis*, 2007; *Deus Ex: Human Revolution*; 2011; *Deus Ex: Mankind Divided*, 2016) and role-playing games.

Carnal Art, a form of sculpture originated by French artist Orlan, uses the body as its medium and plastic surgery as its method. French biological anthropologist Dr. Judith Nicogossian also works on representations of the hybrid body.

Debate

The very notion and prospect of human enhancement and related issues arouse public controversy. Criticisms of transhumanism and its proposals take two main forms: those objecting to the likelihood of transhumanist goals being achieved (practical criticisms) and those objecting to the moral principles or worldview sustaining transhumanist proposals or underlying transhumanism itself (ethical criticisms). Critics and opponents often see transhumanists' goals as posing threats to human values.

Some of the most widely known critiques of the transhumanist program are novels and fictional films. These works of art, despite presenting imagined worlds rather than philosophical analyses, are used as touchstones for some of the more formal arguments. Various arguments have been made to the effect that a society that adopts human enhancement technologies may come to resemble the dystopia depicted in the 1932 novel *Brave New World,* by Aldous Huxley.

On another front, some authors consider that humanity would already be transhuman, because medical advances in recent centuries have significantly altered our species. However, it would not be in a conscious and therefore transhumanistic way.

Feasibility

In a 1992 book, sociologist Max Dublin pointed to many past failed predictions of technological progress and argued that modern futurist predictions would prove similarly inaccurate. He also objected to what he saw as scientism, fanaticism and nihilism by a few in advancing transhumanist causes. Dublin also said that historical parallels existed between Millenarian religions and Communist doctrines.

Although generally sympathetic to transhumanism, public health professor Gregory Stock is skeptical of the technical feasibility and mass appeal of the cyborgization of humanity predicted by Raymond Kurzweil, Hans Moravec and Kevin Warwick. He said that, throughout the 21st century, many humans would find themselves deeply integrated into systems of machines, but would remain biological. Primary changes to their own form and character would arise not from cyberware, but from the direct manipulation of their genetics, metabolism and biochemistry.

In her 1992 book *Science as Salvation*, philosopher Mary Midgley traces the notion of achieving immortality by transcendence of the material human body (echoed in the transhumanist tenet of mind uploading) to a group of male scientific thinkers of the early 20th century, including J. B. S. Haldane and members of his circle. She characterizes these ideas as "quasi-scientific dreams and prophesies" involving visions of escape from the body coupled with "self-indulgent, uncontrolled power-fantasies". Her argument focuses on what she perceives as the pseudoscientific speculations and irrational, fear-of-death-driven fantasies of these thinkers, their disregard for laymen and the remoteness of their eschatological visions.

Another critique is aimed mainly at "algeny" (a portmanteau of *alchemy* and *genetics*), which Jeremy Rifkin defined as "the upgrading of existing organisms and the design of wholly new ones with the intent of 'perfecting' their performance". It emphasizes the issue of biocomplexity and the unpredictability of attempts to guide the development of products of biological evolution. This argument, elaborated in particular by the biologist Stuart Newman, is based on the recognition that cloning and germline genetic engineering of animals are error-prone and inherently disruptive of embryonic development. Accordingly, so it is argued, it would create unacceptable risks to use such methods on human embryos. Performing experiments, particularly ones with permanent biological consequences, on developing humans would thus be in violation of accepted principles governing research on human subjects (see the 1964 Declaration of Helsinki). Moreover, because improvements in experimental outcomes in one species are not automatically transferable to a new species without further experimentation, it is claimed that there is no ethical route to genetic manipulation of humans at early developmental stages.

As a practical matter, however, international protocols on human subject research may not present a legal obstacle to attempts by transhumanists and others to improve their offspring by germinal choice technology. According to legal scholar Kirsten Rabe Smolensky, existing laws would protect parents who choose to enhance their child's genome from future liability arising from adverse outcomes of the procedure.

Transhumanists and other supporters of human genetic engineering do not dismiss practical concerns out of hand, insofar as there is a high degree of uncertainty about the timelines and likely outcomes of genetic modification experiments in humans. However, bioethicist James Hughes suggests that one possible ethical route to the genetic manipulation of humans at early developmental stages is the building of computer models of the human genome, the proteins it specifies and the tissue engineering he argues that it also codes for. With the exponential progress in bioinformatics, Hughes believes that a virtual model of genetic expression in the human body will not be far behind and that it will soon be possible to accelerate approval of genetic modifications by simulating their effects on virtual humans. Public health professor Gregory Stock points to artificial chromosomes as an alleged safer alternative to existing genetic engineering techniques.

ThinkersWikipedia:Manual of Style/Words to watch#Unsupported attributions who defend the likelihood of accelerating change point to a past pattern of exponential increases in humanity's technological capacities. Kurzweil developed this position in his 2005 book *The Singularity Is Near*.

Intrinsic immorality

It has been argued that, in transhumanist thought, humans attempt to substitute themselves for God. The 2002 Vatican statement *Communion and Stewardship: Human Persons Created in the Image of God,* stated that "changing the genetic identity of man as a human person through the production of an infrahuman being is radically immoral", implying, that "man has full right of disposal over his own biological nature". The statement also argues that creation of a superhuman or spiritually superior being is "unthinkable", since true improvement can come only through religious experience and "realizing more fully the image of God". Christian theologians and lay activists of several churches and denominations have expressed similar objections to transhumanism and claimed that Christians attain in the afterlife what radical transhumanism promises, such as indefinite life extension or the abolition of suffering. In this view, transhumanism is just another representative of the long line of utopian movements which seek to create "heaven on earth". On the other hand, religious thinkers allied with transhumanist goals such as the theologians Ronald Cole-Turner and Ted Peters hold that the doctrine of "co-creation" provides an obligation to use genetic engineering to improve human biology.

Other critics target what they claim to be an instrumental conception of the human body in the writings of Marvin Minsky, Hans Moravec and some other transhumanists. Reflecting a strain of feminist criticism of the transhumanist program, philosopher Susan Bordo points to "contemporary obsessions with slenderness, youth and physical perfection", which she sees as affecting both men and women, but in distinct ways, as "the logical (if extreme) manifestations of anxieties and fantasies fostered by our culture." Some critics question other social implications of the movement's focus on body modification. Political scientist Klaus-Gerd Giesen, in particular, has asserted that transhumanism's concentration on altering the human body represents the logical yet tragic consequence of atomized individualism and body commodification within a consumer culture.

Nick Bostrom responds that the desire to regain youth, specifically, and transcend the natural limitations of the human body, in general, is pan-cultural and pan-historical, and is therefore not uniquely tied to the culture of the 20th century. He argues that the transhumanist program is an attempt to channel that desire into a scientific project on par with the Human Genome Project and achieve humanity's oldest hope, rather than a puerile fantasy or social trend.

Loss of human identity

In his 2003 book *Enough: Staying Human in an Engineered Age*, environmental ethicist Bill McKibben argued at length against many of the technologies that are postulated or supported by transhumanists, including germinal choice technology, nanomedicine and life extension strategies. He claims that it would be morally wrong for humans to tamper with fundamental aspects of themselves (or their children) in an attempt to overcome universal human limitations, such as vulnerability to aging, maximum life span and biological constraints on physical and cognitive ability. Attempts to "improve" themselves through such manipulation would remove limitations that provide a necessary context for the experience of meaningful human choice. He claims that human lives would no longer seem meaningful in a world where such limitations could be overcome technologically. Even the goal of using germinal choice technology for clearly therapeutic purposes should be relinquished, since it would inevitably produce temptations to tamper with such things as cognitive capacities. He argues that it is possible for societies to benefit from renouncing particular technologies, using as examples Ming China, Tokugawa Japan and the contemporary Amish.

Figure 82: *In the U.S., the Amish are a religious group probably most known for their avoidance of certain modern technologies. Transhumanists draw a parallel by arguing that in the near-future there will probably be "humanish", people who choose to "stay human" by not adopting human enhancement technologies. They believe their choice must be respected and protected.*

Biopolitical activist Jeremy Rifkin and biologist Stuart Newman accept that biotechnology has the power to make profound changes in organismal identity. They argue against the genetic engineering of human beings because they fear the blurring of the boundary between human and artifact. Philosopher Keekok Lee sees such developments as part of an accelerating trend in modernization in which technology has been used to transform the "natural" into the "artefactual". In the extreme, this could lead to the manufacturing and enslavement of "monsters" such as human clones, human-animal chimeras, or bioroids, but even lesser dislocations of humans and non-humans from social and ecological systems are seen as problematic. The film *Blade Runner* (1982) and the novels *The Boys From Brazil* (1976) and *The Island of Doctor Moreau* (1896) depict elements of such scenarios, but Mary Shelley's 1818 novel *Frankenstein* is most often alluded to by critics who suggest that biotechnologies could create objectified and socially unmoored people as well as subhumans. Such critics propose that strict measures be implemented to prevent what they portray as dehumanizing possibilities from ever happening, usually in the form of an international ban on human genetic engineering.

Science journalist Ronald Bailey claims that McKibben's historical examples are flawed and support different conclusions when studied more closely. For example, few groups are more cautious than the Amish about embracing new technologies, but, though they shun television and use horses and buggies, some are welcoming the possibilities of gene therapy since inbreeding has afflicted them with a number of rare genetic diseases. Bailey and other supporters of technological alteration of human biology also reject the claim that life would be experienced as meaningless if some human limitations are overcome with enhancement technologies as extremely subjective.

Writing in *Reason* magazine, Bailey has accused opponents of research involving the modification of animals as indulging in alarmism when they speculate about the creation of subhuman creatures with human-like intelligence and brains resembling those of *Homo sapiens*. Bailey insists that the aim of conducting research on animals is simply to produce human health care benefits.

A different response comes from transhumanist personhood theorists who object to what they characterize as the anthropomorphobia fueling some criticisms of this research, which science fiction writer Isaac Asimov termed the "Frankenstein complex". For example, Woody Evans argues that, provided they are self-aware, human clones, human-animal chimeras and uplifted animals would all be unique persons deserving of respect, dignity, rights, responsibilities, and citizenship. They conclude that the coming ethical issue is not the creation of so-called monsters, but what they characterize as the "yuck factor" and "human-racism", that would judge and treat these creations as monstrous.

At least one public interest organization, the U.S.-based Center for Genetics and Society, was formed, in 2001, with the specific goal of opposing transhumanist agendas that involve transgenerational modification of human biology, such as full-term human cloning and germinal choice technology. The Institute on Biotechnology and the Human Future of the Chicago-Kent College of Law critically scrutinizes proposed applications of genetic and nanotechnologies to human biology in an academic setting.

Socioeconomic effects

Some critics of libertarian transhumanism have focused on the likely socioeconomic consequences in societies in which divisions between rich and poor are on the rise. Bill McKibben, for example, suggests that emerging human enhancement technologies would be disproportionately available to those with greater financial resources, thereby exacerbating the gap between rich and poor and creating a "genetic divide". Even Lee M. Silver, the biologist and science writer who coined the term "reprogenetics" and supports its applications, has expressed concern that these methods could create a two-tiered society of genetically engineered "haves" and "have nots" if social democratic reforms lag behind implementation of enhancement technologies. The 1997 film *Gattaca* depicts a dystopian society in which one's social class depends entirely on genetic modifications and is often cited by critics in support of these views.

These criticisms are also voiced by non-libertarian transhumanist advocates, especially self-described democratic transhumanists, who believe that the majority of current or future social and environmental issues (such as unemployment and resource depletion) need to be addressed by a combination of political and technological solutions (like a guaranteed minimum income and alternative technology). Therefore, on the specific issue of an emerging genetic divide due to unequal access to human enhancement technologies, bioethicist James Hughes, in his 2004 book *Citizen Cyborg: Why Democratic Societies Must Respond to the Redesigned Human of the Future*, argues that progressives or, more precisely, techno-progressives must articulate and implement public policies (i.e., a universal health care voucher system that covers human enhancement technologies) in order to attenuate this problem as much as possible, rather than trying to ban human enhancement technologies. The latter, he argues, might actually worsen the problem by making these technologies unsafe or available only to the wealthy on the local black market or in countries where such a ban is not enforced.

Sometimes, as in the writings of Leon Kass, the fear is that various institutions and practices judged as fundamental to civilized society would be damaged or destroyed. In his 2002 book *Our Posthuman Future* and in a 2004 *Foreign Policy* magazine article, political economist and philosopher Francis Fukuyama designates transhumanism as the world's most dangerous idea because he believes that it may undermine the egalitarian ideals of democracy (in general) and liberal democracy (in particular) through a fundamental alteration of "human nature". Social philosopher Jürgen Habermas makes a similar argument in his 2003 book *The Future of Human Nature*, in which he asserts that moral autonomy depends on not being subject to another's unilaterally imposed specifications. Habermas thus suggests that the human "species ethic" would be undermined by embryo-stage genetic alteration. Critics such as Kass, Fukuyama and a variety of authors hold that attempts to significantly alter human biology are not only inherently immoral, but also threaten the social order. Alternatively, they argue that implementation of such technologies would likely lead to the "naturalizing" of social hierarchies or place new means of control in the hands of totalitarian regimes. AI pioneer Joseph Weizenbaum criticizes what he sees as misanthropic tendencies in the language and ideas of some of his colleagues, in particular Marvin Minsky and Hans Moravec, which, by devaluing the human organism per se, promotes a discourse that enables divisive and undemocratic social policies.

In a 2004 article in the libertarian monthly *Reason,* science journalist Ronald Bailey contested the assertions of Fukuyama by arguing that political equality has never rested on the facts of human biology. He asserts that liberalism was founded not on the proposition of effective equality of human beings, or *de facto* equality, but on the assertion of an equality in political rights and before the law, or *de jure* equality. Bailey asserts that the products of genetic engineering may well ameliorate rather than exacerbate human inequality, giving to the many what were once the privileges of the few. Moreover, he argues, "the crowning achievement of the Enlightenment is the principle of tolerance". In fact, he says, political liberalism is already the solution to the issue of human and posthuman rights since in liberal societies the law is meant to apply equally to all, no matter how rich or poor, powerful or powerless, educated or ignorant, enhanced or unenhanced. Other thinkers who are sympathetic to transhumanist ideas, such as philosopher Russell Blackford, have also objected to the appeal to tradition and what they see as alarmism involved in *Brave New World*-type arguments.

Specter of coercive eugenicism

Some critics of transhumanism see the old eugenics, social Darwinist, and master race ideologies and programs of the past as warnings of what the promotion of eugenic enhancement technologies might unintentionally encourage. Some fear future "eugenics wars" as the worst-case scenario: the return of coercive state-sponsored genetic discrimination and human rights violations such as compulsory sterilization of persons with genetic defects, the killing of the institutionalized and, specifically, segregation and genocide of *races* perceived as inferior. Health law professor George Annas and technology law professor Lori Andrews are prominent advocates of the position that the use of these technologies could lead to such human-posthuman caste warfare.

The major transhumanist organizations strongly condemn the coercion involved in such policies and reject the racist and classist assumptions on which they were based, along with the pseudoscientific notions that eugenic improvements could be accomplished in a practically meaningful time frame through selective human breeding. Instead, most transhumanist thinkers advocate a "new eugenics", a form of egalitarian liberal eugenics. In their 2000 book *From Chance to Choice: Genetics and Justice*, non-transhumanist bioethicists Allen Buchanan, Dan Brock, Norman Daniels and Daniel Wikler have argued that liberal societies have an obligation to encourage as wide an adoption of eugenic enhancement technologies as possible (so long as such policies do not infringe on individuals' reproductive rights or exert undue pressures on prospective parents to use these technologies) in order to maximize public health and minimize the inequalities that may result from both natural genetic endowments and unequal access to genetic enhancements. Most transhumanists holding similar views nonetheless distance themselves from the term "eugenics" (preferring "germinal choice" or "reprogenetics") to avoid having their position confused with the discredited theories and practices of early-20th-century eugenic movements.

Existential risks

In his 2003 book *Our Final Hour*, British Astronomer Royal Martin Rees argues that advanced science and technology bring as much risk of disaster as opportunity for progress. However, Rees does not advocate a halt to scientific activity. Instead, he calls for tighter security and perhaps an end to traditional scientific openness. Advocates of the precautionary principle, such as many in the environmental movement, also favor slow, careful progress or a halt in potentially dangerous areas. Some precautionists believe that artificial intelligence and robotics present possibilities of alternative forms of cognition that may threaten human life.

Transhumanists do not necessarily rule out specific restrictions on emerging technologies so as to lessen the prospect of existential risk. Generally, however, they counter that proposals based on the precautionary principle are often unrealistic and sometimes even counter-productive as opposed to the technogaian current of transhumanism, which they claim is both realistic and productive. In his television series *Connections*, science historian James Burke dissects several views on technological change, including precautionism and the restriction of open inquiry. Burke questions the practicality of some of these views, but concludes that maintaining the *status quo* of inquiry and development poses hazards of its own, such as a disorienting rate of change and the depletion of our planet's resources. The common transhumanist position is a pragmatic one where society takes deliberate action to ensure the early arrival of the benefits of safe, clean, alternative technology, rather than fostering what it considers to be anti-scientific views and technophobia.

Nick Bostrom argues that even barring the occurrence of a singular global catastrophic event, basic Malthusian and evolutionary forces facilitated by technological progress threaten to eliminate the positive aspects of human society.

One transhumanist solution proposed by Bostrom to counter existential risks is control of differential technological development, a series of attempts to influence the sequence in which technologies are developed. In this approach, planners would strive to retard the development of possibly harmful technologies and their applications, while accelerating the development of likely beneficial technologies, especially those that offer protection against the harmful effects of others.

Further reading

- Maher, Derek F.; Mercer, Calvin, eds. (2009). *Religion and the implications of radical life extension* (1st ed.). New York: Palgrave Macmillan. ISBN 978-0-230-10072-5.
- Cole-Turner, Ronald, ed. (2011). *Transhumanism and transcendence : Christian hope in an age of technological enhancement*. Washington, D.C.: Georgetown University Press. ISBN 978-1-58901-780-1.
- Hansell, Gregory R; Grassie, William, eds. (2011). *H+/-: Transhumanism and Its Critics*. Philadelphia: Metanexus Institute. ISBN 978-1-45681-567-7.
- More, Max; Vita-More, Natasha, eds. (2013). *The transhumanist reader : classical and contemporary essays on the science, technology, and philosophy of the human future* (1.publ. ed.). Hoboken, N.J.: Wiley. ISBN 978-1-118-33429-4.
- Mercer, Calvin; Trothen, Tracy, eds. (2014). *Religion and transhumanism : the unknown future of human enhancement*. Westport, CT: Praeger. ISBN 9781440833250.
- Mercer, Calvin; Maher, Derek, eds. (2014). *Transhumanism and the Body: The World Religions Speak*. New York: Palgrave Macmillan. ISBN 9781137365835.
- Ranisch, Robert; Sorgner, Stefan Lorenz, eds. (2014). *Post- and Transhumanism*. Bruxelles: Peter Lang. ISBN 978-3-631-60662-9.

External links

<indicator name="spoken-icon"> </indicator>

- H+Pedia[1399] Transhumanist Wiki
- Media related to Transhumanism at Wikimedia Commons
- Transhumanism[1400] at Curlie (based on DMOZ)
- What is Transhumanism?[1401]

Appendix

References

[1]//en.wikipedia.org/w/index.php?title=Template:History_of_computing&action=edit
[2]McCorduck 2004.
[3]Kurzweil 2005.
[4]McCorduck 2004, pp. 5–35.
[5](Judah Loew's Golem); (Paracelsus); (Geber's *Takwin*)
[6]McCorduck 2004, pp. 17–25.
[7]Butler 1863.
[8]Nick 2005.
[9]and see also
[10]Quoted in . and discusses sacred statues.
[11]Other important automatons were built by Haroun al-Rashid , Jacques de Vaucanson and Leonardo Torres y Quevedo
[12]Cfr. Carreras Artau, Tomás y Joaquín. *Historia de la filosofía española. Filosofía cristiana de los siglos XIII al XV*. Madrid, 1939, Volume I
[13]Bonner, Anthonny, *The Art and Logic of Ramón Llull: A User's Guide*, Brill, 2007.
[14]Anthony Bonner (ed.), Doctor Illuminatus. A Ramon Llull Reader (Princeton University 1985). Vid. "Llull's Influence: The History of Lullism" at 57–71
[15]17th century mechanism and AI:
[16]Hobbes and AI:
[17]Leibniz and AI:
[18]The Lambda calculus was especially important to AI, since it was an inspiration for Lisp (the most important programming language used in AI).
[19]The original photo can be seen in the article:
[20]The Turing machine: , , and see
[21]Menabrea 1843
[22], , , , .
[23], , , , .
[24], , , and see also
[25], and
[26], , , , . See also
[27]claim that Turing answered all the major objections to AI that have been offered in the years since the paper appeared.
[28]See "A Brief History of Computing" http://www.alanturing.net/turing_archive/pages/Reference%20Articles/BriefHistofComp.html at AlanTuring.net.
[29]Schaeffer, Jonathan. *One Jump Ahead:: Challenging Human Supremacy in Checkers*, 1997,2009, Springer, . Chapter 6.
[30], and
[31]Quoted in and
[32], and
[33]See . Also see where Crevier states "[the proposal] later became known as the 'physical symbol systems hypothesis'". The physical symbol system hypothesis was articulated and named by Newell and Simon in their paper on GPS. It includes a more specific definition of a "machine" as an agent that manipulates symbols. See the philosophy of artificial intelligence.
[34]discusses how the Dartmouth conference alumni dominated the first two decades of AI research, calling them the "invisible college".
[35]"I won't swear and I hadn't seen it before," McCarthy told Pamela McCorduck in 1979. However, McCarthy also stated unequivocally "I came up with the term" in a CNET interview.
[36]writes "the conference is generally recognized as the official birthdate of the new science."
[37]Russell and Norvig write "it was astonishing whenever a computer did anything remotely clever."

[38], and
[39], and
[40]Means-ends analysis, reasoning as search: .
[41]Heuristic: ,
[42]GPS: , ,
[43]and
[44], ,
[45], ,
[46], , and
[47], ,
[48]quoted in . See also
[49]quoted in
[50]quoted in
[51]Minsky strongly believes he was misquoted. See , and .
[52], . McCorduck also notes that funding was mostly under the direction of alumni of the Dartmouth conference of 1956.
[53]and
[54]http://www.humanoid.waseda.ac.jp/booklet/kato_2-j.html
[55]*Robotics and Mechatronics: Proceedings of the 4th IFToMM International Symposium on Robotics and Mechatronics*, page 66 https://books.google.com/books?id=tQqVCgAAQBAJ&pg=PA66
[56]*Robots: From Science Fiction to Technological Revolution* https://books.google.com/books?id=PqQeAQAAIAAJ, page 130
[57]*Handbook of Digital Human Modeling: Research for Applied Ergonomics and Human Factors Engineering*, Chapter 3, pages 1-2 https://books.google.com/books?id=NgLLBQAAQBAJ&pg=SA3-PA1
[58]and
[59], ,
[60], see also : "Early programs were necessarily limited in scope by the size and speed of memory"
[61]. McCarthy has always disagreed with Moravec, back to their early days together at SAIL. He states "I would say that 50 years ago, the machine capability was much too small, but by 30 years ago, machine capability wasn't the real problem." in a CNET interview.
[62]and
[63]; ; ; ;
[64], , and under "Success in Speech Recognition".
[65], , and see also .
[66], , John McCarthy wrote in response that "the combinatorial explosion problem has been recognized in AI from the beginning" in Review of Lighthill report http://www-formal.stanford.edu/jmc/reviews/lighthill/lighthill.html
[67](on whom this account is based). Other views include and under "Success in Speech Recognition".
[68]. Moravec explains, "Their initial promises to DARPA had been much too optimistic. Of course, what they delivered stopped considerably short of that. But they felt they couldn't in their next proposal promise less than in the first one, so they promised more."
[69]under "Shift to Applied Research Increases Investment." While the autonomous tank was a failure, the battle management system (called "DART") proved to be enormously successful, saving billions in the first Gulf War, repaying the investment and justifying the DARPA's pragmatic policy, at least as far as DARPA was concerned.
[70]Lucas and Penrose' critique of AI: , , and see
[71]"Know-how" is Dreyfus' term. (Dreyfus makes a distinction between "knowing how" and "knowing that", a modern version of Heidegger's distinction of ready-to-hand and present-at-hand.)
[72]Dreyfus' critique of artificial intelligence: , , and see , ,
[73]Searle's critique of AI: , , and see
[74]Quoted in
[75]Quoted in

[76]"I became the only member of the AI community to be seen eating lunch with Dreyfus. And I deliberately made it plain that theirs was not the way to treat a human being." Joseph Weizenbaum, quoted in .

[77]. Weizenbaum referred to this text in . Colby and his colleagues later also developed chatterbot-like "computer simulations of paranoid processes (PARRY)" to "make intelligble paranoid processes in explicit symbol processing terms." ()

[78]Weizenbaum's critique of AI: , , and see

[79]showed that people do poorly on completely abstract problems, but if the problem is restated to allow the use of intuitive social intelligence, performance dramatically improves. (See Wason selection task) have shown that people are terrible at elementary problems that involve uncertain reasoning. (See list of cognitive biases for several examples). Eleanor Rosch's work is described in

[80]An early example of McCathy's position was in the journal Science where he said "This is AI, so we don't care if it's psychologically real" , and he recently reiterated his position at the AI@50 conference where he said "Artificial intelligence is not, by definition, simulation of human intelligence" .

[81]Neat vs. scruffy: (who picks up the state of the debate in 1984). (who documents Schank's original use of the term). Another aspect of the conflict was called "the procedural/declarative distinction" but did not prove to be influential in later AI research.

[82], and . Minsky's frame paper: .

[83](Dendral), ,

[84]and

[85], and

[86]Knowledge revolution: ,

[87]Cyc: , , and

[88], , and see also

[89], under "Shift to Applied Research Increases Investment"

[90]. AI winter was first used as the title of a seminar on the subject for the Association for the Advancement of Artificial Intelligence.

[91](who cites institutional reasons for their ultimate failure), (who cites the difficulty of truth maintenance, i.e., learning and updating), (who emphasizes the brittleness and the inability to handle excessive qualification.)

[92], . McCorduck writes "Two and a half decades later, we can see that the Japanese didn't quite meet all of those ambitious goals."

[93]writes: "I am confident that this bottom-up route to artificial intelligence will one date meet the traditional top-down route more than half way, ready to provide the real world competence and the commonsense knowledge that has been so frustratingly elusive in reasoning programs. Fully intelligent machines will result when the metaphorical golden spike is driven uniting the two efforts."

[94]http://people.csail.mit.edu/brooks/papers/elephants.pdf

[95]See, for example,

[96]discusses the fragmentation and the abandonment of AI's original goals.

[97]DARPA Grand Challenge – home page http://www.darpa.mil/grandchallenge/

[98]http://archive.darpa.mil/grandchallenge/

[99]writes that the improvement in computer chess, "according to common wisdom, is governed only by the brute force expansion of computer hardware."

[100]Cycle time of Ferranti Mark 1 was 1.2 milliseconds, which is arguably equivalent to about 833 flops. Deep Blue ran at 11.38 gigaflops (and this does not even take into account Deep Blue's special-purpose hardware for chess). *Very* approximately, these differ by a factor of 10^7.

[101], , where they write: "The whole-agent view is now widely accepted in the field". The intelligent agent paradigm is discussed in major AI textbooks, such as: , ,

[102]Carl Hewitt's Actor model anticipated the modern definition of intelligent agents. Both John Doyle and Marvin Minsky's popular classic *The Society of Mind* used the word "agent". Other "modular" proposals included Rodney Brook's subsumption architecture, object-oriented programming and others.

[103]This is how the most widely accepted textbooks of the 21st century define artificial intelligence. See and
[104]: "As I write, AI enjoys a Neat hegemony."
[105]See
[106]under "Artificial Intelligence in the 90s", and
[107]For the new state of the art in AI based speech recognition, see
[108]"AI-inspired systems were already integral to many everyday technologies such as internet search engines, bank software for processing transactions and in medical diagnosis." Nick Bostrom, quoted in
[109], ,
[110]He goes on to say: "The answer is, I believe we could have ... I once went to an international conference on neural net[s]. There were 40 thousand registrants ... but ... if you had an international conference, for example, on using multiple representations for common sense reasoning, I've only been able to find 6 or 7 people in the whole world."
[111]or see Advanced Human Intelligence http://crnano.typepad.com/crnblog/2005/08/advanced_human_.html where he defines strong AI as "machine intelligence with the full range of human intelligence."
[112]The Age of Artificial Intelligence: George John at TEDxLondonBusinessSchool 2013 http://tedxtalks.ted.com/video/The-Age-of-Artificial-Intellige
[113]//www.worldcat.org/oclc/46890682
[114]https://web.archive.org/web/20070926023314/http://www.aaai.org/AITopics/assets/PDF/AIMag26-04-016.pdf
[115]http://www.aaai.org/AITopics/assets/PDF/AIMag26-04-016.pdf
[116]http://people.csail.mit.edu/brooks/papers/elephants.pdf
[117]//doi.org/10.1016/S0921-8890%2805%2980025-9
[118]http://www.nzetc.org/tm/scholarly/tei-ButFir-t1-g1-t1-g1-t4-body.html
[119]https://exhibits.stanford.edu/feigenbaum/catalog/hk334rq4790
[120]http://i.stanford.edu/pub/cstr/reports/cs/tr/74/457/CS-TR-74-457.pdf
[121]http://www.cnn.com/2006/TECH/science/07/24/ai.bostrom/
[122]http://www.alanturing.net/turing_archive/pages/Reference%20Articles/what_is_AI/What%20is%20AI06.html
[123]//www.worldcat.org/oclc/5056816
[124]http://www.economist.com/science/tq/displaystory.cfm?story_id=9249338
[125]//www.worldcat.org/oclc/61273290
[126]//www.worldcat.org/oclc/48871099
[127]https://web.archive.org/web/20091229084457/http://dli.iiit.ac.in/ijcai/IJCAI-73/PDF/027B.pdf
[128]http://dli.iiit.ac.in/ijcai/IJCAI-73/PDF/027B.pdf
[129]//www.worldcat.org/oclc/225590743
[130]http://www.inf.ed.ac.uk/about/AIhistory.html
[131]http://adsabs.harvard.edu/abs/1982Sci...217.1237K
[132]//doi.org/10.1126/science.217.4566.1237
[133]//www.ncbi.nlm.nih.gov/pubmed/17837639
[134]//www.worldcat.org/oclc/71826177
[135]//www.worldcat.org/oclc/19981533
[136]http://users.ox.ac.uk/~jrlucas/Godel/mmg.html
[137]//doi.org/10.1017/S0031819100057983
[138]https://web.archive.org/web/20081008120238/http://www.engagingexperience.com/2006/07/ai50_ai_past_pr.html
[139]http://www.engagingexperience.com/2006/07/ai50_ai_past_pr.html
[140]https//www.nytimes.com
[141]http://www-formal.stanford.edu/jmc/history/dartmouth/dartmouth.html
[142]http://www-formal.stanford.edu/jmc/mcchay69/mcchay69.html
[143]//www.worldcat.org/oclc/52197627
[144]//doi.org/10.1007/BF02478259
[145]http://www.fourmilab.ch/babbage/sketch.html

[146]//www.worldcat.org/oclc/16924756

[147]http://web.media.mit.edu/~minsky/papers/Frames/frames.html

[148]//www.worldcat.org/oclc/223353010

[149]http://www.ddj.com/hpc-high-performance-computing/197700454?cid=RSSfeed_DDJ_AI

[150]http://www.frc.ri.cmu.edu/users/hpm/project.archive/general.articles/1975/Raw.Power.html

[151]//www.worldcat.org/oclc/245755104

[152]//www.worldcat.org/oclc/246584055

[153]//www.worldcat.org/oclc/246968117

[154]http://www.alshindagah.com/marapr2005/jaziri.html

[155]http://repository.upenn.edu/dissertations/AAI9503804

[156]http://news.cnet.com/Googles-man-behind-the-curtain/2008-1024_3-5208228.html

[157]http://news.cnet.com/Spying-an-intelligent-search-engine/2100-1032_3-6107048.html

[158]//www.worldcat.org/oclc/249625842

[159]http://aima.cs.berkeley.edu/

[160]http://www.cs.ubc.ca/~poole/ci.html

[161]http://domino.research.ibm.com/tchjr/journalindex.nsf/600cc5649e2871db852568150060213c/39a870213169f45685256bfa00683d74?OpenDocument

[162]//doi.org/10.1147/rd.33.0210

[163]https://web.archive.org/web/20071210043312/http://members.aol.com/NeoNoetics/MindsBrainsPrograms.html

[164]//doi.org/10.1017/S0140525X00005756

[165]//doi.org/10.1287/opre.6.1.1

[166]http://news.cnet.com/Getting-machines-to-think-like-us---page-2/2008-11394_3-6090207-2.html?tag=st.next

[167]http://www.bizjournals.com/pittsburgh/stories/2006/08/14/focus3.html?b=1155528000%5E1329573

[168]http://www.abelard.org/turpap2/tp2-ie.asp

[169]//doi.org/10.1112/plms/s2-42.1.230

[170]//doi.org/10.1093/mind/LIX.236.433

[171]//www.worldcat.org/issn/0026-4423

[172]//www.worldcat.org/oclc/10952283

[173] A Thirteenth Century Programmable Robot http://www.shef.ac.uk/marcoms/eview/articles58/robot.html

[174]Please see Mechanical calculator#Calculating clocks: unsuccessful mechanical calculators

[175]Hubert Dreyfus, What Computers Can't Do

[176]Please see: Pascal's calculator#Pascal versus Schickard

[177]Quoted in

[178]Project Gutenberg eBook Erewhon by Samuel Butler.Poes..... http://www.gutenberg.org/etext/1906

[179] and see also The Life and Work of Konrad Zuse https://web.archive.org/web/20100418164050/http://www.epemag.com/zuse

[180]Crevier 1993:22–25

[181]Samuel 1959

[182] Schaeffer, Jonathan. *One Jump Ahead:: Challenging Human Supremacy in Checkers*, 1997,2009, Springer, . Chapter 6.

[183]http://www.cs.york.ac.uk/mlg/MI/mi.html

[184] DART: Revolutionizing Logistics Planning http://doi.ieeecs.org/10.1109/MIS.2002.1005635

[185]http://research.microsoft.com/en-us/projects/vrkinect/

[186]http://research.microsoft.com/en-us/groups/vision/default.aspx

[187] Alibaba's AI Outguns Humans in Reading Test https://www.bloomberg.com/news/articles/2018-01-15/alibaba-s-ai-outgunned-humans-in-key-stanford-reading-test. January 15, 2018

[188]https://web.archive.org/web/20070926023314/http://www.aaai.org/AITopics/assets/PDF/AIMag26-04-016.pdf

[189]http://www.aaai.org/AITopics/assets/PDF/AIMag26-04-016.pdf

[190]http://people.csail.mit.edu/brooks/papers/elephants.pdf

[191] //doi.org/10.1016/S0921-8890%2805%2980025-9
[192] http://www.inf.ed.ac.uk/about/AIhistory.html
[193] http://users.ox.ac.uk/~jrlucas/Godel/mmg.html
[194] https://web.archive.org/web/20070826230310/http://www-formal.stanford.edu/jmc/history/dartmouth/dartmouth.html
[195] http://www-formal.stanford.edu/jmc/history/dartmouth/dartmouth.html
[196] http://www-formal.stanford.edu/jmc/mcchay69.html
[197] //doi.org/10.1007/BF02478259
[198] http://web.media.mit.edu/~minsky/papers/Frames/frames.html
[199] http://www.frc.ri.cmu.edu/users/hpm/project.archive/general.articles/1975/Raw.Power.html
[200] http://www.nap.edu/readingroom/books/far/ch9.html
[201] http://aima.cs.berkeley.edu/
[202] http://www.cs.ubc.ca/~poole/ci.html
[203] http://domino.research.ibm.com/tchjr/journalindex.nsf/600cc5649e2871db852568150060213c/39a870213169f45685256bfa00683d74?OpenDocument
[204] //doi.org/10.1147/rd.33.0210
[205] https://web.archive.org/web/20100118120034/http://www.bbsonline.org/documents/a/00/00/04/84/bbs00000484-00/bbs.searle2.html
[206] //doi.org/10.1017/S0140525X00005756
[207] http://www.bbsonline.org/documents/a/00/00/04/84/bbs00000484-00/bbs.searle2.html
[208] //doi.org/10.1287/opre.6.1.1
[209] http://www.abelard.org/turpap2/tp2-ie.asp
[210] //doi.org/10.1112/plms/s2-42.1.230
[211] https://web.archive.org/web/20080702224846/http://loebner.net/Prizef/TuringArticle.html
[212] //doi.org/10.1093/mind/LIX.236.433
[213] http://loebner.net/Prizef/TuringArticle.html
[214] http://aitopics.org/misc/brief-history
[215] https://www.nytimes.com/interactive/2010/06/25/science/20100625_AI_TIMELINE.html
[216] Russell, Stuart J.; Norvig, Peter (2010), Artificial Intelligence: A Modern Approach (3rd ed.), Upper Saddle River, New Jersey: Prentice Hall, , p. 437-439
[217] Hayes P, Naive physics I: Ontology for liquids. University of Essex report, 1978, Essex, UK.
[218] Davis R, Shrobe H E, Representing Structure and Behavior of Digital Hardware, IEEE Computer, Special Issue on Knowledge Representation, 16(10):75-82.
[219] https://web.archive.org/web/20040110031608/http://citeseer.nj.nec.com/context/177306/0
[220] http://www.lirmm.fr/gbkrbook/
[221] http://citeseer.ist.psu.edu/davis93what.html
[222] https://web.archive.org/web/20070211062250/http://www.informatik.uni-trier.de/~ley/db/conf/er/HainautHEHR96.html
[223] http//14.139.206.50:8080
[224] http://dai.fmph.uniba.sk/~sefranek/kri/handbook/handbook_of_kr.pdf
[225] http://medg.lcs.mit.edu/ftp/psz/k-rep.html
[226] http://www.makhfi.com/KCM_intro.htm
[227] http://www.inf.unibz.it/~franconi/dl/course/
[228] http://www.ccl.kuleuven.ac.be/LKR/html/datr.html
[229] http://www.isi.edu/isd/LOOM/LOOM-HOME.html
[230] http://www.research.att.com/sw/tools/classic/tm/ijcai-95-with-scenario.html
[231] https://web.archive.org/web/20031202202839/http://www.dfki.uni-kl.de/ruleml/
[232] http://nelements.org
[233] Liu, Hugo, and Push Singh. "ConceptNet—a practical commonsense reasoning tool-kit." BT technology journal 22.4 (2004): 211-226.
[234] Yampolskiy, Roman V. "AI-Complete, AI-Hard, or AI-Easy-Classification of Problems in AI." MAICS. 2012.
[235] Andrich, C, Novosel, L, and Hrnkas, B. (2009). Common Sense Knowledge https://pdfs.semanticscholar.org/59e3/495f4f1e70629e17ddbbce06834626b2a363.pdf. Information Search and Retrieval, 2009.

[236]http://www-formal.stanford.edu/cmason/circulation-ws07cmason.pdf

[237]Speer, Robert, and Catherine Havasi. "Representing General Relational Knowledge in ConceptNet 5." LREC. 2012.

[238]http://www.pst.stsci.edu/spss/doc/spss-abs.html

[239]http://www.stsci.edu/resources/software_hardware/spike/

[240]https://web.archive.org/web/20131222165824/http://www.eetn.gr/index.php/eetn-publications/ai-research-in-greece/planning-and-scheduling

[241]http://www.eetn.gr/index.php/eetn-publications/ai-research-in-greece/planning-and-scheduling

[242]http://www.icaps-conference.org/

[243]//en.wikipedia.org/w/index.php?title=Template:Machine_learning_bar&action=edit

[244]Supposedly paraphrased from:
See, for example,

[245]http://www.britannica.com/EBchecked/topic/1116194/machine-learning

[246]R. Kohavi and F. Provost, "Glossary of terms," Machine Learning, vol. 30, no. 2–3, pp. 271–274, 1998.

[247]Honglak Lee, Roger Grosse, Rajesh Ranganath, Andrew Y. Ng. " Convolutional Deep Belief Networks for Scalable Unsupervised Learning of Hierarchical Representations http://citeseerx.ist.psu.edu/viewdoc/download?doi=10.1.1.149.802&rep=rep1&type=pdf" Proceedings of the 26th Annual International Conference on Machine Learning, 2009.

[248]Aharon, M, M Elad, and A Bruckstein. 2006. "K-SVD: An Algorithm for Designing Overcomplete Dictionaries for Sparse Representation." Signal Processing, IEEE Transactions on 54 (11): 4311–4322

[249]Bridge, James P., Sean B. Holden, and Lawrence C. Paulson. " Machine learning for first-order theorem proving https://www.cl.cam.ac.uk/~lp15/papers/Reports/Bridge-ml.pdf." Journal of automated reasoning 53.2 (2014): 141–172.

[250]Loos, Sarah, et al. " Deep Network Guided Proof Search https://arxiv.org/pdf/1701.06972." arXiv preprint arXiv:1701.06972 (2017).

[251]Finnsson, Hilmar, and Yngvi Björnsson. " Simulation-Based Approach to General Game Playing https://vvvvw.aaai.org/Papers/AAAI/2008/AAAI08-041.pdf." AAAI. Vol. 8. 2008.

[252]Sarikaya, Ruhi, Geoffrey E. Hinton, and Anoop Deoras. " Application of deep belief networks for natural language understanding http://www.cs.utoronto.ca/~hinton/absps/ruhijournal.pdf." IEEE/ACM Transactions on Audio, Speech and Language Processing (TASLP) 22.4 (2014): 778–784.

[253]"BelKor Home Page" http://www2.research.att.com/~volinsky/netflix/ research.att.com

[254]When A Machine Learning Algorithm Studied Fine Art Paintings, It Saw Things Art Historians Had Never Noticed https//medium.com, *The Physics at ArXiv blog*

[255]https://odsc.com

[256]http://ai.stanford.edu/people/nilsson/mlbook.html

[257]https://web.archive.org/web/20091110212529/http://www-stat.stanford.edu/~tibs/ElemStatLearn/

[258]http://www.inference.phy.cam.ac.uk/mackay/itila/book.html

[259]http://world.std.com/~rjs/indinf56.pdf

[260]http://machinelearning.org/

[261]https://www.coursera.org/learn/machine-learning

[262]https://see.stanford.edu/Course/CS229

[263]https://mloss.org/

[264]https://developers.google.com/machine-learning/crash-course/

[265]http://www.paripath.com/mida-book

[266]http://www.mlyearning.org/

[267]http://www.deeplearningbook.org/

[268] Implementing an online help desk system based on conversational agent http://ahs.annaisd.org/common/pages/GalleryPhoto.aspx?photoId=11772950&width=180&height=180 Authors: Alisa Kongthon, Chatchawal Sangkeettrakarn, Sarawoot Kongyoung and Choochart Haruechaiyasak. Published by ACM 2009 Article, Bibliometrics Data Bibliometrics. Published

in: Proceeding, MEDES '09 Proceedings of the International Conference on Management of Emergent Digital EcoSystems, ACM New York, NY, USA. ,

[269]Chomskyan linguistics encourages the investigation of "corner cases" that stress the limits of its theoretical models (comparable to pathological phenomena in mathematics), typically created using thought experiments, rather than the systematic investigation of typical phenomena that occur in real-world data, as is the case in corpus linguistics. The creation and use of such corpora of real-world data is a fundamental part of machine-learning algorithms for natural language processing. In addition, theoretical underpinnings of Chomskyan linguistics such as the so-called "poverty of the stimulus" argument entail that general learning algorithms, as are typically used in machine learning, cannot be successful in language processing. As a result, the Chomskyan paradigm discouraged the application of such models to language processing.

[270]Antonio Di Marco - Roberto Navigili, "Clustering and Diversifying Web Search Results with Graph Based Word Sense Induction" http://www.mitpressjournals.org/doi/full/10.1162/COLI_a_00148#.VW4sDc_tmko, 2013

[271]Goldberg, Yoav (2016). A Primer on Neural Network Models for Natural Language Processing https://www.jair.org/media/4992/live-4992-9623-jair.pdf. Journal of Artificial Intelligence Research 57 (2016) 345–420

[272]Ian Goodfellow, Yoshua Bengio and Aaron Courville. http://www.deeplearningbook.org/ Deep Learning]. MIT Press.

[273]Rafal Jozefowicz, Oriol Vinyals, Mike Schuster, Noam Shazeer, and Yonghui Wu (2016). https://arxiv.org/abs/1602.02410 Exploring the Limits of Language Modeling

[274]Do Kook Choe and Eugene Charniak (EMNLP 2016). http://www.aclweb.org/website/old_anthology/D/D16/D16-1257.pdf Parsing as Language Modeling

[275]Vinyals, Oriol, et al. (NIPS2015). https://papers.nips.cc/paper/5635-grammar-as-a-foreign-language.pdf

[276]Mark Johnson. How the statistical revolution changes (computational) linguistics. http://www.aclweb.org/anthology/W09-0103 Proceedings of the EACL 2009 Workshop on the Interaction between Linguistics and Computational Linguistics.

[277]Philip Resnik. Four revolutions. http://languagelog.ldc.upenn.edu/nll/?p=2946 Language Log, February 5, 2011.

[278]Winograd, Terry (1971). Procedures as a Representation for Data in a Computer Program for Understanding Natural Language. http://hci.stanford.edu/winograd/shrdlu/

[279]Roger C. Schank and Robert P. Abelson (1977). Scripts, plans, goals, and understanding: An inquiry into human knowledge structures

[280]Kishorjit, N., Vidya Raj RK., Nirmal Y., and Sivaji B. (2012) " Manipuri Morpheme Identification http://aclweb.org/anthology//W/W12/W12-5008.pdf", Proceedings of the 3rd Workshop on South and Southeast Asian Natural Language Processing (SANLP), pages 95–108, COLING 2012, Mumbai, December 2012

[281]Yucong Duan, Christophe Cruz (2011), *Formalizing Semantic of Natural Language through Conceptualization from Existence http://www.ijimt.org/abstract/100-E00187.htm*. International Journal of Innovation, Management and Technology(2011) 2 (1), pp. 37-42.

[282]" Versatile question answering systems: seeing in synthesis https://www.academia.edu/2475776/Versatile_question_answering_systems_seeing_in_synthesis", Mittal et al., IJIIDS, 5(2), 119-142, 2011.

[283]PASCAL Recognizing Textual Entailment Challenge (RTE-7) https://tac.nist.gov//2011/RTE/

[284]//www.ncbi.nlm.nih.gov/pmc/articles/PMC40721

[285]//doi.org/10.1073/pnas.92.22.9977

[286]http://nlp.stanford.edu/IR-book/

[287]http://www.bmva.org/visionoverview The British Machine Vision Association and Society for Pattern Recognition Retrieved February 20, 2017

[288]https://www.robots.com/blog/viewing/the-future-of-automated-random-bin-picking

[289]O. Russakovsky et al., " ImageNet Large Scale Visual Recognition Challenge https://arxiv.org/abs/1409.0575", 2014.

[290]Barghout, Lauren. "Visual Taxometric Approach to Image Segmentation Using Fuzzy-Spatial Taxon Cut Yields Contextually Relevant Regions." Information Processing and Management of Uncertainty in Knowledge-Based Systems. Springer International Publishing, 2014.

[291] http://www.nada.kth.se/~tony/book.html
[292] http://www.cs.auckland.ac.nz/~rklette/Books/SpringerCV98/Springer98.html
[293] http://www.mas.ecp.fr/vision/Personnel/nikos/paragios-chen-faugeras/
[294] http://www.imagingbook.com/
[295] http://ivt.sourceforge.net/book.html
[296] http://szeliski.org/Book/
[297] https://www.amazon.ca/Algorithms-Image-Processing-Computer-Vision/dp/0470643854
[298] http://iris.usc.edu/Information/Iris-Conferences.html
[299] http://www.cvpapers.com/index.html
[300] http://www.computervisiononline.com/
[301] http://iris.usc.edu/Vision-Notes/bibliography/contents.html
[302] http://homepages.inf.ed.ac.uk/rbf/CVonline/
[303] http://www.bmva.org/
[304] Lowerre, Bruce. "The Harpy Speech Recognition System", Ph.D. thesis, Carnegie Mellon University, 1976
[305] Cravero, M., R. Pieraccini, and F. Raineri. "Definition and evaluation of phonetic units for speech recognition by hidden Markov models." Acoustics, Speech, and Signal Processing, IEEE International Conference on ICASSP'86.. Vol. 11. IEEE, 1986.
[306] Billi, Roberto. "Vector quantization and Markov source models applied to speech recognition." Acoustics, Speech, and Signal Processing, IEEE International Conference on ICASSP'82.. Vol. 7. IEEE, 1982.
[307] Billi, R., Canavesio, F., Ciaramella, A., & Nebbia, L. (1994, September). Interactive voice technology at work: The CSELT experience. In Interactive Voice Technology for Telecommunications Applications, 1994., Second IEEE Workshop on (pp. 43–48). IEEE.
[308] Pirani, Giancarlo, ed. Advanced algorithms and architectures for speech understanding. Vol. 1. Springer Science & Business Media, 2013.
[309] Herve Bourlard and Nelson Morgan, Connectionist Speech Recognition: A Hybrid Approach, The Kluwer International Series in Engineering and Computer Science; v. 247, Boston: Kluwer Academic Publishers, 1994.
[310] Alex Graves, Santiago Fernandez, Faustino Gomez, and Jürgen Schmidhuber (2006). Connectionist temporal classification: Labelling unsegmented sequence data with recurrent neural nets. Proceedings of ICML'06, pp. 369–376.
[311] Santiago Fernandez, Alex Graves, and Jürgen Schmidhuber (2007). An application of recurrent neural networks to discriminative keyword spotting http://www6.in.tum.de/pub/Main/Publications/Fernandez2007b.pdf. Proceedings of ICANN (2), pp. 220–229.
[312] Haşim Sak, Andrew Senior, Kanishka Rao, Françoise Beaufays and Johan Schalkwyk (September 2015): Google voice search: faster and more accurate. http://googleresearch.blogspot.ch/2015/09/google-voice-search-faster-and-more.html
[313] NIPS Workshop: Deep Learning for Speech Recognition and Related Applications, Whistler, BC, Canada, Dec. 2009 (Organizers: Li Deng, Geoff Hinton, D. Yu).
[314] Morgan, Bourlard, Renals, Cohen, Franco (1993) "Hybrid neural network/hidden Markov model systems for continuous speech recognition. ICASSP/IJPRAI"
[315] T. Robinson. (1992) A real-time recurrent error propagation network word recognition system https//www.researchgate.net , ICASSP.
[316] Waibel, Hanazawa, Hinton, Shikano, Lang. (1989) "Phoneme recognition using time-delay neural networks. IEEE Transactions on Acoustics, Speech and Signal Processing."
[317] Sepp Hochreiter (1991), Untersuchungen zu dynamischen neuronalen Netzen http://people.idsia.ch/~juergen/SeppHochreiter1991ThesisAdvisorSchmidhuber.pdf , Diploma thesis. Institut f. Informatik, Technische Univ. Munich. Advisor: J. Schmidhuber.
[318] Y. Bengio (1991). "Artificial Neural Networks and their Application to Speech/Sequence Recognition," Ph.D. thesis, McGill University, Canada.
[319] Keynote talk: Recent Developments in Deep Neural Networks. ICASSP, 2013 (by Geoff Hinton).
[320] Keynote talk: "Achievements and Challenges of Deep Learning: From Speech Analysis and Recognition To Language and Multimodal Processing," Interspeech, September 2014 (by Li Deng).

[321] S. A. Zahorian, A. M. Zimmer, and F. Meng, (2002) "Vowel Classification for Computer based Visual Feedback for Speech Training for the Hearing Impaired," in ICSLP 2002

[322] ICASSP 2013.

[323] Deng L., Li, J., Huang, J., Yao, K., Yu, D., Seide, F. et al. Recent Advances in Deep Learning for Speech Research at Microsoft. ICASSP, 2013.

[324] L. Deng, M. Seltzer, D. Yu, A. Acero, A. Mohamed, and G. Hinton (2010) Binary Coding of Speech Spectrograms Using a Deep Auto-encoder. Interspeech.

[325] http://www.verbyx.com

[326] Friends International Support Group

[327] Forgrave, Karen E. "Assistive Technology: Empowering Students with Disabilities." Clearing House 75.3 (2002): 122–6. Web.

[328] Ciaramella, Alberto. "A prototype performance evaluation report." Sundial workpackage 8000 (1993).

[329] National Institute of Standards and Technology. " The History of Automatic Speech Recognition Evaluation at NIST http://www.itl.nist.gov/iad/mig/publications/ASRhistory/ ".

[330] https://demo.cobaltspeech.com/

[331] https://www.academia.edu/4685517/SpeeG2_A_Speech-_and_Gesture-based_Interface_for_Efficient_Controller-free_Text_Entry

[332] https://curlie.org/Computers/Speech_Technology

[333] Cited by Tao and Tan.

[334] "Affective Computing" http://affect.media.mit.edu/pdfs/95.picard.pdf MIT Technical Report #321 (Abstract http://vismod.media.mit.edu/pub/tech-reports/TR-321-ABSTRACT.html), 1995

[335] Breazeal, C. and Aryananda, L. Recognition of affective communicative intent in robot-directed speech. Autonomous Robots 12 1, 2002. pp. 83–104.

[336]

[337] Lee, C.M.; Narayanan, S.; Pieraccini, R., Recognition of Negative Emotion in the Human Speech Signals, Workshop on Auto. Speech Recognition and Understanding, Dec 2001

[338] "Gaussian Mixture Model" http://cnx.org/content/m13205/latest/. Connexions – Sharing Knowledge and Building Communities. Retrieved 10 March 2011.

[339] Ekman, P. & Friesen, W. V (1969). The repertoire of nonverbal behavior: Categories, origins, usage, and coding. Semiotica, 1, 49–98.

[340] "Facial Action Coding System (FACS) and the FACS Manual" http://face-and-emotion.com/dataface/facs/description.jsp . A Human Face. Retrieved 21 March 2011.

[341] Clever Algorithms. "Bacterial Foraging Optimization Algorithm – Swarm Algorithms – Clever Algorithms" http://www.cleveralgorithms.com/nature-inspired/swarm/bfoa.html. Clever Algorithms. Retrieved 21 March 2011.

[342] "Soft Computing" http://www.softcomputing.net/bfoa-chapter.pdf. Soft Computing. Retrieved 18 March 2011.

[343] Williams, Mark. "Better Face-Recognition Software – Technology Review" http://www.technologyreview.com/Infotech/18796/?a=f. Technology Review: The Authority on the Future of Technology. Retrieved 21 March 2011.

[344] J. K. Aggarwal, Q. Cai, Human Motion Analysis: A Review, Computer Vision and Image Understanding, Vol. 73, No. 3, 1999

[345] Picard, Rosalind (1998). Affective Computing. MIT.

[346] Larsen JT, Norris CJ, Cacioppo JT, "Effects of positive and negative affect on electromyographic activity over zygomaticus major and corrugator supercilii", (September 2003)

[347] Ritendra Datta, Dhiraj Joshi, Jia Li and James Z. Wang, Studying Aesthetics in Photographic Images Using a Computational Approach, Lecture Notes in Computer Science, vol. 3953, Proceedings of the European Conference on Computer Vision, Part III, pp. 288-301, Graz, Austria, May 2006.

[348] AutoTutor http://www.autotutor.org/

[349] Projects in Affective Computing http://affect.media.mit.edu/projects.php

[350] //citeseerx.ist.psu.edu/viewdoc/summary?doi=10.1.1.180.6429

[351] //doi.org/10.1016/s1071-5819%2803%2900047-8

[352] http://affect.media.mit.edu/

[353] http://emotions.usc.edu/
[354] http://sites.google.com/site/memphisemotivecomputing/
[355] http://www.acii2011.org/
[356] http://www.eecs.tufts.edu/~agirou01/workshop/
[357] http://www.igi-global.com/bookstore/titledetails.aspx?TitleId=1144
[358] http://www.computer.org/portal/web/tac
[359] https://dx.doi.org/10.14569/IJACSA.2010.010615
[360] http://opensmile.sourceforge.net/
[361] or see Advanced Human Intelligence http://crnano.typepad.com/crnblog/2005/08/advanced_human_.html where he defines strong AI as "machine intelligence with the full range of human intelligence."
[362] The Age of Artificial Intelligence: George John at TEDxLondonBusinessSchool 2013 http://tedxtalks.ted.com/video/The-Age-of-Artificial-Intellige
[363] . This is the term they use for "human-level" intelligence in the physical symbol system hypothesis.
[364] Encyclopædia Britannica Strong AI, applied AI, and cognitive simulation http://www.britannica.com/eb/article-219086/artificial-intelligence or Jack Copeland What is artificial intelligence? http://www.cs.usfca.edu/www.AlanTuring.net/turing_archive/pages/Reference%20Articles/what_is_AI/What%20is%20AI02.html on AlanTuring.net
[365] The Open University on Strong and Weak AI http://www.open2.net/nextbigthing/ai/ai_in_depth/in_depth.htm
[366] AI founder John McCarthy writes: "we cannot yet characterize in general what kinds of computational procedures we want to call intelligent." (For a discussion of some definitions of intelligence used by artificial intelligence researchers, see philosophy of artificial intelligence.)
[367] This list of intelligent traits is based on the topics covered by major AI textbooks, including: , , and

.
[368] Pfeifer, R. and Bongard J. C., How the body shapes the way we think: a new view of intelligence (The MIT Press, 2007).
[369] deCharms, R. (1968). Personal causation. New York: Academic Press.
[370] Shapiro, Stuart C. (1992). Artificial Intelligence http://www.cse.buffalo.edu/~shapiro/Papers/ai.pdf In Stuart C. Shapiro (Ed.), *Encyclopedia of Artificial Intelligence* (Second Edition, pp. 54–57). New York: John Wiley. (Section 4 is on "AI-Complete Tasks".)
[371] Roman V. Yampolskiy. Turing Test as a Defining Feature of AI-Completeness. In Artificial Intelligence, Evolutionary Computation and Metaheuristics (AIECM) –In the footsteps of Alan Turing. Xin-She Yang (Ed.). pp. 3–17. (Chapter 1). Springer, London. 2013. http://cecs.louisville.edu/ry/TuringTestasaDefiningFeature04270003.pdf
[372] Luis von Ahn, Manuel Blum, Nicholas Hopper, and John Langford. CAPTCHA: Using Hard AI Problems for Security http://www.captcha.net/captcha_crypt.pdf. In Proceedings of Eurocrypt, Vol. 2656 (2003), pp. 294–311.
[373] (unpublished?)
[374] quoted in
[375] Scientist on the Set: An Interview with Marvin Minsky http://mitpress.mit.edu/e-books/Hal/chap2/two1.html
[376] Marvin Minsky to , quoted in .
[377] The Lighthill report specifically criticized AI's "grandiose objectives" and led the dismantling of AI research in England. (;) In the U.S., DARPA became determined to fund only "mission-oriented direct research, rather than basic undirected research". See under "Shift to Applied Research Increases Investment". See also and
[378] , and see also
[379] ; ;
[380] As AI founder John McCarthy writes "it would be a great relief to the rest of the workers in AI if the inventors of new general formalisms would express their hopes in a more guarded form than has sometimes been the case."
[381] "At its low point, some computer scientists and software engineers avoided the term artificial intelligence for fear of being viewed as wild-eyed dreamers."

[382]Goertzel & Pennachin 2006.
[383]. See also with an up-to-date summary and lots of links.
[384]Goertzel 2007.
[385]
[386]. "The basic idea is to take a particular brain, scan its structure in detail, and construct a software model of it that is so faithful to the original that, when run on appropriate hardware, it will behave in essentially the same way as the original brain."
[387]Sandberg & Boström 2008.
[388]Drachman 2005.
[389]Russell & Norvig 2003.
[390]In "Mind Children" 10^{15} cps is used. More recently, in 1997, <> Moravec argued for 10^{8} MIPS which would roughly correspond to 10^{14} cps. Moravec talks in terms of MIPS, not "cps", which is a non-standard term Kurzweil introduced.
[391]Artificial brain '10 years away' 2009 BBC news http://news.bbc.co.uk/1/hi/technology/8164060.stm
[392]University of Calgary news http://gauntlet.ucalgary.ca/story/10343, MSNBC news http://msnbc.msn.com/id/12037941/
[393]. A wide range of views in current research, all of which require grounding to some degree
[394]some links to bee brain studies http://www.setiai.com/archives/cat_honey_bee_brain.html
[395]In Goertzels' AGI book (Yudkowsky 2006), Yudkowsky proposes 5 levels of organisation that must be understood – code/data, sensory modality, concept & category, thought, and deliberation (consciousness) – in order to use the available hardware
[396]"nervous system, human." http://search.eb.com/eb/article-75525 *Encyclopædia Britannica*. 9 January 2007
[397]Azevedo et al. 2009.
[398]Yudkowsky 2006.
[399]Aleksander 1996.
[400]As defined in a standard AI textbook: "The assertion that machines could possibly act intelligently (or, perhaps better, act as if they were intelligent) is called the 'weak AI' hypothesis by philosophers, and the assertion that machines that do so are actually thinking (as opposed to simulating thinking) is called the 'strong AI' hypothesis."
[401]For example: • , • Oxford University Press Dictionary of Psychology http://www.encyclopedia.com/doc/1O87-strongAI.html (quoted in "High Beam Encyclopedia"), • MIT Encyclopedia of Cognitive Science http://www.aaai.org/AITopics/html/phil.html (quoted in "AITopics") • Planet Math http://planetmath.org/encyclopedia/StrongAIThesis.html • Will Biological Computers Enable Artificially Intelligent Machines to Become Persons? http://www.cbhd.org/resources/biotech/tongen_2003-11-07.htm Anthony Tongen
[402]Russell & Norvig 2003, p. 947.
[403]Clocksin 2003.
[404]McCarthy 2003.
[405]Gelernter 2010.
[406]McCarthy 2007.
[407]Holte & Choueiry 2003.
[408]Zucker 2003.
[409]Note that consciousness is difficult to define. A popular definition, due to Thomas Nagel, is that it "feels like" something to be conscious. If we are not conscious, then it doesn't feel like anything. Nagel uses the example of a bat: we can sensibly ask "what does it feel like to be a bat?" However, we are unlikely to ask "what does it feel like to be a toaster?" Nagel concludes that a bat appears to be conscious (i.e. has consciousness) but a toaster does not. See
[410]Müller, V. C., & Bostrom, N. (2016). Future progress in artificial intelligence: A survey of expert opinion. In Fundamental issues of artificial intelligence (pp. 555-572). Springer, Cham.
[411]Hein, Andreas Makoto, and Hélène Condat. "Can Machines Design? An Artificial General Intelligence Approach." https://arxiv.org/abs/1806.02091 arXiv preprint arXiv:1806.02091 (2018).
[412]https://web.archive.org/web/20130606101835/http://www.techcast.org/Upload/PDFs/633615794236495345_TCTheAutomationofThought.pdf

[413]http://www.techcast.org/Upload/PDFs/633615794236495345_TCTheAutomationofThought.pdf
[414]http://www.fhi.ox.ac.uk/Reports/2008-3.pdf
[415]https//www.researchgate.net
[416]//doi.org/10.1002/cne.21974
[417]//www.ncbi.nlm.nih.gov/pubmed/19226510
[418]http://berglas.org/Articles/AIKillGrandchildren/AIKillGrandchildren.html
[419]http://adsabs.harvard.edu/abs/2003RSPTA.361.1721C
[420]//doi.org/10.1098/rsta.2003.1232
[421]//www.ncbi.nlm.nih.gov/pubmed/12952683
[422]//doi.org/10.1212/01.WNL.0000166914.38327.BB
[423]//www.ncbi.nlm.nih.gov/pubmed/15985565
[424]http://www.edge.org/3rd_culture/gelernter10.1/gelernter10.1_index.html
[425]https//web.archive.org
[426]http://people.inf.elte.hu/csizsekp/ai/books/artificial-general-intelligence-cognitive-technologies.9783540237334.27156.pdf
[427]http://sites.google.com/site/narswang/publications/wang-goertzel.AGI_Aspects.pdf?attredirects=1
[428]https://scholar.google.com/scholar?hl=sv&lr=&cluster=15189798216526465792
[429]//doi.org/10.1016/j.artint.2007.10.011
[430]http://www.foresight.org/Conferences/MNT05/Papers/Gubrud/
[431]//doi.org/10.1098/rstb.2003.1317
[432]//www.ncbi.nlm.nih.gov/pmc/articles/PMC1693218
[433]//www.ncbi.nlm.nih.gov/pubmed/12903653
[434]http://www.dai.ed.ac.uk/AI_at_Edinburgh_perspective.html
[435]http://www.cs.unm.edu/~luger/ai-final/tocfull.html
[436]//doi.org/10.1016/j.artint.2007.10.009
[437]http://www.pamelamc.com/html/machines_who_think.html
[438]http://www.frc.ri.cmu.edu/users/hpm/project.archive/general.articles/1975/Raw.Power.html
[439]http://organizations.utep.edu/Portals/1475/nagel_bat.pdf
[440]http://portal.acm.org/citation.cfm?id=360022
[441]//doi.org/10.1145/360018.360022
[442]http://aima.cs.berkeley.edu/
[443]http://www.nap.edu/readingroom/books/far/ch9.html
[444]http://www.cs.ubc.ca/spider/poole/ci.html
[445]https://web.archive.org/web/20100118120034/http://www.bbsonline.org/documents/a/00/00/04/84/bbs00000484-00/bbs.searle2.html
[446]//doi.org/10.1017/S0140525X00005756
[447]http://www.bbsonline.org/documents/a/00/00/04/84/bbs00000484-00/bbs.searle2.html
[448]//doi.org/10.1146/annurev.ne.11.030188.002231
[449]//www.ncbi.nlm.nih.gov/pubmed/3284447
[450]https://web.archive.org/web/20090411050423/http://www.singinst.org/upload/LOGI/LOGI.pdf
[451]//doi.org/10.1146/annurev.psych.49.1.585
[452]//www.ncbi.nlm.nih.gov/pubmed/9496632
[453]http://www.singinst.org/upload/LOGI//LOGI.pdf
[454]//doi.org/10.1098/rstb.2003.1308
[455]//www.ncbi.nlm.nih.gov/pmc/articles/PMC1693211
[456]//www.ncbi.nlm.nih.gov/pubmed/12903672
[457]http://adsabs.harvard.edu/abs/2008gcr..book..303Y
[458]http://sites.google.com/site/narswang/home/agi-introduction
[459]http://genesis.csail.mit.edu/index.html
[460]http://www.opencog.org/
[461]http://academia.wikia.com/wiki/A_Method_for_Simulating_the_Process_of_Logical_Human_Thought
[462]http://gameofid.com

[463]http://www.openphilanthropy.org/focus/global-catastrophic-risks/potential-risks-advanced-artificial-intelligence/ai-timelines

[464]Shapiro, Stuart C. (1992). Artificial Intelligence http://www.cse.buffalo.edu/~shapiro/Papers/ai.pdf In Stuart C. Shapiro (Ed.), *Encyclopedia of Artificial Intelligence* (Second Edition, pp. 54–57). New York: John Wiley. (Section 4 is on "AI-Complete Tasks".)

[465]Roman V. Yampolskiy. Turing Test as a Defining Feature of AI-Completeness . In Artificial Intelligence, Evolutionary Computation and Metaheuristics (AIECM) –In the footsteps of Alan Turing. Xin-She Yang (Ed.). pp. 3-17. (Chapter 1). Springer, London. 2013. http://cecs.louisville.edu/ry/TuringTestasaDefiningFeature04270003.pdf

[466]Luis von Ahn, Manuel Blum, Nicholas Hopper, and John Langford. CAPTCHA: Using Hard AI Problems for Security http://www.captcha.net/captcha_crypt.pdf. In Proceedings of Eurocrypt, Vol. 2656 (2003), pp. 294-311.

[467](unpublished?)

[468]Mueller, Erik T. (1987, March). *Daydreaming and Computation* (Technical Report CSD-870017) ftp://ftp.cs.ucla.edu/tech-report/198_-reports/870017.pdf Ph.D. dissertation, University of California, Los Angeles. ("Daydreaming is but one more *AI-complete* problem: if we could solve any one artificial intelligence problem, we could solve all the others", p. 302)

[469]Raymond, Eric S. (1991, March 22). Jargon File Version 2.8.1 http://catb.org/esr/jargon/oldversions/jarg282.txt (Definition of "AI-complete" first added to jargon file.)

[470]Ide, N., & Veronis, J. (1998). Introduction to the special issue on word sense disambiguation: the state of the art. Computational Linguistics, 24(1), 2–40.

[471]Dafna Shahaf and Eyal Amir (2007) Towards a theory of AI completeness http://www.aaai.org/Papers/Symposia/Spring/2007/SS-07-05/SS07-05-026.pdf. Commonsense 2007, 8th International Symposium on Logical Formalizations of Commonsense Reasoning http://www.ucl.ac.uk/commonsense07.

[472]Heinz von Foerster (1981), 'Observing Systems", Intersystems Publications, Seaside, CA.

[473]Tange, Kenzo (1966) "Function, Structure and Symbol".

[474]*"La cybernétique est l'art de l'efficacité de l'action"* originally a French definition formulated in 1953, lit. "Cybernetics is the art of effective action"

[475]Couffignal, Louis, "Essai d'une définition générale de la cybernétique", *The First International Congress on Cybernetics*, Namur, Belgium, June 26–29, 1956, Paris: Gauthier-Villars, 1958, pp. 46-54.

[476]CYBCON discusstion group 20 September 2007 18:15

[477]Leary, Timothy. "The Cyberpunk: the individual as reality pilot" in Storming the Reality Studio. Duke University Press: 1991.

[478]H.S. Tsien. *Engineering Cybernetics*, Preface vii. McGraw Hill, 1954.

[479]http://people.wku.edu/charles.smith/wallace/S043.htm

[480]*Философский словарь* (Philosophical dictionary), 1954; "Cybernetics", *The Great Soviet Encyclopedia* (1979)

[481]Jean-Pierre Dupuy, "The autonomy of social reality: on the contribution of systems theory to the theory of society" in: Elias L. Khalil & Kenneth E. Boulding eds., *Evolution, Order and Complexity*, 1986.

[482]Peter Harries-Jones (1988), "The Self-Organizing Polity: An Epistemological Analysis of Political Life by Laurent Dobuzinskis" in: *Canadian Journal of Political Science* (Revue canadienne de science politique), Vol. 21, No. 2 (Jun., 1988), pp. 431-433.

[483]Kenneth D. Bailey (1994), *Sociology and the New Systems Theory: Toward a Theoretical Synthesis*, p.163.

[484]//en.wikipedia.org/w/index.php?title=Template:Economic_systems_sidebar&action=edit

[485]Allin Cottrell & W.Paul Cockshott, *Towards a new socialism* http://ricardo.ecn.wfu.edu/~cottrell/socialism_book/index.html (Nottingham, England: Spokesman, 1993). Retrieved: 17 March 2012.

[486]This does not refer to the concept of Racial Memory but to the concept of cumulative adaptation to a particular niche, such as the case of the pepper moth having genes for both light and dark environments.

[487]McClelland, Kent A., and Thomas J. Fararo (Eds.). 2006. Purpose, Meaning, and Action: Control Systems Theories in Sociology. New York: Palgrave Macmillan.

[488]"SSCC (Sinai Sustainability Cybernetics Center)" the 46th team to qualify for this year's MIT semi-finalist round — Naharnet http://www.naharnet.com/stories/en/80769. Naharnet.com (2013-04-25). Retrieved on 2013-11-02.

[489]Pask, G. (1971). A comment, a case history and a plan. In J. Reichardt (Ed.), Cybernetics, art and ideas (pp. 76-99). London: Studio Vista. Fernandez, M. (2009). "Aesthetically-Potent Environments" or How Pask Detourned Instrumental Cybernetics. In P. Brown, C. Gere, N. Lambert & C. Mason (Eds.), White Heat Cold Logic: British Computer Art 1960-1980 Cambridge, Massachusetts: MIT Press.

[490]Cybernetics and Art: Cultural Convergence in the 1960s http://www.artexetra.com/ /CyberneticsArtCultConv.pdf

[491]Upitis, A. (2013). Alexander's Choice: How Architecture avoided Computer Aided Design c. 1962. In A. Dutta (Ed.), A Second Modernism: MIT, Architecture, and the 'Techno-Social' Moment (pp. 474-505). Cambridge, Massachusetts: SA+P Press.

[492]Boyd Davis, S., & Gristwood, S. (2016). The Structure of Design Processes: Ideal and Reality in Bruce Archer's 1968 Doctoral Thesis. In Proceedings of DRS 2016, Design Research Society 50th Anniversary Conference, Brighton, UK. 27–30 June 2016. Retrieved from http://www.drs2016.org/240/

[493]Mathews, S. (2007). From Agit-Prop to Free Space: The Architecture of Cedric Price. London: Black Dog. Isabelle Doucet (University of Manchester, UK), Samantha Hardingham (Architectural Association, London, UK), Tanja Herdt (TU Munich, Germany), Jim Njoo (École Nationale Supérieure d'Architecture de Paris-La Villette, France), Ben Sweeting (University of Brighton, UK). An Afternoon with Cedric Price no. 1, CCA c/o Lisboa http://www.cca.qc.ca/en/explore?event=40500. Panel discussion moderated by Kim Förster, CCA Associate Director, Research. Organised by the Canadian Centre for Architecture, Montreal and Artéria, Lisbon. Held at Barbas Lopes Arquitectos. Part of the Lisbon Architecture Triennale 2016. 22 October 2016.

[494]Furtado Cardoso Lopes, G. M. (2008). Cedric Price's Generator and the Frazers' systems research. Technoetic Arts, 6(1), 55-72.

[495]Glanville, R. (Ed.). (2007). Cybernetics and design. Special double issue of Kybernetes, 36(9/10); Jones, P. H. (2014). Systemic Design Principles for Complex Social Systems. In G. Metcalf (Ed.), Social Systems and Design (pp. 91-128). Tokyo: Springer.

[496]Glanville, R. (2007). Try again. Fail again. Fail better: The cybernetics in design and the design in cybernetics. Kybernetes, 36(9/10), 1173-1206. ; Glanville, R. (2014). How Design and Cybernetics Reflect Each Other. In B. Sevaldson & P. H. Jones (Eds.), Proceedings of Third Symposium of Relating Systems Thinking to Design, Oslo, Norway. October 15–17, 2014. Sweeting, B. (2016). Design Research as a Variety of Second-Order Cybernetic Practice http://www.univie.ac.at/constructivism/journal/11/3/572.sweeting. Constructivist Foundations, 11(3), 572-579.

[497]Schellnhuber, H.-J., Discourse: Earth system analysis - The scope of the challenge, pp. 3-195. In: Schellnhuber, H.-J. and Wenzel, V. (Eds.). 1998. Earth system analysis: Integrating science for sustainability. Berlin: Springer.

[498]Schellnhuber, H.-J., Earth system analysis and the second Copernican revolution. *Nature*, 402, C19-C23. 1999.

[499]http://pespmc1.vub.ac.be/books/IntroCyb.pdf

[500]https//web.archive.org

[501]http://web.mit.edu/slava/homepage/newspeak.htm

[502]http://pespmc1.vub.ac.be/Papers/Cybernetics-EPST.pdf

[503]http://neocybernetics.com/report151/

[504]http://pangaro.com/published/cyber-macmillan.html

[505]http://www.cybsoc.org/gcyb.htm

[506]//doi.org/10.1086/283881

[507]//www.jstor.org/stable/2460822?

[508]ftp://ftp.vub.ac.be/pub/projects/Principia_Cybernetica/Papers_Umpleby/Science-Cybernetics.txt

[509]http://www.stanford.edu/group/SHR/4-2/text/foerster.html

[510]https://books.google.com/books?id=2AKaAAAAIAAJ

[511] http://www.bu.edu/wcp/Papers/Comp/CompJurc.htm
[512] http://bactra.org/notebooks/cybernetics.html
[513] http://pespmc1.vub.ac.be/DEFAULT.html
[514] http://pespmc1.vub.ac.be/ASC/indexASC.html
[515] http://www.gwu.edu/~asc/slide/s1.html
[516] https://web.archive.org/web/20100811013353/http://www.smithsrisca.demon.co.uk/cybernetics.html
[517] http://www.smithsrisca.demon.co.uk/cybernetics.html
[518] https://www.youtube.com/watch?v=_hjAXkNbPfk
[519] http://journal.sonicstudies.org/vol02/nr01/a03
[520] https://www.jstor.org/stable/10.7757/persnewmusi.54.1.0005
[521] http://www.asc-cybernetics.org/
[522] http://www.ieeesmc.org/
[523] http://3rd-street.net/Group/index.php/index.php?topic=68.msg216#msg216
[524] http://www.cybsoc.org
[525] What is computational neuroscience? Patricia S. Churchland, Christof Koch, Terrence J. Sejnowski. in Computational Neuroscience pp.46-55. Edited by Eric L. Schwartz. 1993. MIT Press
[526] Review article
[527] //doi.org/10.1016/j.neuron.2004.08.012
[528] //www.ncbi.nlm.nih.gov/pubmed/15339643
[529] http://www.jphysiol.org/cgi/pmidlookup?view=long&pmid=12991237
[530] //doi.org/10.1113/jphysiol.1952.sp004764
[531] //www.ncbi.nlm.nih.gov/pmc/articles/PMC1392413
[532] //www.ncbi.nlm.nih.gov/pubmed/12991237
[533] https://www.springer.com/mathematics/journal/13408
[534] https://www.springer.com/10827
[535] http://www.mitpressjournals.org/loi/neco
[536] http://www.springerlink.com/content/1871-4099/
[537] http://frontiersin.org/neuroscience/computationalneuroscience/
[538] http://www.ploscompbiol.org/home.action
[539] http://www.frontiersin.org/Journal/specialty.aspx?s=752&name=neuroinformatics&x=y
[540] http://www.cnsorg.org
[541] http://www.nips.cc
[542] https://web.archive.org/web/20070309063503/http://www.iccn2007.org/
[543] http://www.icms.org.uk/workshops/mathneuro
[544] https://web.archive.org/web/20071012155706/http://www.neurocomp.fr/index.php?page=welcome
[545] https://web.archive.org/web/20110429094455/http://www.nncn.de/Aktuelles-en/bernsteinsymposium/Symposium/view?set_language=en
[546] http://www.nncn.de/willkommen-en/view?set_language=en
[547] http://www.areadne.org/index.html
[548] http://www.scholarpedia.org/article/Encyclopedia_of_Computational_Neuroscience
[549] //en.wikipedia.org/w/index.php?title=Template:Evolutionary_biology&action=edit
[550] G. C. Onwubolu and B V Babu,
[551] https://plato.stanford.edu/entries/information-biological/#InfEvo
[552] https://plato.stanford.edu/entries/information-biological/#InfEvo/
[553] https://www.amazon.com/Handbook-Evolutionary-Computation-Thomas-Back/dp/0750303921
[554] https://www.springer.com/computer+science/theoretical+computer+science/foundations+of+computations/book/978-3-540-67353-8
[555] https://www.springer.com/engineering/computational+intelligence+and+complexity/book/978-3-642-23423-1
[556] http://www.sciencedirect.com/science/article/pii/S0020019002002041
[557] https://www.springer.com/computer/theoretical+computer+science/book/978-3-540-40184-1
[558] https://www.springer.com/computer/ai/book/978-3-540-74109-1

[559] https://www.amazon.com/Parameter-Evolutionary-Algorithms-Computational-Intelligence/dp/3642088929/

[560] https://www.springer.com/computer/ai/book/978-3-540-60676-5

[561] https://www.springer.com/computer/theoretical+computer+science/book/978-3-540-22494-5

[562] http://academic.csuohio.edu/simond/EvolutionaryOptimization

[563] https://biodatamining.biomedcentral.com/articles/10.1186/s13040-018-0164-x

[564] https://arxiv.org/pdf/1709.09840.pdf

[565] " The Nature of Mathematical Programming http://glossary.computing.society.informs.org/index.php?page=nature.html ," *Mathematical Programming Glossary*, INFORMS Computing Society.

[566] W. Erwin Diewert (2008). "cost functions," *The New Palgrave Dictionary of Economics*, 2nd Edition Contents http://www.dictionaryofeconomics.com/article?id=pde2008_C000390&edition=current&q=.

[567] Lionel Robbins (1935, 2nd ed.) *An Essay on the Nature and Significance of Economic Science*, Macmillan, p. 16.

[568] A.G. Malliaris (2008). "stochastic optimal control," *The New Palgrave Dictionary of Economics*, 2nd Edition. Abstract http://www.dictionaryofeconomics.com/article?id=pde2008_S000269&edition=&field=keyword&q=Taylor's%20th&topicid=&result_number=1.

[569] From *The New Palgrave Dictionary of Economics* (2008), 2nd Edition with Abstract links:

- " numerical optimization methods in economics http://www.dictionaryofeconomics.com/article?id=pde2008_N000148&edition=current&q=optimization&topicid=&result_number=1" by Karl Schmedders
- " convex programming http://www.dictionaryofeconomics.com/article?id=pde2008_C000348&edition=current&q=optimization&topicid=&result_number=4" by Lawrence E. Blume
- " Arrow–Debreu model of general equilibrium http://www.dictionaryofeconomics.com/article?id=pde2008_A000133&edition=current&q=optimization&topicid=&result_number=20" by John Geanakoplos.

[570] S. Koziel and J.W. Bandler, "Space mapping with multiple coarse models for optimization of microwave components," http://ieeexplore.ieee.org/xpl/abstractCitations.jsp?arnumber=4408469 IEEE Microwave and Wireless Components Letters, vol. 8, no. 1, pp. 1–3, Jan. 2008.

[571] S. Tu, Q.S. Cheng, Y. Zhang, J.W. Bandler, and N.K. Nikolova, "Space mapping optimization of handset antennas exploiting thin-wire models," http://ieeexplore.ieee.org/stamp/stamp.jsp?arnumber=06487389 IEEE Trans. Antennas Propag., vol. 61, no. 7, pp. 3797-3807, July 2013.]

[572] N. Friedrich, "Space mapping outpaces EM optimization in handset-antenna design," http://mwrf.com/software/space-mapping-outpaces-em-optimization-handset-antenna-design microwaves&rf, Aug. 30, 2013.

[573] Juan C. Cervantes-González, J. E. Rayas-Sánchez, C. A. López, J. R. Camacho-Pérez, Z. Brito-Brito, and J. L. Chavez-Hurtado, "Space mapping optimization of handset antennas considering EM effects of mobile phone components and human body," http//onlinelibrary.wiley.com Int. J. RF and Microwave CAE, vol. 26, no. 2, pp. 121-128, Feb. 2016

[574] J.W. Bandler, R.M. Biernacki, S.H. Chen, P.A. Grobelny, and R.H. Hemmers, "Space mapping technique for electromagnetic optimization," http://ieeexplore.ieee.org/xpl/articleDetails.jsp?arnumber=339794 IEEE Trans. Microwave Theory Tech., vol. 42, no. 12, pp. 2536-2544, Dec. 1994.

[575] J.W. Bandler, R.M. Biernacki, S.H. Chen, R.H. Hemmers, and K. Madsen, "Electromagnetic optimization exploiting aggressive space mapping," http://ieeexplore.ieee.org/xpls/abs_all.jsp?arnumber=475649&tag=1 IEEE Trans. Microwave Theory Tech., vol. 43, no. 12, pp. 2874-2882, Dec. 1995.

[576] Piryonesi, S. M., & Tavakolan, M. (2017). A mathematical programming model for solving cost-safety optimization (CSO) problems in the maintenance of structures. KSCE Journal of Civil Engineering, 1-10. https://link.springer.com/article/10.1007/s12205-017-0531-z

[577] Hegazy, T., (1999) Optimization of Resource Allocation and Leveling Using Genetic Algorithms, Journal of Construction Engineering and Management, ASCE, Vol. 125, No. 3, pp 167-175.

[578]http://apmonitor.com/me575/index.php/Main/BookChapters
[579]https://www.springer.com/gp/book/9783319775852
[580]http://www.wellesleycambridge.com/tocs/toc-appl
[581]//www.ams.org/mathscinet-getitem?mr=0870634
[582]//www.ams.org/mathscinet-getitem?mr=1104662
[583]//www.ams.org/mathscinet-getitem?mr=2571910
[584]https://mathscinet.ams.org/mathscinet-getitem?mr=868279
[585]//www.ams.org/mathscinet-getitem?mr=1105099
[586]//www.ams.org/mathscinet-getitem?mr=0544669
[587]https://www.springer.com/mathematics/applications/book/978-3-540-35445-1
[588]//doi.org/10.1007/978-3-540-35447-5
[589]//www.ams.org/mathscinet-getitem?mr=2265882
[590]//www.ams.org/mathscinet-getitem?mr=1756264
[591]http://www.stanford.edu/~boyd/cvxbook/bv_cvxbook.pdf
[592]http://www.ece.northwestern.edu/~nocedal/book/num-opt.html
[593]//www.ams.org/mathscinet-getitem?mr=2199043
[594]http://www.princeton.edu/~rvdb/LPbook/index.html
[595]https://mathscinet.ams.org/mathscinet-getitem?mr=868083
[596]//www.ams.org/mathscinet-getitem?mr=2552933
[597]https://mathscinet.ams.org/mathscinet-getitem?mr=745802
[598]https://books.google.com/books?id=3pL1B7WVYnAC&printsec=frontcover&source=gbs_ge_summary_r&cad=0#v=onepage&q&f=false
[599]https://www.amazon.co.uk/exec/obidos/ASIN/3110145421/qid%3D1000637090/202-7422833-6521417
[600]https://www.springer.com/mathematics/journal/10589
[601]https://www.novapublishers.com/catalog/product_info.php?products_id=6353
[602]http://www.journals.elsevier.com/journal-of-economic-dynamics-and-control/
[603]http://www.siam.org/journals/siopt.php
[604]http://www.siam.org/journals/siopt/policy.php
[605]http://www.siam.org/journals/sicon.php
[606]http://www.siam.org/journals/sicon/policy.php
[607]http://www.coin-or.org/
[608]http://plato.asu.edu/guide.html
[609]http://www.mat.univie.ac.at/%7Eneum/glopt.html
[610]http://glossary.computing.society.informs.org/
[611]http://www.mathprog.org/
[612]https://web.archive.org/web/20020822203831/http://www-fp.mcs.anl.gov/otc/GUIDE/index.html
[613]http://wiki.mcs.anl.gov/neos
[614]http://www.optimization-online.org
[615]http://www2.arnes.si/%7Eljc3m2/igor/links.html
[616]https://web.archive.org/web/20090204025624/http://see.stanford.edu/SEE/courseinfo.aspx?coll=2db7ced4-39d1-4fdb-90e8-364129597c87
[617]http://www.stanford.edu/~boyd/cvxbook
[618]https://www.informs.org/About-INFORMS/History-and-Traditions/OR-Methodologies/Mathematical-Programming
[619]John L. Pollock
[620]C. Hales, Thomas "Formal Proof" http://www.ams.org/notices/200811/tx081101370p.pdf, University of Pittsburgh. Retrieved on 2010-10-19
[621]"Automated Deduction (AD)" http://www.cs.cornell.edu/info/projects/nuprl/Intro/AutoDeduction/autoded.html, [*The Nature of PRL Project*]. Retrieved on 2010-10-19
[622]Martin Davis, "The Prehistory and Early History of Automated Deduction," in Automation of Reasoning, eds. Siekmann and Wrightson, vol. 1, 1-28 at p. 15
[623]"Principia Mathematica" http://plato.stanford.edu/entries/principia-mathematica/, at Stanford University. Retrieved 2010-10-19

[624]"The Logic Theorist and its Children" http://www.cs.swarthmore.edu/~eroberts/cs91/projects/ethics-of-ai/sec1_2.html. Retrieved 2010-10-18

[625]Shankar, Natarajan *Little Engines of Proof http://www.csl.sri.com/~shankar/*, Computer Science Laboratory, SRI International. Retrieved 2010-10-19

[626]https://arxiv.org/abs/1605.00723

[627] *The Boyer- Moore Theorem Prover http://www.cs.utexas.edu/~moore/best-ideas/nqthm/index.html.* Retrieved on 2010-10-23

[628]Harrison, John *HOL Light: an overview http://www.cl.cam.ac.uk/~jrh13/slides/tphols-18aug09/slides.pdf.* Retrieved 2010-10-23

[629] *Introduction to Coq http://coq.inria.fr/a-short-introduction-to-coq.* Retrieved 2010-10-23

[630] *Automated Reasoning http://plato.stanford.edu/entries/reasoning-automated/*, Stanford Encyclopedia. Retrieved 2010-10-10

[631]https://web.archive.org/web/20090310081227/http://www.csc.liv.ac.uk/~konev/iwil2008/

[632]http://www.eprover.org/EVENTS/es_series.html

[633]Beame & Fich 2002, p. 39.

[634]Knuth 1998, §6.5 ("Retrieval on Secondary Keys").

[635]Knuth 1998, §6.1 ("Sequential Searching").

[636]Knuth 1998, §6.2 ("Searching by Comparison of Keys").

[637]Knuth 1998, §6.3 (Digital Searching).

[638]Knuth 1998, §6.4, (Hashing).

[639]http://www.sciencedirect.com/science/article/pii/S0022000002918222

[640]//doi.org/10.1006/jcss.2002.1822

[641]http://sites.google.com/site/hantarto/quantum-computing/unsorted

[642]//en.wikipedia.org/w/index.php?title=Template:Programming_paradigms&action=edit

[643]Cordell Green. **Application of Theorem Proving to Problem Solving http://www.dtic.mil/get-tr-doc/pdf?AD=ADA459656** IJCAI 1969.

[644]J.M. Foster and E.W. Elcock. ABSYS 1: An Incremental Compiler for Assertions: an Introduction, Machine Intelligence 4, Edinburgh U Press, 1969, pp. 423–429

[645]Carl Hewitt. **Planner: A Language for Proving Theorems in Robots** IJCAI 1969.

[646]Pat Hayes. Computation and Deduction. In Proceedings of the 2nd MFCS Symposium. Czechoslovak Academy of Sciences, 1973, pp. 105–118.

[647]Robert Kowalski **Predicate Logic as a Programming Language** http://www.doc.ic.ac.uk/~rak/papers/IFIP%2074.pdf Memo 70, Department of Artificial Intelligence, Edinburgh University. 1973. Also in Proceedings IFIP Congress, Stockholm, North Holland Publishing Co., 1974, pp. 569–574.

[648]Robert Kowalski and Donald and Kuehner **Linear Resolution with Selection Function** http://www.doc.ic.ac.uk/~rak/papers/sl.pdf Artificial Intelligence, Vol. 2, 1971, pp. 227–60.

[649]Also appeared in

[650]Prakken, H. and Sartor, G., 2015. Law and logic: a review from an argumentation perspective http://www.cs.uu.nl/groups/IS/archive/henry/ReviewLogicAndLawRevised.pdf. Artificial Intelligence, 227, 214–245.

[651]Sergot, M.J., Sadri, F., Kowalski, R.A., Kriwaczek, F., Hammond, P. and Cory, H.T., 1986. The British Nationality Act as a logic program http://www.doc.ic.ac.uk/~rak/papers/British%20Nationality%20Act.pdf. Communications of the ACM, 29(5), 370–386.

[652]Shunichi Uchida and Kazuhiro Fuchi **Proceedings of the FGCS Project Evaluation Workshop** Institute for New Generation Computer Technology (ICOT). 1992.

[653]Joshua Hodas and Dale Miller. **Logic Programming in a Fragment of Intuitionistic Linear Logic http://repository.upenn.edu/cgi/viewcontent.cgi?article=1540&context=cis_reports**, Information and Computation, 1994, 110(2), 327–365.

[654]Naoki Kobayashi and Akinori Yonezawa. **Asynchronous communication model based on linear logic**, Formal Aspects of Computing, 1994, 279–294.

[655]http://redwood.cs.ttu.edu/~mgelfond/PAPERS/survey.pdf

[656]//doi.org/10.1016/0743-1066%2894%2990025-6

[657]http://www.doc.ic.ac.uk/~rak/papers/the%20early%20years.pdf

[658]//doi.org/10.1145/35043.35046

[659]http://www.sciencedirect.com/science/article/pii/016800729190068W

[660] https://www.ijcai.org/Proceedings/71/Papers/014%20A.pdf
[661] https://pdfs.semanticscholar.org/9993/ec68770faaab132da6945492b0e4ad07eb7b.pdf
[662] http://cmpe.emu.edu.tr/bayram/courses/531/forpresentation/p374-dantsin.pdf
[663] http://www.ida.liu.se/~ulfni/lpp/
[664] https://web.archive.org/web/20050828194751/http://vl.fmnet.info/logic-prog/
[665] http://liinwww.ira.uka.de/bibliography/LogicProgramming/
[666] http://www.logicprogramming.org/
[667] http://www.cs.kuleuven.be/~dtai/projects/ALP/TPLP/
[668] https://web.archive.org/web/20090108012954/http://www.mpprogramming.com/Cpp/
[669] http://www.mozart-oz.org/documentation/tutorial/node12.html
[670] http://www.pdc.dk/
[671] http://docs.racket-lang.org/racklog/
[672] //en.wikipedia.org/w/index.php?title=Template:Bayesian_statistics&action=edit
[673] M. Scanagatta, G. Corani, C. P. de Campos, and M. Zaffalon. Learning Treewidth-Bounded Bayesian Networks with Thousands of Variables. http://papers.nips.cc/paper/6232-learning-treewidth-bounded-bayesian-networks-with-thousands-of-variables In NIPS-16: Advances in Neural Information Processing Systems 29, 2016.
[674] Russell & Norvig 2003, p. 496.
[675] Russell & Norvig 2003, p. 499.
[676] D. Roth, On the hardness of approximate reasoning http://cogcomp.cs.illinois.edu/page/publication_view/5, IJCAI (1993)
[677] D. Roth, On the hardness of approximate reasoning http://cogcomp.cs.illinois.edu/papers/hardJ.pdf, Artificial Intelligence (1996)
[678] Christos L. Koumenides and Nigel R. Shadbolt. 2012. Combining link and content-based information in a Bayesian inference model for entity search. http://eprints.soton.ac.uk/342220 In Proceedings of the 1st Joint International Workshop on Entity-Oriented and Semantic Search (JIWES '12). ACM, New York, NY, USA, Article 3, 6 pages.
[679] http://www.eng.tau.ac.il/~bengal/BN.pdf
[680] //doi.org/10.1002/9780470061572.eqr089
[681] http://fuzzy.cs.uni-magdeburg.de/books/gm/
[682] http://www.csse.monash.edu.au/~dld/David.Dowe.publications.html#ComleyDowe2003
[683] http://www.csse.monash.edu.au/~dld/David.Dowe.publications.html#ComleyDowe2005
[684] http://www.csse.monash.edu.au/~dld/MML.html
[685] http://www.csse.monash.edu.au/~dld/Publications/2003/Comley+Dowe03_HICS2003_GeneralBayesianNetworksAsymmetricLanguages.pdf
[686] http://www.cambridge.org/9780521884389
[687] http//www.csse.monash.edu.au
[688] http://japan.elsevier.com/products/books/HPS.pdf
[689] http://www.agenarisk.com/resources/apps_bayesian_networks.pdf
[690] https://web.archive.org/web/20070927153751/https://www.dcs.qmul.ac.uk/~norman/papers/Combining%20evidence%20in%20risk%20analysis%20using%20BNs.pdf
[691] https://www.dcs.qmul.ac.uk/~norman/papers/Combining%20evidence%20in%20risk%20analysis%20using%20BNs.pdf
[692] https://books.google.com/books?id=TNYhnkXQSjAC&pg=PA120
[693] https://books.google.com/books?id=TNYhnkXQSjAC
[694] http://research.microsoft.com/research/pubs/view.aspx?msr_tr_id=MSR-TR-95-06
[695] //doi.org/10.1023/A%3A1009730122752
[696] https://web.archive.org/web/20060719171558/http://research.microsoft.com/research/pubs/view.aspx?msr_tr_id=MSR-TR-95-06
[697] http://www.kamran-karimi.com/pubs/khISMIS2000.pdf
[698] //doi.org/10.1007/s10044-004-0214-5
[699] //doi.org/10.1002/sim.3680
[700] //www.ncbi.nlm.nih.gov/pubmed/19630097
[701] http://www.dcs.qmul.ac.uk/~norman/papers/oprisk.pdf
[702] //doi.org/10.1111/j.1539-6924.2005.00641.x
[703] //www.ncbi.nlm.nih.gov/pubmed/16268944

[704] //doi.org/10.1016/0004-3702%2886%2990072-X
[705] http://aima.cs.berkeley.edu/
[706] http://pages.cs.wisc.edu/~dyer/cs540/handouts/charniak.pdf
[707] http://www.niedermayer.ca/papers/bayesian/bayes.html
[708] http://www.dcs.qmw.ac.uk/%7Enorman/BBNs/BBNs.htm
[709] http://princesofserendib.com/
[710] http://robotics.stanford.edu/~nodelman/papers/ctbn.pdf
[711] https://web.archive.org/web/20090923200511/http://wiki.syncleus.com/index.php/DANN%3ABayesian_Network
[712] http://videolectures.net/kdd07_neapolitan_lbn/
[713] http://www.biomedcentral.com/1471-2105/7/514/abstract
[714] http://www.labmedinfo.org/download/lmi339.pdf
[715] //en.wikipedia.org/w/index.php?title=Template:Machine_learning_bar&action=edit
[716] Thad Starner, Alex Pentland. Real-Time American Sign Language Visual Recognition From Video Using Hidden Markov Models http://www.cc.gatech.edu/~thad/p/031_10_SL/real-time-asl-recognition-from%20video-using-hmm-ISCV95.pdf. Master's Thesis, MIT, Feb 1995, Program in Media Arts
[717] B. Pardo and W. Birmingham. Modeling Form for On-line Following of Musical Performances http://www.cs.northwestern.edu/~pardo/publications/pardo-birmingham-aaai-05.pdf. AAAI-05 Proc., July 2005.
[718] Satish L, Gururaj BI (April 2003). " Use of hidden Markov models for partial discharge pattern classification http://ieeexplore.ieee.org/xpl/freeabs_all.jsp?arnumber=212242". *IEEE Transactions on Dielectrics and Electrical Insulation*.
[719] Boudaren et al. http://asp.eurasipjournals.com/content/pdf/1687-6180-2012-134.pdf, M. Y. Boudaren, E. Monfrini, W. Pieczynski, and A. Aissani, Dempster-Shafer fusion of multisensor signals in nonstationary Markovian context, EURASIP Journal on Advances in Signal Processing, No. 134, 2012.
[720] Lanchantin et al. http//ieeexplore.ieee.org, P. Lanchantin and W. Pieczynski, Unsupervised restoration of hidden non stationary Markov chain using evidential priors, IEEE Trans. on Signal Processing, Vol. 53, No. 8, pp. 3091-3098, 2005.
[721] Boudaren et al. http//ieeexplore.ieee.org, M. Y. Boudaren, E. Monfrini, and W. Pieczynski, Unsupervised segmentation of random discrete data hidden with switching noise distributions, IEEE Signal Processing Letters, Vol. 19, No. 10, pp. 619-622, October 2012.
[722] Sipos, I. Róbert. *Parallel stratified MCMC sampling of AR-HMMs for stochastic time series prediction.* In: Proceedings, 4th Stochastic Modeling Techniques and Data Analysis International Conference with Demographics Workshop (SMTDA2016), pp. 295-306. Valletta, 2016. PDF http://1drv.ms/b/s!ApL_0Av0YGDLglwEOv1aYAGbmQeL
[723] R. Paroli. et al., *Poisson hidden Markov models for time series of overdispersed insurance counts* http://actuaries.org/ASTIN/Colloquia/Porto_Cervo/Paroli_Redaelli_Spezia.pdf
[724] http://compbio.mit.edu/ChromHMM/
[725] Sotirios P. Chatzis, "Hidden Markov Models with Nonelliptically Contoured State Densities," IEEE Transactions on Pattern Analysis and Machine Intelligence, vol. 32, no. 12, pp. 2297-2304, Dec. 2010. http://ieeexplore.ieee.org/abstract/document/5551154/
[726]
[727]
[728] Sotirios P. Chatzis, Dimitrios Kosmopoulos, "Visual Workflow Recognition Using a Variational Bayesian Treatment of Multistream Fused Hidden Markov Models," IEEE Transactions on Circuits and Systems for Video Technology, vol. 22, no. 7, pp. 1076-1086, July 2012. http://ieeexplore.ieee.org/document/6164251/
[729] M. Lukosevicius, H. Jaeger (2009) Reservoir computing approaches to recurrent neural network training, Computer Science Review **3**: 127–149.
[730] https://arxiv.org/pdf/1004.5514
[731] //arxiv.org/abs/1004.5514
[732] http://adsabs.harvard.edu/abs/2010JPCM...22O4105T
[733] //doi.org/10.1088/0953-8984/22/41/414105
[734] http://www.cs.sjsu.edu/~stamp/RUA/HMM.pdf

[735] https://web.archive.org/web/20120415032315/http://www.ee.washington.edu/research/guptalab/publications/EMbookChenGupta2010.pdf
[736] http://www.comp.leeds.ac.uk/roger/HiddenMarkovModels/html_dev/main.html
[737] http://www.cs.brown.edu/research/ai/dynamics/tutorial/Documents/HiddenMarkovModels.html
[738] http://jedlik.phy.bme.hu/~gerjanos/HMM/node2.html
[739] http://www.eecis.udel.edu/~lliao/cis841s06/hmmtutorialpart1.pdf
[740] http://www.eecis.udel.edu/~lliao/cis841s06/hmmtutorialpart2.pdf
[741] http://videolectures.net/hltss2010_eisner_plm/video/2/
[742] http://www.cs.jhu.edu/~jason/papers/eisner.hmm.xls
[743] http://www.cs.ubc.ca/~murphyk/Software/HMM/hmm.html
[744] http://htk.eng.cam.ac.uk/
[745] https://cran.r-project.org/web/packages/HMM/index.html
[746] http://birc.au.dk/software/zipHMM
[747] http://www.ghmm.org/
[748] https://github.com/KommuSoft/jahmm
[749] http://www.kanungo.com/software/software.html
[750] http://hackage.haskell.org/cgi-bin/hackage-scripts/package/hmm
[751] http://www.haskell.org/
[752] http://gt2k.cc.gatech.edu/
[753] http://www.lwebzem.com/cgi-bin/courses/hidden_markov_model_online.cgi
[754] http://sourceforge.net/projects/cvhmm/
[755] https://cran.r-project.org/web/packages/depmixS4/index.html
[756] https://adrianulbona.github.io/hmm
[757] http://tuvalu.santafe.edu/~simon/styled-8/
[758] Mohinder S. Grewal and Angus P. Andrews http://ieeecss.org/CSM/library/2010/june10/11-HistoricalPerspectives.pdf
[759] http://www.trs-80.org/interview-jack-crenshaw/
[760] Stratonovich, R. L. (1959). *Optimum nonlinear systems which bring about a separation of a signal with constant parameters from noise*. Radiofizika, 2:6, pp. 892–901.
[761] Stratonovich, R. L. (1959). *On the theory of optimal non-linear filtering of random functions*. Theory of Probability and its Applications, 4, pp. 223–225.
[762] Stratonovich, R. L. (1960) *Application of the Markov processes theory to optimal filtering*. Radio Engineering and Electronic Physics, 5:11, pp. 1–19.
[763] Stratonovich, R. L. (1960). *Conditional Markov Processes*. Theory of Probability and its Applications, 5, pp. 156–178.
[764] Hamilton, J. (1994), *Time Series Analysis*, Princeton University Press. Chapter 13, 'The Kalman Filter'
[765] 2006 Corrected Version http://www.frc.ri.cmu.edu/~alonzo/pubs/reports/kalman_V2.pdf
[766] Three optimality tests with numerical examples are described in
[767] Jingyang Lu. "False information injection attack on dynamic state estimation in multi-sensor systems" http://ieeexplore.ieee.org/abstract/document/6916211/,, Fusion 2014
[768] Bucy, R.S. and Joseph, P.D., *Filtering for Stochastic Processes with Applications to Guidance,* John Wiley & Sons, 1968; 2nd Edition, AMS Chelsea Publ., 2005.
[769] Jazwinski, Andrew H., *Stochastic processes and filtering theory,* Academic Press, New York, 1970.
[770] http://www.intechopen.com/books/smoothing-filtering-and-prediction-estimating-the-past-present-and-future
[771] //doi.org/10.1016/j.automatica.2014.12.044
[772] https://web.archive.org/web/20080529105724/http://www.elo.utfsm.cl/~ipd481/Papers%20varios/kalman1960.pdf
[773] //doi.org/10.1115/1.3662552
[774] http://www.elo.utfsm.cl/~ipd481/Papers%20varios/kalman1960.pdf
[775] http://citeseerx.ist.psu.edu/viewdoc/summary?doi=10.1.1.361.6851
[776] //doi.org/10.1162/089976699300016674
[777] //www.ncbi.nlm.nih.gov/pubmed/9950734

[778] http://academic.csuohio.edu/simond/estimation/
[779] http://www.princeton.edu/~stengel/OptConEst.html
[780] http://www.informaworld.com/index/779885789.pdf
[781] //doi.org/10.1080/00207178708933989
[782] http://www.cs.unc.edu/~welch/media/pdf/scaat.pdf
[783] //doi.org/10.1145/258734.258876
[784] http://www.cs.unc.edu/~welch/kalman/media/pdf/maybeck_ch1.pdf
[785] https://www.researchgate.net/publication/216411106_Methods_for_Estimating_State_and_Measurement_Noise_Covariance_Matrices_Aspects_and_Comparison
[786] //doi.org/10.3182/20090706-3-FR-2004.00061
[787] http://pubs.acs.org/doi/abs/10.1021/ie9018116
[788] //doi.org/10.1021/ie9018116
[789] http://www.cs.unc.edu/~welch/kalman/kalmanPaper.html
[790] https://github.com/rlabbe/Kalman-and-Bayesian-Filters-in-Python
[791] http://www.bzarg.com/p/how-a-kalman-filter-works-in-pictures/
[792] http://www.eng.tau.ac.il/~liptser/lectures1/lect6.pdf
[793] https://www.youtube.com/watch?v=d0D3VwBh5UQ
[794] http://www.cs.unc.edu/~tracker/media/pdf/SIGGRAPH2001_CoursePack_08.pdf
[795] http://www.cs.unc.edu/~welch/kalman/
[796] http://www.siam.org/pdf/news/362.pdf
[797] //doi.org/10.1021/ie034308l
[798] http://netlib.org/a/esl.tgz
[799] http://www.mathworks.com/matlabcentral/fileexchange/32537
[800] http://eia.udg.es/~qsalvi/Slam.zip
[801] https://web.archive.org/web/20160304042652/http://www.cnel.ufl.edu/~weifeng/publication.htm
[802] http://www.mathfinance.cn/kalman-filter-finance-revisited/
[803] http://www.data-assimilation.net/Tools/AssimDemo/?method=KF
[804] //doi.org/10.3390/s110808164
[805] http://www.mathworks.com/discovery/kalman-filter.html
[806] http://blog.sciencenet.cn/home.php?mod=space&uid=1565&do=blog&id=851754
[807] https://www.laas.fr/public/en
[808] P. Del Moral, G. Rigal, and G. Salut. Estimation and nonlinear optimal control : An unified framework for particle solutions
LAAS-CNRS, Toulouse, Research Report no. 91137, DRET-DIGILOG- LAAS/CNRS contract, April (1991).
[809] P. Del Moral, G. Rigal, and G. Salut. Nonlinear and non Gaussian particle filters applied to inertial platform repositioning.
LAAS-CNRS, Toulouse, Research Report no. 92207, STCAN/DIGILOG-LAAS/CNRS Convention STCAN no. A.91.77.013, (94p.) September (1991).
[810] P. Del Moral, G. Rigal, and G. Salut. Estimation and nonlinear optimal control : Particle resolution in filtering and estimation. Experimental results.
Convention DRET no. 89.34.553.00.470.75.01, Research report no.2 (54p.), January (1992).
[811] P. Del Moral, G. Rigal, and G. Salut. Estimation and nonlinear optimal control : Particle resolution in filtering and estimation. Theoretical results
Convention DRET no. 89.34.553.00.470.75.01, Research report no.3 (123p.), October (1992).
[812] P. Del Moral, J.-Ch. Noyer, G. Rigal, and G. Salut. Particle filters in radar signal processing : detection, estimation and air targets recognition.
LAAS-CNRS, Toulouse, Research report no. 92495, December (1992).
[813] P. Del Moral, G. Rigal, and G. Salut. Estimation and nonlinear optimal control : Particle resolution in filtering and estimation.
Studies on: Filtering, optimal control, and maximum likelihood estimation. Convention DRET no. 89.34.553.00.470.75.01. Research report no.4 (210p.), January (1993).
[814] http://web.maths.unsw.edu.au/~peterdel-moral/mprfs.pdf
[815] https://www.springer.com/us/book/9780387202686#reviews
[816] https://www.crcpress.com/product/isbn/9781466504059

[817] http://www.people.fas.harvard.edu/~junliu/TechRept/98folder/liu&chen98_2.pdf
[818] //doi.org/10.1080/01621459.1998.10473765
[819] http://www.people.fas.harvard.edu/~junliu/TechRept/94folder/klw94.pdf
[820] //doi.org/10.1080/01621459.1994.10476469
[821] http://www.people.fas.harvard.edu/~junliu/TechRept/95folder/liu&chen95_s.pdf
[822] //doi.org/10.2307/2291068
[823] http://www.cs.ubc.ca/%7Earnaud/doucet_johansen_tutorialPF.pdf
[824] http://www.springerlink.com/content/q6452k2x37357l3r/
[825] //doi.org/10.1023/A%3A1008935410038
[826] http://ieeexplore.ieee.org/xpls/abs_all.jsp?arnumber=978374
[827] http://adsabs.harvard.edu/abs/2002ITSP...50..174A
[828] //doi.org/10.1109/78.978374
[829] //doi.org/10.1109/JPROC.2007.893250
[830] //doi.org/10.2307/1390750
[831] //www.jstor.org/stable/1390750
[832] http://www.mitre-corporation.net/work/tech_papers/tech_papers_05/05_0211/05_0211.pdf
[833] https://www.questia.com/PM.qst?a=o&se=gglsc&d=5002321997
[834] //doi.org/10.2307/2670179
[835] //www.jstor.org/stable/2670179
[836] http://ieeexplore.ieee.org/xpls/abs_all.jsp?arnumber=210672
[837] //doi.org/10.1049/ip-f-2.1993.0015
[838] //citeseerx.ist.psu.edu/viewdoc/summary?doi=10.1.1.107.7415
[839] //doi.org/10.1109/tpami.2007.1081
[840] http://web.maths.unsw.edu.au/~peterdel-moral/simulinks.html
[841] http://www-sigproc.eng.cam.ac.uk/smc/
[842] https://web.archive.org/web/20060612210237/http://www.cs.washington.edu/ai/Mobile_Robotics/mcl/
[843] http://blogs.oregonstate.edu/hess/code/particles/
[844] http://www.jstatsoft.org/v30/i06/
[845] http://www.oursland.net/projects/particlefilter/
[846] https://zhouyan.github.io/vSMC/
[847] https://www.youtube.com/watch?v=bO_GajDgGJ4/
[848] Steele, Katie and Stefánsson, H. Orri, "Decision Theory", The Stanford Encyclopedia of Philosophy (Winter 2015 Edition), Edward N. Zalta (ed.), URL = http://plato.stanford.edu/archives/win2015/entries/decision-theory
[849] Hansson, Sven Ove. "Decision theory: A brief introduction." (2005) Section 1.2: A truly interdisciplinary subject.
[850] For instance, see:
[851] Crozier, M. & Friedberg, E. 1995. "Organization and Collective Action. Our Contribution to Organizational Analysis" in Bacharach S.B, Gagliardi P. & Mundell P. (Eds). Research in the Sociology of Organizations. Vol. XIII, Special Issue on European Perspectives of Organizational Theory, Greenwich, CT: JAI Press.
[852] Rosenburgh, D. 2011. "Decision Theory with its Applications in Wireless Communication" in Zhang, Y. (Ed.), GUIZANI, M. (Ed.). (2011). Game Theory for Wireless Communications and Networking. Boca Raton: CRC Press.
[853] //www.ams.org/mathscinet-getitem?mr=0804611
[854] //www.ams.org/mathscinet-getitem?mr=1274699
[855] https://web.archive.org/web/20060705052730/http://www.infra.kth.se/~soh/decisiontheory.pdf
[856] http://www.infra.kth.se/~soh/decisiontheory.pdf
[857] //doi.org/10.1016/0028-3932%2885%2990022-3
[858] //www.ncbi.nlm.nih.gov/pubmed/4022303
[859] //doi.org/10.1016/0028-3932%2885%2990023-5
[860] //www.ncbi.nlm.nih.gov/pubmed/4022304
[861] //doi.org/10.1109/TSSC.1968.300114
[862] //doi.org/10.1007/0-387-71599-1

[863]//www.ams.org/mathscinet-getitem?mr=1835885
[864]http://psychclassics.yorku.ca/Peirce/small-diffs.htm
[865]https://web.archive.org/web/20030605070939/http://cepa.newschool.edu/het/texts/ramsey/ramsess.pdf
[866]http://www.numdam.org/item?id=AIHP_1937__7_1_1_0
[867]//en.wikipedia.org/w/index.php?title=Template:Utilitarianism&action=edit
[868]Von Neumann, J., Morgenstern, O. (1953). Theory of games and economic behavior, 3rd ed., Princeton University Press.
[869]Kahneman, D., Tversky, A. (1979). Prospect theory: An analysis of decision under risk. Econometrica, 47, 263-291.
[870]Grechuk, B., Zabarankin, M. (2016). Inverse portfolio problem with coherent risk measures https://www.researchgate.net/publication/283985485_Inverse_Portfolio_Problem_With_Coherent_Risk_Measures , European Journal of Operational Research 249(2), 740-750.
[871]Castagnoli, E. and M. LiCalzi. "Expected Utility Theory without Utility." Theory and Decision, 1996
[872]Bordley, R. and M. LiCalzi. "Decision Analysis with Targets instead of Utilities," Decisions in Economics and Finance. 2000.
[873]Bordley,R. And C.Kirkwood. Multiattribute preference analysis with Performance Targets. Operations Research. 2004.
[874]Paul H. Rubin and C. Monica Capra. The evolutionary psychology of economics. In
[875]//doi.org/10.2307/1891094
[876]//www.jstor.org/stable/1891094
[877]//doi.org/10.2307/1907266
[878]//www.jstor.org/stable/1907266
[879]http://www.investopedia.com/terms/u/utility.asp#axzz1eGQlDNsr
[880]http://www2.hawaii.edu/~fuleky/anatomy/anatomy.html
[881]http://www2.hawaii.edu/~fuleky/anatomy/anatomy2.html
[882]http://www.investopedia.com/university/economics/economics5.asp
[883]http://philpapers.org/rec/BRAMOO-2
[884]http://smallbusiness.chron.com/four-types-utility-marketing-24420.html
[885]http://www.businessdictionary.com/definition/form-utility.html
[886]http://www.businessdictionary.com/definition/place-utility.html
[887]http://www.businessdictionary.com/definition/time-utility.html
[888]http://www.allbusiness.com/glossaries/possession-utility/4957292-1.html
[889]http://www.witiger.com/marketing/5utilities.htm
[890]//en.wikipedia.org/w/index.php?title=Template:Machine_learning_bar&action=edit
[891]Fisher R.A. (1936) " The use of multiple measurements in taxonomic problems", *Annals of Eugenics*, 7, 179–188
[892]Fisher R.A. (1938) " The statistical utilization of multiple measurements", *Annals of Eugenics*, 8, 376–386
[893]Gnanadesikan, R. (1977) *Methods for Statistical Data Analysis of Multivariate Observations*, Wiley. (p. 83–86)
[894]Rao, C.R. (1952) *Advanced Statistical Methods in Multivariate Analysis*, Wiley. (Section 9c)
[895]Anderson, T.W. (1958) *An Introduction to Multivariate Statistical Analysis*, Wiley.
[896]Binder, D.A. (1978) "Bayesian cluster analysis", *Biometrika*, 65, 31–38.
[897]Binder, D.A. (1981) "Approximations to Bayesian clustering rules", *Biometrika*, 68, 275–285.
[898]Har-Peled, S., Roth, D., Zimak, D. (2003) "Constraint Classification for Multiclass Classification and Ranking." In: Becker, B., Thrun, S., Obermayer, K. (Eds) *Advances in Neural Information Processing Systems 15: Proceedings of the 2002 Conference*, MIT Press.
[899]//en.wikipedia.org/w/index.php?title=Template:Machine_learning_bar&action=edit
[900]J. Weng, N. Ahuja and T. S. Huang, " Cresceptron: a self-organizing neural network which grows adaptively http://www.cse.msu.edu/~weng/research/CresceptronIJCNN1992.pdf," *Proc. International Joint Conference on Neural Networks*, Baltimore, Maryland, vol I, pp. 576–581, June, 1992.

[901] J. Weng, N. Ahuja and T. S. Huang, " Learning recognition and segmentation of 3-D objects from 2-D images http://www.cse.msu.edu/~weng/research/CresceptronICCV1993.pdf," *Proc. 4th International Conf. Computer Vision*, Berlin, Germany, pp. 121–128, May, 1993.
[902] J. Weng, N. Ahuja and T. S. Huang, " Learning recognition and segmentation using the Cresceptron http://www.cse.msu.edu/~weng/research/CresceptronIJCV.pdf," *International Journal of Computer Vision*, vol. 25, no. 2, pp. 105–139, Nov. 1997.
[903] Dominik Scherer, Andreas C. Müller, and Sven Behnke: " Evaluation of Pooling Operations in Convolutional Architectures for Object Recognition https://www.ais.uni-bonn.de/papers/icann2010_maxpool.pdf," *In 20th International Conference Artificial Neural Networks (ICANN)*, pp. 92–101, 2010.
[904] S. Hochreiter., " Untersuchungen zu dynamischen neuronalen Netzen http://people.idsia.ch/~juergen/SeppHochreiter1991ThesisAdvisorSchmidhuber.pdf," *Diploma thesis. Institut f. Informatik, Technische Univ. Munich. Advisor: J. Schmidhuber*, 1991.
[905] J. Schmidhuber., "Learning complex, extended sequences using the principle of history compression," *Neural Computation*, 4, pp. 234–242, 1992.
[906] 2012 Kurzweil AI Interview http://www.kurzweilai.net/how-bio-inspired-deep-learning-keeps-winning-competitions with Jürgen Schmidhuber on the eight competitions won by his Deep Learning team 2009–2012
[907] Graves, Alex; and Schmidhuber, Jürgen; *Offline Handwriting Recognition with Multidimensional Recurrent Neural Networks http://www.idsia.ch/~juergen/nips2009.pdf*, in Bengio, Yoshua; Schuurmans, Dale; Lafferty, John; Williams, Chris K. I.; and Culotta, Aron (eds.), *Advances in Neural Information Processing Systems 22 (NIPS'22), 7–10 December 2009, Vancouver, BC*, Neural Information Processing Systems (NIPS) Foundation, 2009, pp. 545–552.
[908] J. Weng, " Why Have We Passed 'Neural Networks Do not Abstract Well'? http://www.cse.msu.edu/~weng/research/WhyPass-Weng-NI-2011.pdf," *Natural Intelligence: the INNS Magazine*, vol. 1, no.1, pp. 13–22, 2011.
[909] Z. Ji, J. Weng, and D. Prokhorov, " Where-What Network 1: Where and What Assist Each Other Through Top-down Connections http://www.cse.msu.edu/~weng/research/ICDL08_0077.pdf," *Proc. 7th International Conference on Development and Learning (ICDL'08)*, Monterey, CA, Aug. 9–12, pp. 1–6, 2008.
[910] X. Wu, G. Guo, and J. Weng, " Skull-closed Autonomous Development: WWN-7 Dealing with Scales http://www.cse.msu.edu/~weng/research/WWN7-Wu-ICBM-2013.pdf," *Proc. International Conference on Brain-Mind*, July 27–28, East Lansing, Michigan, pp. 1–9, 2013.
[911] Eiji Mizutani, Stuart Dreyfus, Kenichi Nishio (2000). On derivation of MLP backpropagation from the Kelley-Bryson optimal-control gradient formula and its application. Proceedings of the IEEE International Joint Conference on Neural Networks (IJCNN 2000), Como Italy, July 2000. Online http://queue.ieor.berkeley.edu/People/Faculty/dreyfus-pubs/ijcnn2k.pdf
[912] Arthur E. Bryson (1961, April). A gradient method for optimizing multi-stage allocation processes. In Proceedings of the Harvard Univ. Symposium on digital computers and their applications.
[913] Seppo Linnainmaa (1970). The representation of the cumulative rounding error of an algorithm as a Taylor expansion of the local rounding errors. Master's Thesis (in Finnish), Univ. Helsinki, 6–7.
[914] Paul Werbos (1974). Beyond regression: New tools for prediction and analysis in the behavioral sciences. PhD thesis, Harvard University.
[915] Eric A. Wan (1993). "Time series prediction by using a connectionist network with internal delay lines." In *Proceedings of the Santa Fe Institute Studies in the Sciences of Complexity*, **15**: p. 195. Addison-Wesley Publishing Co.
[916] ESANN. 2009
[917] Ting Qin, et al. "A learning algorithm of CMAC based on RLS." Neural Processing Letters 19.1 (2004): 49–61.
[918] Ting Qin, et al. "Continuous CMAC-QRLS and its systolic array." Neural Processing Letters 22.1 (2005): 1–16.
[919] LeCun *et al.*, "Backpropagation Applied to Handwritten Zip Code Recognition," *Neural Computation*, 1, pp. 541–551, 1989.
[920] Yann LeCun (2016). Slides on Deep Learning Online https://indico.cern.ch/event/510372/

[921] D. Graupe, "Principles of Artificial Neural Networks.3rd Edition", World Scientific Publishers, 2013, pp. 203–274.
[922] D. Graupe, "Principles of Artificial Neural Networks.3rd Edition", World Scientific Publishers", 2013, pp. 253–274.
[923] //www.google.com/patents/US5920852
[924] S. Das, C.L. Giles, G.Z. Sun, "Learning Context Free Grammars: Limitations of a Recurrent Neural Network with an External Stack Memory," Proc. 14th Annual Conf. of the Cog. Sci. Soc., p. 79, 1992.
[925] Grefenstette, Edward, et al. "Learning to Transduce with Unbounded Memory." https://arxiv.org/pdf/1506.02516.pdf (2015).
[926] Graves, Alex, Greg Wayne, and Ivo Danihelka. "Neural Turing Machines." (2014).
[927] Salakhutdinov, Ruslan, and Geoffrey Hinton. "Semantic hashing." http://www.utstat.toronto.edu/~rsalakhu/papers/sdarticle.pdf International Journal of Approximate Reasoning 50.7 (2009): 969–978.
[928] Weston, Jason, Sumit Chopra, and Antoine Bordes. "Memory networks." (2014).
[929] Sukhbaatar, Sainbayar, et al. "End-To-End Memory Networks." (2015).
[930] Bordes, Antoine, et al. "Large-scale Simple Question Answering with Memory Networks." (2015).
[931] Vinyals, Oriol, Meire Fortunato, and Navdeep Jaitly. "Pointer networks." (2015).
[932] Kurach, Karol, Andrychowicz, Marcin and Sutskever, Ilya. "Neural Random-Access Machines." (2015).
[933] Cho, Kyunghyun, Aaron Courville, and Yoshua Bengio. "Describing Multimedia Content using Attention-based Encoder–Decoder Networks." (2015).
[934] D. J. Felleman and D. C. Van Essen, " Distributed hierarchical processing in the primate cerebral cortex http://cercor.oxfordjournals.org/content/1/1/1.1.full.pdf+html," *Cerebral Cortex*, 1, pp. 1–47, 1991.
[935] J. Weng, " Natural and Artificial Intelligence: Introduction to Computational Brain-Mind https://www.amazon.com/Natural-Artificial-Intelligence-Introduction-Computational/dp/0985875720," BMI Press, , 2012.
[936] NASA – Dryden Flight Research Center – News Room: News Releases: NASA NEURAL NETWORK PROJECT PASSES MILESTONE http://www.nasa.gov/centers/dryden/news/NewsReleases/2003/03-49.html. Nasa.gov. Retrieved on 2013-11-20.
[937] Roger Bridgman's defence of neural networks http://members.fortunecity.com/templarseries/popper.html
[938] Sun and Bookman (1990)
[939] http://www.msm.cam.ac.uk/phase-trans/abstracts/neural.review.pdf
[940] //doi.org/10.2355/isijinternational.39.966
[941] https://www.worldcat.org/oclc/33101074
[942] //www.worldcat.org/oclc/33101074
[943] https://www.worldcat.org/oclc/76538146
[944] //www.worldcat.org/oclc/76538146
[945] https://books.google.com/books?id=4RtVAAAAMAAJ&pg=PA303
[946] http://actcomm.dartmouth.edu/gvc/papers/approx_by_superposition.pdf
[947] https://www.worldcat.org/oclc/35558945
[948] //www.worldcat.org/oclc/35558945
[949] https://www.worldcat.org/oclc/41347061
[950] //www.worldcat.org/oclc/41347061
[951] //doi.org/10.1016/S0031-3203%2801%2900178-9
[952] http://www.cs.iastate.edu/~honavar/fahlman.pdf
[953] https://www.worldcat.org/oclc/37875698
[954] //www.worldcat.org/oclc/37875698
[955] https://www.worldcat.org/oclc/38908586
[956] //www.worldcat.org/oclc/38908586
[957] https://www.worldcat.org/oclc/21522159
[958] //www.worldcat.org/oclc/21522159
[959] //www.worldcat.org/oclc/52377690

[960]https://www.worldcat.org/oclc/837524179
[961]//www.worldcat.org/oclc/837524179
[962]https://www.worldcat.org/oclc/32179420
[963]//www.worldcat.org/oclc/32179420
[964]http://www.inference.phy.cam.ac.uk/itprnn/book.pdf
[965]https://www.worldcat.org/oclc/29877717
[966]//www.worldcat.org/oclc/29877717
[967]https://books.google.com/books?id=m12UR8QmLqoC
[968]https://pdfs.semanticscholar.org/861e/de32115d157e1568622b153e7ed3dca28467.pdf
[969]//doi.org/10.1016/0304-3975%2894%2990178-3
[970]https://www.worldcat.org/oclc/27145760
[971]//www.worldcat.org/oclc/27145760
[972]https://www.worldcat.org/oclc/27429729
[973]//www.worldcat.org/oclc/27429729
[974]http://www.dkriesel.com/en/science/neural_networks
[975]http://deeplearning4j.org/neuralnet-overview.html
[976]http://people.revoledu.com/kardi/tutorial/NeuralNetwork/index.html
[977]https://www.youtube.com/watch?v=q0pm3BrIUFo
[978]https://www.academia.edu/25708860/A_Concise_Introduction_to_Machine_Learning_with_Artificial_Neural_Networks
[979]https://www.coursera.org/course/neuralnets
[980]http://www.deeplearningbook.org/
[981]https://nnplayground.com
[982]http://numerentur.org/redes-neuronales-artificiales/
[983]https://plato.stanford.edu/entries/connectionism/
[984]//en.wikipedia.org/w/index.php?title=Connectionism&action=edit
[985]http://www.blutner.de/NeuralNets/Texts/Medler.pdf
[986]http://philosophy.uwaterloo.ca/MindDict/connectionism.html
[987]https://plato.stanford.edu/entries/connectionism/
[988]http://srsc.ulb.ac.be/pdp/iac/IAC.html
[989]http://www.iep.utm.edu/connect
[990]//en.wikipedia.org/w/index.php?title=Template:Machine_learning_bar&action=edit
[991]Jürgen Schmidhuber (2015). Deep Learning. Scholarpedia, 10(11):32832. Online http://www.scholarpedia.org/article/Deep_Learning
[992]Balázs Csanád Csáji (2001). Approximation with Artificial Neural Networks; Faculty of Sciences; Eötvös Loránd University, Hungary
[993]Rina Dechter (1986). Learning while searching in constraint-satisfaction problems. University of California, Computer Science Department, Cognitive Systems Laboratory. Online https://www.researchgate.net/publication/221605378_Learning_While_Searching_in_Constraint-Satisfaction-Problems
[994]Igor Aizenberg, Naum N. Aizenberg, Joos P.L. Vandewalle (2000). Multi-Valued and Universal Binary Neurons: Theory, Learning and Applications. Springer Science & Business Media.
[995]Co-evolving recurrent neurons learn deep memory POMDPs. Proc. GECCO, Washington, D. C., pp. 1795-1802, ACM Press, New York, NY, USA, 2005.
[996]Seppo Linnainmaa (1970). The representation of the cumulative rounding error of an algorithm as a Taylor expansion of the local rounding errors. Master's Thesis (in Finnish), Univ. Helsinki, 6-7.
[997]LeCun *et al.*, "Backpropagation Applied to Handwritten Zip Code Recognition," *Neural Computation*, 1, pp. 541–551, 1989.
[998]J. Weng, N. Ahuja and T. S. Huang, " Cresceptron: a self-organizing neural network which grows adaptively http://www.cse.msu.edu/~weng/research/CresceptronIJCNN1992.pdf," *Proc. International Joint Conference on Neural Networks*, Baltimore, Maryland, vol I, pp. 576-581, June, 1992.
[999]J. Weng, N. Ahuja and T. S. Huang, " Learning recognition and segmentation of 3-D objects from 2-D images http://www.cse.msu.edu/~weng/research/CresceptronICCV1993.pdf," *Proc. 4th International Conf. Computer Vision*, Berlin, Germany, pp. 121-128, May, 1993.

[1000] J. Weng, N. Ahuja and T. S. Huang, " Learning recognition and segmentation using the Cresceptron http://www.cse.msu.edu/~weng/research/CresceptronIJCV.pdf," *International Journal of Computer Vision*, vol. 25, no. 2, pp. 105-139, Nov. 1997.

[1001] S. Hochreiter., " Untersuchungen zu dynamischen neuronalen Netzen http://people.idsia.ch/~juergen/SeppHochreiter1991ThesisAdvisorSchmidhuber.pdf," *Diploma thesis. Institut f. Informatik, Technische Univ. Munich. Advisor: J. Schmidhuber*, 1991.

[1002] Santiago Fernandez, Alex Graves, and Jürgen Schmidhuber (2007). An application of recurrent neural networks to discriminative keyword spotting. Proceedings of ICANN (2), pp. 220–229.

[1003] G. E. Hinton., " Learning multiple layers of representation http://www.csri.utoronto.ca/~hinton/absps/ticsdraft.pdf," *Trends in Cognitive Sciences*, 11, pp. 428–434, 2007.

[1004] Yann LeCun (2016). Slides on Deep Learning Online https://indico.cern.ch/event/510372/

[1005] D. Yu, L. Deng, G. Li, and F. Seide (2011). "Discriminative pretraining of deep neural networks," U.S. Patent Filing.

[1006] Chellapilla, K., Puri, S., and Simard, P. (2006). High performance convolutional neural networks for document processing. International Workshop on Frontiers in Handwriting Recognition.

[1007] "Toxicology in the 21st century Data Challenge"

[1008] Ting Qin, et al. "A learning algorithm of CMAC based on RLS." Neural Processing Letters 19.1 (2004): 49-61.

[1009] Ting Qin, et al. "Continuous CMAC-QRLS and its systolic array." Neural Processing Letters 22.1 (2005): 1-16.

[1010] *TIMIT Acoustic-Phonetic Continuous Speech Corpus* Linguistic Data Consortium, Philadelphia.

[1011] Nvidia Demos a Car Computer Trained with "Deep Learning" http://www.technologyreview.com/news/533936/nvidia-demos-a-car-computer-trained-with-deep-learning/ (2015-01-06), David Talbot, *MIT Technology Review*

[1012] Brocardo ML, Traore I, Woungang I, Obaidat MS. " Authorship verification using deep belief network systems http://onlinelibrary.wiley.com/doi/10.1002/dac.3259/full". Int J Commun Syst. 2017. doi:10.1002/dac.3259

[1013] "An Infusion of AI Makes Google Translate More Powerful Than Ever." Cade Metz, WIRED, Date of Publication: 09.27.16. https://www.wired.com/2016/09/google-claims-ai-breakthrough-machine-translation/

[1014] S. Blakeslee., "In brain's early growth, timetable may be critical," *The New York Times, Science Section*, pp. B5–B6, 1995.

[1015] Miller, G. A., and N. Chomsky. "Pattern conception." Paper for Conference on pattern detection, University of Michigan. 1957.

[1016] //en.wikipedia.org/w/index.php?title=Template:Machine_learning_bar&action=edit

[1017] Page 150 ff demonstrates credit assignment across the equivalent of 1,200 layers in an unfolded RNN.

[1018] Bo Fan, Lijuan Wang, Frank K. Soong, and Lei Xie (2015). Photo-Real Talking Head with Deep Bidirectional LSTM. In Proceedings of ICASSP 2015.

[1019] Seppo Linnainmaa (1970). The representation of the cumulative rounding error of an algorithm as a Taylor expansion of the local rounding errors. Master's Thesis (in Finnish), Univ. Helsinki, 6-7.

[1020] Cruse, Holk; *Neural Networks as Cybernetic Systems* http://www.brains-minds-media.org/archive/615/bmm615.pdf, 2nd and revised edition

[1021] W. Maass, T. Natschläger, and H. Markram. A fresh look at real-time computation in generic recurrent neural circuits. Technical report, Institute for Theoretical Computer Science, TU Graz, 2002.

[1022] Sepp Hochreiter (1991), Untersuchungen zu dynamischen neuronalen Netzen http://people.idsia.ch/~juergen/SeppHochreiter1991ThesisAdvisorSchmidhuber.pdf, Diploma thesis. Institut f. Informatik, Technische Univ. Munich. Advisor: J. Schmidhuber.

[1023] C.L. Giles, C.B. Miller, D. Chen, H.H. Chen, G.Z. Sun, Y.C. Lee, "Learning and Extracting Finite State Automata with Second-Order Recurrent Neural Networks" https://clgiles.ist.psu.edu/pubs/NC1992-recurrent-NN.pdf, Neural Computation, 4(3), p. 393, 1992.

[1024] C.W. Omlin, C.L. Giles, "Constructing Deterministic Finite-State Automata in Recurrent Neural Networks" http://citeseerx.ist.psu.edu/viewdoc/download?doi=10.1.1.32.2364&rep=rep1&type=pdf Journal of the ACM, 45(6), 937-972, 1996.

[1025] https://pytorch.org/

[1026] http://caffe.berkeleyvision.org/

[1027] http://deeplearning4j.org/

[1028] https://github.com/deeplearning4j/libnd4j

[1029] http://dynet.io/

[1030] http://www.torch.ch/

[1031] A. Graves, J. Schmidhuber. Offline Handwriting Recognition with Multidimensional Recurrent Neural Networks. Advances in Neural Information Processing Systems 22, NIPS'22, pp 545–552, Vancouver, MIT Press, 2009.

[1032] M. Baccouche, F. Mamalet, C Wolf, C. Garcia, A. Baskurt. Sequential Deep Learning for Human Action Recognition. 2nd International Workshop on Human Behavior Understanding (HBU), A.A. Salah, B. Lepri ed. Amsterdam, Netherlands. pp. 29–39. Lecture Notes in Computer Science 7065. Springer. 2011

[1033] https://github.com/zhongkaifu/RNNSharp

[1034] http://www.idsia.ch/~juergen/rnn.html

[1035] http://jsalatas.ictpro.gr/weka

[1036] http://deeplearning4j.org/recurrentnetwork

[1037] https://gammastorm.github.io/TheBrain/Brain.html

[1038] AI set to exceed human brain power http://www.cnn.com/2006/TECH/science/07/24/ai.bostrom/ CNN.com (July 26, 2006)

[1039] under "Artificial Intelligence in the 90s"

[1040] Approximate year AI started beating top human experts

[1041] Watson beats Jeopardy grand-champions. https://www.nytimes.com/2011/02/17/science/17jeopardy-watson.html

[1042] Computer bridge#Computers versus humans

[1043] Proverb: The probabilistic cruciverbalist. By Greg A. Keim, Noam Shazeer, Michael L. Littman, Sushant Agarwal, Catherine M. Cheves, Joseph Fitzgerald, Jason Grosland, Fan Jiang, Shannon Pollard, and Karl Weinmeister. 1999. In Proceedings of the Sixteenth National Conference on Artificial Intelligence, 710-717. Menlo Park, Calif.: AAAI Press.

[1044] Antol, Stanislaw, et al. "Vqa: Visual question answering." Proceedings of the IEEE International Conference on Computer Vision. 2015.

[1045] Brynjolfsson, E., & Mitchell, T. (2017). What can machine learning do? Workforce implications. Science, 358(6370), 1530-1534.

[1046] There are several ways of evaluating machine translation systems. People competent in a second language frequently outperform machine translation systems but the average person is often less capable. Some machine translation systems are capable of a large number of languages, like google translate, and as a result have a broader competence than most humans. For example, very few humans can translate from Arabic to Polish *and* French to Swahili *and* Armenian to Vietnamese. When comparing over several languages machine translation systems will tend to outperform humans.

[1047] Müller, V. C., & Bostrom, N. (2016). Future progress in artificial intelligence: A survey of expert opinion. In Fundamental issues of artificial intelligence (pp. 555-572). Springer, Cham.

[1048]

[1049] Muehlhauser, L., & Salamon, A. (2012). Intelligence explosion: Evidence and import. In Singularity Hypotheses (pp. 15-42). Springer, Berlin, Heidelberg.

[1050] https://aiimpacts.org/miri-ai-predictions-dataset/

[1051] United States, National Science and Technology Council – Committee on Technology. Executive Office of the President. (2016). Preparing for the future of artificial intelligence.

[1052] AIDA Homepage http://www.kbs.twi.tudelft.nl/Research/Projects/AIDA/. Kbs.twi.tudelft.nl (April 17, 1997). Retrieved on 2013-07-21.

[1053] The Story of Self-Repairing Flight Control Systems. http://crgis.ndc.nasa.gov/crgis/images/c/c9/88798main_srfcs.pdf NASA Dryden. (April 2003). Retrieved on 2016-08-25.

[1054] http://www.cse.unsw.edu.au/~claude/papers/ilp92.pdf

[1055] https://www.kensho.com/#/
[1056] Marco Costantino, Paolo Coletti, Information Extraction in Finance, Wit Press, 2008.
[1057] https://www.eyediagnosis.net/single-post/2018/04/12/FDA-permits-marketing-of-IDx-DR-for-automated-detection-of-diabetic-retinopathy-in-primary-care
[1058] http://www.pomato.com/
[1059] https://www.cooperativecomputing.com/blog/ai-taking-world/
[1060] https://www.hirevue.com
[1061] US Patent #7696426 https://www.google.com/patents/US7696426
[1062] SACEM Database, https://repertoire.sacem.fr/resultats?filters=parties&query=aiva&nbWorks=20)
[1063] business intelligence solutions http://www.narrativescience.com/solutions.html . Narrative Science. Retrieved on 2013-07-21.
[1064] https://archive.is/20131116104250/http://yseop.com/EN/home.php
[1065] Implementing an online help desk system based on [[conversational agent http://portal.acm.org/citation.cfm?id=1643823.1643908]] Authors: Alisa Kongthon, Chatchawal Sangkeet-trakarn, Sarawoot Kongyoung and Choochart Haruechaiyasak. Published by ACM 2009 Article, Bibliometrics Data Bibliometrics. Published in: Proceeding, MEDES '09 Proceedings of the International Conference on Management of Emergent Digital EcoSystems, ACM New York, NY, USA. ,
[1066] https://digitalgenius.com/
[1067] http://www.ipsoft.com/
[1068] https://www.inbenta.com/en
[1069] Success Stories http://www.theorsociety.com/Science_of_Better/htdocs/prospect/can_do/success_stories/dwsbt.htm .
[1070] http://aima.cs.berkeley.edu/
[1071] //www.worldcat.org/oclc/246584055
[1072] http://aitopics.org/topic/applications
[1073] http://www.buzzle.com/articles/applications-of-artificial-intelligence.html
[1074] //en.wikipedia.org/w/index.php?title=Lethal_autonomous_weapon&action=edit
[1075] http://www.newsweek.com/russia-military-challenge-us-china-missile-own-decisions-639926
[1076] https://www.rbth.com/defence/2017/08/09/comrade-in-arms-russia-is-developing-a-freethinking-war-machine_819686
[1077] https://www.rbth.com/defence/2017/06/06/rise-of-the-machines-a-look-at-russias-latest-combat-robots_777480
[1078] https://www.rbth.com/science_and_tech/2016/02/10/is-terminator-back-russians-make-major-advances-in-artificial-intelligence_566553
[1079] https://www.rbth.com/news/2017/05/15/virtual-trainer-for-robots-and-drones-developed-in-russia_763016
[1080] http://www.jpost.com/Israel-News/Politics-And-Diplomacy/Kara-I-wasnt-revealing-state-secrets-about-the-robots-482616
[1081] https://www.cnbc.com/2017/03/17/mini-nukes-and-inspect-bot-weapons-being-primed-for-future-warfare.html
[1082] https://sputniknews.com/asia/201807221066595641-china-large-AI-umanned-subs/
[1083] http://philpapers.org/rec/MLLAKR
[1084] https://www.cs.berkeley.edu/~russell/research/LAWS.html
[1085] http://philpapers.org/rec/SIMJWA
[1086] "Robot Television Bomber" https://books.google.com/books?id=19kDAAAAMBAJ&pg=PA805 *Popular Mechanics* June 1940
[1087] Mikoyan-Gurevitch Skat in Aviation Week http://www.aviationweek.com/aw/generic/story.jsp?id=news/MIG082307.xml&headline=First%20Look%20At%20MiG%20Skat%20UCAV%20%5BUpdated%5D&channel=null
[1088] Farmer, Ben (6 February 2014) Successful test flight for Taranis stealth drone https://www.telegraph.co.uk/news/uknews/defence/10619378/Successful-test-flight-for-Taranis-stealth-drone.html Daily Telegraph, Page 12

[1089]"Pentagon Sets Plan For New Bomber, Terminates J-UCAS Program" http://www.globalsecurity.org/org/news/2006/060113-j-ucas-terminated.htm, by Jason Sherman, GlobalSecurity.org, 13 January 2006

[1090]"Carrier UCAVs: The Return of UCAS" http://www.defenseindustrydaily.com/cv-ucavs-the-return-of-ucas-03557/, Defense Industry Daily, 7 February 2010

[1091]Editorial https://www.nytimes.com/2011/01/02/opinion/02sun2.html?hp, *The New York Times*, January 1, 2011 (January 2, 2011 p. WK7 NY ed.). Retrieved 2011-01-02.

[1092](14 July 2014) Israel downs Gaza drone as suspects in Palestinian's killing appear in court https://www.theglobeandmail.com/news/world/israel-gaza/article19583865/ The Globe and Mail/ Associated Press, Retrieved 18 July 2014

[1093]Zitun, Yoav (14 July 2014) Hamas claims multiple UAVs launched into Israel http://www.ynetnews.com/articles/0,7340,L-4543077,00.html YNet News, Retrieved 18 July 2014

[1094]Karayel UCAV MAM L Firing – Vestel Defence Industry https://www.youtube.com/watch?v=5NBS04ckgps 16.06.2016: Vestel UCAV's firing test video 1

[1095]Karayel UCAV MAM-L Firing Test – Vestel Defence Industry https://www.youtube.com/watch?v=yyOtd7jd7ic 25.06.2016: Vestel UCAV's firing test video 2

[1096]*The Guardian*, 23 March 2009. "Cut to pieces: the Palestinian family drinking tea in their courtyard: Israeli unmanned aerial vehicles – the dreaded drones – caused at least 48 deaths in Gaza during the 23-day offensive." https://www.theguardian.com/world/2009/mar/23/gaza-war-crimes-drones Retrieved on 3 August 2009.

[1097]Anatomy of an Afghan war tragedy http://www.latimes.com/news/nationworld/world/la-fg-afghanistan-drone-20110410,0,2818134,full.story, David S. Cloud, *Los Angeles Times*, April 10, 2011

[1098]Drone operators blamed in airstrike that killed Afghan civilians in February https://www.washingtonpost.com/wp-dyn/content/article/2010/05/29/AR2010052901390.html, Karin Brulliard, *The Washington Post*, Sunday, May 30, 2010

[1099]http://www.dawn.com/wps/wcm/connect/dawn-content-library/dawn/news/world/13+drones+kill+10+civilians+for+one+militant-za-09, Dawn (newspaper), 21 July 2009

[1100]*Newsweek*, 8 July 2009. Anita Kirpalani, "Drone On. Q&A: A former Pakistani diplomat says America's most useful weapon is hurting the cause in his country" http://www.newsweek.com/id/205725 Retrieved on 3 August 2009.

[1101]Michael Isikoff: Ex-Pentagon official has 'heavy heart' over US teen's inadvertent killing by drone http://investigations.nbcnews.com/_news/2013/07/18/19546292-ex-pentagon-official-has-heavy-heart-over-us-teens-inadvertent-killing-by-drone

[1102][Yemeni] Official: U.S. drone attack in Yemen kills wedding guests http://www.cbsnews.com/news/official-us-drone-attack-in-yemen-kills-wedding-guests/ – AP, December 12, 2013

[1103]Cole, Jim and Chris Wright. "Drone Wars UK". January 2010. http://dronewarsuk.wordpress.com/aboutdrone/

[1104]Salama, Vivian. "Death From Above: How American Drone Strikes Are Devastating Yemen". April 2014. https://www.rollingstone.com/politics/news/death-from-above-how-american-drone-strikes-are-devastating-yemen-20140414#ixzz3YocHjBIz

[1105]Chatterjee, Pratap (March 2015). Is Drone Warfare Fraying at the Edges? http://www.tomdispatch.com/post/175964/tomgram%3A_pratap_chatterjee%2C_is_drone_warfare_fraying_at_the_edges/. "Are Pilots Deserting Washington's Remote-Control War? A New Form of War May Be Producing a New Form of Mental Disturbance."

[1106]Fairleigh Dickinson University's PublicMind, (February 7, 2013). Public says it's illegal to target Americans abroad as some question CIA drone attacks http://www.publicmind.fdu.edu/2013/drone/ (press release)

[1107]Allen, Greg, and Taniel Chan. "Artificial Intelligence and National Security." Report. Harvard Kennedy School, Harvard University. Boston, MA (2017).

[1108]Antebi, Liran. "Who Will Stop the Robots?." Military and Strategic Affairs 5.2 (2013).

[1109]Shulman, C., & Armstrong, S. (2009, July). Arms control and intelligence explosions. In 7th European Conference on Computing and Philosophy (ECAP), Bellaterra, Spain, July (pp. 2-4).

[1110]Coiera, E. (1997). Guide to medical informatics, the Internet and telemedicine. Chapman & Hall, Ltd..

[1111] Power, B. (March 19, 2015). Artificial Intelligence Is Almost Ready for Business. Retrieved from https://hbr.org/2015/03/artificial-intelligence-is-almost-ready-for-business
[1112] Bloch-Budzier, S. (November 22, 2016). NHS using Google technology to treat patients. Retrieved from https://www.bbc.com/news/health-38055509
[1113] Lorenzetti, L. (April 5, 2016). Here's How IBM Watson Health is Transforming the Health Care Industry. Retrieved from http://fortune.com/ibm-watson-health-business-strategy/
[1114] CB Insights Artificial Intelligence report. (June 28, 2016). Retrieved from https://www.cbinsights.com/reports/CB-Insights-Artificial-Intelligence-Webinar.pdf
[1115] Lindsay, R. K., Buchanan, B. G., Feigenbaum, E. A., & Lederberg, J. (1993). DENDRAL: a case study of the first expert system for scientific hypothesis formation. Artificial intelligence, 61(2), 209-261.
[1116] Clancey, W. J., & Shortliffe, E. H. (1984). Readings in medical artificial intelligence: the first decade. Addison-Wesley Longman Publishing Co., Inc..
[1117] Bruce, G., Buchanan, B. G., & Shortliffe, E. D. (1984). Rule-based expert systems: the MYCIN experiments of the Stanford Heuristic Programming Project.
[1118] Duda, R. O., & Shortliffe, E. H. (1983). Expert systems research. Science, 220(4594), 261-268.
[1119] Miller, R. A. (1994). Medical diagnostic decision support systems—past, present, and future. Journal of the American Medical Informatics Association, 1(1), 8-27.
[1120] Adlassnig KP. A fuzzy logical model of' computer-assisted medical diagnosis. Methods Inf Med. 1980; 19:14
[1121] Reggia, J. A., & Peng, Y. (1987). Modeling diagnostic reasoning: a summary of parsimonious covering theory. Computer methods and programs in biomedicine, 25(2), 125-134.
[1122] Baxt, W. G. (1991). Use of an artificial neural network for the diagnosis of myocardial infarction. Annals of Internal Medicine, 115(11), 843-848.
[1123] Maclin, P. S., Dempsey, J., Brooks, J., & Rand, J. (1991). Using neural networks to diagnose cancer. Journal of Medical Systems, 15(1), 11-19.
[1124] Koomey, J., Berard, S., Sanchez, M., & Wong, H. (2011). Implications of historical trends in the electrical efficiency of computing. IEEE Annals of the History of Computing, 33(3), 46-54.
[1125] Dinov, I. D. (2016). Volume and value of big healthcare data. Journal of medical statistics and informatics, 4.
[1126] Barnes, B., & Dupré, J. (2009). Genomes and what to make of them. University of Chicago Press.
[1127] Jha, A. K., DesRoches, C. M., Campbell, E. G., Donelan, K., Rao, S. R., Ferris, T. G., ... & Blumenthal, D. (2009). Use of electronic health records in US hospitals. New England Journal of Medicine, 360(16), 1628-1638.
[1128] Banko, M., & Brill, E. (2001, July). Scaling to very very large corpora for natural language disambiguation. In Proceedings of the 39th annual meeting on association for computational linguistics (pp. 26-33). Association for Computational Linguistics.
[1129] Dougherty, G. (2009). Digital image processing for medical applications. Cambridge University Press.
[1130] "Artificial Intelligence and Machine Learning for Healthcare" Sigmoidal, December 21, 2017. https://sigmoidal.io/artificial-intelligence-and-machine-learning-for-healthcare/.
[1131] Cohn, Jonathan. "The Robot Will See You Now." The Atlantic, March 2013. https://www.theatlantic.com/magazine/archive/2013/03/the-robot-will-see-you-now/309216/.
[1132] Spear, Andrew. "From Cancer to Consumer Tech: A Look Inside IBM's Watson Health Strategy." Fortune, April 5, 2016. http://fortune.com/ibm-watson-health-business-strategy/.
[1133] Bass, Dina. "Microsoft Develops AI to Help Cancer Doctors Find the Right Treatments." Bloomberg, September 20, 2016. https://www.bloomberg.com/news/articles/2016-09-20/microsoft-develops-ai-to-help-cancer-doctors-find-the-right-treatments.
[1134] Knapton, Sarah. "Microsoft Will 'Solve' Cancer within 10 Years by 'Reprogramming' Diseased Cells." The Telegraph, September 20, 2016. https://www.telegraph.co.uk/science/2016/09/20/microsoft-will-solve-cancer-within-10-years-by-reprogramming-dis/.
[1135] Bloch-Budzier, Sarah. "NHS Using Google Technology to Treat Patients." BBC News, November 22, 2016. https://www.bbc.com/news/health-38055509.
[1136] Lee, Chris Baraniuk, Dave. "Google DeepMind Targets NHS Head and Neck Cancer Treatment." BBC News, August 31, 2016. https://www.bbc.com/news/technology-37230806.

[1137] Primack, Dan. "Intel Capital Cancels $1 Billion Portfolio Sale." Fortune, May 26, 2016. http://fortune.com/2016/05/26/intel-capital-cancels-1-billion-portfolio-sale/.

[1138] https://www.eyediagnosis.net/idx-dr

[1139] https://en.wikipedia.org/wiki/Michael_Abramoff

[1140] https://www.eyediagnosis.net/single-post/2018/04/12/FDA-permits-marketing-of-IDx-DR-for-automated-detection-of-diabetic-retinopathy-in-primary-care

[1141] https://www.eyediagnosis.net/

[1142] Medvice Official Website http://www.medvice.eu

[1143] Hernandez, Daniela. "Artificial Intelligence Is Now Telling Doctors How to Treat You." WIRED, June 2, 2014. https://www.wired.com/2014/06/ai-healthcare/.

[1144] Felten, Ed. "Preparing for the Future of Artificial Intelligence." Whitehouse.gov, May 3, 2016. https://www.whitehouse.gov/blog/2016/05/03/preparing-future-artificial-intelligence.

[1145] "The National Artificial Intelligence Research and Development Strategic Plan." Office of Science and Technology Policy, October 16, 2016. https://www.nitrd.gov/PUBS/national_ai_rd_strategic_plan.pdf.

[1146] Office of the Press Secretary. "At Cancer Moonshot Summit, Vice President Biden Announces New Actions to Accelerate Progress Toward Ending Cancer As We Know It." Whitehouse.gov, June 28, 2016. https://www.whitehouse.gov/the-press-office/2016/06/28/fact-sheet-cancer-moonshot-summit-vice-president-biden-announces-new.

[1147] Office of the Press Secretary. "President Obama's Precision Medicine Initiative." Whitehouse.gov, January 30, 2015. https://www.whitehouse.gov/the-press-office/2015/01/30/fact-sheet-president-obama-s-precision-medicine-initiative.

[1148] *European Roadmap Smart Systems for Automated Driving* http://www.smart-systems-integration.org/public/documents/publications/EPoSS%20Roadmap_Smart%20Systems%20for%20Automated%20Driving_2015_V1.pdf, European Technology Platform on Smart Systems Integration (EPoSS), 2015.

[1149] SAE International

[1150] http://ieeecss.org/sites/ieeecss.org/files/documents/IoCT-Part4-13VehicleToVehicle-HR.pdf

[1151] Natasha Merat and A. Hamish Jamson. " HOW DO DRIVERS BEHAVE IN A HIGHLY AUTOMATED CAR? http://drivingassessment.uiowa.edu/DA2009/072_MeratJamson.pdf " Institute for Transport Studies University of Leeds. Quote: "Drivers' response to all critical events was found to be much later in the automated driving condition, compared to manual driving."

[1152] Self-driving Car Logs More Miles http://googleblog.blogspot.hu/2012/08/the-self-driving-car-logs-more-miles-on.html, googleblog

[1153] //en.wikipedia.org/w/index.php?title=Autonomous_car&action=edit

[1154] O'Toole, Randal, Policy Implications of Autonomous Vehicles (18 September 2014). Cato Institute Policy Analysis No. 758. Available at SSRN:

[1155] Litman, Todd. "Autonomous vehicle implementation predictions." *Victoria Transport Policy Institute* 28 (2014).

[1156] Global Automotive Regulation page related to automated driving https://web.archive.org/web/20171201050545/https://globalautoregs.com/rules/157-1968-vienna-convention-on-road-traffic

[1157] Ana Valdes (July 5, 2012). Florida Embraces Self-Driving Cars http://www.wptv.com/dpp/news/state/florida-embraces-self-driving-cars-as-engineers-and-lawmakers-prepare-for-the-new-technology Retrieved March 31, 2013.

[1158] John Oram (9-27-2012). Governor Brown Signs California Driverless Car Law at Google HQ http://www.brightsideofnews.com/news/2012/9/27/governor-brown-signs-california-driverless-car-law-at-google-hq-.aspx Retrieved March 31, 2013.

[1159] CDA Press (July 8, 2014). Aye, robot: Cd'A City Council approves robot ordinance http://www.cdapress.com/news/local_news/article_c6e1b189-66da-5b53-83fd-91cf8a431f84.html

[1160] http://jalopnik.com/how-the-cars-of-logan-grappled-with-the-very-real-futur-1793099275

[1161] https://books.google.com/books?id=DW2QAAAAQBAJ

[1162] http://www.inf.ed.ac.uk/publications/thesis/online/IM110982.pdf

[1163] https://www.technologyreview.com/s/520431/driverless-cars-are-further-away-than-you-think/

[1164] http://onlinepubs.trb.org/onlinepubs/nchrp/nchrp_lrd_069.pdf

[1165]http://theloadstar.co.uk/the-driving-forces-behind-what-would-be-the-next-revolution-in-the-haulage-sector/

[1166]http://recode.net/2015/10/27/meet-the-companies-building-self-driving-cars-for-google-and-tesla-and-maybe-apple/

[1167]http://ntl.bts.gov/lib/57000/57000/57076/Review_FMVSS_AV_Scan.pdf

[1168]http://knowledgecenter.csg.org/kc/system/files/CR_automomous.pdf

[1169]https://www.npr.org/2015/07/31/427990392/remembering-when-driverless-elevators-drew-skepticism

[1170]http://www.rand.org/content/dam/rand/pubs/research_reports/RR400/RR443-2/RAND_RR443-2.pdf

[1171]define the philosophy of AI as consisting of the first two questions, and the additional question of the ethics of artificial intelligence. writes "In the current literature, philosophy has two chief roles: to determine whether or not such machines would be conscious, and, second, to predict whether or not such machines are possible." The last question bears on the first two.

[1172]This is a paraphrase of the essential point of the Turing test. , , ,

[1173]. This assertion was printed in the program for the Dartmouth Conference of 1956, widely considered the "birth of AI."also

[1174]and

[1175]This version is from , and is also quoted in . Searle's original formulation was "The appropriately programmed computer really is a mind, in the sense that computers given the right programs can be literally said to understand and have other cognitive states." . Strong AI is defined similarly by : "The assertion that machines could possibly act intelligently (or, perhaps better, act as if they were intelligent) is called the 'weak AI' hypothesis by philosophers, and the assertion that machines that do so are actually thinking (as opposed to simulating thinking) is called the 'strong AI' hypothesis."

[1176]See , where they make the distinction between *acting* rationally and *being* rational, and define AI as the study of the former.

[1177]Saygin 2000.

[1178]and see , where they call his paper "famous" and write "Turing examined a wide variety of possible objections to the possibility of intelligent machines, including virtually all of those that have been raised in the half century since his paper appeared."

[1179]under "The Argument from Consciousness"

[1180]Russell and Norvig would prefer the word "rational" to "intelligent".

[1181]consider a thermostat a simple form of intelligent agent, known as a reflex agent. For an in-depth treatment of the role of the thermostat in philosophy see "4. Is Experience Ubiquitous?" subsections *What is it like to be a thermostat?*, *Whither panpsychism?*, and *Constraining the double-aspect principle*.

[1182]. Also see and . The most extreme form of this argument (the brain replacement scenario) was put forward by Clark Glymour in the mid-1970s and was touched on by Zenon Pylyshyn and John Searle in 1980

[1183]Hubert Dreyfus writes: "In general, by accepting the fundamental assumptions that the nervous system is part of the physical world and that all physical processes can be described in a mathematical formalism which can, in turn, be manipulated by a digital computer, one can arrive at the strong claim that the behavior which results from human 'information processing,' whether directly formalizable or not, can always be indirectly reproduced on a digital machine." . John Searle writes: "Could a man made machine think? Assuming it possible produce artificially a machine with a nervous system, ... the answer to the question seems to be obviously, yes ... Could a digital computer think? If by 'digital computer' you mean anything at all that has a level of description where it can be correctly described as the instantiation of a computer program, then again the answer is, of course, yes, since we are the instantiations of any number of computer programs, and we can think."

[1184]Searle writes "I like the straight forwardness of the claim."

[1185]Gödel, Kurt, 1951, *Some basic theorems on the foundations of mathematics and their implications* in Solomon Feferman, ed., 1995. *Collected works / Kurt Gödel, Vol. III*. Oxford University Press: 304-23. - In this lecture, Gödel uses the incompleteness theorem to arrive at the following disjunction: (a) the human mind is not a consistent finite machine, or (b) there exist Diophantine equations for which it cannot decide whether solutions exist. Gödel finds (b) implausible, and thus seems to have believed the human mind was not equivalent to a finite machine, i.e., its power exceeded that of any finite machine. He recognized that this was only a conjecture, since one could never disprove (b). Yet he considered the disjunctive conclusion to be a "certain fact".

[1186] , ,

[1187] Mark Colyvan. An introduction to the philosophy of mathematics. Cambridge University Press, 2012. From 2.2.2, 'Philosophical significance of Gödel's incompleteness results': "The accepted wisdom (with which I concur) is that the Lucas-Penrose arguments fail."

[1188] LaForte, G., Hayes, P. J., Ford, K. M. 1998. Why Gödel's theorem cannot refute computationalism. Artificial Intelligence, 104:265-286, 1998.

[1189] They point out that real machines with finite memory can be modeled using propositional logic, which is formally decidable, and Gödel's argument does not apply to them at all.

[1190] According to , this statement was first proposed by C. H. Whiteley

[1191] , , under "The Argument from Mathematics" where he writes "although it is established that there are limitations to the powers of any particular machine, it has only been stated, without sort of proof, that no such limitations apply to the human intellect."

[1192] , , . See also , and

[1193] under "(8) The Argument from the Informality of Behavior"

[1194] See and

[1195] under "(4) The Argument from Consciousness". See also , where they identify Searle's argument with Turing's "Argument from Consciousness."

[1196] "[P]eople always tell me it was very hard to define consciousness, but I think if you're just looking for the kind of commonsense definition that you get at the beginning of the investigation, and not at the hard nosed scientific definition that comes at the end, it's not hard to give commonsense definition of consciousness." The Philosopher's Zone: The question of consciousness http://www.abc.net.au/rn/philosopherszone/stories/2006/1639491.htm. Also see

[1197] For example, John Searle writes: "Can a machine think? The answer is, obvious, yes. We are precisely such machines."

[1198] . See also , , and

[1199] under "1. The Systems Reply (Berkeley)", , , . Among those who hold to the "system" position (according to Cole) are Ned Block, Jack Copeland, Daniel Dennett, Jerry Fodor, John Haugeland, Ray Kurzweil and Georges Rey. Those who have defended the "virtual mind" reply include Marvin Minsky, Alan Perlis, David Chalmers, Ned Block and J. Cole (again, according to)

[1200] ascribes this position to Ned Block, Daniel Dennett, Tim Maudlin, David Chalmers, Steven Pinker, Patricia Churchland and others.

[1201] under "2. The Robot Reply (Yale)". ascribes this position to Margaret Boden, Tim Crane, Daniel Dennett, Jerry Fodor, Stevan Harnad, Hans Moravec and Georges Rey

[1202] Quoted in

[1203] under "3. The Brain Simulator Reply (Berkeley and M.I.T.)" ascribes this position to Paul and Patricia Churchland and Ray Kurzweil

[1204] under "5. The Other Minds Reply", . makes this reply under "(4) The Argument from Consciousness." Cole ascribes this position to Daniel Dennett and Hans Moravec.

[1205] under "(5) Arguments from Various Disabilities"

[1206] Quoted in

[1207] under "(6) Lady Lovelace's Objection"

[1208] under "(5) Argument from Various Disabilities"

[1209] The Coming Technological Singularity: How to Survive in the Post-Human Era https://web.archive.org/web/20070101133646/http://www-rohan.sdsu.edu/faculty/vinge/misc/singularity.html, by Vernor Vinge, Department of Mathematical Sciences, San Diego State University, (c) 1993 by Vernor Vinge.

[1210]

[1211] Call for debate on killer robots http://news.bbc.co.uk/2/hi/technology/8182003.stm, By Jason Palmer, Science and technology reporter, BBC News, 8/3/09.

[1212] Science New Navy-funded Report Warns of War Robots Going "Terminator" http://www.dailytech.com/New%20Navyfunded%20Report%20Warns%20of%20War%20Robots%20Going%20Terminator/article14298.htm, by Jason Mick (Blog), dailytech.com, February 17, 2009.

[1213] Navy report warns of robot uprising, suggests a strong moral compass https://www.engadget.com/2009/02/18/navy-report-warns-of-robot-uprising-suggests-a-strong-moral-com/, by Joseph L. Flatley engadget.com, Feb 18th 2009.

[1214] AAAI Presidential Panel on Long-Term AI Futures 2008-2009 Study http://research.microsoft.com/en-us/um/people/horvitz/AAAI_Presidential_Panel_2008-2009.htm, Association for the Advancement of Artificial Intelligence, Accessed 7/26/09.

[1215] Article at Asimovlaws.com http://www.asimovlaws.com/articles/archives/2004/07/why_we_need_fri_1.html, July 2004, accessed 7/27/09.

[1216] under "(1) The Theological Objection", although it should be noted that he also writes "I am not very impressed with theological arguments whatever they may be used to support"

[1217] https://philpapers.org/browse/philosophy-of-artificial-intelligence/

[1218] https://www.pt-ai.org

[1219] http://www.sophia.de

[1220] http://people.csail.mit.edu/brooks/papers/elephants.pdf

[1221] //doi.org/10.1016/S0921-8890%2805%2980025-9

[1222] http://plato.stanford.edu/archives/fall2004/entries/chinese-room/

[1223] http://cogprints.org/4023/1/searlbook.htm

[1224] http://plato.stanford.edu/entries/computational-mind/

[1225] http://users.ox.ac.uk/~jrlucas/Godel/mmg.html

[1226] https://web.archive.org/web/20080930164306/http://www-formal.stanford.edu/jmc/history/dartmouth/dartmouth.html

[1227] http://www-formal.stanford.edu/jmc/history/dartmouth/dartmouth.html

[1228] http://www.psych.utoronto.ca/~reingold/courses/ai/cache/mcdermott.html

[1229] https://web.archive.org/web/20081007162810/http://www.rci.rutgers.edu/~cfs/472_html/AI_SEARCH/PSS/PSSH4.html

[1230] http://www.rci.rutgers.edu/~cfs/472_html/AI_SEARCH/PSS/PSSH4.html

[1231] http://aima.cs.berkeley.edu/

[1232] https://web.archive.org/web/20150923204614/http://www.class.uh.edu/phil/garson/MindsBrainsandPrograms.pdf

[1233] //doi.org/10.1017/S0140525X00005756

[1234] http://www.class.uh.edu/phil/garson/MindsBrainsandPrograms.pdf

[1235] //www.worldcat.org/oclc/231867665

[1236] //doi.org/10.1093/mind/LIX.236.433

[1237] //www.worldcat.org/issn/0026-4423

[1238] http://lyceumphilosophy.com/Lyceum-5-1.pdf

[1239] The American Heritage Dictionary of the English Language, Fourth Edition

[1240] The Big Question: Should the human race be worried by the rise of robots? https//web.archive.org, Independent Newspaper,

[1241] Loebner Prize Contest Official Rules — Version 2.0 http://loebner03.hamill.co.uk/docs/LPC%20Official%20Rules%20v2.0.pdf The competition was directed by David Hamill and the rules were developed by members of the Robitron Yahoo group.

[1242] Saudi Arabia bestows citizenship on a robot named Sophia https://techcrunch.com/2017/10/26/saudi-arabia-robot-citizen-sophia/

[1243] Joseph Weizenbaum, quoted in

[1244] Hibbard, Bill (2014): "Ethical Artificial Intelligence" https://arxiv.org/abs/1411.1373.

[1245] Open Source AI. http://www.ssec.wisc.edu/~billh/g/hibbard_agi_workshop.pdf Bill Hibbard. 2008 proceedings of the First Conference on Artificial General Intelligence, eds. Pei Wang, Ben Goertzel and Stan Franklin.

[1246] OpenCog: A Software Framework for Integrative Artificial General Intelligence. http://citeseerx.ist.psu.edu/viewdoc/download?doi=10.1.1.366.621&rep=rep1&type=pdf David Hart and Ben Goertzel. 2008 proceedings of the First Conference on Artificial General Intelligence, eds. Pei Wang, Ben Goertzel and Stan Franklin.

[1247] Inside OpenAI, Elon Musk's Wild Plan to Set Artificial Intelligence Free https://www.wired.com/2016/04/openai-elon-musk-sam-altman-plan-to-set-artificial-intelligence-free/ Cade Metz, Wired 27 April 2016.

[1248] Call for debate on killer robots http://news.bbc.co.uk/2/hi/technology/8182003.stm, By Jason Palmer, Science and technology reporter, BBC News, 8/3/09.

[1249] Robot Three-Way Portends Autonomous Future https://www.wired.com/dangerroom/2009/08/robot-three-way-portends-autonomous-future/, By David Axe wired.com, August 13, 2009.

[1250] New Navy-funded Report Warns of War Robots Going "Terminator" http://www.dailytech.com/New%20Navyfunded%20Report%20Warns%20of%20War%20Robots%20Going%20Terminator/article14298.htm, by Jason Mick (Blog), dailytech.com, February 17, 2009.

[1251] Navy report warns of robot uprising, suggests a strong moral compass https://www.engadget.com/2009/02/18/navy-report-warns-of-robot-uprising-suggests-a-strong-moral-com/, by Joseph L. Flatley engadget.com, Feb 18th 2009.

[1252] http://search.proquest.com/docview/1372020233

[1253] Evolving Robots Learn To Lie To Each Other http://www.popsci.com/scitech/article/2009-08/evolving-robots-learn-lie-hide-resources-each-other, Popular Science, August 18, 2009

[1254] Science New Navy-funded Report Warns of War Robots Going "Terminator" http://www.dailytech.com/New%20Navyfunded%20Report%20Warns%20of%20War%20Robots%20Going%20Terminator/article14298.htm, by Jason Mick (Blog), dailytech.com, February 17, 2009.

[1255] Navy report warns of robot uprising, suggests a strong moral compass https://www.engadget.com/2009/02/18/navy-report-warns-of-robot-uprising-suggests-a-strong-moral-com/, by Joseph L. Flatley engadget.com, Feb 18th 2009.

[1256] AAAI Presidential Panel on Long-Term AI Futures 2008-2009 Study http://research.microsoft.com/en-us/um/people/horvitz/AAAI_Presidential_Panel_2008-2009.htm, Association for the Advancement of Artificial Intelligence, Accessed 7/26/09.

[1257] The Coming Technological Singularity: How to Survive in the Post-Human Era http://www-rohan.sdsu.edu/faculty/vinge/misc/singularity.html, by Vernor Vinge, Department of Mathematical Sciences, San Diego State University, (c) 1993 by Vernor Vinge.

[1258] Article at Asimovlaws.com http://www.asimovlaws.com/articles/archives/2004/07/why_we_need_fri_1.html , July 2004, accessed 7/27/09.

[1259]

[1260] Muehlhauser, Luke, and Louie Helm. 2012. "Intelligence Explosion and Machine Ethics" https://intelligence.org/files/IE-ME.pdf. In Singularity Hypotheses: A Scientific and Philosophical Assessment, edited by Amnon Eden, Johnny Søraker, James H. Moor, and Eric Steinhart. Berlin: Springer.

[1261] Bostrom, Nick. 2003. "Ethical Issues in Advanced Artificial Intelligence" http://www.nickbostrom.com/ethics/ai.html. In Cognitive, Emotive and Ethical Aspects of Decision Making in Humans and in Artificial Intelligence, edited by Iva Smit and George E. Lasker, 12–17. Vol. 2. Windsor, ON: International Institute for Advanced Studies in Systems Research / Cybernetics.

[1262] Yudkowsky, Eliezer. 2011. "Complex Value Systems in Friendly AI" https://intelligence.org/files/ComplexValues.pdf. In Schmidhuber, Thórisson, and Looks 2011, 388–393.

[1263] "Partnership on Artificial Intelligence to Benefit People and Society". N.p., n.d. 24 October 2016.

[1264] Fiegerman, Seth. "Facebook, Google, Amazon Create Group to Ease AI Concerns". CNNMoney. n.d. 4 December 2016.

[1265] http://philpapers.org/browse/ethics-of-artificial-intelligence/

[1266] http://www.nature.com/news/robotics-ethics-of-artificial-intelligence-1.17611

[1267] http://news.bbc.co.uk/1/hi/sci/tech/1809769.stm

[1268] http://www.dasboot.org/thorisson.htm

[1269] https://web.archive.org/web/20080418122849/http://www.southernct.edu/organizations/rccs/resources/research/introduction/bynum_shrt_hist.html

[1270]

[1271] I.J. Good, "Speculations Concerning the First Ultraintelligent Machine" http://commonsenseatheism.com/wp-content/uploads/2011/02/Good-Speculations-Concerning-the-First-Ultraintelligent-Machine.pdf (HTML http://www.acceleratingfuture.com/pages/ultraintelligentmachine.html), *Advances in Computers*, vol. 6, 1965.

[1272] A M Turing, *Intelligent Machinery, A Heretical Theory http://philmat.oxfordjournals.org/content/4/3/256.full.pdf*, 1951, reprinted *Philosophia Mathematica* (1996) 4(3): 256–260 "once the machine thinking method has started, it would not take long to outstrip our feeble powers. .. At some stage therefore we should have to expect the machines to take control, in the way that is mentioned in Samuel Butler's *Erewhon*"

[1273] Eden, Amnon H., et al. "Singularity hypotheses: An overview." Singularity Hypotheses. Springer Berlin Heidelberg, 2012. 1-12.

[1274] Scientists Worry Machines May Outsmart Man https://www.nytimes.com/2009/07/26/science/26robot.html?_r=1&ref=todayspaper By JOHN MARKOFF, NY Times, July 26, 2009.

[1275] Syndicated http://www.businessinsider.com/intelligent-machines-and-human-life-2014-8 at Business Insider

[1276] Yudkowsky, E. (2013). Intelligence explosion microeconomics. Machine Intelligence Research Institute.

[1277] Yampolskiy, Roman V. "Analysis of types of self-improving software." Artificial General Intelligence. Springer International Publishing, 2015. 384-393.

[1278] Yudkowsky, E. (2011, August). Complex value systems in friendly AI. In International Conference on Artificial General Intelligence (pp. 388-393). Springer, Berlin, Heidelberg.

[1279] Haidt, Jonathan; Kesebir, Selin (2010) "Chapter 22: Morality" In Handbook of Social Psychology, Fifth Edition, Hoboken NJ, Wiley, 2010, pp. 797-832.

[1280] Yudkowsky, Eliezer. "Complex value systems in friendly AI." In Artificial general intelligence, pp. 388-393. Springer Berlin Heidelberg, 2011.

[1281]

[1282]

[1283] Waser, Mark. "Rational Universal Benevolence: Simpler, Safer, and Wiser Than 'Friendly AI'." Artificial General Intelligence. Springer Berlin Heidelberg, 2011. 153-162. "Terminal-goaled intelligences are short-lived but mono-maniacally dangerous and a correct basis for concern if anyone is smart enough to program high-intelligence and unwise enough to want a paperclip-maximizer.

[1284] Cohen, Paul R., and Edward A. Feigenbaum, eds. The handbook of artificial intelligence. Vol. 3. Butterworth-Heinemann, 2014.

[1285] Harvnb|Simon|1965|p=96 quoted in Harvnb|Crevier|1993|p=109

[1286] Müller, V. C., & Bostrom, N. (2016). Future progress in artificial intelligence: A survey of expert opinion. In Fundamental issues of artificial intelligence (pp. 555-572). Springer, Cham.

[1287] Artificial Intelligence Alarmists Win ITIF's Annual Luddite Award https://itif.org/publications/2016/01/19/artificial-intelligence-alarmists-win-itif%E2%80%99s-annual-luddite-award, ITIF Website, January 19, 2016

[1288] Elliott, E. W. (2011). Physics of the Future: How Science Will Shape Human Destiny and Our Daily Lives by the Year 2100, by Michio Kaku. *Issues in Science and Technology*, 27(4), 90.

[1289] via http://lukemuehlhauser.com/hillary-clinton-on-ai-risk/

[1290] " Tech Luminaries Address Singularity http://spectrum.ieee.org/computing/hardware/tech-luminaries-address-singularity, IEEE Spectrum. Special Report: The Singularity (June 2008).

[1291] http://intelligence.org/files/AIFoomDebate.pdf

[1292] https://web.archive.org/web/20050826094312/http://www.gslis.utexas.edu/~palmquis/courses/reviews/amy.htm

[1293] https://web.archive.org/web/20050508173416/http://www.smeed.org/1735

[1294] http://www.ilmarefilm.org/W_E_2.htm

[1295] http://chaosradio.ccc.de/25c3_m4v_3052.html

[1296] https://www.nytimes.com/2008/03/13/world/europe/13weizenbaum.html?_r=1&oref=slogin

[1297] http://www.plugandpray-film.com/en/''Plug

[1298] Eliezer Yudkowsky (2008) in *Artificial Intelligence as a Positive and Negative Factor in Global Risk http://intelligence.org/files/AIPosNegFactor.pdf*

[1299] Omohundro, S. M. (2008, February). The basic AI drives. In AGI (Vol. 171, pp. 483-492).

[1300] Chapter 7: The Superintelligent Will.

[1301]' How Skynet Might Emerge From Simple Physics http://io9.com/how-skynet-might-emerge-from-simple-physics-482402911, *io9, Published 2013-04-26.*
[1302]Goertzel, Ben. " Should Humanity Build a Global AI Nanny to Delay the Singularity Until It's Better Understood? https//web.archive.org", Journal of consciousness studies 19.1-2 (2012): 1-2.
[1303]Pernar, Stefan. " The Evolutionary Perspective - a Transhuman Philosophy http://rationalmorality.info/wp-content/uploads/2015/04/TranshumanPhilosophy_formatted.pdf", 8th Conference on Artificial General Intelligence in Berlin, July 22–25, 2015
[1304]Mason, Cindy. " Engineering Kindness - Giving Machines Compassionate Intelligence https://pdfs.semanticscholar.org/02da/b0b1022c940788c016792ab3f08ad1d88cac.pdf", International Journal of Synthetic Emotions, 6(1), June – December 2015.
[1305]Armstrong, S., Sandberg, A., & Bostrom, N. (2012). Thinking inside the box: Controlling and using an oracle ai. Minds and Machines, 22(4), 299-324.
[1306]Kornai, András. " Bounding the impact of AGI http://www.kornai.com/Papers/agi12.pdf". Journal of Experimental & Theoretical Artificial Intelligence ahead-of-print (2014): 1-22. "...the essence of AGIs is their reasoning facilities, and it is the very logic of their being that will compel them to behave in a moral fashion... The real nightmare scenario (is one where) humans find it advantageous to strongly couple themselves to AGIs, with no guarantees against self-deception."
[1307]http://intelligence.org/files/AIPosNegFactor.pdf
[1308]http://citeseerx.ist.psu.edu/viewdoc/download?doi=10.1.1.393.8356&rep=rep1&type=pdf
[1309]https://aaai.org/Papers/Workshops/2008/WS-08-07/WS08-07-023.pdf
[1310]http://www.nickbostrom.com/ethics/ai.html
[1311]https://intelligence.org/ie-faq/#WhatIsFriendlyAI
[1312]https://intelligence.org/files/CFAI.pdf
[1313]http://www.ssec.wisc.edu/~billh/g/SIAI_critique.html
[1314]http://www.optimal.org/peter/siai_guidelines.htm
[1315]http://www.thenewatlantis.com/publications/the-problem-with-friendly-artificial-intelligence
[1316]*Artificial Intelligence: A Modern Approach* includes the philosophical foundations of AI including the questions of consciousness http://aima.cs.berkeley.edu/contents.html, Russell, Stuart J., Norvig, Peter, 2003, Upper Saddle River, New Jersey: Prentice Hall,
[1317]Artificial consciousness: Utopia or real possibility? http://ieeexplore.ieee.org/xpl/login.jsp?tp=&arnumber=933500&url=http%3A%2F%2Fieeexplore.ieee.org%2Fiel5%2F2%2F20203%2F00933500 Buttazzo, Giorgio, July 2001, Computer, ISSN 0018-9162
[1318]Loebner Prize Contest Official Rules — Version 2.0 http://loebner03.hamill.co.uk/docs/LPC%20Official%20Rules%20v2.0.pdf The competition was directed by David Hamill and the rules were developed by members of the Robitron Yahoo group.
[1319]Joëlle Proust in *Neural Correlates of Consciousness*, Thomas Metzinger, 2000, MIT, pages 307-324
[1320]Christof Koch, *The Quest for Consciousness*, 2004, page 2 footnote 2
[1321]Tulving, E. 1985. Memory and consciousness. Canadian Psychology 26:1-12
[1322]Franklin, Stan, et al. "The role of consciousness in memory." Brains, Minds and Media 1.1 (2005): 38.
[1323]Franklin, Stan. "Perceptual memory and learning: Recognizing, categorizing, and relating." Proc. Developmental Robotics AAAI Spring Symp. 2005.
[1324]Shastri, L. 2002. Episodic memory and cortico-hippocampal interactions. Trends in Cognitive Sciences
[1325]Kanerva, Pentti. Sparse distributed memory. MIT press, 1988.
[1326]Aleksander 1995
[1327]Takeno - Archive No... http://www.rs.cs.meiji.ac.jp/Takeno_Archive.html
[1328]The world first self-aware robot and The success of mirror image cognition https://web.archive.org/web/20090203193304/http://www.mimed.mw.tum.de/06-12-18_MIMED_LUETH-002918.PDF, Takeno
[1329]A Robot Succeeds in 100% Mirror Image Cognition http://www.s2is.org/Issues/v1/n4/papers/paper4.pdf, Takeno, 2008
[1330]Aleksander I (1996) *Impossible Minds: My Neurons, My Consciousness*, Imperial College Press

[1331] Thaler, S.L., " Device for the autonomous generation of useful information https//web.archive.org"
[1332] Roque, R. and Barreira, A. (2011). "O Paradigma da "Máquina de Criatividade" e a Geração de Novidades em um Espaço Conceitual," 3º Seminário Interno de Cognição Artificial - SICA 2011 – FEEC – UNICAMP.
[1333] Minati, G. and Vitiello, G (2006) Mistakemaking machines,"inSyst.Emer.:Res.Develop. (Springer, NewYork), pp.67-68.
[1334] Thaler, S. L. (2013) The Creativity Machine Paradigm, Encyclopedia of Creativity, Invention, Innovation, and Entrepreneurship http://www.springerreference.com/docs/html/chapterdbid/358097.html, (ed.) E.G. Carayannis, Springer Science+Business Media
[1335] Thaler, S. L. (2011). "The Creativity Machine: Withstanding the Argument from Consciousness," APA Newsletter on Philosophy and Computers
[1336] Thaler, S. L. (1995). Death of a gedanken creature, *Journal of Near-Death Studies*, 13(3), Spring 1995
[1337] Thaler, S. L. (1996). Is Neuronal Chaos the Source of Stream of Consciousness? In Proceedings of the World Congress on Neural Networks, (WCNN'96), Lawrence Erlbaum, Mawah, NJ.
[1338] Mayer, H. A. (2004). A modular neurocontroller for creative mobile autonomous robots learning by temporal difference http://citeseerx.ist.psu.edu/viewdoc/download?doi=10.1.1.89.5420&rep=rep1&type=pdf, Systems, Man and Cybernetics, 2004 IEEE International Conference(Volume:6)
[1339] Mapping the Landscape of Human-Level Artificial General Intelligence http://web.eecs.utk.edu/~itamar/Papers/AI_MAG_2011.pdf
[1340] http://www.conscious-robots.com/consscale
[1341] "Consciousness". In Honderich T. The Oxford companion to philosophy. Oxford University Press.
[1342] http://iase.info
[1343] https://web.archive.org/web/19970302014628/http://www.ee.ic.ac.uk/research/neural/publications/iwann.html
[1344] https://web.archive.org/web/20110721234802/http://www.conscious-robots.com/raul/papers/Arrabales_ICCI09_preprint.pdf
[1345] http://www.conscious-robots.com/raul/papers/Arrabales_ICCI09_preprint.pdf
[1346] https://web.archive.org/web/20040816183746/http://vesicle.nsi.edu/users/baars/BaarsConsciousnessBook1988/
[1347] http://vesicle.nsi.edu/users/baars/BaarsConsciousnessBook1988
[1348] http://philpapers.org/archive/BOYAQI.pdf
[1349] //www.worldcat.org/issn/2012-3477
[1350] https//web.archive.org
[1351] http://www.conscious-robots.com/en/conscious-machines/theories-of-consciousness/pentti-haikonens-architecture-for-conscious-mac.html
[1352] //doi.org/10.1016/j.concog.2007.04.005
[1353] //doi.org/10.1080/00048407212341301
[1354] //doi.org/10.1016/j.neunet.2013.03.011
[1355] http://cogprints.org/5941/1/ASLAB%2DA%2D2007%2D011.pdf
[1356] //doi.org/10.1016/j.neunet.2007.09.012
[1357] //www.ncbi.nlm.nih.gov/pubmed/17936581
[1358] //doi.org/10.1016/j.concog.2005.11.005
[1359] //doi.org/10.1006/ccog.1999.0405
[1360] //www.ncbi.nlm.nih.gov/pubmed/10600249
[1361] //doi.org/10.1016/S1389-0417%2800%2900013-9
[1362] http://ieeexplore.ieee.org/xpl/freeabs_all.jsp?arnumber=1554325
[1363] //doi.org/10.1109/CIRA.2005.1554325
[1364] http://srsc.ulb.ac.be/axcWWW/papers/pdf/01-AXCLJ.pdf
[1365] https://web.archive.org/web/20151223105456/http://j-cs.org/gnuboard/bbs/download.php?bo_table=__vol012i4&wr_id=1&no=0
[1366] http://j-cs.org/gnuboard/bbs/download.php?bo_table=__vol012i4&wr_id=1&no=0
[1367] //doi.org/10.1016/s1364-6613%2803%2900056-1

[1368]http://www.brains-minds-media.org/archive/150

[1369]http://www-formal.stanford.edu/jmc/generality/generality.html

[1370]http://www.d6.dion.ne.jp/~takenof/SAR-LAL-Takeno.pdf

[1371]http://ccsl.mae.cornell.edu/sites/default/files/GECCO09_Zagal.pdf

[1372]http//www.slideserve.com

[1373]https://web.archive.org/web/20110719195048/http://www.aco.gatech.edu/conference/focs-aco/abstracts.html

[1374]http://www.Conscious-Robots.com

[1375]Cadwalladr, Carole (2014). " Are the robots about to rise? Google's new director of engineering thinks so... https://www.theguardian.com/technology/2014/feb/22/robots-google-ray-kurzweil-terminator-singularity-artificial-intelligence" *The Guardian*. Guardian News and Media Limited.

[1376]Müller, V. C., & Bostrom, N. (2016). Future progress in artificial intelligence: A survey of expert opinion. In Fundamental issues of artificial intelligence (pp. 555–572). Springer, Cham.

[1377]"The World's Technological Capacity to Store, Communicate, and Compute Information" http://www.sciencemag.org/content/332/6025/60, Martin Hilbert and Priscila López (2011), Science, 332(6025), 60–65; free access to the article through here: martinhilbert.net/WorldInfoCapacity.html

[1378]*The Singularity Is Near*

[1379]John R. Searle, "What Your Computer Can't Know", *The New York Review of Books*, 9 October 2014, p. 54.

[1380]Korotayev A. The 21st Century Singularity and its Big History Implications: A re-analysis https://www.academia.edu/36810724/The_21st_Century_Singularity_and_its_Big_History_Implications_A_re-analysis. *Journal of Big History* 2/3 (2018): 71 - 118.

[1381]LePoire D. Potential Economic and Energy Indicators of Inflection in Complexity. *Evolution* 3 (2013): 108–118.

[1382]Nordhaus, W. D. (2007). Two Centuries of Productivity Growth in Computing. The Journal of Economic History, 67(1), 128–159. http://doi.org/10.1017/S0022050707000058

[1383]Schmidhuber, Jürgen. "New millennium AI and the convergence of history." Challenges for computational intelligence. Springer Berlin Heidelberg, 2007. 15–35.

[1384]William D. Nordhaus, "Why Growth Will Fall" (a review of Robert J. Gordon, *The Rise and Fall of American Growth: The U.S. Standard of Living Since the Civil War*, Princeton University Press, 2016, , 762 pp., $39.95), *The New York Review of Books*, vol. LXIII, no. 13 (August 18, 2016), p. 68.

[1385]

[1386]*The Singularity Is Near*, p.215,

[1387]*The Singularity is Near*, p.216.

[1388]Sandberg, Anders. "An overview of models of technological singularity." Roadmaps to AGI and the Future of AGI Workshop, Lugano, Switzerland, March. Vol. 8. 2010.

[1389]//en.wikipedia.org/w/index.php?title=Template:Transhumanism&action=edit

[1390]Lin (2010), p. 24

[1391]I.J. Good, "Speculations Concerning the First Ultraintelligent Machine" https://web.archive.org/web/20090420061605/http://www.aeiveos.com/~bradbury/Authors/Computing/Good-IJ/SCtFUM.html (HTML http://www.acceleratingfuture.com/pages/ultraintelligentmachine.html), *Advances in Computers*, vol. 6, 1965.

[1392]http://genity.org/

[1393]Pellissier, Hank. "Do all Transhumanists Want Immortality? No? Why Not?" Futurist 46.6 (2012): 65-. Web.

[1394]Human Purpose and Transhuman Potential: A Cosmic Vision of Our Future Evolution, Ted Chu 2014

[1395]The thinker's guide to ethical reasoning, Linda Elder and Richard Paul 2013

[1396]How to Think about Weird Things: Critical Thinking for a New Age Theodore Schick

[1397]Ten Billion Tomorrows: How Science Fiction Technology Became Reality and Shapes the Future, Brian Clegg 2015

[1398]//en.wikipedia.org/w/index.php?title=Template:Humanism&action=edit

[1399]https://hpluspedia.org/

[1400] https://curlie.org/Society/Future/Transhumanism
[1401] http://whatistranshumanism.org/

Article Sources and Contributors

The sources listed for each article provide more detailed licensing information including the copyright status, the copyright owner, and the license conditions.

History of artificial intelligence *Source:* https://en.wikipedia.org/w/index.php?oldid=850692791 *License:* Creative Commons Attribution-Share Alike 3.0 *Contributors:* 22merlin, Adavidb, Aircorn, Allenchue, Amccann421, AmericanLemming, AoV2, Arado, Audiodude, AustralianMelodrama, Axem Titanium, Ayyu, BD2412, Bender235, Bgwhite, Blueclaw, CharlesGillingham, Chomsky1, Chris the speller, Chris55, Chrissymad, Christian75, Citation bot 1, ClueBot NG, Cordes∼enwiki, Cyanidethistles, Daniel.Cardenas, Darker Dreams, Delusion23, Devourer09, Dispenser, Dl2000, Edward, Froth, Funandtrvl, Gaius Cornelius, General Ization, Gilliam, GioCM, Grafen, Helper70, HitchieH, Hmains, Ilovescratch, Itinerant1, Ivanvector, Jackson Peebles, Jamelan, JaneStillman, Jim1138, Js229, Junling, KGirlTrucker81, KeithRagoonaden, Ketiltrout, Kirbanzo, Knight1993, Koavf, Kothlover, Kwamikagami, Lechgar, Lightmouse, Liontooth, M3tainfo, Malleus Fatuorum, Materialscientist, Mblumber, MegMaker, Mild Bill Hiccup, Mosaic9134, MrsHudson, Mschribr, Mtpaley, MusikAnimal, Myscrnnm, Narky Blert, Nealmcb, Neelix, NewEnglandYankee, Nneonneo, No Swan So Fine, Ohconfucius, Omnipaedista, Onel5969, Open2universe, Oshwah, Pgr94, Philip Trueman, Pinethicket, Pkbwcgs, Powolniak, Primefac, Quentin X, Qwertyus, Riceissa, Rich Farmbrough, Rjwilmsi, Rootxploit, SchreyP, Serouj, Shellwood, Shoessss, SimonTrew, Skittleys, SkyWarrior, SoxBot III, Spencerk, StAnselm, Sunshinehit, Supdiop, Syncategoremata, TAnthony, Tassedethe, TedColes, The Mol Man, The Transhumanist, Trappist the monk, Vadmium, Vivohobson, VjiaoBlack, WOSlinker, Wa3frp, Wavelength, Widr, Woohookitty, WordSeventeen, Xentyr, Yahya Abdal-Aziz, Yuriz, Zy26, 151 anonymous edits 1

Timeline of artificial intelligence *Source:* https://en.wikipedia.org/w/index.php?oldid=852325808 *License:* Creative Commons Attribution-Share Alike 3.0 *Contributors:* Ad48, Adavidb, Albertzeyer, Alistairdavidson, Arado, Ashleytway, Austintate, AvalerionV, Awesomeo99, BD2412, BenKovitz, Bender235, Bilby, Blueclaw, Bluekiller79, Bugone, Catlemur, Ceyockey, CharlesGillingham, Chris the speller, Chris55, Citation bot 1, ClueBot NG, Coder m, Curious1i, DSmurf, DangerousJXD, Daniel Mietchen, Davechatting, Dawnseeker2000, Dekimasu, Denisarona, Dewritech, Dialectric, Dissident93, Dmcq, Dr.apricot, Drbobpgh, Electricmaster, Eric Le Bigot, Ezrdr, Faizhaider, Florianagrasso, Florrat, Fratrep, Funandtrvl, Fuxx, Gnubo, Goooshlrific, Graham87, Gregbard, GünniX, Hmains, Huon, Iridescent, J04n, JackofOz, Jagged 85, JamesBWatson, Johannahrodgers, Jonesey95, Joshualouie711, Jutta, Kennethhurst, Ketiltrout, Khazar2, Kumioko (renamed), Letter 7, Lotje, M2545, MSRCambridge, Mandarax, Manning Bartlett, Mcibula, Meredyth, Metatinara, Mikael Häggström, Mrwojo, Mschribr, NQ, Neo-Jay, Nwerneck, Onel5969, Pcoucke, Philip Trueman, Physikite, Qetuth, QuartzReload, R'n'B, Rjwilmsi, Rolf h nelson, Rootxploit, Roshiof, Rror, Sacerdoti, SchreiberBike, Shellwood, Shuimujianghu, Smallpond, T3dkjn89q00vl02Cxp1kqs3x7, TAnthony, TedColes, The Transhumanist, The1337gamer, Tony1, Tpbradbury, Trappist the monk, UnsungKing123, Vipul, VoABot II, WRK, Yeedle, 94 anonymous edits 31

Knowledge representation and reasoning *Source:* https://en.wikipedia.org/w/index.php?oldid=849216379 *License:* Creative Commons Attribution-Share Alike 3.0 *Contributors:* 1ForTheMoney, 4th-otaku, Adrian.walker, Aeternus, Aleksd, Amy8492, Ancheta Wis, Andrewspencer, AndriesVanRenssen, Andycjp, Architectchao, Arjayay, Arodichevski, Auminski, Avicennasis, Beauty School Dropout, Beland, Biogeographist, Blarley, BrianPansky, Bruyninc∼enwiki, C.Koltzenburg, Cattus, CharlesGillingham, Charvest, Clairebw1, ClamDip, ClueBot NG, Comp.arch, Corb3126, Curb Chain, Cygri, D0t, D6, Davidwilliam.csn, Deeptrivia, Dewritech, Dfass, Dinhvan5481, Dkgoodman, Don4of4, Dpv, Drunken Pirate, Duncancumming, Dustinsmith, El C, Emoritz2017, Equilux, Erik Sandberg, Faizan, Felixlover4ever, Frap, Fraxtil, Frietjes, GeekyEnki, George100, GerryWolff, Gewild, Giraffedata, Gorbag42, Graham87, Grantbow, Gregbard, Grshutt, Haakon, He Who Is, Helenjjstanford, Hirzel, Hobsonlane, HolgerK∼enwiki, ILikeThings, Indon, Isotope23, Ivan Štambuk, Ivanche, JCLately, January2009, Jarble, Jeblad, Jeff3000, Jefferythomas, JimBrander, Jjpcondor, JoBrodie, Joeoettinger, JonDCurtis, Joymmart, Julian Mendez, Jvstein, Kai-Hendrik, Kiwibird, Kku, Klbrain, Kpmiyapuram, Lambiam, Levineps, Lh389, Lheuer, Linas, LongAgedUser, M.O.X, MadScientistX11, Magioladitis, Materialscientist, MaxEnt, Mdd, Medmrt2008, Michael Hardy, Mike Rosoft, Mild Bill Hiccup, Modify, Mogism, Mpeisenbr, NOrbeck, Nabeth, Nick Number, Nono64, Ohnoitsjamie, Olivier Auber, Omnipaedista, Openman, Owl-syme, Parkerdr, Paul Magnussen, Phuzion, Poor Yorick, R'n'B, Rds1970, Rhododendrites, RichardF, Rjwilmsi, Robotczar, Robotwisdom, Samlyn.josfyn, Satori Son, Sderose, Semantia, Shermanmonroe, Sina2, Smimram, Smoken Flames, Snowmanradio, Spencerk, StraboVarenius, Sugandha2901, Sun Creator, Supdiop, Tdbreaux, TiagoDias, Tizio, ToBeFree, Toimild, Trappist the monk, Tremilux, VanishedUserABC, Waldir, Widr, Williamchen16, Woohookitty, Xyzzyplugh, Yzeng, 夙, 142 anonymous edits 45

Commonsense knowledge (artificial intelligence) *Source:* https://en.wikipedia.org/w/index.php?oldid=842359929 *License:* Creative Commons Attribution-Share Alike 3.0 *Contributors:* 4th-otaku, Alai, Alynna Kasmira, AtzaMan, BryanD, CharlesGillingham, Cmouse, Daniele Gallesio, ErnestSDavis, Etm, Feydey, Forbsey, Geogene, Jaibe, Jefferson2011, JonHarder, MBParker, MadScientistX11, Mindpixel, Peak, Rolf h nelson, Spencerk, Tom.Reding, Waldir, Widefox, WikiHannibal, Zouxiaohui, 16 anonymous edits 54

Automated planning and scheduling *Source:* https://en.wikipedia.org/w/index.php?oldid=852878746 *License:* Creative Commons Attribution-Share Alike 3.0 *Contributors:* "alyosha", AST3, AlanUS, Albertzeyer, Allan McInnes, Apparition11, BAxelrod, Chaosdruid, CharlesGillingham, Chris.hames, Citation bot 1, D-s-weld, DGG, Danim, Dcirovic, Discospinster, Dougher, Dsnpost, Erel Segal, Faizhaider, Fayenatic london, Gaius Cornelius, Galoubet, Gonesbuyo, Greenrd, Hectorpal, Highbrownknoll, Hxp0402, Iv, IJULEN92, Jaibe, Jamelan, JamesBWatson, Jarble, JessicaGibbs, Kappa, LilHelpa, Linas, Magioladitis, Mdd, Mild Bill Hiccup, Moxon, MrOllie, Nono64, Pgr94, PrologFan, Qwertyus, Rijkbenik, Saxifrage, Simeon, TAnthony, The Transhumanist, Tizio, Trixie05, Trylks, Woodshed, Yaris678, 38 anonymous edits 57

Machine learning *Source:* https://en.wikipedia.org/w/index.php?oldid=853049718 *License:* Creative Commons Attribution-Share Alike 3.0 *Contributors:* ATPhosphate, Acfleiss, Agamemnonc, Aios3837, Ajav13, Alex-jprof, AlexandreHagen, Allforrous, Amir Rahat, An Inconvenient Truth, Aomatveev, Arijit.chkbr, Auguel, AvnishIT, Avoided, Bbarmadillo, Beland, Ben j Scott, Bilbo571, Blankshot, Blythwood, Bobbybobbie, BunnyShark, Caoanexo, Carlosayam, Cedar101, Chire, Chris55, Clarevoyance, Clarities, Clenchedteeth, Clipname, ClueBot NG, Cnnsbs, CobraL0rd, Count Count, Cowprophet, Crobbins5, Danhash, Dataminerengine, Datasci007, David.moreno72, Dewritech, Dhruvh, Diego Moya, Docurbs, Donner60, Dorianinou, DougBoost, DrStrauss, Emily ly, Emoody17, Erel Segal, Erukhhirhsr318, Ettrig, Faustino.josemar, Fcastro2, Feelmybrainwaves, Flyer22 Reborn, Fmadd, Galois fu, GeneSobolev, GerryWolff, Ghost30, Gilliam, Graham87, Grayfell, Gufosowa, Headbomb, HelpUsStopSpam, Ian (Wiki Ed), Imersion, JR Munoz, Jackjackjackbruce, Jacob.stein, Jarble, Jim1138, Joel B. Lewis, John of Reading, JohnBlackburne, Johnbandler, Johnnyballs454545, Jowlstirs, Jytdog, Kaartic, Keith arano, Kojikawano, Lamb99, Lenov, Letsstopwikidictators, Like.liberation, LilHelpa, Llorenzi, MageTeo, Magioladitis, ManU0710, Master of Time, Materialscientist, MathKeduor7, Mean as custard, Ml 771, Moltean, MrOllie, NatGertler, Nbro, Ncsny, Netimpacter, Neutrality, Newestin, Nitpicking polish, Noah Kastin, Olexa Riznyk, Olivia482, Onel5969, Oshwah, Otto bek, Pcsontos, Peterwittek, Physicalmathematics, Pineapple2018, PouyaRZ, Praaaks, Premeditated Chaos, Quercus solaris, RachelTan, Randallbritten, Rhiever, RichardWeiss, Roxy the dog, Rstats8108, SMcCandlish, Schellluri, SecurityPanther, Serols, Shubh24aug, SinPantuflas, Somda, StanLeeP, Staszek Lem, Stuart Feffer, Suriyakrish, Systemal, Sytelus, Techiemntr, The Anome, Tony1, Tordek ar, TuCo, Ubba abba, User8178, VeniVidiVicipedia, Wang Zichen, WikiUser2211, Xl2085, 161 anonymous edits 61

Natural language processing *Source:* https://en.wikipedia.org/w/index.php?oldid=852787341 *License:* Creative Commons Attribution-Share Alike 3.0 *Contributors:* 098760sam, 0x0077BE, 7804j, Aios3837, AllenAkhaumere, Allforrous, Alvations, Ambertk, Andrewrosenberg, Antikithera, Apesteilen, Ayoopdog, BD2412, Basit1122, BeachComber1972, Bear-rings, Beland, Benoît Sagot, Billinghurst, Biografer, BlackVegetable, Blahma, Bonadea, Bumm13, Cantaloupe2, Celct, Chester Markel, Chris the speller, ChrisGualtieri, Clockery, ClueBot NG, Coenni, CognitiveAI, Crisco 1492, Cs california, Csmaltes, DVdm, Dakaplo, Daniel Hershcovich, Daniel.Cardenas, David Eppstein, Dawit semere, Dawnseeker2000, Diego Moya, DivineAlpha, Dmwpowers, Doc Daneeka, Donner60, Drmihaelavan, Dtunkelang, Dudu90, EurekaDS, Fellytone84, Fixuture, Flyer22 Reborn, Fmadd, Forris0129, Francopoulo, Gelingvistoj, GermanJoe, Get Learnt, Graham87, Greenrd, Gwil, Haael, HelpUsStopSpam, Hyksandra, I dream of horses, ILikeThings, Icarusgenius, Ifnord, Iicvx, Ish ishwar, Jahub, Jarble, Jdcomix, Jdstout11, JimPark, Joel B. Lewis, Jona, Jonpatterns, Jytdog, Karkand23, Keith arano, Kk7nc, Kwamikagami, Leondz, Lesbar2, Linguist111, LoMaPh, LordOfPens, Luraatcelct, Maghnus, Marco Ciaramella, Marina Dorosh, Materialscientist, Matthew V. Milone, Mattiasostmar, Mcguire, Me, Myself, and I are Here, Mikael Häggström, Miko Yeoh, Mindmatrix, Mortense, Mquantin, MrOllie, Mx. Granger, NLPGuy, Ncasale, Newwikieditor678, Niceguyedc, Nkishorjitsingh, Ocelot8, Operator873, Pairadonis, Partha lal, Phamnhatkhanh, PhnomPencil, Pleasantville, Prakash Nadkarni, Primefac, Prinr, ProprioMe OW, PullJosh, Rcblanke, Reatlas, RenamedUser01302013, Renepick, Rjwilmsi, Ronz, Royote, Sahraoui17, Sandstein, Sangeenjan, Sawalha.majdi, Sawwie, Schrieveslaach, Sderose, Senu, Stateditor, The Transhumanist, Thüringer, Tillander, Titodutta, Transcendence, UKoch, Unbuttered Parsnip, VeniVidiVicipedia, Versageek, Waggie, Wavelength, WilliamJennings1989, Yucong Duan, Zparcell, 185 anonymous edits 76

Machine perception *Source:* https://en.wikipedia.org/w/index.php?oldid=847836913 *License:* Creative Commons Attribution-Share Alike 3.0 *Contributors:* Alai, Alexser16, Anaxial, Bhny, CafyFab, Dcirovic, Deathsythe, Dicklyon, GregorB, Hayman30, Headbomb, HendrixEesti, HombreDHojalata, Jarble, Jfishel, Kku, Kristinah 2710, Malcolma, Martagallegorodenas, Mcld, Mild Bill Hiccup, Neøn, Ninjakannon, QuiteUnusual, RCtheViking, Shaded0, 11 anonymous edits 86

Computer vision *Source:* https://en.wikipedia.org/w/index.php?oldid=852880582 *License:* Creative Commons Attribution-Share Alike 3.0 *Contributors:* A Train, AGUWMT, Aas213, Ahmadh84, Ancheta Wis, Animalparty, Arado, Arsi Warrior, Arthur Rubin, Atiqahad, BD2412, Babitaarora, Beland, Bender235, Blueclaw, Bob1960evens, Breno, Bths83Cu87Aiu06, Casablancateam, Chcameron, Cinchtray, ClueBot NG, Comp.arch, Cuarta dimension, DMahalko, David Eppstein, DavidCary, Dcirovic, Dewritech, Dicklyon, Dolotta, Don4of4, Download, DrSchlagger, Eidenberger, Enterprisey, Falconiere, Fgnievinski, Fmadd, Gengiskanhg, Georgopa, GerryWolff, GorillaWarfare, Graham87, Graphium, Graywash, HaeB, Hardshiv, HelpUsStopSpam, Hidasri,

Hu12, Hydrargyrum, Iineko, Ijuander, Imareaver, Isacdaavid, Iztwoz, JaGa, JacobDang, Jarble, Jdaloner, Jenna yan, JimVC3, JinOuKim, Joe Gazz84, Johnbandler, Josh3580, Jprg1966, KYN, Kamykowsari, Kendall-K1, Kethrus, Kku, Lamro, Lashlees, Laxmikanthpujari, Lebeda Karel, Leondz, Liridon, Ltomlinjr, Mariella.dimiccoli, Materialscientist, Maurice Carbonaro, Me, Myself, and I are Here, Mean as custard, Mikew72324, Mild Bill Hiccup, MishoMiqo, MrOllie, Myconix, Niceguyedc, NoToleranceForIntolerance, North8000, Novaseminary, PatrickMaeder, Pawn number zero, Phamnhatkhanh, Pierreinformatique, Primefac, Prinsipe Ybarro, Profesoratecnologia, Ram Charitra Dwivedi, Redxine, Rgdboer, Rich Farmbrough, RippleSax, Rolf h nelson, Royote, Samratsubedi, Scribbleink, Severin.stalder, SideSlapd, Smallman12q, StanLeeP, Sunshine-Moonlight, Superegz, SyteAI, Tedickey, The Transhumanist, Toussaint, TylerLangan, UnterLink, Vishnuvg15, WolgangW, Yes deeper, Yobert, Ὁ οἶστρος, 104 anonymous edits 87

Speech recognition *Source:* https://en.wikipedia.org/w/index.php?oldid=852863524 *License:* Creative Commons Attribution-Share Alike 3.0 *Contributors:* 23krot, 2bfrankmchugh, 2ndjpeg, Adcasaa, Andrewpmk, Atomota, Bbrafman, Beland, Bender235, Bhny, Blueclaw, Bobbygalaxy, BoldLuis, Bomazi, Chess, Chris the speller, ChristopheS, ClueBot NG, Compfreak7, Cyberbot II, Danali600, Dane, Dbabbitt, Diallord, Diannaa, Diego Moya, Dithridge, Dmwpowers, Doc James, Download, Drchriswilliams, Drewmutt, Editor2020, ElleLee, FeatherPluma, Fixuture, Fmadd, Francopoulo, Fulldecent, Ggatsby, Gilo1969, Giraffedata, Golopotw, Gracefoo, GünniX, Hajasu, Headbomb, HelpUsStopSpam, Hyksandra, Ijablokov, Jack Greenmaven, JaconaFrere, Jahid.Hamid, Jak86, Jarble, Jcgabin, Jeankowkow, Jim1138, Jmertel23, John of Reading, Jonathanharoldadams, Jonnyglobal, Julian Herzog, Junling, Justin Ormont, KH-1, Kanguole, Kku, Kolbasz, Lairzoqs, Liao, LilHelpa, Macvishnu1983, Magioladitis, Marco Ciaramella, Martarius, Martinevans123, Materialscientist, Mathpriest, Mattramzzz, Mattramzzz1971, Me, Myself, and I are Here, Mean as custard, Mild Bill Hiccup, Mistakefinder, Mr ezhov, MrOllie, Mrt doulaty, NaBUru38, Narky Blert, Nev1, Ngocminh.oss, Nitpicking polish, Norman21, Onel5969, Pauli133, Phamnhatkhanh, Phil Boswell, Pintoch, Pkiesele, Prakash Nadkarni, Praveentawade, Primefac, Pst9, Quebec99, Rajivshah01, Renzoy16, Rjwilmsi, Rodw, Ronz, ScotXW, Selichm, Serols, Sharma vasu55, SheriffIsInTown, Shoaib 075, Skyerise, Sonic Purity, Sundayclose, TAnthony, Tabletop, Techpeace, The Anome, Thnidu, Uncomp, Velella, Wbm1058, Widefox, Wikid77, Yikkayaya, Yobert, Yogainyosemite, Zerothspeller, エマニュエル・マクロン, 153 anonymous edits 103

Affective computing *Source:* https://en.wikipedia.org/w/index.php?oldid=846969548 *License:* Creative Commons Attribution-Share Alike 3.0 *Contributors:* A284093825730, ATren, Aeternus, Ainlina, Anon126, Anonymous but Registered, Areldyb, ArielGold, Arru, Awaterl, Beland, Bender235, Bissinger, Bonadea, Brandmeister, Brennon, Cambriaerik, Candiedbug, Celestial Phineas, CharlesGillingham, Chpapa~enwiki, Christianjones, Citation bot 1, Ckatz, ClueBot NG, Commander Keane, Ctacebu4, Czar, DThomas33, Dammit, Dcirovic, Diego Moya, Dimtet, DreamGuy, Drjheise, Earthtodorian, Ehheh, Eric Mathews Technology, Eroesch~enwiki, Filll, Firefly, Gamer chick21, Grafen, Guettarda, Hamish59, Hanifbbz, Headbomb, HelpUsStopSpam, HendrixEesti, Hrafn, Initial B, Interiot, Jac16888, Jim62sch, John of Reading, K.Nevelsteen, Kappa, Keith arano, LamaHippoOtter, Lamro, MZMcBride, Magioladitis, Matteo004, Mattisse, Mcmillin24, Me, Myself, and I are Here, Merciadriluca, Merzul, Michael Hardy, Mmkmisp, Moulton, Mr. Magoo and McBarker, MrOllie, Myopic Bookworm, Niceguyedc, Nishantrai18, Nixeagle, Oddity-, Omnipaedista, Orangemarlin, Ownlyanangel, Patrick, Paul Erik, Petelonsdale, Pgr94, Phil Boswell, PhnomPencil, Pintoch, Potokiwi, QuackGuru, Qwertyus, Rich Farmbrough, Ringbang, Rjanag, Rjwilmsi, Robert Daoust, Robofish, S L Happy, S33us00n, SEVEREN, Saiorg, SebastianHelm, Shitalnpatil, Sidney.dmello, SimonP, SiobhanHansa, Snaily, Snowdog, Sphinxer, Stamptrader, Susann edits, TAnthony, Technocratic, Tgeller, Thiseye, Trappist the monk, Utcursch, Wavelength, ZayZayEM, Zeppomedio, 96 anonymous edits 125

Artificial general intelligence *Source:* https://en.wikipedia.org/w/index.php?oldid=850034056 *License:* Creative Commons Attribution-Share Alike 3.0 *Contributors:* 0x0077BE, AManWithNoPlan, Aetheling, AlvoMaia, Anrnusna, Apokrif, Aprock, Arado, Asdklf;, AskUNICE, BD2412, Beland, Ben Standeven, Bender235, Bgwhite, BigDwiki, BioMeow, CFCF, CSimons, Chalst, CharlesGillingham, Chealer, Chris the speller, Citation bot 1, ClueBot NG, Cnwilliams, Comp-heur-intel, Curious1i, Daniel.Cardenas, Darker Dreams, Dawnseeker2000, Dcirovic, Dewritech, Dhart, Ds13, Edenplib2, Ego White Tray, EricEnfermero, ErikHaugen, Estevezj, FT2, Fixuture, Flyer22 Reborn, Foobarnix, FrankCarroll, G6cid, Gap9551, Geodejerry, GerryWolff, GliderMaven, Greensburger, Groupuscule, Gubrud, Hisashiyarouin, IjonTichyIjonTichy, Imka7, Inimino, Iztwoz, Jarble, Jeff Muscato, Jhertel, Jmcgnh, Kaeslin, KenTancwell, Kerrick Staley, Kmzurn, Kri, Lamro, Legobot II, Leszek Jańczuk, Lynndunn, Magioladitis, Magyar25, Mais oui!, Mattflaschen, Maximillion Pegasus, McGeddon, Melab-1, Mightyname, Mindbound, Mithoron, Mmortal03, Monterege, Mortense, MrLinkinPark333, MrOllie, Myconix, Nbarth, Nbro, Niceguyedc, Nickchmura, OlEnglish, Omnipaedista, Paolo Lipparini, Paul2520, Pgr94, Phantomket5, Phuzion, Pintoch, Quinton Feldberg, R'n'B, RP88, Renamed user 2560613081, Rich Farmbrough, Rjwilmsi, Rolf h nelson, RomanSpa, Ronsams, Rovercaps, SchreyP, Sean6668881, Sevakprime, Silence, Simul, Slippingspy, Smk65536, Tassedethe, Tga.D, The Transhumanist, Timwi, Tom Edwards, Tom.Reding, Tomazz2, Tomzx, Topbanana, Trappist the monk, Tuntable, Ueutyi, Viriditas, Vx1119, Waldir, Wavelength, Wikijoby, WilliamH, Winterst, Wjhonson, Wonderfl, 101 anonymous edits 138

AI-complete *Source:* https://en.wikipedia.org/w/index.php?oldid=849232842 *License:* Creative Commons Attribution-Share Alike 3.0 *Contributors:* 0, 144.132.75.xxx, A3nm, Adamdaley, Auntof6, Beland, Bethcarey, Bryan Derksen, Byelf2007, CarlHewitt, Chalst, CharlesGillingham, Chimpex, Citation bot 1, Cristi215, Darker Dreams, Datenverlust, Davehi1, David290, Dcoetzee, Dlrohrer2003, EagerToddler39, ErikHaugen, Etm3, FakirNL, Ferrara p, GregorB, Hairy Dude, HelpUsStopSpam, JakeVortex, Jamelan, Jim1138, Karada, Kevin Saff, Last Avenue, Magioladitis, Michael Hardy, MisterHand, Ngocminh.oss, Pgr94, Phil Boswell, PhilKnight, Pintoch, Rich Farmbrough, Rizzardi, Rjwilmsi, RobinK, Serols, Sophus Bie, Stephenb, Tenured, The Transhumanist, Tobias Hoevekamp, Twri, Untalker, WikHead, Yuzz, 42 anonymous edits 154

Cybernetics *Source:* https://en.wikipedia.org/w/index.php?oldid=851520163 *License:* Creative Commons Attribution-Share Alike 3.0 *Contributors:* Aaronshaw, Aeternus, AgnosticPreachersKid, Agoramachina, Akhil999in, Aliensyntax, Altenmann, Anotherwellwisher, Arjayay, Artimis Cat, Asukara, AugPi, BD2412, Battlecry, Bear-rings, Bender235, Bgwhite, Biogeographist, BorgQueen, Bowest2, Bulwersator, Byelf2007, CanadianLinuxUser, Chaosdruid, Chris the speller, Citobun, ClueBot NG, Codwiki, Comp-heur-intel, Compfreak7, Crau999, Cricobr, Crzer07, Csc2.71, DamiaanLVDW, Danim, Dankonikolic, DaveApter, David Ludwig, Davidssonus, Deadclever23, Delusion23, DerBorg, DevOhm, DigDugDogDoog, EMBViki, Ed Shanken, Emurphy42, EricEnfermero, Fgnievinski, Fitmoos, François Robere, Frze, Fuper, Fylbecatulous, Fynsnman, Georgopa, Gilliam, Gisling, GliderMaven, Groupuscule, Harbingyr, Headbomb, Heraclites, Hmains, Hyperbolick, Ifny, J 1982, JacobTrue, Jcc, Jeraphine Gryphon, Jmcgnh, Jonesey95, Jonpatterns, Jprg1966, Kaecyy, Klassica, Kmzurn, Kocabas, KosMal, Leutha, LieutenantLatvia, Lily1721, Liwangyan, Loraof, Madalibi, Magioladitis, Mark viking, Martarius, Master of Time, Materialscientist, Mato, Mblumber, Mcarmier, Mdd, Me, Myself, and I are Here, MeanMotherJr, Mecorbin08, Molisi, Moveo ergo sum, My Chemistry romantic, NMaia, Nagualdesign, Neko-chan, Nemo bis, Nick Green, Nigellwh, Nimbosa, Ninjoust, NoiselessJoe, Omnipaedista, Ondixonhill, OrsanS, Parkwells, Paulpangaro, Phantomsnake, PhilDWraight, Pinethicket, Pjrich, Polmandc, Psheld, Psyus, Pumpernickel, Pyrsmis, Qwertyus, R'n'B, Randy Kryn, Ranepa4, Raphaele7, RecoveringAddict, RekishiEJ, Rememberway, Rich Farmbrough, Roisterer, Sah111-666, Samf4u, Seaphoto, Settdigger, Solarra, Tanmay.jain39, Tassedethe, Teapeat, The Transhumanist, ThePlaz, Theresa knott, Trevithj, Twirligig, Wbm1058, WereSpielChequers, Wikipelli, Wildman2000, Woodlot, Woohookitty, Xanzzibar, Xcedrinod, Yobol, Zati.orzo, Zujine, ZuluPapa5, 176 anonymous edits 157

Computational neuroscience *Source:* https://en.wikipedia.org/w/index.php?oldid=849272277 *License:* Creative Commons Attribution-Share Alike 3.0 *Contributors:* ASmartKid, ATPhosphate, Aaron Schulz, Allforrous, Alterego, Andorin, ArguMentor, Ariangiovanni, Arne Heise, Astro Reeves, BUCompNeuro, Bayle Shanks, Bender235, Benhocking, BertSeghers, Bkroeger, Brainboychik, CBM, Cardoso44, Cardosodeoliveira, CarstenCarsten, CasperEACClemence, Chardar, Chealer, Chrissymad, Chzz, Citation bot 1, Clicketyclack, ClueBot NG, Colonies Chris, Conejo23, CortexSurfer, Cuaxdon, Cyberfay, Dankonikolic, David Eppstein, Davidcofer73, Dcirovic, Dendritic33, Depthgr8, Destexhe, Dicklyon, Djhbrown, Docu, DrPhoenix, Duagloth, Ebegoli, Eerosim, ErikDS59, Fixuture, Gal chechik, Gcastellanos, Gewaltig, Gianluca.baldassarre, Haeinous, Headbomb, Hiturik, Jarble, Jean-Francois Gariepy, Jgarridoalcazar, Jmencisom, Jonesey95, Josephus78, Jvstone, Jwmillerusa, KNakamura, Kazkaskazkasako, Keith D, Kernsters, Kghose, Krunchyman, LaurentPerrinet, Leondz, Lfrench, LittleOldMe, Logical Gentleman, Looie496, Lordvolton, Lova Falk, Mange01, Mangst, Mark viking, MartinSpacek, Mattisse, MaxEnt, Me, Myself, and I are Here, Memming, Meteoralv, Michael Hardy, Mosquitopsu, N d s f, Naiaratoker, Necrotranson, Nemo bis, Neurocompute, Neuromusic, Neuropsychology, Ngoddard, Niarb, Numedeon, Obankston, Occhanikov, Omnipaedista, Orvilleduck, Pascal.Tesson, Pgr94, Phamnhatkhanh, Pillai ajay, Pleasantville, Pointer2948, Polarwind13, Primefac, Psyklic, Pulga0907, RDBrown, Rbrette, Rich Farmbrough, Richard.decal, Rjanag, Rjwilmsi, Rnwatanabe, Robclewley, Rwalker, Ryly, STBotD, SamuelRiv, Sardanaphalus, Scott Illini, Sjs2008, Sluox, Smartprogrammer93, SparkOfCreation, Specdude, Srikanth rswamy, Sun Creator, Sîc S. Drêwalla, Talanvalash, Teemu08, Theneuroscientist123, TjeerdB, Toddphen, Torcini, Transient cns, Trna.michal, Tryptofish, U+003F, Umar yahaya, UninvitedCompany, Whitelet, Xurtio, Yurigerhard, Zdx3578, Zfountas, 175 anonymous edits 174

Symbolic artificial intelligence *Source:* https://en.wikipedia.org/w/index.php?oldid=841805702 *License:* Creative Commons Attribution-Share Alike 3.0 *Contributors:* Aik, Aios3837, Algoprof, Angela, CharlesGillingham, ChrisGriswold, Clemmcculloch, DASHBotAV, Darker Dreams, Dysprosia, Frugal fisherman, Gray hat, Gulumeemee, Jaibe, Jarble, Jytdog, Kku, Kokiri, Me, Myself, and I are Here, Meters, Misterwindupbird, Nbarth, OccultZone, Ogai, Pgr94, Primefac, Qwertyus, Reciprocal, Rursus, The Anome, Tizio, WorldsApart, 24 anonymous edits 182

Evolutionary computation *Source:* https://en.wikipedia.org/w/index.php?oldid=846584032 *License:* Creative Commons Attribution-Share Alike 3.0 *Contributors:* Adoniscik, Aiesparcia, Alex Kosorukoff, Anrnusna, Antonielly, Arkenflame, BD2412, Babajobu, Bahriyebasturk, Baldy Bill, Bgwhite, Biblbroks, Bilby, Boky90, Calltech, Ccottap, CharlesGillingham, ChrisGualtieri, Classicwiki, ClueBot NG, DMacks, Dggreen, Duoduoduo, Dzkd, Ehasl, Emir of Wikipedia, Epipelagic, Epktsang, Ernesto.benini, Ettrig, FerzenR, Fheyligh, Fryed-peach, Gary, George100, Gusz Eiben, HRV, Harizotoh9, Headbomb, Horsman, Inego~enwiki, Intelcompu, Jdaloner, Jeanpmartins, Jiy, Jjmerelo~enwiki, Jlaire, Joe Decker, JohnBlackburne, JonHarder, Jr271, Juckele, JustinTime55, Jwojt, Kamitsaha, Karada, Kevin Dewitt, Lexor, LilHelpa, Lordvolton, Lotje, Luís Felipe Braga, MBazzy, Marco castellani 1965, Mark viking, Mdorigo, Me, Myself, and I are Here, Mercere99, Michael Allan, Michael Hardy, Midas02, MikiWiki, Mneser, Mohdavary, Moxon, Muaafa, NaujPierluigi, Nbro, Obscurans, Oli Filth, Onel5969, Oritnk, Paskornc, Pdcook, Peterdjones, Phamnhatkhanh, Power~enwiki, Primefac, Pruetboonma~enwiki, R1D1, RKUrsem, RSRV0306, Rayman60, RichardF77, Ritchy, Rjwilmsi, Rodw, Ronz, Ruud Koot, Salih, Samsara, Sean.hoyland, Sergey539, Solomon7968, Tense, TheAMmollusc, TheProfessor, Theelepeltje, TimVickers, Togelius, Tom.Reding, Tony1, Trappist the monk, Trbiyanto, Vrenator, Wavelength, WeShareKnowledge, Wo.luren, Woohookitty, Xzesey, Yafrani, Yintan, Zach Winkler, Zilupe, Zwgeem, Илтиорел, 187 anonymous edits 183

Mathematical optimization *Source:* https://en.wikipedia.org/w/index.php?oldid=849809465 *License:* Creative Commons Attribution-Share Alike 3.0 *Contributors:* Albert triv, Allforrous, Amaury, Auntof6, Avisnacks, BD2412, Bender235, Bfinn, Bonadea, Brycehughes, Buzzzman, CBM, Centathlete, Chester Markel, ChrisGualtieri, Cic, ClueBot NG, Cpt Wise, Cubism44, David Eppstein, David R. Ingham, Delafé, Denmartines, Diego Moya, DonSiano, Download, Dsol, Dufrin, Duoduoduo, Eiro06, ElPimiento1231, EncMstr, Epistemenical, Erel Segal, FiveColourMap, Frietjes, Gabortherman, Geogene, Ggbil2, Graham87, Hakkinen, Hamsterlopithecus, Headbomb, HelpUsStopSpam, Heroestr, Heron, Hossein6785, InverseHypercube, Isheden, JHunterJ, JIP, JPRBW, Jarble, Joeyjoon, Johnbandler, Jonesey95, Jowa fan, Ju12358, Kabanov Dmitry, Kiefer.Wolfowitz, Kimoscholar, Kiril Simeonovski, Kleuske,

Kolbasz, Kri, Krystofer, Lazars53, Lbertolotti, Leonardo61, Limit-theorem, LokiClock, Loraof, Lycurgus, MSchlueter, MaNeMeBasat, Magioladitis, Malcops, Marcocapelle, Mcld, Mdann52, Me, Myself, and I are Here, Mehr86, Meldraft, Micemug, Michael Hardy, Michael X Zhou, Mike Rosoft, Miklomp, Miraclexix, Mitch Ames, Mitsisko, MoutazRamadan, Mr Stephen, MrOllie, Nageh, Newbyman, Nicoguaro, Nihiltres, Nojhan, Omnipaedista, Onlybumblebeescanfertilizeredcloverplants, Oshwah, Osiris, P99am, Paolo.dL, Paul August, Philip Trueman, PiotrGregorczyk, Pirehelokan, Pkalczynski, Quebec99, Qwfp, RandallZ, Rinconsoleao, Rjwilmsi, RobbieIanMorrison, Robbiemorrison, Robiminer, Ruud Koot, Rxnt, Saiteki-ka, SakeUPenn, Salih, Sardanaphalus, Sepehr ebrahimi, Serols, Simplexity22, Sofia Koutsouveli 2005, Solomon Douglas, Stevan White, Stochastics, SwanseaBen, Thennicke, This, that and the other, Thiscomments voice, Thomasmeeks, Trappist the monk, TureTrueTrue, Tyrneaeplith, Vermont, Waterfall mg, Wavelength, Wcherowi, Widefox, Woohookitty, Yuanfangdelang, Zemminamia, Zfeinst, Βελτιστοποίησης, Іванко1, Гармонический Мир, 133 anonymous edits 189

Automated reasoning *Source:* https://en.wikipedia.org/w/index.php?oldid=852748433 *License:* Creative Commons Attribution-Share Alike 3.0 *Contributors:* Adavidb, Alai, Antonielly, Arapajoe, ArtemisOne, Brian.davis475, CBM, Caesura, Cek, Charles Matthews, ClueBot NG, DaYZman, Dell Adams, Doezxcty, DrStrauss, FleetCommand, From Adam, Gaius Cornelius, Grantbow, Gregbard, HairyWombat, IjonTichyIjonTichy, JamesMazur22, Jarble, Jason Davies, JenniferForUnity, Jochem V, Jochen Burghardt, Kinaro7, Kingpin13, Lesser Cartographies, Lumingz, Matěj Grabovský, Mdebellis, Mets501, Moe Epsilon, Monkey Bounce, Mr.Z-man, Myasuda, Oatsy, Obradovic Goran, OlEnglish, Pacerier, Pathoschild, Phamnhatkhanh, Primefac, Pwjb, R'n'B, RekishiEJ, Rhododendrites, Robertbowerman, Sanchitw, Search4Lancer, Serketan, Shamp00, Simon04, SimonP, Stephan Schulz, TaBOT-zerem, Tea2min, The Transhumanist, Waldir, Witbrock, 40 anonymous edits 208

Search algorithm *Source:* https://en.wikipedia.org/w/index.php?oldid=843200672 *License:* Creative Commons Attribution-Share Alike 3.0 *Contributors:* !dea4u, Adavidb, Adityapasalapudi, Akhil999in, Alex.g, Alexius08, Alexmorter, Altenmann, AngleWyrm, AnotherPerson 2, Barneca, Baruneju, Bbik, Beland, Bogdangiusca, Boleslav Bobcik, Bonadea, Bsilverthorn, Buchibaba, CRGreathouse, CYD, Canopus Grandiflora, Capricorn42, CharlesGillingham, Chris G, Christian75, Cjsmith.us, ClansOfIntrigue, Cleancutkid, ClueBot NG, Colin Barrett, Conversion script, Cspooner, DHN-bot∼enwiki, Dariopy, Daveagp, Dcdanny, Dcoetzee, Decrypt3, DevastatorIIC, Dggreen, Diego Moya, Doc glasgow, Ericvalero2, Esquivalience, Famtaro, Finnusertop, Flandidlydanders, Flordeneu, Fmicaeli, Frikle, GPHemsley, Gattom, Gene.arboit, Giftlite, Gregman2, HamburgerRadio, Horcrux, JNW, Jacobjames10, Jarble, Jim1138, Jitse Niesen, Jorge Stolfi, Jschnur, Kareekacha, Kc03, Kdakin, Kenyon, Kgeza7, KingPuppy, Kku, Kragen, Lamdk, Leprof 7272, Ligulem, MEOGLOBAL, Materialscientist, MattGiuca, Me, Myself, and I are Here, Meredyth, Mezzanine∼enwiki, Michael Hardy, Mikeblas, Mild Bill Hiccup, Mironearth, Mlpkr, Mundhenk, Nabla, Nanshu, Neil Smithline, Nik-t, Nixdorf, Nohat, Northamerica1000, Nyq, Ohnoitsjamie, Omicronpersei8, Onel5969, Oshwah, Pgr94, Phils, Pjoef, Plasticup, Ptbotgourou, Quadrescence, R'n'B, RW Marloe, Ramzysamman, Rgoodermote, RichF, Richardj311, RichiH, Robinh, Rrburke, Ruud Koot, SLi, Saaska, Sam Hocevar, SemiHypercube, Serols, Shuroo, SiobhanHansa, Snowolf, Spidern, Stefano∼enwiki, TaBOT-zerem, Taral, Taw, Tcncv, Tea2min, The12game, TheDragonFire, ThomHImself, Tide rolls, Tijfo098, Tomaxer, Vassloff, Widr, Wtmitchell, Ww, Xijiahe, Xonqnopp, Zzedar, ㅂㄱㅇ, 199 anonymous edits 212

Logic programming *Source:* https://en.wikipedia.org/w/index.php?oldid=850501944 *License:* Creative Commons Attribution-Share Alike 3.0 *Contributors:* 2ndMouse, Afterthewar∼enwiki, Allforrous, Arthur Rubin, Ashmont42, Balrog, Beland, Bellerophon5685, CBM, Carbo1200, Catamorphism, Cdrdata, Cedar101, CharlesGillingham, Chris55, ChrisGualtieri, Classicalecon, ClueBot NG, Comp.arch, Compulogger, Diego Moya, Disavian, Dkwebsub, Docurbs, Don4of4, EdJohnston, Elizabeth Safro, EnOreg, EngineerScotty, Eriake, Eyu100, Fayenatic london, FluxCapacitor, Furrykef, GoingBatty, Gregbard, Izkala, JFreeman, JIP, Jacinto.davila, Janburse, Jarble, Jim Carnicelli, Jmlk17, Jpbowen, Kcho, Linas, LinguistAtLarge, Logicistestpeople, Logperson, Lysy, Maarten van Emden, Magioladitis, Magnus7lab, Mahanga, Mdebellis, Mesoderm, MicahDCochran, Michaelkifer, MontecitoSequoia, MrOllie, Naikrosh, Nathan Johnson, Nejko, Neparis, Nima Baghaei, O.Koslowski, Obscuranym, OlEnglish, Orcrist, PeterSchachte, Pgr94, Pinar, Pleclech, Qwertyus, R'n'B, Radagast3, Raquel Baranow, Reedy, Renku, Rich Farmbrough, Rjwilmsi, Robert Kowalski, Ronchet, Ruud Koot, Shaddim, SimonHova, Skunk44, Slaniel, Smallclone2, Stanislav Nowak∼enwiki, StarmanW, StephenKJohnson, Sun Creator, TakuyaMurata, The Thing That Should Not Be, The Transhumanist, Thepigdog, Thierry Le Provost, Thomas Linder Puls, Toccata quarta, Toussaint, Tresiden, Una Smith, Unquietwiki, Untalker, Vincent Lextrait, Vpab15, Waldir, Wasell, Wavelength, Widr, WikiWinters, Winston Chuen-Shih Yang, Wywin, Zootalures, 127 anonymous edits 216

Bayesian network *Source:* https://en.wikipedia.org/w/index.php?oldid=852435850 *License:* Creative Commons Attribution-Share Alike 3.0 *Contributors:* A3nm, Abacenis, Abbastaher, AgarwalSumeet, Akseli.palen, Aleksanndar1995, AndreasWittenstein, AndrewHZ, Andrewaskew, Angelferrer, Anrnusna, Arrandale, Ausaen, BD2412, Balizarde, Bender235, Benwing, Bgwhite, Biogeographist, Boghog, Brews ohare, C. lorenz, Camrn86, Ceyockey, Chafe66, Chalst, CharlesGillingham, Chris troutman, Citation bot 1, Ckling, ClueBot NG, Colonies Chris, Converge on truth, Cskudzu, Dairy501, Dandinpantelimon, Daniel347x, Daniel5Ko, Danielkorzekwa, David Eppstein, Dcirovic, Destynova, Djsyclik, Druzdzel, Dzmanto, Edward, Enterprisey, Ereator, Erget2005, Erreip, Escribiente, Foerstj, Fpetitjean∼enwiki, Freewol, Frietjes, Garde, GermanJoe, Gidonb, Giliev, Gliall, Grsmvg, Habil zare, Headbomb, Hous21, Hyacinth, Ibsen 13, Illinoisnois, Irad Ben-Gal, Iridescent, IronGargoyle, JGS2010, Jarekt, Jim.Callahan,Orlando, Jodosma, John "Hannibal" Smith, John of Reading, JohnBlackburne, Jonesey95, Jordi Burguet Castell, Jpbowen, Jwatson89, Kercker, Kku, Kolyma, Kurtan∼enwiki, Kvihill, Kyle Cronan, Laeciosantosbsb, Lambiam, Latex-yow, Lebd, Leonariso, Linas, Loskana, Magioladitis, MarcoLittel, Marcocapelle, Mark viking, Mathewk1300, Mattg82, Matthiaspaul, MaxEnt, Melcombe, Mesospheric, Michael Hardy, Mikayé, Mild Bill Hiccup, Mindmatrix, Mogism, MrFeroz91, MrOllie, Mtanti, Neko-chan, Nemo bis, Notsofeo, Ohconfucius, Olexa Riznyk, Omnipaedista, Paheld, PapaJupe, Papadim.G, Paul2520, Pedram Masoudi, Peripattikos, Ph.eyes, Pknad, Priyankp87, Probinf, Puhfyn, Qqq06, Qwertyus, Qwfp, RDBrown, RaphaelQS, Raspabill, Rich Farmbrough, Rjwilmsi, Robin klein, Rxnt, Salix alba, Shaider iba, Siddharth.mv, Sieste, Smartse, Sonarpulse, Stamptrader, Sulabh.avatar, Tassedethe, Tatome, Technokratisch, Terrycojones, The Whispering Wind, The enemies of god, Thomas Kist, Trappist the monk, Trig-ger, Trioculite, Turgidson, Twexcom, Vycl1994, Vyepesp, Wavelength, Widianpear, Wikid77, Wrongfilter, Zanetu, Zariski83, 150 anonymous edits 230

Hidden Markov model *Source:* https://en.wikipedia.org/w/index.php?oldid=852959646 *License:* Creative Commons Attribution-Share Alike 3.0 *Contributors:* AManWithNoPlan, Aberdeen01, Aednichols, Alexander E Ross, Amkilpatrick, Amorymeltzer, Andresmoreira, Anshuldby, Aresgram, Arkanosis, ArtfulVampire, Aymatth2, Balenko, Bang55555bang, Bear-rings, Bender235, Benwing, Boudaren, BubbaRich, Captainfranz, Cedar101, Chewings72, Chire, Chrisjj3, Christian.fritz, ChristopherKingChemist, Ciphers, Citation bot 1, ClueBot NG, Clusternote, David z 1, Dcirovic, Dhirajjoshi16, Dratman, Duncan.Hull, Edeveien, Eflatmajor7th, Eitt, Elaz85, Exeunt, FindingEllipsoids, Frederic Y Bois, Galyet, Gareth Jones, Giftlite, Gioto, Glopk, Gtfjbl, Guslacerda, Hanif Al Husaini, Headbomb, HelpUsStopSpam, Hilvar, Iazyges, InverseHypercube, Izno, J kabudian, J.delanoy, JA(000)Davidson, Jamie7687, Jay Page, Jesse V., Jldurrieu, John of Reading, Jonkerz, K6ka, Klbrain, Lolmid, Lourakis, Luke Maurits, Magioladitis, Marek69, Materialscientist, Maximilianh, McGucket, MelbourneStar, MichaK, Michael Hardy, Minimac, Mmernex, Mnemosyne89, MrOllie, MrSampson, Mv240, Nbro, Nealmcb, Neo1942, Novangelis, OhGodItsSoAmazing, Olexa Riznyk, Otakuftf, Pcompeau, Pintaio, PlyrStar93, ProprioMe OW, Prosperosity, Quantling, Qwfp, Razorbliss, Rcsprinter123, Reposit, Rich Farmbrough, Richwiss, Rjwilmsi, Rolf h nelson, Schmock, SciCompTeacher, Sergmain, Simenheg, Skittleys, Skoch3, Smh.oloomi, Sminthopsis84, Soeding, Sraybaud, Surement, TeaDrinker, Terrycojones, Thorwald, Tobo∼enwiki, Tonyszedlak, Trappist the monk, Tymewalk, Uncle Dick, Urhixidur, Variance3, Virgil disgr4ce, Vivekrs, Vuerqex, Waldir, Wavelength, Widr, Wiki2016edit, WikiLaurent, Wkretzsch, Zixu1986, 212 anonymous edits 246

Kalman filter *Source:* https://en.wikipedia.org/w/index.php?oldid=852434461 *License:* Creative Commons Attribution-Share Alike 3.0 *Contributors:* 14GTR, AManWithNoPlan, Acheck10, AlexCornejo, Avi.nehemiah, Babitaarora, BarrelProof, Bender235, Bytesock, CBM, Certes, ChrisGualtieri, Clemlaflemme, Colorete87, Dalemcl, Davmre, DmitTrix, Don Reba, Dough34, Eitt, Emmanuel Battesti, Excirial, FantasyJXF, Ferrofield, Fgnievinski, Fiacobelli, Frietjes, Funandtrvl, Gpanders, GünniX, Headbomb, HelpUsStopSpam, Hikenstuff, IjonTichyIjonTichy, Ilyushka88, Impsswoon, InterCity(IC), Iridescent, Jamal Abdalla, Jehrlich, Jensschulz, Jerrykorulla, Jevansen, John of Reading, Johnrpenner, Jordi Burguet Castell, Khaosoahk, KorgBoy, Kotika98, Krauss, Kri, L T T H U, Laserhao, Lfstevens, Lipkenszaal, Liwangyan, Loraof, Luigidigiacomo, Lujingyang1029, Mac Cathmhaoil, Markdask, Me, Myself, and I are Here, Michael Hardy, Mintz l, MisterSheik, Mmeijeri, MrOllie, Mtpaley, MultiPoly, Myraeri, Mzhang88, Navraj42, Nbarth, Neko-chan, Niceguyedc, Nikevich, Offsure, Oldlab, Olexa Riznyk, Olli Niemitalo, Onel5969, Oruanaidh, Pdecalculus, Petteri Aimonen, Pezcore2014, Pglotov, Potwang, Przemek-Majewski, Ptsneves, Purefel, Rainald62, Rjwilmsi, Rkinch, Robbbb, Rothhapm, SA 13 Bro, Saung Tadashi, Serols, Sohale, Spacetime123, Stamptrader, Stickee, Stochastics, ThomasTC, Tomhallward, Trappist the monk, Turgidson, Viktor O. Ledenyov, Vishalv2050, Vivic2000, Wikitidy, Xenxax, Yazdandaste, Yongqi lee, צורטק, 219 anonymous edits 266

Particle filter *Source:* https://en.wikipedia.org/w/index.php?oldid=851846700 *License:* Creative Commons Attribution-Share Alike 3.0 *Contributors:* A5, Adoniscik, Allforrous, Amit man, Arnauddoucet, Ashishdhiman12, Avrahamleib, BAxelrod, BeenAroundAWhile, Benlansdell, Benwing, Bgwhite, Billjefferys, Chaosdruid, Chris53516, Chrismartinet, Citation bot 1, Cs california, Danim, DrSparkle713, Drewnoakes, Edcolins, Ego White Tray, Eleschinski2000, Ema–or, Engineeru, Fblasqueswiki, Feyrauth, Forwardmeasure, Giftlite, Gilliam, Gurch, Headbomb, I dream of horses, ILikeThings, Irshadali1, J04n, JSHuisman, Javad.rezvani, Jiuguang Wang, Jmath666, John of Reading, Jonesey95, Jose278, KYN, Kfriston, LindsayH, Loopy30, Magioladitis, Malosse, Marcocapelle, Maugu, Mcld, Mebden, Melcombe, Memming, Michael Hardy, Minusix, MistWiz, Monfera, Muhammad Hamza, Myrbock, NaBUru38, Niceguyedc, Palfrey, Patschy, Patstuart, Pdelmoral, Phil Boswell, Phmresearch, Pierre-delmoral, Pintoch, Quantling, Redgecko, Rich Farmbrough, Rjwilmsi, Rkrish67, Rmcantin, Robertmacl, Robomanx, Rodkaroma117, Ronz, Sanho Kakiherma, Sharkir hussain, Skybugs, Tetu-s, The Anome, Tom.Reding, Trappist the monk, Vema123, Vrenator, Vudro, Vzaliva, Wavelength, Winterstein, Wolverenstein, Wzyuan, XochitlSpinoza, Yoderj, 162 anonymous edits 298

Decision theory *Source:* https://en.wikipedia.org/w/index.php?oldid=852989542 *License:* Creative Commons Attribution-Share Alike 3.0 *Contributors:* 3mta3, 4RugbyRd, 7&6=thirteen, Adoniscik, Alan Dawrst, Andeggs, AndriusKulikauskas, Anonyhole, Arthur Rubin, Awesomekoala, BD2412, Baking Soda, Barek, Bender235, Bennose, BillieBC, Billjefferys, Biogeographist, Bjoram11@yahoo.co.in, Bordley, Bracton, Byelf2007, CRGreathouse, Canadaduane, ChangChienFu, Charledl, Chewings72, Chris the speller, Ciphergoth, Citation bot 1, ComputerGeezer, Cretog8, Curious1i, Cyan, Cyrus noto3at bulaga, D6, DRE, Dandv, Danielpcox, David Eppstein, Dcirovic, Dcljr, Decthe, Denisarona, DoubleBlue, Duoduoduo, Edsix, Edward, Epeefleche, Finn Krogstad, Flaviusvulso, FletchFFletch16, Forich, Frymaster, GESICC, Gandalf61, Gandtedi, Giftlite, Gilliam, Gis72, Glumundrung, GoingBatty, Gomm, Goochelaar, Gregbard, Gurchzilla, Gustronico, Hactuary, Headbomb, Helgus, HelpUsStopSpam, Hisabness, Hubbardaie, Infarom, Ixfd64, J04n, Jackdavinci, Jamelan, Jbusemey, Jeff3000, Jimmaths, Jiuguang Wang, Jmath666, John Quiggin, John V John, Jon Awbrey, Josephina Obama, KSchutte, Keith D, KenKendall, Kiefer.Wolfowitz, Kmbuki, Kpmiyapuram, Kvng, Kzollman, LeadSongDog, Loraof, Lova Falk, Läkarstudenten, Magioladitis, Magmi, Mandarax, Matdrodes, Matt.voroney, Matthias Stein HOU, Mdd, Melcombe, Michael Hardy, Michael Slone, Mike Rosoft, Miraclexix, Miskin, Mmmarilyn, MrOllie, Nbarth, Neelix, NeoJustin, NewEconomist, NoychoH, Nrlsouza, Oeconomica, Ohnoitsjamie, Orpheus Lummis, Pa68, Paulscrawl, Pbech, Penbat, Phiwum, Phronetic, Pm master, Polmandc, Posiebers, PxPz, RDBrown, RJBurkhart3, Radford Neal, Regregex, Rich Farmbrough, RichardF,

Rjwilmsi, Rlsheehan, Rolf h nelson, SUNDARZEET, Sadeghd, Samf4u, Sardanaphalus, SilverLocust, Single-cell, Skr15081997, Smellyfarts001, Stephensuleeman, Superegz, Surveyofipsych, Tassedethe, TenPoundHammer, Terra Novus, Tesseract2, The Gnome, Themepark, Themusicgod1, Thomasmeeks, Three-quarter-ten, Tom.Reding, Trade2tradewell, Trialsanderrors, Twins Too!, U89djt, Volunteer Marek, Vonscheel, Waterfall mg, WayneToms, WeakTrain, WikiWikiWilson, Wikih101, Winterfors, Xyzzyplugh, Zvika, Владимир Паронджанов, Гармонический Мир, 157 anonymous edits 321

Utility *Source:* https://en.wikipedia.org/w/index.php?oldid=849624262 *License:* Creative Commons Attribution-Share Alike 3.0 *Contributors:* 1ForTheMoney, A Macedonian, ASCIIn2Bme, Acadēmica Orientālis, Adrin10, Agent032125, Agor153, Andyjsmith, Avi Harel, BD2412, Bender235, Bfinn, BigK HeX, Blaylockjam10, BobKawanaka, Bowei Huang 2, Breno, Bubby33, Buldri, Calmer Waters, Canadaduane, Caramsay, Charles Matthews, Chealer, Clementi, Clemwang, ClueBot NG, Clément Pillias, CompliantDrone, Cretog8, Crow, D.S. Cordoba-Bahle, Dave Muscato, DemocraticLuntz, Dewritech, Dhan111, Dl2000, Donner60, Dreispt, Dreslough, Duoduoduo, E to the Pi times i, Eb00kie, Ejw wiki editor, Ellisun, EmpireOfWisdom, Erel Segal, Ezhiki, Forich, Frietjes, Fules, Furthermost, Gilliam, Gioto, GoingBatty, Goochelaar, Hammersbach, Histrion, Hve, HybridGirl, Icseaturtles, Jackelfive, Jackftwist, Jackzhp, Jarble, Jayanta Sen, Jbening, Jerome Charles Potts, Jonkerz, Kalinin Nikita, Kayau, Kellysa, Kfsung, Kndimov, Kobusdg, L Kensington, Lamro, Lawrencekhoo, Lbertolotti, Likebox, LilHelpa, Loraof, Lugia2453, Luis Goslin, MER-C, Magic5ball, Makdisse, Marcocapelle, Mark Renier, Masalih, Matthew Proctor, Mediran, MelbourneStar, Melcombe, Michael Hardy, Miguel Meireles, Mild Bill Hiccup, Mrevan, Mrkonium7, Noel Streatfield, Nrlsouza, Nubeli, O18, O5o7, Ohconfucius, Oliverlyc, Orenburg1, Oshwah, Owen Ambur, PLA y Grande Covián, Pacerier, Patrick, Penbat, Pischinas George, Reaper Eternal, Renamed user 5695569576f6b340, Rinconsoleao, Rjwilmsi, RoyalTS, Ruby Murray, Rupertgatti, Sfan00 IMG, SheriffIsInTown, Shiftchange, SlamDiego, Sniedo, Stimulieconomy, Sugarfoot1001, Sylvain1972, THEN WHO WAS PHONE?, The Banner Turbo, The Master of Mayhem, Themightyrambo, Titodutta, UKoch, Uppalapadu27, Vapour, Vasil', Vatsal4592, Velella, Volunteer Marek, Vramasub, Wdspann, Widr, Wikimorphism, William3576, Windharp, Woohookitty, Wragge, Yolkyal, Zfeinst, Δ, 168 anonymous edits 328

Statistical classification *Source:* https://en.wikipedia.org/w/index.php?oldid=852260700 *License:* Creative Commons Attribution-Share Alike 3.0 *Contributors:* Abc-wtf-def, Agor153, AlexNet22, Allforrous, Anthony Appleyard, Ashershow1, BD2412, Beetstra, Beland, Benwing, Bgwhite, Billgaitas@hotmail.com, Bkkbrad, Black Kite, Bohunk, Buntworthy, BustYourMyth, CAPTAIN RAJU, Can't sleep, clown will eat me, Chire, Chris the speller, Chyvve, Classifier1234, ClueBot NG, Cybercobra, Datakeeper, David Eppstein, Deacon Vorbis, Denoir, Dtrebbien, Favonian, Ferrarisailor, Fgnievinski, Fly by Night, Francisbach, GTBacchus, Gdupont, Giftlite, Giggly37, Gire 3pich2005, Girish280, HelpUsStopSpam, Hike395, Hu12, I Less than3 Maths, Imphil, Jamelan, Jarble, Joel B. Lewis, Jonkerz, Jrbouldin, Juansempere, Junling, Jérôme, Kierano, Klilidiplomus, LdyBruin, Loraof, Mack2, Mark viking, Mcld, Me, Myself, and I are Here, Melcombe, Memming, Michael Hardy, Microfries, Movado73, Nbarth, Nervexmachina, Object01, Oleg Alexandrov, Oleginger, Peteymills, Prolog, Quocminh9, Qwertyus, R.davtalab, RJASE1, Richard001, Rmashhadi, Robotman1974, Roboto de Ajvol, Rod57, Sigma0 1, Swsboarder366, Telfordbuck, The Anome, ThomHImself, Tiffanicita, Tobi Kellner, Trauber, Verbum Veritas, Violetriga, VoABot II, WikiBotas, Xfthhxk, 136 anonymous edits 337

Artificial neural network *Source:* https://en.wikipedia.org/w/index.php?oldid=852861901 *License:* Creative Commons Attribution-Share Alike 3.0 *Contributors:* 72, ATPhosphate, Abou Ben Adhem, Adityauj, Alex4532, AllyD, BD2412, Bikingdog, Billinghurst, Bjordan555, Botatao, Breezeway77, BronHiggs, CaraniS, Certes, Chrissymad, ClueBot NG, ColeDeanShepherd, Dgpop, Dhruvh, Dividingbyzerofordummies, Dmt137, Dschotze, Ettrig, EyedMoon, Frosztbyte, Get Learnt, Ghost30, GünniX, Headbomb, HelpUsStopSpam, IAFSK, Iluminame, Jarble, Jessicapierce, Jim1138, Joeykai, Jwilleke, Keministi, Kri, L8 ManeValidus, Lenov, Lepsyleon, Leschnei, Lfstevens, Lindsay658, Ma ra pa pe, Magioladitis, Mattythewhite, MaynardClark, Mbisgaier, Memento Mori 4, MfortyoneA, Michaelmalak, Miraclexix, Monadial, MrOllie, MrRK, Nada bin Alshaikh, Narky Blert, Nbarth, Nbro, Neko-chan, Neo-Jay, Newslinger, Nitpicking polish, Olexa Riznyk, Omnipaedista, Onco p53, Onel5969, Optate823, Oranjelo100, Oxinabox, ParliamentFunk, Patriotmouse, PcPrincipal, Pgan002, Physicalmathematics, Pkbwcgs, Plastikspork, PointyOintment, Quasar G., RichardWeiss, Rjwilmsi, Rodw, Sandstein, Seppi333, Softtest123, Sophiemaywilson, Stefan Weil, The Quixotic Potato, Tom.Reding, Trappist the monk, Vbrm, VeniVidiVicipedia, Wbm1058, Wcherowi, Zidong Du, طاها, 91 anonymous edits 343

Connectionism *Source:* https://en.wikipedia.org/w/index.php?oldid=846751415 *License:* Creative Commons Attribution-Share Alike 3.0 *Contributors:* AKA MBG, ATPhosphate, Abigbluetable, AgamemnonZ, Aldux, Andycjp, Ariel Pontes, Arjayay, Asbestos, Aubreybardo, Boblayton, Brian Sayrs, Brinerustle, Bscotland, Byelf2007, CBM, Cadr, Chaos, Charles Matthews, Christian.B, Cyrus Grisham, DARTH SIDIOUS 2, Dankonikolic, Deodar∼enwiki, Dicklyon, Djfrost711, DorisH, Dpleibovitz, Dreiche2, Drphilharmonic, Ewlyahoocom, Fadyone, Flyhighplato, Gene s, Gerocus, Givekeeps1109, Gregbard, Gzornenplatz, Helgus, Helwr, Ignacio Icke, Iztwoz, JMBrown, Jandalhandler, Jeblad, Jm34harvey, Jonathan.s.kt, Junes, Jwdietrich2, Khazar2, Kku, Kzollman, Lestrade, Libcub, Lizia7, Lova Falk, Lumos3, Male1979, Mangst, McGeddon, Mdd, Michael Hardy, Mikeblas, Mnaini, Mneser, Mogism, Mosquitopsu, Moyamensing2253, Mporch, Mr. Stradivarius, Nesbit, Netlad, Noticket, Omnipaedista, P.r.newman, PBS-AWB, Paine Ellsworth, Paul August, PaulTanenbaum, Pgan002, Physicalmathematics, Prari, Ps29, Quadell, Quota, Randykitty, Reedbeta, Reyk, Rhododendrites, Rich Farmbrough, RichardF, Robin klein, Robotczar, RoyBoy, Saikiri, Sandman2007, SchreiberBike, Shellwood, Skiera1987, SlipperyN, Solomon7968, SparkOfCreation, Spoon!, StephenReed, Thatotherpoetryguy, Thomas Kist, Thrilway, Timwi, Trappist the monk, Tremilux, Truecobb, Tyrell turing, Vacio, Velvel2, WIZLOR, Whoiamudunno12, Wknight94, Wmahan, WurmWoode, Zeno Gantner, Zinjixmaggir, 169 anonymous edits 377

Deep learning *Source:* https://en.wikipedia.org/w/index.php?oldid=852565513 *License:* Creative Commons Attribution-Share Alike 3.0 *Contributors:* A Train, ADMIS Walker, ATPhosphate, Ahiijny, Alex Cohn, Alvations, Amkgp, Amkilpatrick, Ancheta Wis, Andreponcedeleon, Andyx96, AnoopBadhe, Avoided, Billinghurst, Black Kite, Bluepenguin14, Boghog, Breezeway77, BustYourMyth, Camberleyb, Chris the speller, Classicwiki, ClueBot NG, Comp.arch, Davemck, Deeplearningdude, Deville, Dharmablues, Diannaa, DrTonyR, Edechter87, Elitestocks, Equinox, Ettrig, EvilxFish, FelixF1iX, Finlay McWalter, Fionaz6954, Fluda3, Fuyuko Yukimura, Fxpena, George100, GoingBatty, GünniX, Headbomb, HelpUsStopSpam, Imonban, Ira Leviton, Isingness, Jarble, Jhertel, Jim1138, Johngraybosch, Josve05a, Jpopmo, Julietdeltalima, Junkyardprince, Justeditingtoday, Jytdog, Kamykowsari, Keith D, Kk7nc, Kku, Kndimov, Koertefa, Lalzary, Larry.europe, Lfstevens, Lgfg86, Like.liberation, Linstancy, LittleWalrus, Lojbanist, Lotje, Ma ra pa pe, MaNeMeBasat, Magioladitis, Malangali, Marcelo.brocardo, Mathpriest, MatthiasWalter, Me, Myself, and I are Here, Mean as custard, Mehdi.moh, Mellbourn, Microbiomeman, Narky Blert, Nbarth, Nbro, Neo-Jay, NewEnglandYankee, Ni7hinkz, Nuclearduck, PaulTanenbaum, Person54, Peterl, Pineapple2018, Prakash Nadkarni, Privatewikiuser, Quebec99, Quercus solaris, RasF, Retimuko, RichardWeiss, Rjwilmsi, Rolf h nelson, Rubbadub42, Rupendra.sh, Sadegh3333, Seazzy, Shaded0, ShesLostControlAgain, StanLeeP, Stesmo, Synchronist, SyteAI, Szo4, Telempe, The Anome, TheRealPascalRascal, Three-quarter-ten, Tjcpl, Tlinse, Tom.Reding, Tony Dou, Tpbradbury, Trougnouf, Vaishnavi Agrawal, VeniVidiVicipedia, Yoderj, Zyumbik, 99 anonymous edits 385

Recurrent neural network *Source:* https://en.wikipedia.org/w/index.php?oldid=852792764 *License:* Creative Commons Attribution-Share Alike 3.0 *Contributors:* Aaronbrick, Achler, Acx01b, Adrian Lange, Alai, Arado, AugustusSFXvanDusen, Banus, Barak∼enwiki, Bender235, Bhny, Certes, Charivari, Charles Matthews, Chris.addy, Chrisk3141, Citation bot 1, ClueBot NG, Comp.arch, Curdeius, DASHBotAV, Daniel Hershcovich, DavidFarmbrough, Daytona2, Dcirovic, Dicklyon, Dithridge, Djfrost711, Dmitry St, DoubleDr, Doug4, Eczanne, Epsiloner, Fadyone, Fhhdak, Flewis, Fmadd, Frze, Fyedernoggersnodden, Gdupont, Glenn, Glovacki, GünniX, Headbomb, Headlessplatter, Incfk8, Itb3d, Ivan Ukhov, JaGa, Jamalex∼enwiki, Jamesmfr, Jeblad, Jennica, Jmander, John of Reading, Jtaylor100, Jungkanji, Justin Wabscott, Justinyap88, Jwray, Keith D, Kissaki0, Kjerish, KylieTastic, Lazaronijunior, Lfstevens, Like.liberation, Lupesko, Magioladitis, Male1979, Mathpriest, Me, Myself, and I are Here, Mean as custard, MichaelGasser, Minky76, Mister Mormon, Mknjbhvg, Moxon, Mreczko, NTheZone, Nehalem, Nikolatesla90, OhGodItsSoAmazing, Olexa Riznyk, Omnipaedista, Philiptkd, Primefac, Qwertyus, RJFJR, Randykitty, RichardF77, RichardTowers, Rjwilmsi, Roux, Rwalker, Selichm, Seliopou, Serols, Shugah, Slashdottir, SlaterDeterminant, Slowfun, Snowycats, Sunnydreamrain, Terry Bollinger, That Guy, From That Show!, The Anome, Tom.Reding, Toni 001, TowerOfBricks, Trappist the monk, Tyrell turing, Ukpg, Urhixidur, VeniVidiVicipedia, Wiki aurel, Winterstein, Yume149∼enwiki, طاها, 110 anonymous edits 402

Progress in artificial intelligence *Source:* https://en.wikipedia.org/w/index.php?oldid=852483261 *License:* Creative Commons Attribution-Share Alike 3.0 *Contributors:* Acyclic, Ad48, Arado, BD2412, BenKovitz, CharlesGillingham, Chris55, ChrisGualtieri, Citation bot 1, ClueBot NG, Danim, Esrogs, Fixuture, Gap9551, Hyperforin, IO Device, Inx272, JanAson, Jon Kolbert, Koko90, KyleH87, LarryJeff, LeJuifErrant, Marcocapelle, Maveridze, Mikael Häggström, Mr Minderbinder, Mschribr, N7403118k, Narky Blert, O.mangold, Onel5969, Pgr94, Pintoch, Rallen7753, Rangoon11, Rjwilmsi, Rolf h nelson, Sandegud, Shadowjams, Silence, Tedsanders, The Transhumanist, Tom.Reding, Wikid77, Xavier Combelle, 28 anonymous edits 416

Applications of artificial intelligence *Source:* https://en.wikipedia.org/w/index.php?oldid=852792557 *License:* Creative Commons Attribution-Share Alike 3.0 *Contributors:* 1337pino, A*-search, AManWithNoPlan, APMcM, Achowat, Alex Silverio, Alviniac, Ameya text recruit, Annaseg, Apparition11, Arthur Rubin, Ashleytway, Auric, Ayomawdb, Beland, Birchfaber, Bmancl, Bquast, Brandmeister, Canzhiye, Cbdorsett, CharlesGillingham, Chase me ladies, I'm the Cavalry, Citation bot 1, ClueBot NG, Coco Briaval, Comp.arch, Compfreak7, Countchoc, Cpgupta9460, Cresdajv, Csc300student, CuriousBrendan, Diannaa, Diego Moya, Diyguy13, DkimEYE, EinZwei, ElPao2, Elperzon, Fathomer, Ferrara p, Flyer22 Reborn, Flytbase, Fouindor, Fuddle, Gh23, GraffDecors, Grayfell, Guy Harris, HaeB, Hannahmaxtra, HarlandQPitt, HelpUsStopSpam, Hestelars, Hyksandra, Hyperforin, IEat, IO Device, IjonTichyIjonTichy, JHP, JamesKotecki, JamieWilson99, Jarble, Jdaloner, JeremiahY, Jim1138, Jiuguang Wang, John of Reading, Jones585, Julaei, KAP03, KH-1, Karunamon, Kesara, Kku, Kuru, Kvsbharath, LakesideMiners, Landon1980, LaurenWakeling, Laurenmcq, LeadSongDog, Lizfarabee, Llorenzi, MER-C, Magioladitis, Makecat, Markicus, Materialscientist, Mawiyakaram, Me, Myself, and I are Here, Mean as custard, Mikael Häggström, MilborneOne, MosheEmes, Mpritham, MrOllie, Nepenthes, NicoScribe, Nicoaraeradu, Nikkimaria, Omnipaedista, Onel5969, Orchidplays, Oshwah, PL290, Panacheola, Pgr94, Philip Trueman, Pinethicket, Quipa, Randykitty, Rangoon11, ReformatMe, Rich Farmbrough, Rikki233752, Rjwilmsi, Robofish, Rocketrod1960, RogerWilco13, Rolf h nelson, Sandstein, Skr15081997, Soler97, Speed57, Steel1943, SuperSmashBros.Brawl777, Surlycyborg, Susanmurray84, Svick, Tankst, Telfordbuck, The Transhumanist, TheEducatedPlatypus, Thine Antique Pen, Tnordlan, Tocretpa, Ugog Nizdast, Vasiliy Faronov, Vcwatcher, Velella, Veveapril, Vinita.m, Wangxinningcandice, Wassname, Wavelength, Woohookitty, Wtmitchell, Xamuel, 146 anonymous edits 425

Lethal autonomous weapon *Source:* https://en.wikipedia.org/w/index.php?oldid=852189875 *License:* Creative Commons Attribution-Share Alike 3.0 *Contributors:* Antepali, Arado, Audacity, Baerentp, Beland, BilCat, Biscuittin, Bosonghan, Ciphers, ClueBot NG, CommanderOzEvolved, Compassionate727, Fixuture, FockeWulf FW 190, GoDaddyWilliams52759, Gubrud, Hagen82, Hcobb, IHarris, Ira Leviton, Kbog, Keith D, LilHelpa, Lopifalko, Madreterra, Mhazard, Onel5969, Oranjelo100, Rolf h nelson, Serols, Sonacarry, Widr, 19 anonymous edits 439

Unmanned combat aerial vehicle *Source:* https://en.wikipedia.org/w/index.php?oldid=852463408 *License:* Creative Commons Attribution-Share Alike 3.0 *Contributors:* AHMED XIV, ATLStudent19, Aaqibobaid, AdRock, Adamgerber80, Adûnâi, Afernand74, Afrachel, Ahpook, And16kmsouth, Andrew Dalby, Anna Frodesiak, Arado, Ashuyadav974, Azcolvin429, BD2412, Beland, Bender235, Bgwhite, BilCat, Binksternet, Blacktiger87, Boris Lewis, C0mmunistdaugther, CAPTAIN RAJU, CensoredScribe, CheChe, ClueBot NG, Cojo21, Coolinu, Cyberbot II, DASonnenfeld, Dana Paymardi,

Daniel.Cardenas, Danrok, Davemillman, David Biddulph, Dawnseeker2000, Dcirovic, DexDor, Diblidabliduu, Dingsuntil, Dougmcdonell, DreamLiner, Edit44441, Edsurridge, Edward321, Ekem, Epicgenius, Esmoen00, Ex nihil, Faizan, Faizankhan913, Faizannehal, Flayer, Flyer22 Reborn, Fnlayson, FreeRangeFrog, Fyrael, Galant Khan, Garuda28, Gazkthul, Gire 3pich2005, Handball-Linus, Hebrides, Hmains, Hohum, I dream of horses, Iamramizansari, Icewhiz, IjonTichyIjonTichy, Ita140188, JMitakon, Jack Greenmaven, Jasonanaggie, Je.est.un.autre, Jim1138, John of Reading, Julesd, Justlettersandnumbers, KConWiki, KH-1, KSFT, KTo288, Kidburla, Kihaduwege, Kintaro, Kku, Kuldeepsingh567, Lfstevens, Librsh, LuigiPortaro29, MER-C, MRD2014, Magioladitis, Marko75, Materialscientist, Mimarx, Mozzie, Muruga86, NDPlume, Newbreeder, Nicky mathew, Nickyfrenchdoc, Nikeunltd, NikhilReddy8320, Nikolas Ojala, OnBeyondZebrax, POONSLAYER59, Pawyilee, Prabu Muthusamy, Professortimithy, Ravinder3790, Rei, Rjwilmsi, Roberticus, SHCarter, Samw, ScienceApe, Scotteaton92, Seanker, Simplytoodope, Sp33dyphil, Starless∼enwiki, Stesmo, SylviaStanley, Tatedengu, ToBeFree, Travers, Unreal7, Vianello, Volunteer2017, Webmgr, Wiki User 10262016, WikiBulova, Wikideas1, Woodgridge, X28BZ1, XX8BZ3!, YSSYguy, Yourmomsfun1789, Zion 1991, ZombieSteve, سامي الرحيلي, 青天白日中国 , 160 anonymous edits 444

Artificial intelligence arms race *Source:* https://en.wikipedia.org/w/index.php?oldid=851272666 *License:* Creative Commons Attribution-Share Alike 3.0 *Contributors:* Dawnseeker2000, Diptanshu Das, GoingBatty, Jefflayman, Rodw, Rolf h nelson, Wakari07, X1\, 2 anonymous edits 459

Artificial intelligence in healthcare *Source:* https://en.wikipedia.org/w/index.php?oldid=852878773 *License:* Creative Commons Attribution-Share Alike 3.0 *Contributors:* Andrescrodriguez, Apparition11, Bremcdee, DkimEYE, EllaLewisTexas, Haraldmmueller, Headbomb, Jarble, Katharineamy, Keith D, Kku, Laurenmcq, LilHelpa, Magioladitis, Me, Myself, and I are Here, Microbiomeman, Mikael Häggström, North8000, Pawn number zero, Retinarecorder, Robjokcs, Seacactus 13, Vonaurum, 5 anonymous edits 464

Autonomous car *Source:* https://en.wikipedia.org/w/index.php?oldid=852701618 *License:* Creative Commons Attribution-Share Alike 3.0 *Contributors:* 98r-b, A Train, Absolutelypuremilk, Adueppen, AidanMStokes, Aisteco, AkselA, Alexf, Alpha3031, AntiCompositeNumber, Avernarius, Avg W, Azcolvin429, BMoore OCC, Beland, Berria1, Besselfunctions, Bloger, Bongwarrior, BrightR, Camofury, Canol, Chensiyuan, ClueBot NG, ColinClark, Cordz, DMacks, DadaNeem, Daniel.Cardenas, Darrelljon, Dawnseeker2000, Ddabest, DeFacto, Deli nk, Diannaa, Dicklyon, Dl2000, Doctor Hobnob, DriveByWriter, EatePurple, Ehrenkater, El C, Elwwod, Entranced98, Frmorrison, Gareth Griffith-Jones, Gmarmstrong, GoingBatty, Greshthegreat, Greyjoy, Groiglery1217, Habari, Hacker445, Hawa098, Headbomb, Hummerrocket, I dream of horses, Iamthomas nine, Imreadytodie, Ita140188, Iuliiai, Jac16888, Jantangring, Jaywubba1887, Jimwilcox79, Jlr3001, Joe Moore Jr., Jytdog, Kashmiri, Kge013, Khayatian, KinkyLipids, Kintetsubuffalo, Kku, Krashlandon, KylieTastic, Lfstevens, Librsh, Lklundin, Lmccanna, Madison.d, Matt Whyndham, Merxistan, Mickgusty, Mikeblas, MikeyTay, Moaz786, Mohanpawar1425, Mpawluk, MrVhyte, Mschribr, Nathanbypark, Neils51, NicolasJz, One passer by, Onel5969, Oshwah, Owen214, Pbritti, Petersons79, Pmoney6, Pmt, Radiphus, Randfal2, Retimuko, RexPatricius, Rishabhsharma1895, Rjwilmsi, Rodney Baggins, Ryan778, SaintWilliamLaSalle, Samdoj, Sanchb1, Sarrotrkux, Scrabbler94, Shellwood, Sir Beluga, Smeagol 17, Sonictrey, SophieWhitton, Staalij, Staszek Lem, Stickee, TheConnorMan, Theinstantmatrix, Theroadislong, ThomasFTomczyk, Toasted Meter, Tortskiel, Trickshot12, Twillisjr, V29, Vlatkosh, Wakari07, Widefox, Winsoft∼enwiki, Xjdjdksnsjsekemrk, Ywqz228, Ziounclesi, 166 anonymous edits 469

Philosophy of artificial intelligence *Source:* https://en.wikipedia.org/w/index.php?oldid=848364986 *License:* Creative Commons Attribution-Share Alike 3.0 *Contributors:* Aeternus, Alan smithee, Aliensyntax, Andy Dingley, Arne Heise, BD2412, Bamyers99, Bender235, Bensin, Bgwhite, CForcino, Chaosdruid, CharlesGillingham, Chiswick Chap, Chlebashořčicí, Chris Hallquist, Chris55, Citation bot 1, ClueBot NG, Coffee2theorems, Cpiral, David Ludwig, DavidCary, Derek Ross, Diletante, Drbogdan, Drunken Pirate, Dschwen, Edokter, Edward, Ego White Tray, Erianna, Favonian, FelixRosch, Foryourinfo, Gregbard, Ian.aaron, Ilkali, Imminent77, Ioverka, JaGa, JamesBWatson, Jamestown Easy Slider, Jnorthup, Jochen Burghardt, Jonesey95, KenTancwell, KindOfHuman, Kumioko, LilHelpa, Liz, Lordvolton, Mandruss, Markhurd, Maurice Carbonaro, McGeddon, Me, Myself, and I are Here, Meat.machine, Michael Hardy, Mimihitam, MrRK, Nbro, Nihil novi, No Swan So Fine, NoToleranceForIntolerance, OccultZone, Omnipaedista, Onel5969, Paine Ellsworth, Paradoctor, Phil.wasag, Raffethefirst, Ramaksoud2000, Ran4, Rjwilmsi, Robertinventor, Rolf h nelson, Roy W. Wright, Ruud Koot, Schaefer, SchreyP, Sfan00 IMG, ShelfSkewed, Silence, Skilmer, Slippingspy, Sm8900, Sophus Bie, Stefanomione, TeacherofJewishEthics, The Blade of the Northern Lights, The Transhumanist, TheTito, TiagoTiago, To-2098, Tom.Reding, Tommy2010, Tuntable, User000name, VerticalNexus, VerticleNexus, WRK, Wejer, Woohookitty, Wvbailey, ZeroOne, Zhang dapeng, 108 anonymous edits 507

Ethics of artificial intelligence *Source:* https://en.wikipedia.org/w/index.php?oldid=850204271 *License:* Creative Commons Attribution-Share Alike 3.0 *Contributors:* AManWithNoPlan, Aeternus, Alan Dawrst, Aliensyntax, Alieseraj, Alimony, Aloisdimpflmoser, Altenmann, Antepali, Antimatter15, Asfan95, BD2412, Beland, Blurryman, Boindil∼enwiki, Brian Tomasik, BritishWatcher, BurritoBazooka, Byelf2007, CharlesGillingham, Cheesefondue, Chucks91, Citation bot 1, ClueBot NG, ComputerGeezer, Computerboy0, Crzer07, Danim, Datboi69696, David Gerard, Download, Dr.enh, Ekips39, EvanProdromou, Fama Clamosa, Fd1193, FelixRosch, Fixuture, Gap9551, Gilliam, IjonTichyIjonTichy, Jamesontai, Jarble, Jason P Doherty, Jdaloner, Jknacnud, Jules.lt, Kai-Hendrik, Keith D, KenTancwell, Kijuhy, Kku, Kpooji74, Kumioko (renamed), Lamro, Langchri, Lightbound, Lustrousstasia, MBlume, Markhurd, Me, Myself, and I are Here, Meneth, MichalKwiatkowski, Motherhalo, MrOllie, Omnipaedista, Oshwah, Owlsmcgee, Polar Mermaid, Pratyya Ghosh, RA0808, Ragwafire, RememberingLife, Riceissa, Rjwilmsi, RogueMedkit, Rolf h nelson, S.Saab77, Seigi Shugoshin, Seniorexpat, Shadowfall8, Sidhray, Silence, SingleUse278, Skilmer, Sm8900, Somewhatdazed, Spectrum629, Staszek Lem, Telfordbuck, Terraflorin, The Transhumanist, Thomas Kist, Tideflat, Tompop888, Turn on a Dime, Viriditas, Virtualerian, Vluczkow, Whibbard, Yakri, 88 anonymous edits 523

Existential risk from artificial general intelligence *Source:* https://en.wikipedia.org/w/index.php?oldid=852347178 *License:* Creative Commons Attribution-Share Alike 3.0 *Contributors:* 06Dom,TJ,Alex ext2015, Aisteco, AntHerder, Apollo The Logician, Arado, Bgwhite, Bosonghan, BriefAeon, CataracticPlanets, Catlemur, Ceosad, Chiswick Chap, Clevera, ClueBot NG, Curious1i, DanielVallstrom, Darko A7, Davearthurs, David Gerard, Debresser, Deku-shrub, Dewritech, Dirtyfilthy, Djkutmasterl, Doulph88, Dqeswn, Dweller, Edokter, Eldorado234, Exa∼enwiki, Fixuture, Fuddle, Gap9551, GermanJoe, GliderMaven, Greenrd, GünniX, Hddty., Headbomb, I dream of horses, IO Device, IjonTichyIjonTichy, Jadixg, Jeoresearch, Jeraphine Gryphon, Jhertel, Kbog, KeanuMoowgliie, KenTancwell, LaundryPizza03, LithiumFlash, Madreterra, Marblespolice, Marcocapelle, Monolithica, Niceguyedc, Nicolap16, Nikolaiho, Nirinsanity, Ohnoitsjamie, Omnipaedista, Onel5969, PaleoNeonate, Paperclip Maximiser, Phantomket5, Polar Mermaid, Quantanew, Rebroad, RedViking, Riceissa, Rinaku, Rjwilmsi, Robert the Devil, Rolf h nelson, S'Trololol, Silence, Spectra239, The Transhumanist, TheEyesOfNye, Tom.Reding, Tri400, Twilight Magic, Ugog Nizdast, Vianello, Vluczkow, Whibbard, WurmWoode, 65 anonymous edits 531

Computer Power and Human Reason *Source:* https://en.wikipedia.org/w/index.php?oldid=848808114 *License:* Creative Commons Attribution-Share Alike 3.0 *Contributors:* 666achmedia, Aloisdimpflmoser, CharlesGillingham, Conscious, Fadesga, GregorB, HMSSolent, Janks, Kai-Hendrik, Kku, Lesser Cartographies, Liz, NW7US, Pegship, Protonk, Ra2007, Randy Kryn, Rayanenadiem, Seaphoto, Superversive, 19 anonymous edits 550

Friendly artificial intelligence *Source:* https://en.wikipedia.org/w/index.php?oldid=845501436 *License:* Creative Commons Attribution-Share Alike 3.0 *Contributors:* "alyosha", AThing, Absinf, Adavidb, AipediaFridemar, Alfio, AllenZh, Angela, AntHerder, Apollo The Logician, Arokha, Asbestos, Auric, Autrijus, Bcooper87, Brian Tomasik, C133, CharlesGillingham, Chris the speller, ChrisGualtieri, Ciphergoth, Civic Cat, ClueBot NG, CosineKitty, Curious1i, Cyberbot II, Cybercobra, CyliCircueta, Danim, Dastract, Davearthurs, David Gerard, Davidcpearce, Dcarroll9999, Dcoetzee, Deego, Deodar∼enwiki, Diza, Dmytrobilan, Doopokko, Edward, Eequor, Emesee, Eric Herboso, ErikHaugen, Esprqii, Ferrara p, Fixuture, Frap, GGordonWorleyIII, Gabbe, Gap9551, Gmcomp, Graham87, Greenrd, Gregbard, GregorB, HLieberman, Hawkseraph, HelpUsStopSpam, Jackfork, Kbog, Kingturtle, Krunchyman, Lambiam, Lamro, Legobot II, Lexor, Lightbound, Lordvolton, Lrieber, Ludwigs2, MBlume, Mark Waser, Mathiastck, Mayumashu, Me, Myself, and I are Here, Meegs, Meelar, Mike Schwartz, Mporter, MusikAnimal, Mz7, Natg 19, NeatGrey, Nickjhay, Oashi, Omnipaedista, Orenburg1, Peterdjones, ProfitlineInc, Quinton Feldberg, Rbarreira, RedWolf, Rich Farmbrough, RichardF, Rjwilmsi, RobinK, Rolf h nelson, Rrococo, Schaefer, Silence, Snori, SoledadKabocha, Spectra239, Sta064, Sunrise, Tawker, The Transhumanist, TheWanderer1357, Tuntable, Ulric1313, Vluczkow, Waldir, WilliamH, Xyzzyplugh, Yodin, 130 anonymous edits 551

Artificial consciousness *Source:* https://en.wikipedia.org/w/index.php?oldid=841118099 *License:* Creative Commons Attribution-Share Alike 3.0 *Contributors:* Asdklf;, AtosiOBP, BD2412, Bender235, Biogeographist, Brian Tomasik, Bronze2018, ClueBot NG, Comp-heur-intel, CooperDenn, Cynulliad, David.moreno72, Deanhorak, Diego Moya, Epipelagic, Faolin42, Favonian, Fixuture, Flarn2006, Greensburger, Harizotoh9, Headbomb, Hyksandra, IdreamofJeanie, Ignatios2000, Jarble, Jasper Deng, Jim1138, Jimhanks289, Keith D, Mdtrotter, Me, Myself, and I are Here, Medgirl131, Michael2118, Motivação, Naresh Choudhary, Nbro, Niceguyedc, Nick Moyes, Nobobo, Omnipaedista, Periksson28, Pmyxjwdm, Qwertyus, RainFall, Rjwilmsi, SJ Defender, Scgtrp, SkyWarrior, Staszek Lem, The Transhumanist, Tkorrovi, Trappist the monk, UltraBee, Usa60527, Virtuosonic, Widr, Zero Thrust, 110 anonymous edits 557

Technological singularity *Source:* https://en.wikipedia.org/w/index.php?oldid=852378572 *License:* Creative Commons Attribution-Share Alike 3.0 *Contributors:* Abcjake, Abierma3, AddressOKX5, Allan428, Anastrophe, Andre Baerfuss, Apollo The Logician, Arado, ArtKocsis, Aviartm, Bachware, Bancron, Bender235, Chienlit, Chiswick Chap, ChrisHigh, Cleeder, ClueBot NG, ColRad85, Comp.arch, Cristi215, Cwobeel, Cyberbot II, Dan Koehl, Daniel Chmielnicki, David Gerard, Dawid2009, Dawnseeker2000, Dcbritt1, Deepred6502, Deez123456, DigitalDev, Earcanal, Editor plus, Editor2020, Edoe, Emmanuel Korir, Ericbe, Favonian, Favre1fan93, Fixuture, Furrykef, Gap9551, Garli, Gendelev, Geranim0, Ghmyrtle, Gilliam, Gilo1969, Gizziiusa, Gracefool, Grayfell, GreenIn2010, Harizotoh9, Headbomb, InfoCmplx, Intgr, Ita140188, Jakeman123, Jamesmcmahon0, Jarble, John of Reading, Jonesey95, Joosh77, Jorgelpt, Junbibi, K6ka, Kapil.xerox, Kbog, Keri, Kkmurray, Koustave, LionMans Account, Lovelove17243875683, Marcocapelle, Materialscientist, Me, Myself, and I are Here, Microplant, MjolnirPants, Modanung, Narky Blert, Nihil novi, Obsidian Soul, Osndok, Ost316, Paradoctor, Philip02003, Pincrete, Pitchcapper, Plandu, Plumbago, Prinsgezinde, QuinnLopi, Quinton Feldberg, Qwertyus, Raymond Leonard, Reason is Immortal, Rivrdansr, Rodw, Rolf h nelson, Ronsams, Sandegud, Silence, Smotezoned, Sofcap, Sorenlouv, Spectra239, Stephenb, Susurrus, TDurr1, The Transhumanist, Tim333, Tom.Reding, Twilight Magic, Ubiquity, Usa60527, ValentinLageard, Widefox, Widr, Wiki-uk, WingedPig, Wiseparticle, Wolfdog, WurmWoode, Zamaster4536, Zarex, Zupanto, 128 anonymous edits 571

Transhumanism *Source:* https://en.wikipedia.org/w/index.php?oldid=852438366 *License:* Creative Commons Attribution-Share Alike 3.0 *Contributors:* 49ersBelongInSanFrancisco, 84user, AKS.9955, Abierma3, Alex Spade, Arado, Archon 2488, Arosophos, Asamind, Auferu, Bender235, Brandmeister, Bri, BrightonC, Cassandra3001, Chris Howard, Chris the speller, Christopherdiogenes, Cishumanist, Citydoctor, Clarinetguy097, ClueBot NG, CommonsDelinker, Crocutacrocultist, Curious1i, Cyberbot II, Daniel G Rego, Danish maggot, David Gerard, DavidJKelley, Davidcpearce, Davlujan, Dawnseeker2000, Dcirovic, Deku-shrub, DieBuche, Dnaber, Doncram, Dr.K., Edward321, Eelizabethb, EricPeden, Evolution and evolvability, Examen Intelligentia, Faolin42, Fences and windows, Ferret, Fixuture, Flash401, Flashpoint2589, Florian Blaschke, Frietjes, Galobtter, Gap9551, Geniac, Gilliam, GoingBatty, Grayfell, Habitmelon, Headbomb, Henderman312, Heuh0, Hwbehrens, Ian.thomson, ImHere2015, Infrasonik, Ioshe, IronGargoyle,

Isambard Kingdom, JJMC89, JackRonan1, Jan Winnicki, Jarble, Jiklo1569, Jim1138, Jmcgnh, John of Reading, Jonesey95, Joseperez22, Julietdeltalima, KailaReed, Kbog, Kevin Dewitt, Kind Tennis Fan, Kjerish, KsaveroEO, Lalaithion, Laurenfroud, Lentower, Longvietdang, Loremaster, LuigiLTC, Lycurgus, Magioladitis, Maximajorian Viridio, Maximumcoolbeans, Mckburton, Me, Myself, and I are Here, Mechanic1c, MellowMurmur, Metamagician3000, Michaelanthonyalberta, Michling'd, Miracle Pen, Mlitwa1, Monolithica, Musegaze2, Neon Meadows, Neutrality, NewEnglandYankee, Nhinchey, Niceguyedc, NoToleranceForIntolerance, Nobody Too, Nomespot, Nøkkenbuer, Ogress, Ojref, Oluwa2Chainz, Omnipaedista, Openlydialectic, PC-XT, Palosirkka, Publiceditz, PvOberstein, Pwjb, Ratemonth, Rayman60, Reagle, Realmdee, Rhododendrites, Ringtail Jack, Robot psychiatrist, Roy McCoy, RyanisaQT, SQGibbon, Saradelic, ScienceDawns, ShankTim, Sheila Ki Jawani, Silence, Sir Paul, SkyWarrior, Slightsmile, Slowlythereagain, Smk65536, SoWhy, Somethingwickedly, StN, Steel1943, Stopquiningeverything, SunnyMiao, Sunrise, TenPoundHammer, The Transhumanist, Thevideodrome, Thewolfchild, Timtempleton, Trappist the monk, Truyopx, Twilight Magic, UniversityofPi, Usa60527, Vote again to coincide with the general election., Waters.Justin, WhatWeKnow123, Whoosit, Widr, WikiEditorial101, Wikigoodnews, William Myers Bugbee forgot his password, WyldPhyr, Ydde2015, Yurivict, Z11o22, 136 anonymous edits 585

Image Sources, Licenses and Contributors

The sources listed for each image provide more detailed licensing information including the copyright status, the copyright owner, and the license conditions.

Image *Source:* https://en.wikipedia.org/w/index.php?title=File:Folder_Hexagonal_Icon.svg *License:* GNU Free Documentation License *Contributors:* Anomie, Jo-Jo Eumerus, Mifter 3

Figure 1 *Source:* https://en.wikipedia.org/w/index.php?title=File:Al-jazari_robots.jpg *License:* Public Domain *Contributors:* Clusternote, Cwbm (commons), File Upload Bot (Magnus Manske), Infrogmation, Man vyi, Nikkimaria, Quibik, Shakko, Takeaway, WikipediaMaster 5

Figure 2 *Source:* https://en.wikipedia.org/w/index.php?title=File:Gottfried_Wilhelm_von_Leibniz.jpg *License:* Public Domain *Contributors:* AndreasPraefcke, Auntof6, Beria, Bernd Schwabe in Hannover, Beyond My Ken, Boo-Boo Baroo, Chenspec, Cirt, Davidlud, Ecummenic, Eusebius, Factumquintus, FalconL, Gabor~commonswiki, Jan Arkesteijn, Jianhui67, Jochen Burghardt, Leyo, Lorry, Luestling~commonswiki, Mattes, Missvain, Rettinghaus, Rilegator, Schaengel89~commonswiki, Shakko, Svencb, Tomisti, Trijnstel, Trzęsacz, Túrelio, 13 anonymous edits 6

Figure 3 *Source:* https://en.wikipedia.org/w/index.php?title=File:Classic_shot_of_the_ENIAC.jpg *License:* Public Domain *Contributors:* Unidentified U.S. Army photographer 7

Figure 4 *Source:* https://en.wikipedia.org/w/index.php?title=File:BRL61-IBM_702.jpg *License:* Public Domain *Contributors:* User RTC on en.wikipedia 8

Figure 5 *Source:* https://en.wikipedia.org/w/index.php?title=File:Semantic_Net.svg *License:* Public Domain *Contributors:* Bjankuloski06en, Konstable~commonswiki, Vlasenko D, Wst, 1 anonymous edits 11

Figure 6 *Source:* https://en.wikipedia.org/w/index.php?title=File:Hopfield-net.png *License:* GNU Free Documentation License *Contributors:* A3nm, BotMultichill, Deditos, File Upload Bot (Magnus Manske), Glenn, MGA73bot2, Magog the Ogre, MartinThoma, OgreBot 2, ŠJů, 1 anonymous edits 19

Image *Source:* https://en.wikipedia.org/w/index.php?title=File:Symbol_support_vote.svg *License:* Public Domain *Contributors:* Anomie, Fastily, Jo-Jo Eumerus 30

Image *Source:* https://en.wikipedia.org/w/index.php?title=File:Commons-logo.svg *License:* logo *Contributors:* Anomie, Callanecc, CambridgeBayWeather, Jo-Jo Eumerus, RHaworth 54

Image *Source:* https://en.wikipedia.org/w/index.php?title=File:Kernel_Machine.svg *Contributors:* User:Alisneaky, User:Zirguezi 61

Image *Source:* https://en.wikipedia.org/w/index.php?title=File:Portal-puzzle.svg *License:* Public Domain *Contributors:* Anomie, Jo-Jo Eumerus, Topbanana 61

Figure 7 *Source:* https://en.wikipedia.org/w/index.php?title=File:Svm_max_sep_hyperplane_with_margin.png *License:* Public Domain *Contributors:* Cyc 63

Figure 8 *Source:* https://en.wikipedia.org/w/index.php?title=File:Automated_online_assistant.png *License:* Attribution *Contributors:* Mikael Häggström 77

Image *Source:* https://en.wikipedia.org/w/index.php?title=File:Lock-green.svg *License:* Creative Commons Zero *Contributors:* User:Trappist the monk 85

Figure 9 *Source:* https//en.wikipedia.org *Contributors:* User:Arsi Warrior 92

Figure 10 *Source:* https://en.wikipedia.org/w/index.php?title=File:DARPA_Visual_Media_Reasoning_Concept_Video.ogv *Contributors:* FAEP, Simisa, Smallman12q 93

Figure 11 *Source:* https://en.wikipedia.org/w/index.php?title=File:NASA_Mars_Rover.jpg *License:* Public Domain *Contributors:* NASA/JPL/Cornell University, Maas Digital LLC 94

Figure 12 *Source:* https://en.wikipedia.org/w/index.php?title=File:SRT_Shape_Recognition_Technology.png *License:* Creative Commons Attribution-Sharealike 3.0 *Contributors:* User:QueSera4710 97

Figure 13 *Source:* https://en.wikipedia.org/w/index.php?title=File:Em-face-2.png *License:* Public Domain *Contributors:* Auric, S33us00n, 1 anonymous edits 134

Figure 14 *Source:* https://en.wikipedia.org/w/index.php?title=File:Gsrplot.svg *Contributors:* User:Celestial Phineas 135

Figure 15 *Source:* https://en.wikipedia.org/w/index.php?title=File:Estimations_of_Human_Brain_Emulation_Required_Performance.svg *License:* Creative Commons Attribution-Sharealike 3.0 *Contributors:* User:Tga.D 143

Figure 16 *Source:* https://en.wikipedia.org/w/index.php?title=File:Ideal_feedback_model.svg *License:* Public Domain *Contributors:* Me (Intgr) 159

Figure 17 *Source:* https://en.wikipedia.org/w/index.php?title=File:James_Watt.jpg *License:* Public Domain *Contributors:* BotMultichill, Shizhao, Voyager 160

Figure 18 *Source:* https://en.wikipedia.org/w/index.php?title=File:JohnvonNeumann-LosAlamos.gif *License:* Public Domain *Contributors:* LANL 163

Image *Source:* https://en.wikipedia.org/w/index.php?title=File:Emblem-money.svg *License:* GNU General Public License *Contributors:* Ain92, Allforrous, Arugula999, Genium, Jdcollins13, Perhelion, Rocket000, SamJohnston, Thomas Linard 165

Figure 19 *Source:* https://en.wikipedia.org/w/index.php?title=File:Honda_ASIMO_Walking_Stairs.JPG *License:* Public Domain *Contributors:* User:Indolences 166

Figure 20 *Source:* https://en.wikipedia.org/w/index.php?title=File:JARVIK_7_artificial_heart.jpg *License:* Public Domain *Contributors:* National Heart, Lung and Blood Institute (NHLBI) 168

Image *Source:* https://en.wikipedia.org/w/index.php?title=File:Tree_of_life.svg *License:* GNU Free Documentation License *Contributors:* User:Samsara 183

Image *Source:* https://en.wikipedia.org/w/index.php?title=File:Open_book_nae_02.svg *License:* Public Domain *Contributors:* Claus Obana, Liftarn, Rocket000, Thomas Linard, VIGNERON 183

Figure 21 *Source:* https://en.wikipedia.org/w/index.php?title=File:Max_paraboloid.svg *Contributors:* Allforrous, Geek3, IkamusumeFan, Sarang, Shiyu Ji 190

Figure 22 *Source:* https://en.wikipedia.org/w/index.php?title=File:Nelder-Mead_Simionescu.gif *Contributors:* User:nicoguaro 190

Figure 23 *Source:* https://en.wikipedia.org/w/index.php?title=File:Hash_table_3_1_1_0_1_0_0_SP.svg *License:* Creative Commons Attribution-Sharealike 3.0 *Contributors:* Jorge Stolfi 213

Figure 24 *Source:* https://en.wikipedia.org/w/index.php?title=File:SimpleBayesNetNodes.svg *License:* Public Domain *Contributors:* AnAj ..232

Image *Source:* https://en.wikipedia.org/w/index.php?title=File:Bayes_icon.svg *License:* Creative Commons Zero *Contributors:* MIKHAIL RYAZANOV 230

Image *Source:* https://en.wikipedia.org/w/index.php?title=File:Fisher_iris_versicolor_sepalwidth.svg *License:* Creative Commons Attribution-Sharealike 3.0 *Contributors:* en:User:Qwfp (original); Pbroks13 (talk) (redraw) 231

Figure 25 *Source:* https://en.wikipedia.org/w/index.php?title=File:SimpleBayesNet.svg *License:* Public domain *Contributors:* AnAj 232

Figure 26 *Source:* https://en.wikipedia.org/w/index.php?title=File:HiddenMarkovModel.svg *License:* Creative Commons Attribution 3.0 *Contributors:* Tdunning vectorization: 248

Image *Source:* https://en.wikipedia.org/w/index.php?title=File:Hmm_temporal_bayesian_net.svg *License:* Public Domain *Contributors:* Clusternote, O (bot), Qef, Sarang 250

Figure 27 *Source:* https://en.wikipedia.org/w/index.php?title=File:HMMsequence.svg *License:* Creative Commons Attribution-Sharealike 3.0 *Contributors:* Hakeem.gadi 251

Image *Source:* https://en.wikipedia.org/w/index.php?title=File:HMMGraph.svg *License:* Public Domain *Contributors:* Terencehonles 254

Figure 28 *Source:* https://en.wikipedia.org/w/index.php?title=File:Basic_concept_of_Kalman_filtering.svg *License:* Creative Commons Zero *Contributors:* User:Petteri Aimonen 267

Figure 29 *Source:* https://en.wikipedia.org/w/index.php?title=File:Kalman_filter_model_2.svg *License:* Public Domain *Contributors:* User:Headlessplatter 271

Figure 30 *Source:* https://en.wikipedia.org/w/index.php?title=File:Kalman.png *Contributors:* User:Lipkenszaal 275

Image *Source:* https://en.wikipedia.org/w/index.php?title=File:HMM_Kalman_Filter_Derivation.svg *License:* Public Domain *Contributors:* Qef 281

Figure 31 *Source:* https://en.wikipedia.org/w/index.php?title=File:Colored_neural_network.svg *License:* Creative Commons Attribution-Sharealike 3.0 *Contributors:* Erik Streb, Glosser.ca, MartinThoma, Olexa Riznyk, Sarang 344

Figure 32 *Source:* https://en.wikipedia.org/w/index.php?title=File:Ann_dependency_(graph).svg *License:* Creative Commons Attribution-Sharealike 3.0 *Contributors:* Glenn, Glosser.ca, Sarang 350

License

Index